FT
2/14/05
135.00

QM 501 C75 2005
Atlas of the sensory organs
CD stored separately from book
CD/TEXT QM 501 C75 2005

JUL 09
JUL X X 2015

CUMBERLAND COUNTY COLLEGE LIBRARY
PO BOX 1500
VINELAND, NJ 08362-1500

WITHDRAWN

Atlas of the Sensory Organs

ATLAS OF THE SENSORY ORGANS

FUNCTIONAL AND CLINICAL ANATOMY

ANDRÁS CSILLAG, MD, PhD, DSc

Department of Anatomy, Histology, and Embryology
Semmelweis University, Budapest, Hungary

HUMANA PRESS
TOTOWA, NEW JERSEY

© 2005 Humana Press Inc.
999 Riverview Drive, Suite 208
Totowa, New Jersey 07512

humanapress.com

For additional copies, pricing for bulk purchases, and/or information about other Humana titles, contact Humana at the above address or at any of the following numbers: Tel.: 973-256-1699; Fax: 973-256-8341, E-mail: humana@humanapr.com; or visit our Website: www.humanapress.com

All rights reserved.

No part of this book may be reproduced, stored in a retrieval system, or transmitted in any form or by any means, electronic, mechanical, photocopying, microfilming, recording, or otherwise without written permission from the Publisher.

All articles, comments, opinions, conclusions, or recommendations are those of the author(s), and do not necessarily reflect the views of the publisher.

Due diligence has been taken by the publishers, editors, and authors of this book to assure the accuracy of the information published and to describe generally accepted practices. The contributors herein have carefully checked to ensure that the drug selections and dosages set forth in this text are accurate and in accord with the standards accepted at the time of publication. Notwithstanding, since new research, changes in government regulations, and knowledge from clinical experience relating to drug therapy and drug reactions constantly occur, the reader is advised to check the product information provided by the manufacturer of each drug for any change in dosages or for additional warnings and contraindications. This is of utmost importance when the recommended drug herein is a new or infrequently used drug. It is the responsibility of the treating physician to determine dosages and treatment strategies for individual patients. Further, it is the responsibility of the health care provider to ascertain the Food and Drug Administration status of each drug or device used in their clinical practice. The publishers, editors, and authors are not responsible for errors or omissions or for any consequences from the application of the information presented in this book and make no warranty, express or implied, with respect to the contents in this publication.

This publication is printed on acid-free paper. ∞
ANSI Z39.48-1984 (American Standards Institute) Permanence of Paper for Printed Library Materials.
Production Editor: Robin B. Weisberg
Cover Illustration: Blood vessels of the uveal tract of the eye
Cover artwork by Csaba Piller
Cover design by Patricia F. Cleary

Photocopy Authorization Policy:
Authorization to photocopy items for internal or personal use, or the internal or personal use of specific clients, is granted by Humana Press Inc., provided that the base fee of US $25.00 per copy is paid directly to the Copyright Clearance Center at 222 Rosewood Drive, Danvers, MA 01923. For those organizations that have been granted a photocopy license from the CCC, a separate system of payment has been arranged and is acceptable to Humana Press Inc. The fee code for users of the Transactional Reporting Service is: [1-58829-412-9/05 $25.00].

Printed in China. 10 9 8 7 6 5 4 3 2 1

eISBN 1-59259-849-8

Library of Congress

Csillag, András (András Laszlo), 1949-
  Atlas of the sensory organs : functional and clinical anatomy / András Csillag.
    p. ; cm.
  Includes bibliographical references and index.
  ISBN 1-588-412-9 (alk. paper)
  1. Sense organs--Atlases.
  [DNLM: 1. Ear--anatomy & histology--Atlases. 2. Eye--anatomy & histology--Atlases. 3. Nose--anatomy & histology--Atlases. 4. Sensation--physiology--Atlases. 5. Skin--anatomy & histology--Atlases. 6. Tongue--anatomy & histology--Atlases. WV 17 C958a 2005] I. Title.
  QM501.C75 2005
    611'.8--dc22
                                                                    20040442382

Preface

Sensory organs constitute an elaborate and demanding chapter of the anatomical curriculum. The structures are confined to a small area, yet have to be discussed to a degree of detail that is unmatched in human anatomy except for neuroanatomy. Partly owing to tradition and partly because of a real practical significance, there is an unprecedented employment of nomenclature in the field of sensory organs, requiring such a comprehensive knowledge that even medically qualified people may find it difficult. Given the great clinical importance of the topic, in particular with reference to the eye and the ear, the time allocated for the study of sensory organs is remarkably limited in most anatomical curricula, with the organs usually being discussed in the framework of (or as a supplement to) the nervous system. Diseases of the organs of vision and hearing are an everyday occurrence in the work of the general practitioner. Furthermore, the latter are dealt with by separate medical specialties: ophthalmology and oto-rhinolaryngology (ENT). Novel diagnostic and surgical results as well as new questions that are raised serve to widen the knowledge of modern anatomy and may also prompt, and rightfully expect, new answers from macroscopic and microscopic anatomy.

Specialized and comprehensive studies covering all sensory organs are rarely encountered in the medical literature. Sufficiently detailed descriptions can only be found in large and expensive handbooks, consisting of several volumes, and often as a chapter of neuroanatomy. Shorter, more concise editions confined to the topic are either at the level of popular science, or deal with the subject from a specific aspect such as comparative anatomy, evolution, etc. Moreover, the different sensory organs are usually discussed in separate volumes. The current atlas is an attempt to present all major sensory systems together with their neural pathways, from primary sensation all the way to the brain.

The morphology of sensory organs is often described separately from their function and also from the results and requirements of clinical research. However, both the surgery and diagnostics of ophthalmology and ENT have undergone an impressive development in the recent years.

In the case of ophthalmic surgery, laser treatment of retinal detachment, lens implantation, corneal grafts, and surgical treatment of ocular lesions or orbital tumors are well known interventions. Modern diagnostic methods have revealed more refined anatomical details in the living patient (e.g., mapping of retinal blood vessels by fluorescent dye markers, FLAG). Electrical signals of the retina (ERG) are indicative of functional disorders. Fine details of the anatomy of the living eye can be observed by using modern diagnostic imaging methods (e.g., keratometry, ultrasonography). Experimental research and medical applications have brought an impressive development in the analysis of the retina, visual pathway and cortical visual field. Retinal photoreceptors of specific function and chemical nature can now be demonstrated by fluorescent immunohistochemistry. Precise projections of the human visual pathway have been described using postmortem pathway degeneration and tract tracing methods (formerly, such data was only available from animal experiments). Further techniques of interest comprise the monitoring of a mitochondrial enzyme cytochrome oxidase histochemistry. The functionally active elements (and potential disorders) of the visual cortex of wakeful patients can now be detected by the most up-to-date diagnostic methods, such as event-related potentials (ERP), visually evoked potentials (VEP), magnetoencephalography (MEG) and direct intraoperative microelectrode recordings, all characterized by good time resolution. Further techniques of high spatial resolution (hence of particular anatomical relevance) comprise regional cerebral blood flow (rCBF) measurements using single photon emission spectroscopy (SPECT) or positron emission spectroscopy (PET). Apart from measurement of local blood perfusion, oxygen and glucose uptake, the latter spectroscopic methods enable also the regional analysis of neurotransmitters and receptors.

Concerning the anatomy of the ear, important progress has been made by the use of modern surgical and endoscopic techniques. This prompted a reappraisal and led to a renaissance of previously existing refined preparatory methods demonstrating the highly complicated microanatomical relations of the organs of hearing and equilibrium. The size of their structural elements verges on the border of visibility, therefore the visualization of details requires microphotography and glass fiber endoscopy of fresh cadaver tissue. Modern diagnostic methods are available also for functional studies of the auditory and vestibular systems. Such methods comprise audiometry, oto-acoustic emission (clinical investigation relevant to the integrity of cochlea), and recordings from the brainstem (BAEP). An important part of this chapter includes CT and MR images, demonstrating the osseous parts and soft tissues of the organs in different section planes.

The other special sensory organs—taste, olfaction and tactile sensation—also have clinical relevance and their investigation is rather difficult. However, they certainly have a place in a comprehensive study focused on the sensory organs. In *Atlas of the Sensory Organs: Functional and Clinical Anatomy*, the material is based primarily upon light and electron microscopic preparations supplemented by a few special semi-macroscopic anatomical preparations (e.g. microdissection specimens demonstrating the Pacinian corpuscles of the skin). All of the chapters contain a considerable amount of original light and electron microscopic specimens and, in some cases, experimental studies, such as immunohistochemistry, have also been included.

The majority of original specimens and recordings are accompanied by schematic explanatory drawings. Furthermore, overview figures and tables assisting the understanding of the material are also included.

Each chapter begins with a detailed anatomical overview, covering also the development (ontogeny) of sensory organs, the functional aspect of sensory mechanisms. This introductory part is followed by original images from macroscopic, microscopic and functional/clinical materials, and a description of the central pathways relevant to the respective sensory organ. As a rule, only healthy and intact organs or tissues are shown, since the diseases fall beyond the scope of the work. The only exception can be if pathology is directly relevant to the understanding of normal structure and function.

The projected audience for this book includes undergraduate students majoring in medicine, dentistry, animal and human biology; graduate students of biomedical courses covering sensory organs or functions; general practitioners, particularly those wanting to specialize in the fields of ophthalmology, ENT, neurology, psychiatry and plastic surgery; optometrists, and radiographers. Given the level of information offered by the book, the main target groups would likely be medical students doing human anatomy, neuroanatomy or neurology courses; young clinicians, residents and post-docs.

András Csillag, MD, PhD, DSc

Acknowledgments

First of all, the authors thank our expert reviewers, Dr. Tarik F. Massoud, Department of Radiology, University of Cambridge, UK, and Professor Ágoston Szél, Head of Department of Human Morphology and Developmental Biology, Semmelweis University, Budapest, Hungary, for helpful comments and suggestions pertaining to both the style and substance of the work.

Professional help by the following contributors and advisors is gratefully acknowledged: Pál Röhlich, Professor in Anatomy, Semmelweis University of Medicine, Budapest, and Department of Human Morphology and Developmental Biology (anatomy and histology of the eye); Ildikó Süveges, Professor and Head, István Gábriel, Senior Lecturer, Ágnes Farkas, Senior Lecturer and József Györy, Lecturer, Semmelweis University of Medicine, Department of Ophthalmology (clinical investigation of the eye); Balázs Gulyás, Professor, Karolinska Institute, Stockholm, Sweden, and Stephanie Clarke, Professor, Department of Neurophysiology, Hopital Nestlé, Lausanne, Switzerland (structure and functional imaging of visual cortex); Dr. Jordi Llorens, Unitat de Fisiologia, Departament de Ciencies Fisiologiques II Universitat de Barcelona (scanning electron microscopy of the internal ear); Lajos Patonay, Lecturer, and Károly Hrabák, Lecturer, Semmelweis University of Medicine, Department of Oto-Rhino-Laryngology (clinical investigation and diagnostic imaging of the ear); Kinga Karlinger, Senior Research Fellow, and Erika Márton, Senior Lecturer, Semmelweis University of Medicine, Department of Radiology and Oncotherapy, (CT and MR imaging studies); Professors Miklós Palkovits, Kálmán Majorossy and Árpád Kiss, Department of Anatomy, Semmelweis University of Budapest, Hungary (experimental histology of the auditory pathway); András Iványi, Consultant Pathologist (Szent István Hospital, Budapest, immunohistochemistry of human skin); Jenö Páli, PhD student, Semmelweis University, Faculty of Physical Education and Sport Sciences (experimental histology of tactile organs); József Takács, Senior Research Fellow, Center for Neurobiology, Hungarian Academy of Sciences, (experimental histology of the vomeronasal organ and the visual pathway).

The authors express their gratitude to Csaba Piller for preparing the majority of the drawings and paintings presented in the work, and to Balázs Kis for the graphic design of selected illustrations.

Most of the original microscopic images were taken using an Olympus Vanox research microscope or an Olympus BX51 microscope equipped with a DP50 digital camera. For endoscopic photography, Olympus Selphoscope systems (4 mm diameter, 0°, and 1.7 mm diameter, 30°) were used. The authors are indebted to Olympus Corporation for valuable advice in the use of their advanced digital microscopic systems, and for providing financial support that facilitated publication of this book.

We wish to thank Dr. Mark Eyre for correcting the English in this manuscript. The expert technical assistance of Mária Szász (electron microscopy), Mária Bakó (light microscopy), Albert Werlesz (scanning electron microscopy), and the devoted editorial assistance of Ágota Ádám and János Barna are gratefully acknowledged.

Contents

PREFACE .... *V*
ACKNOWLEDGMENTS .... *VII*
CONTRIBUTORS .... *XI*
USING THIS BOOK AND THE COMPANION CD .... *XII*

1 THE ORGAN OF HEARING AND EQUILIBRIUM 1
Miklós Tóth and András Csillag

Anatomical Overview of the Ear 1
External Ear 2
Middle Ear 4
Inner Ear 7
Physiology of Hearing 12
Physiology of Equilibrium 14
Chronological Summary of the Development
of the Temporal Bone 15
Atlas Plates I (Embryology) 16
Atlas Plates II (Microdissection and Endoscopy) 28
Atlas Plates III (Radiology) 45
Atlas Plates IV (Histology) 63
The Auditory Pathway 73
Vestibular Pathways 81
Recommended Readings 83

2 THE ORGAN OF VISION 85
András Csillag

Anatomical Overview of the Eye 85
The Globe 86
Accessory Visual Apparatus 103
Development of the Eye 116
Atlas Plates I (Embryology) 119
Atlas Plates II (Clinical Investigations) 121
Atlas Plates III (Histology) 130
The Visual Pathway 147
Recommended Readings 164

3 THE ORGAN OF OLFACTION 165
András Csillag

- Anatomical Overview of the Organ of Olfaction 165
- Development of the Nasal and Oral Cavities in Relation to Olfactory and Taste Sensation 168
- Atlas Plates 173
- Olfactory Pathways 179
- Recommended Readings 185

4 THE ORGAN OF TASTE 187
Andrea D. Székely and András Csillag

- Anatomical Overview of the Organ of Taste 187
- The Development of Tongue 190
- Atlas Plates 191
- Gustatory Pathways 193
- Making Sense of the Texture of Food: *Periodontal Sensation* 194
- Recommended Readings 198

5 THE SKIN AND OTHER DIFFUSE SENSORY SYSTEMS 199
Mihály Kálmán and András Csillag

- Anatomical Overview of the Skin and Other Diffuse Sensory Systems 199
- Receptors 199
- The Skin and Its Appendages 207
- Neural Correlates of Tactile Sensation 215
- Atlas Plates 219
- Mystacial Vibrissae and Somatosensory Pathways 239
- Recommended Readings 243

INDEX 245

Contributors

ANDRÁS CSILLAG, MD, PhD, DSc • *Department of Anatomy, Histology, and Embryology, Semmelweis Univesity, Budapest, Hungary*

MIHÁLY KÁLMÁN, MD, PhD, DSc • *Department of Anatomy, Histology, and Embryology, Semmelweis University, Budapest, Hungary*

ANDREA D. SZÉKELY, DMD, PhD • *Department of Anatomy, Histology, and Embryology, Semmelweis University, Budapest, Hungary*

MIKLÓS TÓTH, MD • *Department of Human Morphology and Developmental Biology, Semmelweis University, Budapest, Hungary*

Using This Book and the Companion CD

The book in its entirety is on the companion CD. To view color figures, please refer to the companion CD. The images are best viewed on a high-resolution (1280 x 1280) color (24 bit or higher true color) computer monitor.

1 The Organ of Hearing and Equilibrium

Miklós Tóth and András Csillag

ANATOMICAL OVERVIEW OF THE EAR

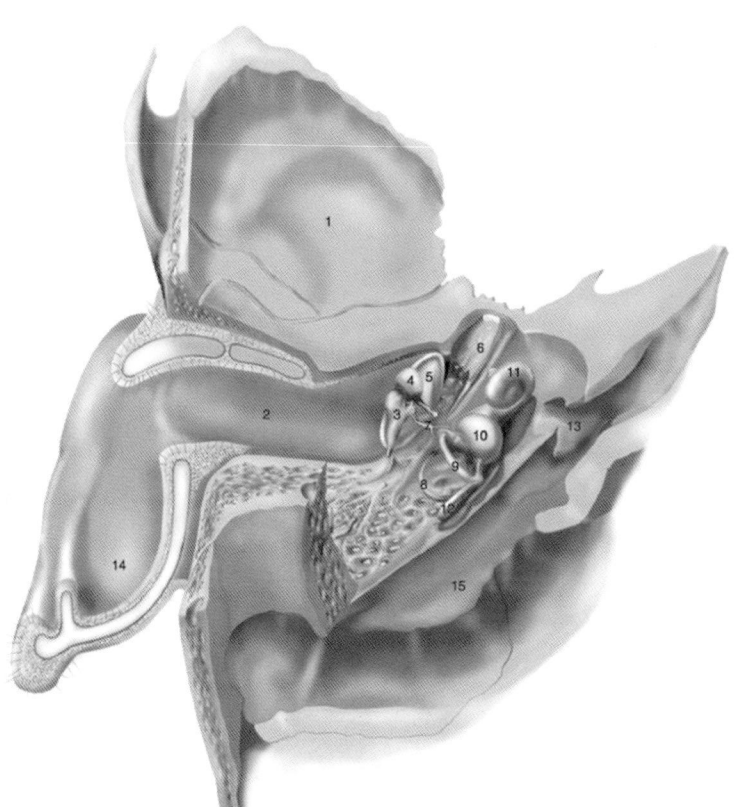

Fig. 1.1

Overview of the ear.

1. Squamous temporal
2. External auditory meatus
3. Malleus
4. Incus
5. Tympanic membrane
6. Pharyngotympanic tube
7. Stapes
8. Lateral semicircular canal
9. Anterior semicircular canal
10. Vestibule
11. Cochlea
12. Posterior semicircular canal
13. Internal auditory meatus
14. Pinna
15. Groove for sigmoid sinus

The human ear is an evolutionary derivative of the lateral line canals of early aquatic vertebrates such as fishes. Both the organs of hearing and equilibrium are based on an "internalized" system of fluid-containing membrane-bound spaces embedded in the petrous part of the temporal bone. Movements of fluid within these ducts owing either to oscillations of atmospheric air (hearing) or to postural changes (balance, equilibrium) elicit specific sensations through the action of highly specialized receptors termed hair cells. The anatomical structures representing both sensory modalities develop from a common ectodermal primordium, the otic placode, surrounded by mesenchyme of the otic capsule. The ectodermal anlage gives rise to a vesicle (otocyst), which later is subdivided into an upper portion (labyrinth, organ of balance) and a lower portion (cochlea, organ of hearing), forming parts of the inner ear. Whereas the former is fully operational alone, the latter requires additional systems for transduction of mechanical energy (sound waves) into bioelectric signals. These systems are situated in the anatomical units called external and middle ear.

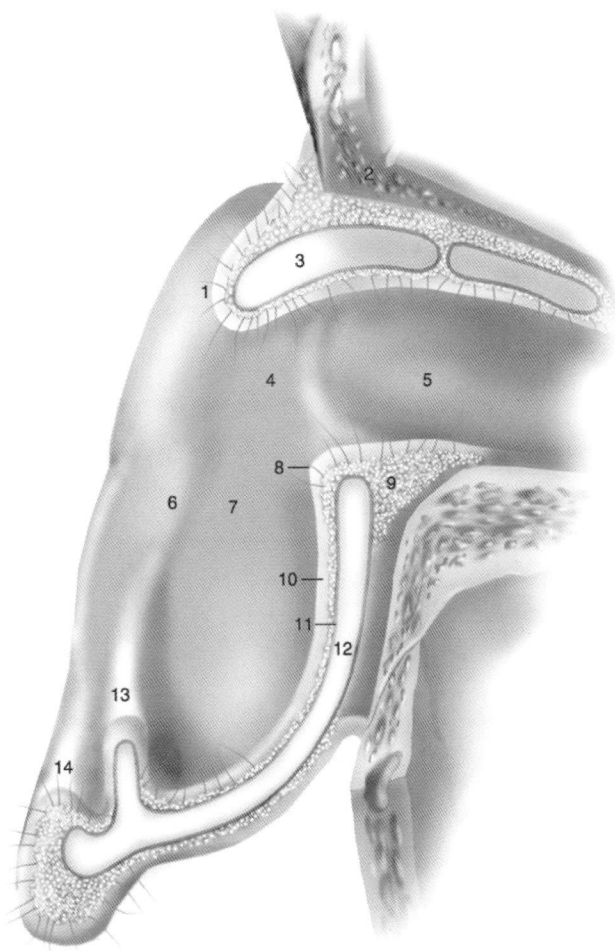

Fig. 1.2

External ear

1. Tragus
2. Osseous part of meatus
3. Cartilagineous part of meatus
4. External acoustic pore
5. External acoustic meatus
6. Antitragus
7. Auricle
8. Auricular hair (tragi)
9. Subcutaneous fat
10. Skin
11. Perichondrium
12. Auricular cartilage
13. Antihelix
14. Helix

External Ear (Figs. 1.1; 1.2; 1.3; 1.4; 1.31; 1.34; 1.42; 1.67; 1.68)

The external ear comprises the visible ear (pinna, auricle) and the external acoustic pore and meatus.

PINNA (FIG. 1.3)

A representative element of the face, this part of the external ear is shaped to collect sound waves from the exterior. Most mammals are capable of directing the auricle at the source of sound but this faculty was virtually lost in humans. A late reminder of an ancient function is a group of rudimentary muscles attached to the external ear and innervated by short branches of the facial nerve.

The pinna is a complicated skinfold reinforced by elastic cartilage and dense connective tissue. The soft and thin skin has a distinctive subcutaneous layer on the posterior surface only. The relief of the inner concave surface of the pinna is defined by numerous prominences and depressions. The posterior free margin of the auricle is called the helix. A second ridge parallel to the helix is termed the antihelix, starting with two limbs, crura anthelicis, above the external acoustic pore. The anterior prominence in front of the pore (tragus) faces another prominence in the lower part of the antihelix, known as the antitragus. The tragus and antitragus are separated by a deep notch, intertragic incisure, pointing toward the soft, cartilage-free earlobe (lobulus auricularis). The groove between the helix and antihelix is called scaphoid fossa, whereas the space delineated by the crura anthelicis is known as the fossa triangularis. The deepest depression inside the pinna is termed cavum conchae, whereas the space surrounded by the crus of helix is known as the cymba conchae.

Vascular supply of the pinna is from the posterior auricular branch of external carotid artery (mainly to the cranial surface), the superficial temporal artery (to the lateral surface), and a branch from the occipital artery.

Apart from the external auricular muscles innervated by the facial nerve, the pinna is innervated by the great auricular and lesser occipital nerves of the cervical plexus, auriculotemporal nerve from the mandibular division of the trigeminal nerve and a small auricular branch of the vagus nerve (see also below). The role of the facial nerve in the cutaneous innervation of the pinna is still a matter of debate.

EXTERNAL ACOUSTIC PORE AND MEATUS (FIGS 1.1; 1.67; 1.68)

Composed of cartilage (the external one-third) and bone (the internal two-thirds), the external acoustic meatus is attached diagonally to the pinna in a seamless fashion. It follows a gentle S-shaped curve that can be partially straightened by pulling the superior margin of the pinna superiorly and posteriorly. The latter manipulation is recommended for examination with an otoscope of the external acoustic meatus and tympanic membrane. The bony part of the canal belongs to the tympanic (posteriorly and inferiorly) and squamous (superiorly) parts of temporal bone. The meatus is lined by skin tightly attached to the underlying perichondrium or periosteum. The integument layer contains massive sebaceous glands (ceruminous glands) that produce ear wax, an agent that protects against damage from moisture trapped inside the canal. Coarse hairs (tragi) can be found at the opening of

CHAPTER 1 / THE EAR

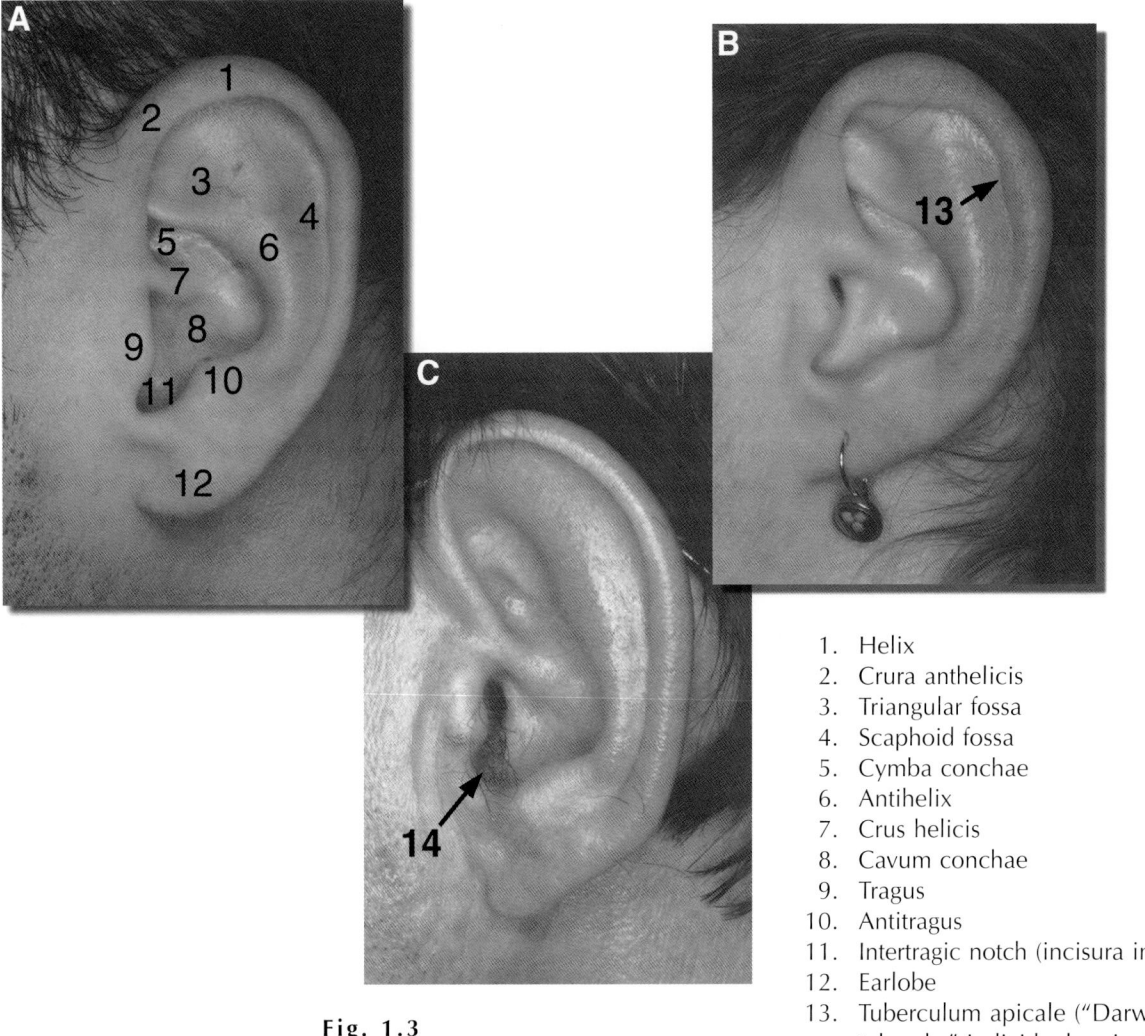

Fig. 1.3
Photographs of human auricles. A – young male; B – young female; C – middle-aged male. Note the individual differences in the shape and size of the earlobe.

1. Helix
2. Crura anthelicis
3. Triangular fossa
4. Scaphoid fossa
5. Cymba conchae
6. Antihelix
7. Crus helicis
8. Cavum conchae
9. Tragus
10. Antitragus
11. Intertragic notch (incisura intertragica)
12. Earlobe
13. Tuberculum apicale ("Darwin's tubercle," individual variant formerly considered "atavistic")
14. Coarse hairs (tragi) surrounding the external acoustic pore

the canal (external acoustic pore). The role of these hairs is to prevent the entrance of insects or other foreign objects. As with other parts of facial hair, the growth of tragi is particularly prominent in old age.

TYMPANIC MEMBRANE (EARDRUM) (FIGS. 1.4; 1.31; 1.34; 1.42; 1.67; 1.68)

The tympanic membrane is a funnel-shaped disk, whose concave side faces the exterior, obturates the external acoustic meatus, fully separating its base from the adjacent anatomical unit, the middle ear, and in particular the tympanic cavity. The membrane is inclined at an approx 50-degree angle; that is, it is not perpendicular to the axis of the auditory canal and its inferior margin is farther than is its superior margin from the external acoustic pore. Notably, in the newborn, the tympanic membrane still occupies a near-horizontal position. The rim of the tympanic membrane (fibrocartilagineous ring, annulus fibrocartilagineus) is anchored in a corresponding semicircular recess of the tympanic part of temporal bone (sulcus tympanicus). A major part of tympanic membrane is taut (pars tensa), whereas the remaining pie-shaped segment adjacent to the squamous temporal above is lax (pars flaccida). The latter area is delineated by thin bands, the anterior and posterior mallear folds (plicae malleares anterior and posterior). A further notable part of the tympanic membrane is the "cone of light," a small inferior area of the membrane nearly perpendicular to the axis of the external acoustic meatus, where a bright reflection of light is visible at otoscopic examination. The tympanic membrane is associated with two processes of the hammer. First, the handle (manubrium) is adherent to the inner side to

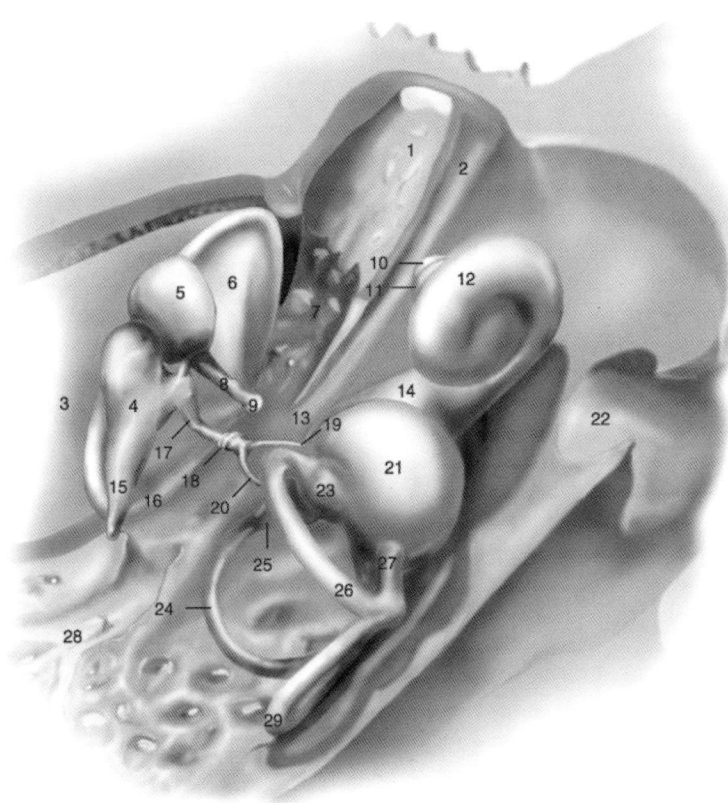

Fig. 1.4

The middle ear in relation to the inner ear.

1. Pharyngotympanic tube (osseous part)
2. Semicanal of tensor tympani muscle
3. External auditory meatus
4. Incus
5. Malleus
6. Tympanic membrane
7. Hypotympanum
8. Manubrium of malleus
9. Umbo
10. Third turn of cochlea
11. Second turn of cochlea
12. First turn of cochlea
13. Cochleariform process
14. Basal turn of cochlea
15. Posterior crus of incus
16. Annulus fibrocartilagineus
17. Long crus of incus
18. Incudostapedial joint
19. Stapes, anterior crus
20. Stapes, posterior crus
21. Vestibule
22. Internal auditory meatus
23. Ampulla of anterior semicircular canal
24. Lateral semicircular canal
25. Ampulla of lateral semicircular canal
26. Anterior semicircular canal
27. Crus commune
28. Air cells of mastoid antrum
29. Posterior semicircular canal

form the stria mallearis and its inferior end corresponding to the umbo. Second, the lateral process of malleus causes a slight bulge at the border between the pars tensa and pars flaccida, the mallear prominence (prominentia mallearis).

The core structure of the tympanic membrane is a layer of dense connective tissue with radially oriented fibers (lamina propria or fibrous stratum). The external side of the eardrum is covered by thin stratified squamous keratinizing epithelium continuous with that of the external auditory canal (cuticular stratum). On the internal side the epithelium is continuous with that of the tympanic cavity (mucous stratum) and is usually reduced to simple squamous nonciliated epithelium.

The outer surface of the tympanic membrane is innervated by the auriculotemporal nerve (from V/3) and a few branches of the vagus nerve (rami auriculares). Involvement of the vagus nerve in the sensory innervation of the external ear explains why, in some individuals, stimulation of the region elicits gagging or vomiting. The internal surface of the tympanic membrane is innervated by branches of the tympanic plexus (from the glossopharyngeal nerve).

Middle Ear

To this region belong the tympanic cavity with the auditory ossicles, the antrum with the mastoid air cells, and the pharyngotympanic tube.

TYMPANIC CAVITY (FIGS. 1.4; 1.5; 1.32; 1.33; 1.35; 1.36; 1.38; 1.39; 1.40; 1.50-1.55; 1.64; 1.67; 1.68)

This air-containing space resembles a drum ("tympanon") lying sideways inside the petrous temporal bone, with the tympanic membrane ("eardrum") directed laterally and the base of the drum corresponding to the medial wall. The main part of the cavity (tympanic cavity proper or mesotympanum) lies opposite the tympanic membrane, whereas other major divisions are the protympanum anterior to it (essentially the bony part of the auditory tube), an upper evagination above this level is termed epitympanum, and the space below the level of the tympanic membrane is known as the hypotympanum.

For an easier description of the otherwise irregular cavity, it is customary to define six walls, as if the cavity had the shape of a rectangular prism. The lateral wall

CHAPTER 1 / THE EAR

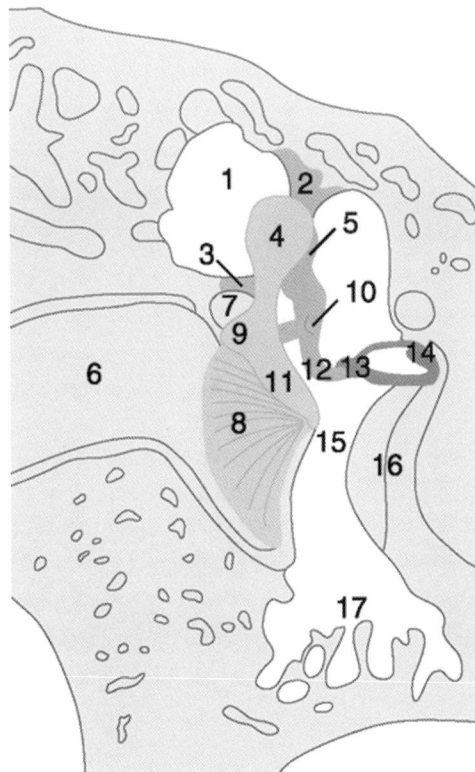

Fig. 1.5

Schematic drawing of the tympanic cavity and the auditory ossicles.

1. Epitympanic recess
2. Superior ligament of malleus
3. Lateral ligament of malleus
4. Head of malleus
5. Head of incus
6. External acoustic meatus
7. Space of Prussak
8. Tympanic membrane
9. Lateral process of malleus
10. Insertion of the tendon of tensor tympani muscle
11. Manubrium of malleus
12. Long limb of incus
13. Incudostapedial joint
14. Footplate of stapes in the stapedial fossa
15. Mesotympanum
16. Promontory
17. Hypotympanum

(paries membranaceus) is mainly composed of the tympanic membrane, supplemented by a part of the squamous temporal bone (scutum). The medial wall has a prominent elevation, the promontory, representing the basal turn of cochlea. Below the promontory, the round window (fenestra cochleae), obturated by the secondary tympanic membrane, is visible. The other important opening leading to the internal ear is the oval window (fenestra vestibuli) at the superior edge of the promontory. This accomodates the footplate of the stapes (meaning: stirrup). In front of the promontory is the opening of the bony canal containing the tensor tympani muscle (semicanal for tensor tympani muscle). The floor of this canal bulges laterally to form a spoon-shaped hook, the cochleariform process. The bony partition between the tensor tympani muscle and the underlying pharyngotympanic tube (semicanal for auditory tube) is incomplete.

The upper wall (roof, paries tegmentalis) of tympanic cavity is composed of the tegmen tympani of petrous temporal bone. The posterior wall is noted for a large opening above (aditus to the mastoid antrum) leading to the antrum and the mastoid air cells. The opening is bordered by prominences of the lateral semicircular canal (above) and the facial nerve canal (below). The pyramidal eminence enclosing the tiny stapedius muscle lies in front of the antrum. The lower wall (floor, paries jugularis) is elevated by the base of styloid process and the jugular bulb. In front of the latter, a small opening passes the tympanic nerve (a branch of glossopharyngeal nerve), which enters the tympanic cavity to form the tympanic plexus ramifying on the promontory for sensory innervation of the middle ear. The nerve also contains parasympathetic preganglionic fibers destined for the otic ganglion (secretory innervation of the parotid gland). The anterior wall (paries caroticus) is closed only in its lower portion, separating it from the internal carotid artery. Above, the wall is perforated by two openings, one for the tensor tympani muscle, whose tendon hooks around the cochleariform process and, after a sharp bend, inserts at the neck of malleus. The other opening leads to the pharyngotympanic (auditory, Eustachian) tube, connecting the middle ear to the nasopharynx. The latter is important for proper ventilation and drainage of excess mucus. Obstruction of this flow can lead to a build-up of pressure inside the middle ear, accompanied by painful sensations, transmitted mainly by the glossopharyngeal nerve. A combination of rapid changes of atmospheric pressure and obstruction of the auditory tube explains the earache and sore throat experienced by some passengers during and after flights. Because the pharyngeal tonsil (adenoid) is well developed in young children, it is more likely to cause obstruction of the auditory tube. Such children are particularly prone to inflammation of the tympanic mucosa (otitis media).

Ossicles (Figs. 1.4; 1.5; 1.32; 1.35; 1.36; 1.38; 1.39; 1.40; 1.43; 1.44; 1.45; 1.61; 1.62; 1.63)

The chain of small bones called malleus, incus and stapes serves the mechanical transmission and amplification of sound energy from the tympanic membrane to the inner ear. The bones are linked to each other by synovial joints, secured in position by ligaments and covered by mucosa. The malleus (hammer) is composed of head, neck, and manubrium (handle). The head lies in the epitympanic recess (evagination of tegmen tympani) and contains a cartilage-covered facet for the incus. The downward-directed manubrium is attached to the umbo of tympanic membrane. Further projections of malleus are the anterior and lateral processes. Three mallear ligaments can be distinguished: the superior ligament anchors the head of malleus in the epitympanic recess, the anterior ligament connects the anterior process with the petrotympanic fissure, and the lateral ligament passes from the neck of malleus to the tympanic notch (incisura tympanica). The incus (anvil) is situated in the epitympanic recess, its body forming the incudomallear joint with the head of malleus. The short limb (crus breve) pointing backward is near horizontal and connected to the fossa incudis, a depression in the posterolateral wall of the tympanic cavity, via the posterior ligament. The long limb (crus longum) courses vertically and articulates with the stapes. The superior ligament of the incus forms a connection between the body and the epitympanic recess. The stapes (stirrup) is composed of a small head (caput), articulated with the long process of incus (incudostapedial joint), two limbs or crura (anterior limb or crus rectilineum and posterior limb or crus curvilineum), and a footplate lodged in the oval window and held in place by the annular ligament. The latter ensures a limited but sufficient mobility of the stapes with respect to the inner ear. Calcification of the annular ligament (otosclerosis) leads to diminished ossicular movements, a common cause of hearing defects, particularly in old age. The space between the limbs and footplate of stapes is occupied by the stapedial membrane.

Muscles of the Middle Ear (Figs. 1.34; 1.38; 1.39; 1.43; 1.45)

The tensor tympani muscle originates in the semicanal above the auditory tube and its tendon takes a sharp turn by the cochleariform process to insert at the base of the manubrium of malleus. Contraction of this muscle combined with a pulley-like mechanism of its tendon diminishes the pressure exerted by the tympanic membrane on the malleus. Thus, the ossicular vibration can be effectively dampened. The muscle is innervated by a branch of the mandibular (V/3) nerve. The stapedius muscle emerges from a cavity inside the pyramidal eminence and is attached to the head of stapes. Retraction of the stapes allows this muscle to attenuate the movements of stapes inside the oval window. The stapedius muscle is innervated by a branch of facial neve. Both muscles of the middle ear are instrumental in the protection of the chain of ossicles and the inner ear from excessive vibration owing to loud sounds. Paralysis of the stapedius muscle (not an infrequent complication of fractures of the skull base damaging the facial nerve) may lead to enhanced sensitivity to sound (hyperacusis).

Pharyngotympanic (Auditory, Eustachian) Tube (Figs. 1.4; 1.32; 1.69)

The auditory tube begins with a trumpet-like opening (hence its other name: salpinx) in the nasopharynx near to the choana (pharyngeal opening, ostium pharyngeum). The prominent elevation posterior to this opening is called torus tubarius, the anterior portion of which forms the posterior lip of the Eustachian tube orifice. The canal is almost 4 cm long and it follows a diagonal course at an angle of approx 45 degrees in a posterolateral direction. The medial two-thirds of the tube are composed of cartilage, whereas the lateral one-third (nearest the tympanic cavity) is bony. The cartilaginous segment is a trough facing downward and laterally, in which the mucosal tube rests. The latero-inferior wall of this part is membranous and adjacent to the (mainly longitudinal) levator veli palatini muscle and the (mainly perpendicular) fibers of tensor veli palatini muscle. The osseous segment with a triangular cross sectional profile (semicanalis tubae auditivae) lies in the petrous temporal bone lateral to the carotid canal. Its tympanic opening (ostium tympanicum) leads to the tympanic cavity but does not allow easy drainage of exsudate because of an elevated rim on the bottom. The auditory tube is normally constricted in the cartilagineous part, especially at the junction with the osseous part (isthmus), and can only be opened by the action of the tensor and levator veli palatini muscles (swallowing). Furthermore, contraction of the salpingopharyngeus muscle (e.g., by yawning) opens up the ostium pharyngeum of the auditory tube, thereby permitting equalization of pressure between the tympanic cavity and the pharynx.

The Facial Nerve in Relation to the Middle Ear (Figs. 1.33; 1.34; 1.35; 1.44; 1.45; 1.48 1.53; 1.57; 1.59; 1.65)

The seventh cranial nerve (facial nerve) passes through the fundus of the internal acoustic meatus and travels in the facial nerve canal in the vicinity of, but never inside, the tympanic cavity. The canal has three segments, labyrinthine, tympanic (horizontal), and mastoid (descending).

The first part of the canal lies on the roof of the vestibule, between the cochlea and the semicircular canals, after which it takes a sharp turn in a posterolateral direction. This bend is called the anterior genu of the facial nerve, hence the ganglion at this site, containing the cell bodies of taste fibers for the anterior two-thirds of the tongue, is known as the geniculate ganglion. A branch of the facial nerve, the greater petrosal nerve, arises from the geniculum and emerges on the anterior surface of petrous temporal bone in its own groove. The peripheral processes of taste neurons leave the facial nerve via another branch, the chorda tympani nerve, which passes through the middle ear cavity and the petrotympanic fissure to join the lingual nerve heading for the tongue. Continuing its course after the geniculate bend, the facial nerve canal raises a small elevation in the angle formed by the roof and medial wall of the tympanic cavity. It then turns sharply downward (this bend is termed the posterior genu) and descends behind the tympanic cavity to leave the cranium at the stylomastoid foramen.

Chorda Tympani Nerve and Mucosa of the Middle Ear (Figs. 1.31; 1.32; 1.35; 1.36; 1.39; 1.41)

An intracranial branch of the facial nerve, the chorda tympani nerve emerges from the facial nerve canal in its descending segment. Enwrapped in a mucosal fold (plica chordae tympani), which can be subdivided into anterior and posterior mallear folds, the chorda tympani nerve traverses the entire tympanic cavity in a superiorly convex arch, passing between the long process of the incus and the neck of the malleus. The nerve leaves the tympanic cavity through the petrotympanic fissure and emerges in the infratemporal fossa, where it joins the lingual nerve. Apart from the taste fibers already mentioned, the chorda tympani nerve also contains preganglionic parasympathetic fibers for secretory innervation of the submandibular and sublingual glands. The pockets formed by the anterior or posterior mallear folds (medially) and the tympanic membrane (laterally) are known as the anterior or posterior tympanic recesses, respectively. The superior tympanic recess (also termed the space of Prussak) is bounded by the head and neck of the malleus and the pars flaccida of the tympanic membrane. The incus and stapes are connected by mucosal folds (plica incudis, plica stapedis) to the wall of tympanic cavity.

Blood Supply and Nerves of the Middle Ear (Figs. 1.42; 1.49)

Four main tympanic arteries supply the tympanic cavity: anterior (from the maxillary artery, entering via the petrotympanic fissure), superior (from the middle meningeal artery, entering via the groove for the lesser petrosal nerve), posterior (from the stylomastoid artery, accompanying the chorda tympani nerve), and inferior (from the ascending pharyngeal artery, accompanying the tympanic nerve). Further small (caroticotympanic) branches derive directly from the internal carotid artery. Venous drainage is to the pterygoid and pharyngeal venous plexuses and to the intracranial venous sinuses, mainly the superior petrosal sinus, which represents a dangerous portal for meningeal infection in young children, as long as the roof of the middle ear cavity remains unclosed.

Lymphatic drainage of the middle ear is mainly to the parotideal, retropharyngeal, superficial and deep cervical, and mastoid lymph nodes.

Sensory innervation of the tympanic cavity is from the tympanic plexus, with contribution from the glossopharyngeal and facial nerves. The terminal branch of the tympanic nerve (lesser petrosal nerve) leaves the tympanic cavity through an opening just lateral to that of the greater petrosal nerve, on the anterior surface of petrous bone. This branch transmits preganglionic parasympathetic fibers for innervation of the parotid, via the otic ganglion. Sympathetic innervation of the tympanic cavity is from the carotid plexus, branches of which pass through the caroticotympanic canaliculi.

Inner Ear (Figs. 1.4; 1.6)

This region comprises a system of membranous ducts containing the receptor structures for hearing and balance (membranous labyrinth). The ducts are enclosed within a matching system of osseous canals made of compact bone, embedded in the petrous temporal bone.

MEMBRANOUS LABYRINTH (FIG. 1.6)

The collective term refers to a continuous system of endolymph-containing ducts, derived from the otic vesicle (otocyst). This is surrounded by a second sleeve-like connective tissue space containing perilymph. The membranous labyrinth is divided into two parts, vestibular (with three semicircular ducts, utricle and saccule) and cochlear (with the cochlear duct). The two sections communicate via the ductus reuniens (often obliterated in adults).

OSSEOUS LABYRINTH (FIGS. 1.1; 1.4; 1.46; 1.47)

A crude correspondent of the membranous labyrinth, the osseous labyrinth encapsulates the membranous ducts. The semicircular ducts are surrounded by the respective semicircular canals, the utricle and saccule by the vestibule and the cochlear duct by the cochlea. Only the vestibule is described here, other parts of the osseous labyrinth are discussed together with their membranous contents.

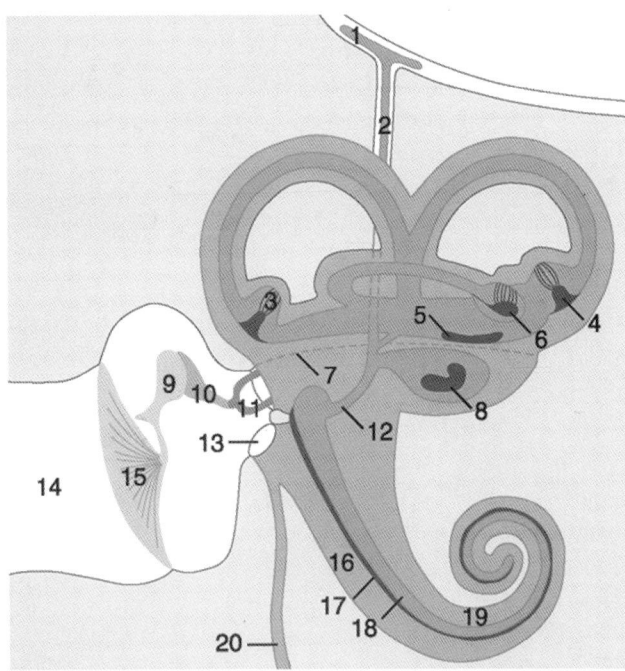

Fig. 1.6

Schematic drawing demonstrating the main components of the internal ear in relation to the tympanic cavity. After Rohen (modified).

1. Endolymphatic sac
2. Endolymphatic duct
3. Ampulla of posterior semicircular duct
4. Ampulla of anterior semicircular duct
5. Macula of utricle
6. Ampulla of lateral semicircular duct
7. Membrana limitans
8. Macula of saccule
9. Malleus
10. Incus
11. Stapes and oval window
12. Ductus reuniens
13. Round window
14. External auditory meatus
15. Tympanic membrane
16. Scala tympani
17. Organ of Corti
18. Cochlear duct
19. Scala vestibuli
20. Perilymphatic duct in cochlear canaliculus

VESTIBULE (FIGS. 1.52; 1.54; 1.55; 1.58; 1.61; 1.62; 1.66)

A central hallway of the osseous labyrinth, the vestibule is connected anteriorly with the cochlea and posteriorly with the semicircular canals. Because of its topographic relationship, the term "vestibular" is used to designate the system of equilibrium sensation. An elliptical orifice on the anterior wall leads to the scala vestibuli of the cochlea. On the lateral wall of the ovoid-shaped cavity is the opening called fenestra vestibuli. The medial wall contains a small depression anteriorly (spherical recess), and behind it an oblique vestibular crest. Two inferior limbs of the latter enclose a small space called cochlear recess. A further depression posterosuperior to the vestibular crest is called the elliptical recess. Below this is the opening of the vestibular aqueduct, a bony canal leading to the posterior surface of the petrous bone. Five openings for the semicircular canals are situated in the posterior region of the vestibule.

COCHLEA AND THE ORGAN OF CORTI (FIGS. 1.7; 1.46; 1.47; 1.55; 1.56; 1.57; 1.58; 1.59; 1.60; 1.70; 1.71; 1.72; 1.73)

The cochlea of the osseous labyrinth truly resembles a snail shell, with two-and-a-half turns of a spiral canal winding around a conical axis, the modiolus. A similarly winding bony ledge (osseous spiral lamina) protruding from the modiolus serves as anchoring for the basilar membrane. The latter spans the width of spiral canal and is attached to the spiral (cochlear) ligament opposite, a fibrous elevation of the endosteum of the cochlear duct. The endolymph-containing cochlear duct is triangular in cross section, the floor being the basilar membrane, the roof corresponding to the thin vestibular membrane (of Reissner) and the lateral wall is an elevation of stratified epithelium, the stria vascularis, followed by the spiral prominence (also richly vascularized) below. The stria vascularis is thought to secrete the endolymph. The cochlear duct has both ends closed, the cecum vestibulare below and the cecum cupulare above. The width of the basilar membrane increases closer to the apex of cochlea. The specialized band of neuroepithelium, termed the spiral organ (of Corti), resting on the basilar membrane, protrudes into the cochlear duct. Lateral and medial borders of the organ are the sulcus spiralis externus and sulcus spiralis internus, respectively. The latter is covered by the tectorial membrane, a gelatinous process in which the hairs of receptor cells are embedded, arising from the labium limbi vestibulare.

Two basic types (sustentacular and receptor) cells are present in the organ of Corti. Bordering the sulcus spiralis internus is the innermost sustentacular cells (of Held), followed by a single row of inner phalangeal cells and the inner hair cells supported by these. The next characteristic structure is a pair of tunnel cells (pillar cells, rods) resem-

CHAPTER 1 / THE EAR

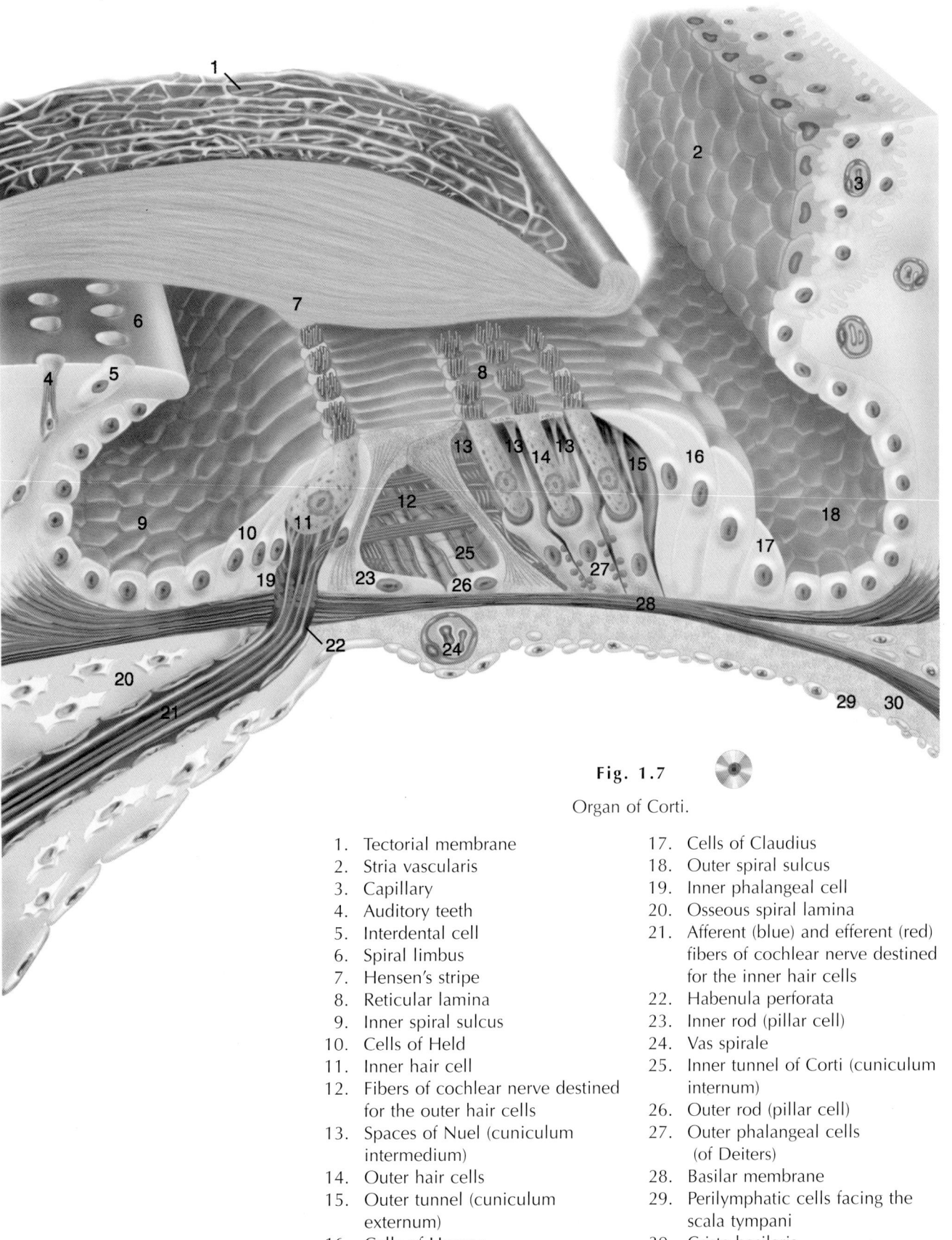

Fig. 1.7

Organ of Corti.

1. Tectorial membrane
2. Stria vascularis
3. Capillary
4. Auditory teeth
5. Interdental cell
6. Spiral limbus
7. Hensen's stripe
8. Reticular lamina
9. Inner spiral sulcus
10. Cells of Held
11. Inner hair cell
12. Fibers of cochlear nerve destined for the outer hair cells
13. Spaces of Nuel (cuniculum intermedium)
14. Outer hair cells
15. Outer tunnel (cuniculum externum)
16. Cells of Hensen
17. Cells of Claudius
18. Outer spiral sulcus
19. Inner phalangeal cell
20. Osseous spiral lamina
21. Afferent (blue) and efferent (red) fibers of cochlear nerve destined for the inner hair cells
22. Habenula perforata
23. Inner rod (pillar cell)
24. Vas spirale
25. Inner tunnel of Corti (cuniculum internum)
26. Outer rod (pillar cell)
27. Outer phalangeal cells (of Deiters)
28. Basilar membrane
29. Perilymphatic cells facing the scala tympani
30. Crista basilaris

bling a house of cards, bordering the inner tunnel (cuniculum internum). Lateral to the tunnel, we find the outer phalanx cells (of Deiters), accomodating three to five rows of outer hair cells. The space of Nuel (cuniculum intermedium) is situated between the outer tunnel cell and the most medial outer phalanx cell. This space (essentially a system of narrow canals), maintains communication between the inner and outer tunnels (the latter, also termed cuniculum externum, is situated just lateral to the outermost hair cell) in the form of an isolated fluid compartment identical with, or very similar to perilymph (cortilymph), presumably optimized for the outer hair cells. Farther laterally, two groups of sustentacular cells, named after Hensen and Claudius, are visible. The latter spread across the sulcus spiralis externus. A special group called the cells of Boettcher is found in the basal turn of the cochlea, below the layer of Hensen's cells. The hair cells are held in position by a perforated reticular lamina overlying the organ of Corti and connected with the flat apical parts of pillar cells. The hairs, of which multiple "stereocilia" (actually microvilli) can be distinguished, project out of the gaps of reticular lamina and are embedded in the jelly (essentially a form of glycocalyx) of the tectorial membrane. This is certainly the case with the outer hair cells, whereas the hairs of the inner hair cells just touch the tectorial membrane.

The cochlear duct is bounded internally (near the modiolus) by an endosteal connective tissue elevation of the osseous spiral lamina, the spiral limbus. The tympanic lip (labium limbi tympanicum) of this structure continues into the basilar membrane, whereas the vestibular lip (labium limbi vestibulare) supports the tectorial membrane. The collagen fibers of the limbus follow a characteristic vertical arrangement known as the "auditory teeth" (of Huschke). The cells lying between these (interdental cells) secrete the substance of the tectorial membrane.

The basilar membrane has a thin inner zone (zona arcuata) stretching from the limbus to the outer rods and a thick outer zone (zona pectinata) between the outer rods and the spiral ligament. The former contains small collagenoid fibers, mainly radially oriented, whereas the latter is composed of three layers: upper with transversely oriented fibers, lower with longitudinal fibers and an intermediate structureless layer sandwiched between these. The inferior surface of the basilar membrane is covered by perilymphatic cells overlying a vascular connective tissue. One prominent blood vessel (vas spirale) is situated below the tunnel of Corti. The peripheral nerve fibers (afferents from the spiral ganglion and efferents from the brainstem) pass to their targets through a zone of perforations (foramina nervosa, habenula perforata) along the inner border of the inner pillar cells. From there the fibers reach the inner hair cells directly, whereas those destined for the outer hair cells traverse the floor of the inner tunnel diagonally before terminating.

Two perilymphatic spaces, the scala tympani and scala vestibuli, accompany the cochlear duct from below and above, respectively. The two scalae communicate with each other in the apex of the cochlea (helicotrema). The scala vestibuli is connected with the vestibule, from which the oval window (fenestra vestibuli), obturated by the footplate of stapes, opens to the tympanic cavity. The scala tympani has communication with the tympanic cavity via the round window (fenestra cochleae), obturated by the secondary tympanic membrane. The perilymphatic space of scala tympani is connected to the subarachnoid space near the superior bulb of the internal jugular vein through the perilymphatic duct enclosed by the cochlear canalicle.

ORGAN OF EQUILIBRIUM (FIGS. 1.48; 1.66; 1.73; 1.74; 1.75; 1.76)

The vestibular part of membranous labyrinth is composed of a system of communicating endolymph-containing vesicles and ducts, including two larger swellings termed saccule and utricle, and three semicircular ducts. The latter are continuous with the utricle, each semicircular duct possessing an ampullar swelling (ampulla membranacea) corresponding to the respective parts of osseous semicircular canals. The receptor structures of the vestibular system (the organ of balance) comprise the maculae of the utricle and saccule, as well as the cristae ampullares of the semicircular ducts. The endolymph-containing duct system lies eccentrically within the osseous labyrinth, surrounded by very loose connective tissue corresponding to the perilymphatic space.

Saccule (Figs. 1.73; 1.74)

This vesicle of about 3 mm diameter, situated in the spherical recess of the vestibule, is connected via the ductus reuniens and the utriculosaccular duct with the cochlear duct and the utricle, respectively. A side branch of the utriculosaccular duct, the endolymphatic duct and its vesicular terminal swelling, the endolymphatic sac, pass to the posterior surface of petrous temporal bone with a blind end beneath the dura mater. These structures, enclosed by the vestibular aqueduct, ensure drainage of the endolymph and equilibration of pressure between the endolymphatic and subarachnoid spaces.

Utricle (Fig. 1.73)

This elongated vesicle, with a size roughly similar to that of the saccule, is situated in the elliptic recess of the vestibule. The utricle is connected with an elaborate system of semicircular ducts.

Semicircular Canals and Ducts (Figs. 1.6; 1.48; 1.61; 1.66; 1.75)

The bony semicircular canals constitute sleeves made of compact bone around the membranous semicircular ducts, with swellings (ampullae) matching those of the membranous ducts. Therefore, only the latter are discussed in detail.

Forming about two-thirds of a complete circular arch, the three semicircular ducts are arranged in three different planes, which are roughly perpendicular to each other. However, the planes do not correspond to any of the principal body planes of an upright-standing individual, with the exception of the lateral semicircular duct, which lies in a near-horizontal plane. The planes of the anterior and posterior semicircular ducts close an angle of about 45° with the frontal and sagittal planes, respectively. The anterior and posterior semicircular ducts have a common limb (crus membranaceum commune), whereas the ampullar limbs (crus membranaceum ampullare) open separately in the utricle. The lateral semicircular duct has its ampullar swelling in the front and a simple posterior limb (crus membranaceum simplex) joining the utricle behind.

Receptor Structures of the Vestibular System (Figs. 1.73; 1.74; 1.75; 1.76)

The maculae of the saccule and utricle are shallow elevations of neuroepithelium, covered by a gelatinous mass (cupula). The composition of the latter is similar to that of the tectorial membrane (see above). The epithelium contains receptor (hair) cells and supporting (sustentacular) cells. The former have many cytoplasmic protrusions ("stereocilia") and one true clilium ("kinocilium") embedded in the cupula. Tonic excitation of the maculae is elicited by calcium containing deposits (statoconia, otoliths) exerting continuous pressure on the hair cells by means of gravity. The macula of the utricle (of 2 × 3 mm size) lies horizontally in the bottom of the utricle. This receptor can therefore aptly detect static body position or linear acceleration with respect to the direction of gravity (vertical). Conversely, the macula of the saccule (of 1.5 mm diameter) is vertically oriented. Although this structure could also participate in the detection of changes in the body position and linear acceleration in a lateral direction, recent findings indicate that the macula of the saccule also represents an accessory hearing system in humans.

The ampullar crests (cristae ampullares) of the semicircular ducts are more prominent crescent-shaped elevations of connective tissue covered by neuroepithelium. Similar to the maculae, this epithelium contains receptor (hair) cells and sustentacular cells, and the receptor hairs are likewise embedded in the gelatinous mass of the cupula. Here, the latter is suspended rather like a swing door that is pushed in or out by the flow of endolymph. The adequate stimulus of hair cells is a deflection of hairs in the direction of kinocilia. Unlike the maculae, the cristae ampullares are sensitive to angular velocity (i.e., rotational movements) in the plane defined by the semicircular duct. Given the three-dimensional arrangement of the three semicircular ducts, movements of any direction will stimulate a specific combination of ampullar receptors.

Vestibulocochlear Nerve (Fig. 1.65)

The eighth cranial nerve is subdivided into cochlear and vestibular parts for hearing and balance, respectively. The cochlear part arises from afferent nerve fibers emerging from the longitudinal canals of the modiolus. These nerves constitute the centrally directed neurites of bipolar cells located in the spiral ganglion (of Corti), whose peripherally directed neurites synapse with the cochlear hair cells. The vestibular part is further subdivided into utriculoampullar nerve (for innervation of the lateral and anterior ampullar crests and the macula of the utricle), saccular nerve (for innervation of the macula of the saccule), and posterior ampullar nerve, innervating the posterior ampullar crest. The vestibulocochlear nerve as well as the vestibular ganglion (of Scarpa) lie near the bottom (fundus) of the internal acoustic meatus.

Summarizing the topographic relations of the fundus, this discoid area consists of four quadrants of unequal size, formed by the intersection of the crista falciformis (transverse bony ridge) with Bill's bar (a verical fibrous ridge). The anterosuperior quadrant is the beginning of the facial nerve canal (see above). The posterosuperior quadrant (superior vestibular area) passes the utriculoampullar nerve, whereas the posteroinferior quadrant (inferior vestibular area) accomodates the saccular nerve and the posterior ampullar nerve (foramen singulare). The cochlear nerve fibers are situated in the anteroinferior quadrant, passing through a helical array of pores (tractus spiralis foraminosus).

Blood Supply of the Inner Ear

The single artery supplying the inner ear is the labyrinthine artery, which arises from the basilar (occasionally the anterior inferior cerebellar) artery and passes through the internal acoustic meatus, with profuse branches for the cochlea and the vestibular organ. Because of its vulnerability, this artery is often implicated in vascular diseases of the inner ear. Venous drainage of the region is via

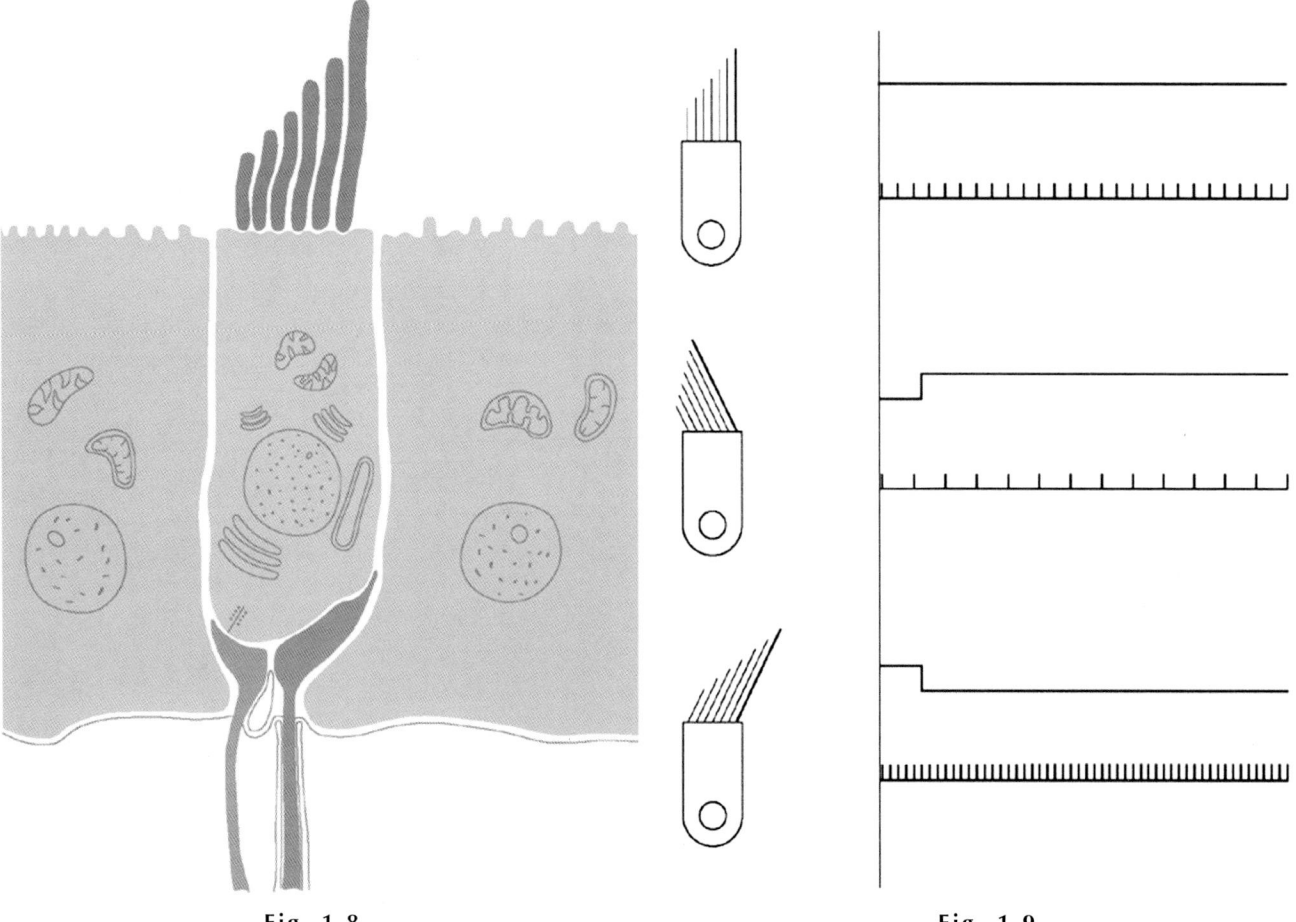

Fig. 1.8

Schematic drawing of a hair cell interposed between two sustentacular cells. The longest apical process represents the kinocilium. The afferent nerve terminal (below left) forms a ribbon synapse, whereas the efferent nerve terminal (below right) has a simple synapse on the basal part of the cell body.

Fig. 1.9

Cartoon illustrating the effect of hair deflection on the membrane potential (top line) and firing frequency (bottom line) of the hair cell.

the labyrinthine veins, returning blood to the inferior petrosal sinus or directly to the internal jugular vein.

Physiology of Hearing

An essential mechanism underlying both hearing and equilibrium sensation is the action of the hair cell. An ancient mechanoreceptor already present in early Agnathan fish, typical hair cells rest on a basement membrane and have numerous cytoplasmic protrusions on their free surface covered by the cuticular plate.

Most of these protrusions are called "stereocilia," which is a misnomer because although they contain actin and myosin filaments, microtubules and basal bodies characteristic for genuine cilia are not present. The longest of the protrusions is termed "kinocilium" and this is the only process to deserve its name because both microtubules and basal bodies can be found. The stereocilia of each hair cell are arranged in a row of increasing length, the kinocilium being the longest of all, rather like an inverted panpipe. The bulbous tips of all cilia are linked together by a fine filamentous material (tip links). The hairs narrow to a slender neck inserted in the cuticular plate. The free surface of the hairs is bathed in endolymph, which is rich in K^+. The whole structure is highly sensitive to deflection of the group of hairs toward or away from the kinocilium (but not in any other direction). Movement of the stereocilia in a direction of the kinocilium causes depolarization of the hair cell with an increased impulse frequency of the sensory nerve, whereas movement in an opposite direction brings about hyperpolarization and a decreased impulse frequency of the sensory nerve.

These near instantaneous responses are the result of gated ionic channels present in the tips of the hairs. Such

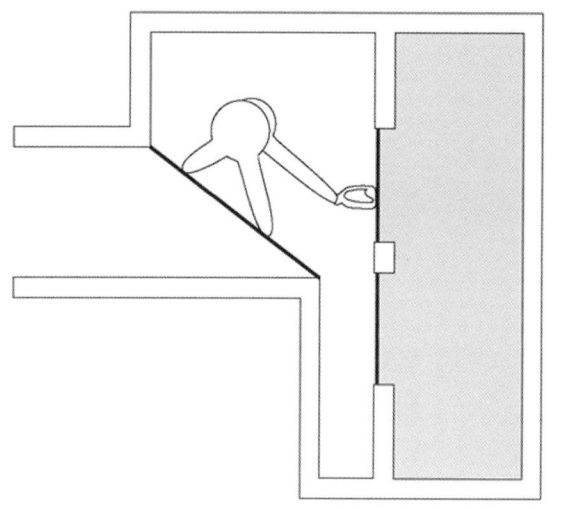 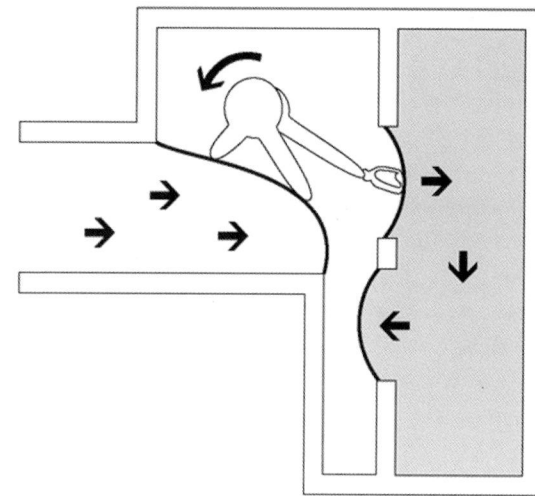

Fig. 1.10

Cartoon illustrating the action of auditory ossicles.

channels pass primarily K+ and Ca²⁺ and are controlled by the mechanical action via the tip links. At rest, the channels are just leaky enough to allow for some baseline activity. When a mechanical stimulus moves the group of stereocilia toward the kinocilium there will be an increasing number of open channels and an enhanced influx of K+ and Ca²⁺ ions, which induces a depolarization of the hair cell membrane. This initial depolarization is called the generator potential. This process is triggered primarily by the influx of Ca²⁺ ions via K+-sensitive (mechanotransduction) channels of stereocilia but, at a later stage, it is further enhanced by the activity of voltage-dependent Ca²⁺ channels of the cuticular plate. Accumulation of Ca²⁺ ions elicits the release of neurotransmitter substance from the hair cell, triggering an action potential that travels along the sensory nerve.

Having tackled the molecular and cellular mechanism of the receptor structure in general, we can now summarize how such a mechanism is operational in hearing. First, sound waves arriving from the exterior hit the tympanic membrane (similarly to the membrane of an old type microphone). Oscillation of the membrane (eardrum) is transmitted via the chain of auditory ossicles to the footplate of stapes and then the fluid compartment of the scala vestibuli. Amplification of signal is achieved by a difference between the area of the tympanic membrane and that of the stapedial footplate (the latter being approx 20 times smaller), whereby the force per unit area is about 20 times greater in the oval window than in the tympanic membrane. Further amplification is to the result of a lever action based on the difference in length of two processes, the manubrium of malleus and crus longum of incus, whose movements are tightly coupled.

As the membrane covering the oval window moves inward it exerts pressure first on the fluid space of the scala vestibuli. This is transmitted on to the vestibular membrane and then the basilar membrane (alternatively, the pressure wave traveling in the perilymphatic space may reach the basilar membrane also via the helicotrema). As the basilar membrane gives way, the subsequent increase in pressure in the scala tympani is released by the outward movement of secondary tympanic membrane covering the round window. The key element is therefore a deflection of the basilar membrane, supporting the organ of Corti. Owing to the inertia of the freely floating tectorial membrane, the hairs of the auditory hair cells embedded in it are always deflected whenever the basilar membrane moves up or down. These hair cells, arranged in one inner row and three to five outer rows, lack kinocilia but the centriole is present in its place. The stereocilia are lined up as the letters W or U with the base directed toward this centriole. Concerted movement of stereocilia in this direction elicits a generator potential on the base of the hair cells. Although the inner hair cells make contact with up to 10 afferent fibers, the outer hair cells are less richly innervated. However, they possess more efferent terminals that apparently can set the sensitivity of the system. The afferent terminals are contacted by characteristic "ribbon" synapses.

Whereas the detection of the intensity of sound can be relatively easily explained by the number of active units and the firing frequency of the afferent cochlear nerve fibers, the perception of sound frequency (pitch) has long been a matter of debate. The human ear can detect frequencies between 20 Hz and 20 kHz, a fair but unimpressive feat when compared to some examples from the animal kingdom (e.g., bats). Having noted the increasing width of the human basilar membrane from the round window end (100 μm) to the helicotrema (500 μm), Hermann von Helmholtz, back in the 19th century, put forward a theory according to which "string-like" segments of the basilar membrane would represent tuned resonators. Low-frequency tones (resonating the "long strings") would cause maximum perturbance of the basilar membrane near the helicotrema, whereas high-frequency tones (resonating the "short strings") would have the same effect near the round window. A simple experiment to demonstrate the effect of tuned resonators is done by opening a grand piano, holding the pedal pressed and loudly singing a tone at close distance to the strings. After cessation of singing, one can hear the faint echo of the same frequency tone from the resonating string of the piano. Although the presence of isolated "strings" within the basilar membrane was not verified, Helmholtz's "place theory" of frequency discrimination gained immense support by the work of the Hungarian-born George Békésy, Nobel laureate. According to this, a complex waveform travels along the basilar membrane, and the point of maximum amplitude is related to the frequency of sound. Recent studies have shown that the hair cells themselves are also tuned to respond to certain frequencies. This is primarily a faculty of the outer hair cells. The size and flexibility of the stereocilia differ in the hair cells, being small and stiff at the round window end and large and flexible at the helicotrema end. By such differences, as well as further differences in the composition of ionic channels each hair cell has an optimum response frequency that can be tuned by efferent stimuli. Thus, the classical place-related tuning of the basilar membrane can now be combined with and modified by an elaborate electromechanical tuning of the individual hair cell. Furthermore, the precise nature of tuning (pitch placement) is somewhat debatable: there is evidence to suggest that, at least in the gerbil cochlea, pitch placement may be based on a sharp cut-off phase (rather than the peak) of cochlear response.

A further impressive mechanism to increase the sharpness of tuning is the ability of outer hair cells to change the length of their hairs on depolarization (shortening) or hyperpolarization (elongation). Increased tension between stereocilia and tectorial membrane (by depolarization) leads to a greater opening probability of the mechanosensitive channels on the stereocilia. This process may also be operational in reverse: spontaneous contractions of hair cells may elicit movements of the basilar membrane and, ultimately, the tympanic membrane. Such movements can be detected in the form of otoacoustic emissions, an important clinical sign of cochlear function.

Physiology of Equilibrium

Adequate stimulus of the hair cells is a deflection of the stereocilia toward the kinocilia. The maculae of the utricle and saccule detect linear acceleration of the head in space. When the head is stationary, linear acceleration of the otoliths caused by gravity exerts pressure on the hairs either perpendicular to the hair cells (utricle) or as a sideways shearing force (saccule). When the head is bent in any direction (or the whole body is accelerating), the hair cells of the maculae will also be bent accordingly, eliciting an appropriate impulse combination in the afferent nerve.

The ampullar receptors of semicircular canals are specialized in detecting angular acceleration of the head. Rotation of the head in space sets the endolymph in motion inside the semicircular canal that is closest to the plane of head movement. First, the endolymph lags behind owing to its inertia (eliciting a stimulus by bending the hairs of cristae ampullares in one direction) but then, as the fluid gradually catches up with the duct wall, the stimulus is decreased. When the rotation is suddenly halted, the endolymph keeps rotating inside the duct, eliciting a stimulus by bending the hairs of cristae ampullares in an opposite direction. This is often accompanied by dizziness and even nausea, especially with the eyes closed (lacking visual cues the patient's orientation is based solely on vestibular input), and a characteristic pattern of eye movements called nystagmus. Movements of endolymph can be triggered also by heat (injection of warm water into the external auditory canal), an experiment that led Robert Bárány to his pioneering study of "caloric nystagmus," awarded with the Nobel prize in the early 20th century. In all cases, the adequate stimulus is transmitted to the hair cells by the cupula, in which the hairs are embedded, floating in the endolymph. Deflection of the hairs will then elicit nerve impulses by a mechanism discussed above.

CHRONOLOGICAL SUMMARY OF THE DEVELOPMENT OF THE TEMPORAL BONE

Fetal week 3	Neuroectoderm and ectoderm lateral to the first branchial groove condense to form the otic placode.
Fetal week 4	Otic placode invaginates to become the otic pit. Then the surface epithelium fuses and the otic pit is detached forming the otocyst or otic vesicle.
Fetal week 5	Wide dorsal and slender part of the otic vesicle appears. The dorsal part becomes the vestibular part of labyrinth and the ventral part forms the cochlea.
Fetal week 6	Condensation of the mesoderm of first and second arches into the hillocks of His. Malleus and incus appear as a single mass. Semicircular canals appear.
Fetal week 7	Stapes ring emerges around the stapedial artery. Further development of the basal turn of cochlea. Maculae differentiate into sensory and supporting cells.
Fetal week 8	The surface ectoderm of first branchial groove thickens and grows as an epithelial core toward the middle ear. The malleus and incus are separated and the incudomallear joint is formed. The semicircular canals and utricle are fully developed.
Fetal week 10	Pneumatization of the middle ear begins. Stapes transforms from ring shape into stirrup shape. Macular cell types are apparent and the otolithic membrane is under development. By week 11 the vestibular end-organs are formed and by week 16 all macular structures are developed and similar to the adult situation.
Fetal week 12	Hillocks fuse to form the auricle. Scala tympani develops its space. The perichondrium of otic capsule appears.
Fetal week 15	Tympanic ring almost fully developed. Formation of the membranous labyrinth is complete without the end-organ. First of the 14 centers of ossification can be identified. Scala vestibuli develops its space.
Fetal week 16	Auricular components recognizable but bulky. Ossicles reach adult size. Ossification appears at the long limb of the incus.
Fetal week 18	Ossification of stapes begins at the obturator surface of the stapedial base.
Fetal week 20	Stria vascularis and tectorial membrane are completed. Last of the 14 centers of ossification appears.
Fetal week 21	Epithelial core begins to resorb to form external acoustic meatus.
Fetal week 22	Antrum appears. Fissula ante fenestram begins ossification.
Fetal week 23	Ossification of otic capsule completed. Membranous and bony labyrinths are adult size excepting the endolymphatic sac, which continues to grow until adulthood.
Fetal week 26	Tunnel of Corti and spaces of Nuel are formed.
Fetal week 28	External acoustic meatus is patent (resorption complete). Eardrum appears. Ossification of stapes is complete except for the vestibular surface of footplate.
Fetal week 30	Excavation of the tympanic cavity is complete.
Birth	The shape of the auricle is definitive but it continues to grow until year 9. The external acoustic meatus is not ossified, ossification is complete at about year 3. The middle ear is well formed, it enlarges only slightly after birth. The mastoid process appears at year 1 and is not fully formed until about 3 years of age. The tympanic ring and external acoustic meatus are ossified also by year 3. Eustachian tube is 17 mm at birth and grows to 36 mm. Part of the manubrium of malleus remains cartilagineous and never ossifies.

ATLAS PLATES I (EMBRYOLOGY)

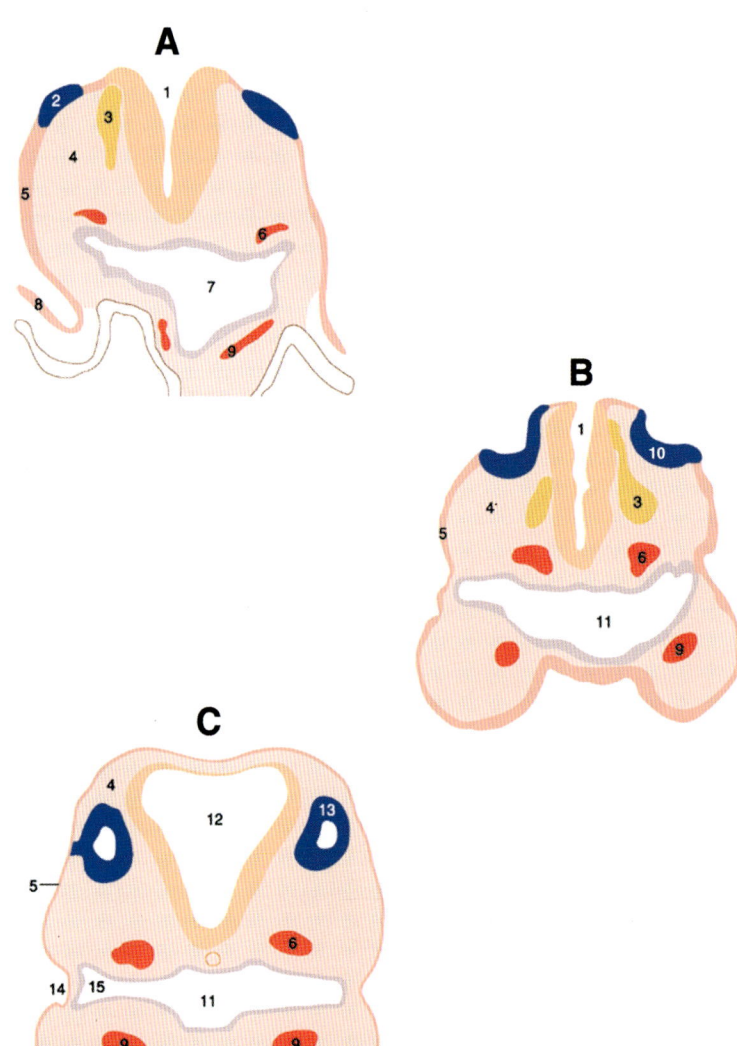

Fig. 1.11

Development of the otic vesicle. Transverse sections of human embryos of 9 somites (A) and 16 somites (B) stages, and approx 4 mm length (C) (after Arey).

1. Neural groove
2. Otic placod
3. Neural crest
4. Mesenchyme
5. Ectoderm
6. Dorsal (primitive) aorta
7. Foregut
8. Amnion
9. Ventral (primitive) aorta
10. Otic pit
11. Pharynx
12. Neural tube
13. Otic vesicle
14. First branchial groove
15. First pharyngeal pouch

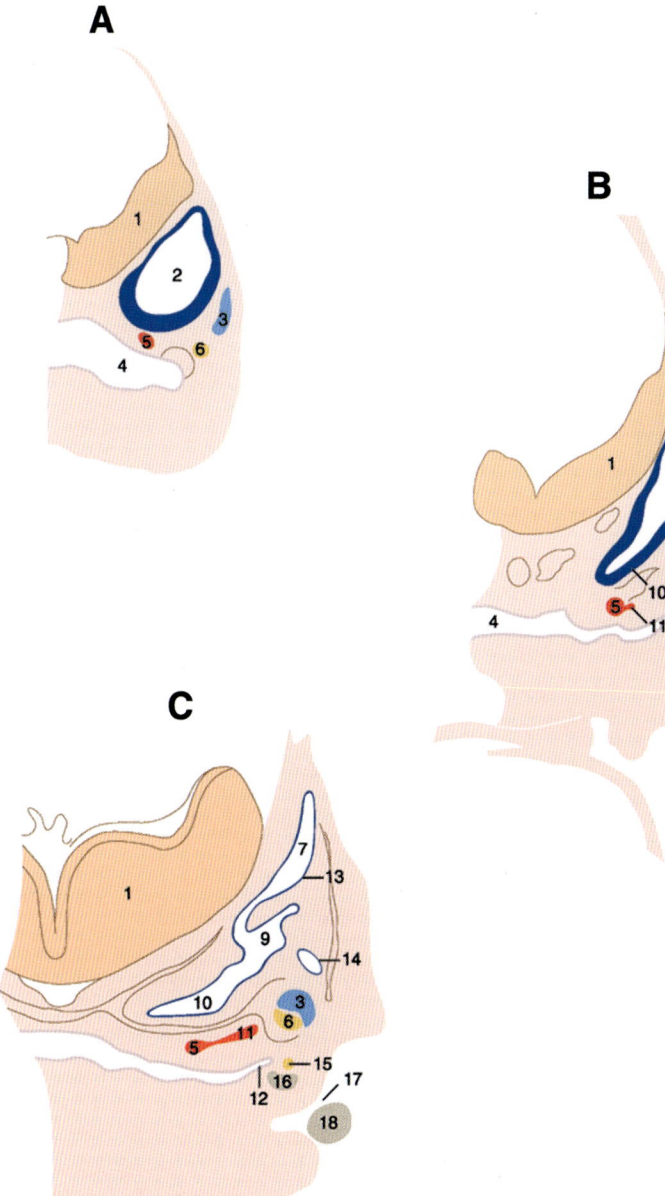

Fig. 1.12

Development of the ear. Transverse sections of human embryos of 7 mm (4 weeks of age, A), 10.5 mm (4½ weeks, B) and 15 mm (5½ weeks, C) (after Siebenmann).

1. Rhombencephalon
2. Otic vesicle
3. Internal jugular vein
4. Pharynx
5. Internal carotid artery
6. Facial nerve
7. Endolymphatic duct
8. Duct portion of otic vesicle
9. Utriculosaccular portion of otic vesicle
10. Cochlear portion of otic vesicle
11. Stapedial artery
12. Primitive tympanic cavity
13. Anterior semicircular duct
14. Lateral semicircular duct
15. Chorda tympani nerve
16. Manubrium of malleus (primordial cartilage)
17. First branchial groove (future external acoustic meatus)
18. Tragus

ATLAS—EAR

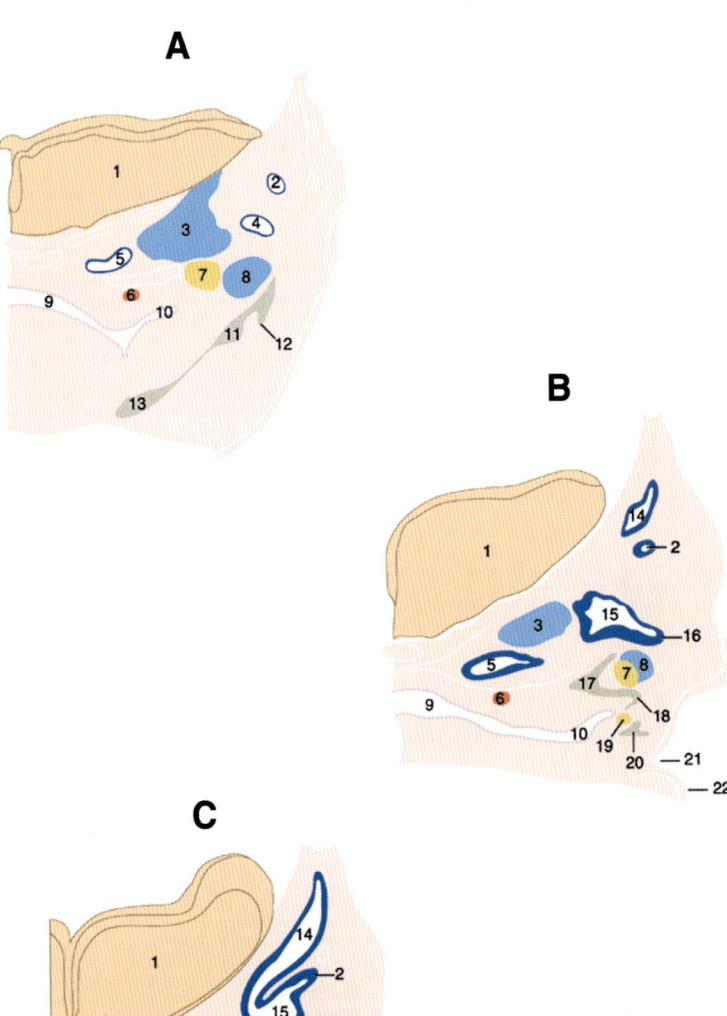

Fig. 1.13

Development of the ear. Consecutive transverse sections of a human embryo of 15 mm length (5½ weeks), drawing A representing most cranial and drawing C most caudal levels (after Siebenmann).

1. Rhombencephalon
2. Anterior semicircular duct
3. Acousticofacial ganglion
4. Ampulla of anterior semicircular duct
5. Cochlea
6. Internal carotid artery
7. Facial nerve
8. Internal jugular vein
9. Pharynx
10. Primitive tympanic cavity
11. Head of malleus (primordial cartilage)
12. Body of incus (primordial cartilage)
13. Meckel's cartilage
14. Endolymphatic duct
15. Vestibule
16. Ampulla of semicircular duct
17. Stapes (primordial cartilage)
18. Crus longum (long limb) of incus (primordial cartilage)
19. Chorda tympani nerve
20. Manubrium of malleus (primordial cartilage)
21. First branchial groove
22. Ear hillock
23. Stapedial artery

ATLAS—EAR

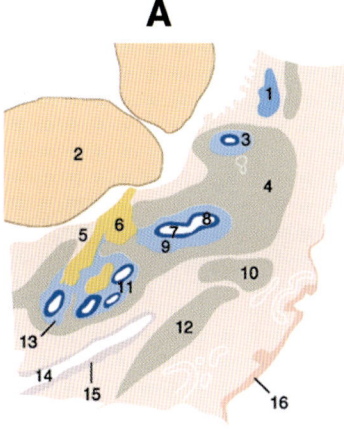

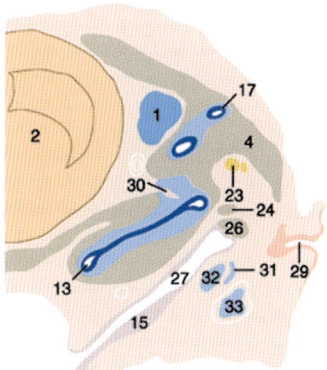

Fig. 1.14

Development of the ear. Consecutive transverse sections of a human embryo of 4.5 cm length (2½ months of age, A–B) and another embryo of 8 cm length (early fourth month of age, C), (after Siebenmann)

1. Transverse sinus
2. Hindbrain
3. Anterior semicircular duct
4. Otic capsule
5. Cochlear nerve
6. Vestibular nerve
7. Ampulla of anterior semicircular duct
8. Ampulla of lateral semicircular duct
9. Anlage of perilymphatic space
10. Body of incus (primordial cartilage)
11. Second and third turn of cochlear duct
12. Malleus continuous with Meckel's cartilage
13. First turn of cochlear duct
14. Pharyngotympanic tube
15. Tubotympanic epithelium
16. Epidermis
17. Posterior semicircular duct
18. Endolymphatic duct
19. Utricle
20. Saccule
21. Lateral semicircular duct
22. Stapes
23. Facial nerve
24. Crus longum (long limb) of incus
25. External acoustic meatus, non-canalized part
26. Manubrium of malleus
27. Primitive tympanic cavity
28. Tympanic membrane
29. External acoustic meatus, canalized part
30. Internal opening of cochlear canaliculus
31. Anterior part of tympanic ring
32. Anterior process of malleus
33. Mandible

ATLAS—EAR

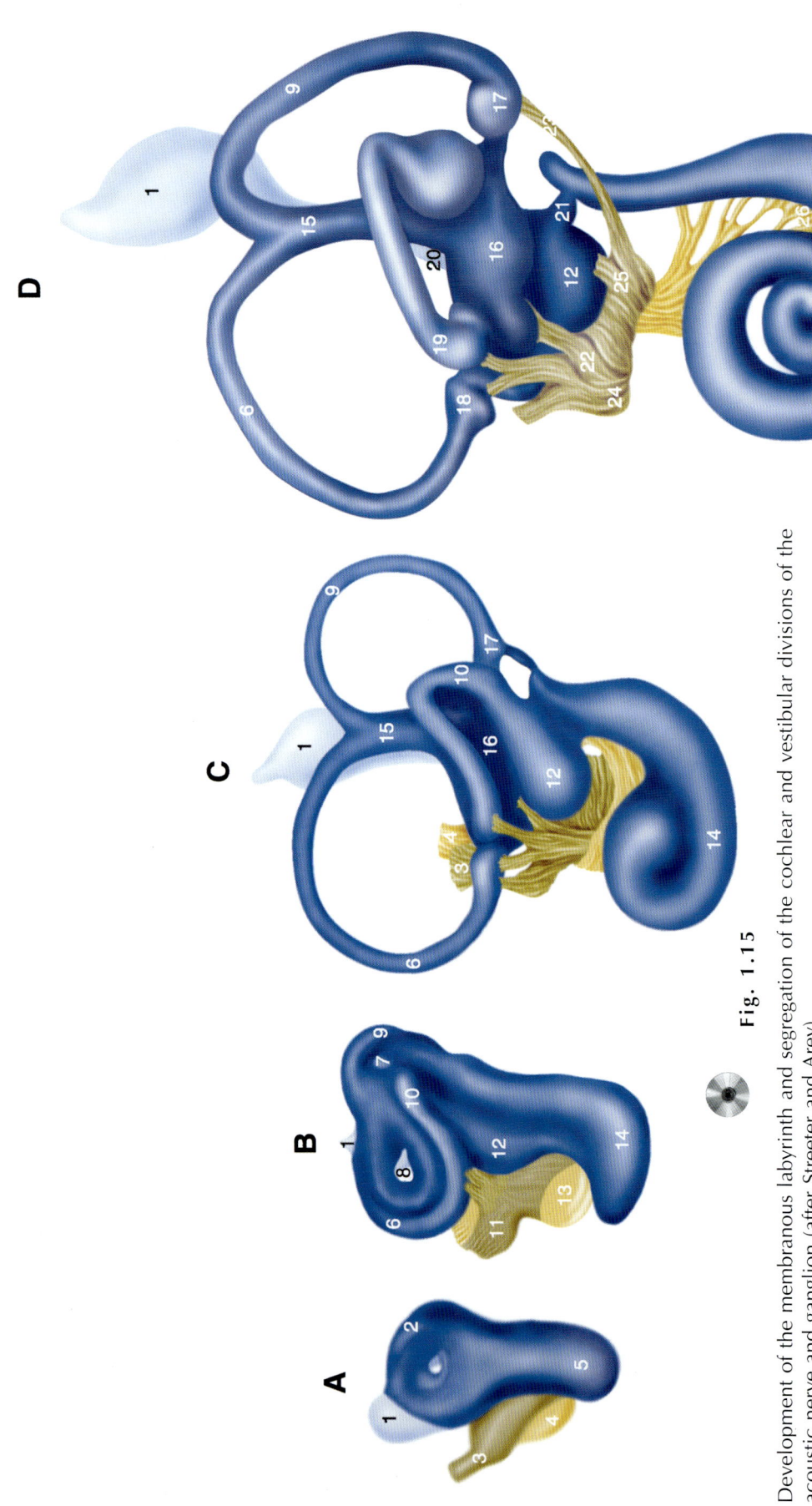

Fig. 1.15

Development of the membranous labyrinth and segregation of the cochlear and vestibular divisions of the acoustic nerve and ganglion (after Streeter and Arey).

A at 6.6 mm   B at 13 mm   C at 20 mm   D at 30 mm body length.

1. Endolymphatic sac
2. Vestibular pouch
3. Vestibular nerve
4. Cochlear nerve
5. Cochlear pouch
6. Anterior semicircular duct
7. Absorption focus for posterior semicircular duct
8. Absorption focus for anterior semicircular duct
9. Posterior semicircular duct
10. Lateral semicircular duct
11. Vestibular nerve
12. Saccule
13. Cochlear nerve
14. Cochlea
15. Crus commune
16. Utricle
17. Ampulla of posterior semicircular duct
18. Ampulla of anterior semicircular duct
19. Ampulla of lateral semicircular duct
20. Endolymphatic duct
21. Ductus reuniens
22. Utriculoampullar nerve
23. Posterior ampullar nerve
24. Vestibular ganglion (of Scarpa)
25. Saccular nerve
26. Spiral ganglion (of Corti)

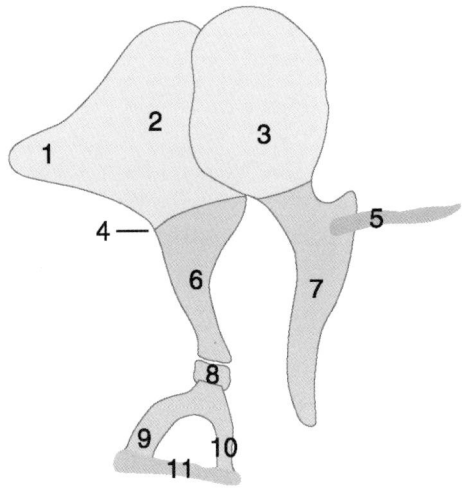

Fig. 1.16

Development of the auditory ossicles. Schematic figure according to Anson and Bast. The constituents marked by the indexes 1–3 belong to the Meckel's cartilage, whereas those marked 6–10 belong to the Reichert's cartilage. The part labeled 11 derives from the otic capsule and that labeled 5 develops by independent intramembranous ossification.

1. Crus posterius of incus
2. Body of incus
3. Head of malleus
4. Developmental border representing the course of chorda tympani nerve
5. Anterior process of malleus
6. Crus longum of incus
7. Manubrium of malleus
8. Head of stapes
9. Crus posterius of stapes
10. Crus anterius of stapes
11. Footplate of stapes

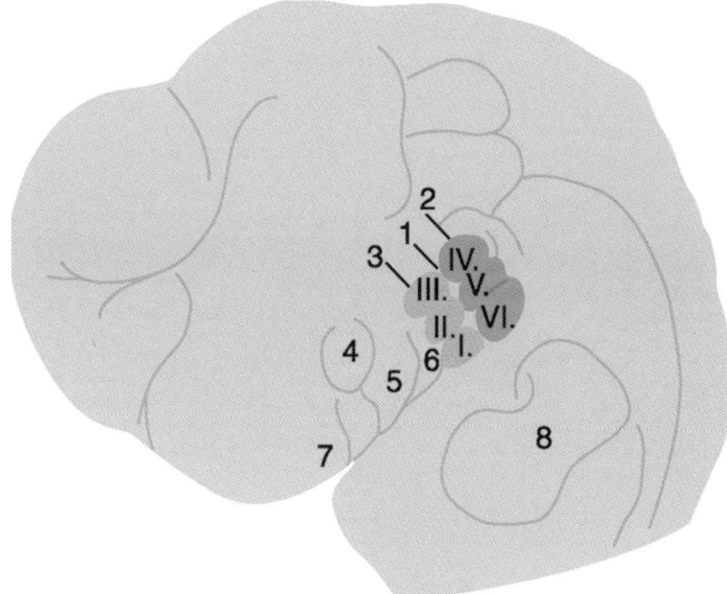

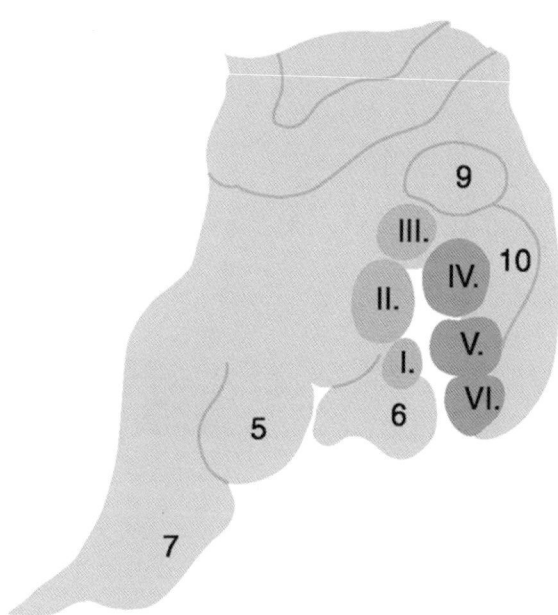

Fig. 1.17

Development of the auricle. Early second month of embryonic age (after Schwalbe). The ear hillocks are numbered by Roman numerals. I. Tragus, II-III. Helix, IV-V. Anthelix, VI - Antitragus.

1. First branchial groove
2. Second branchial (hyoid) arch
3. First branchial (mandibular) arch
4. Primordium of eye
5. Maxillary process
6. Mandibular process
7. Frontal prominence
8. Upper limb bud
9. Otic vesicle
10. Auricular fold

Development of the Internal Ear

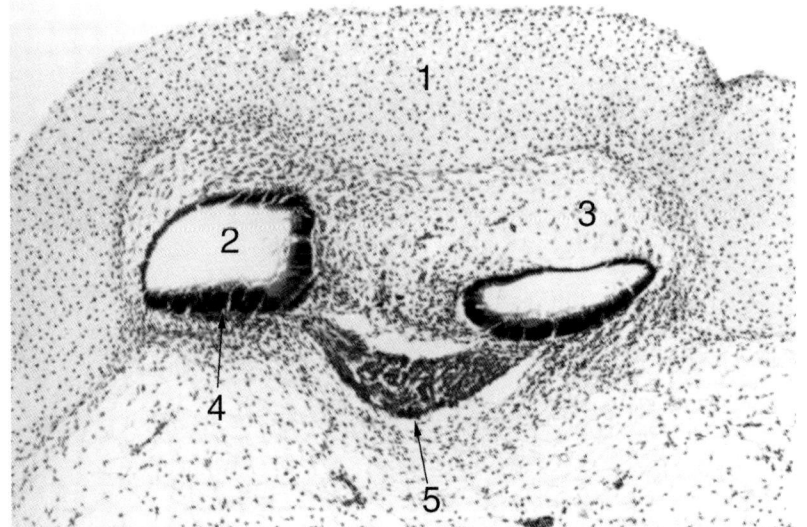

Fig. 1.18

Histological specimen of an 8½-week-old human embryo, depicting the developing cochlea. HE staining.

1. Precartilage of the otic capsule
2. Cochlear duct
3. Embryonic mesenchyme in the otic capsule
4. Neuroepithelium
5. Cochlear nerve

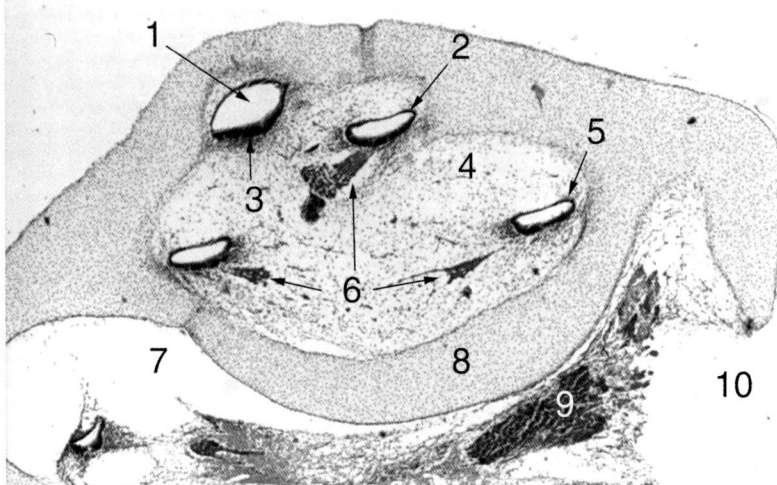

Fig. 1.19

Histological specimen of an 8½-week-old human embryo, depicting the developing cochlea. HE staining.

1. Cochlear duct
2. Third turn of the cochlea
3. Neuroepithelium
4. Embryonic mesenchyme in the otic capsule
5. Second turn of the cochlea
6. Spiral ganglion
7. First turn of the cochlea
8. Precartilage of the otic capsule
9. Cochlear nerve
10. Internal acoustic meatus

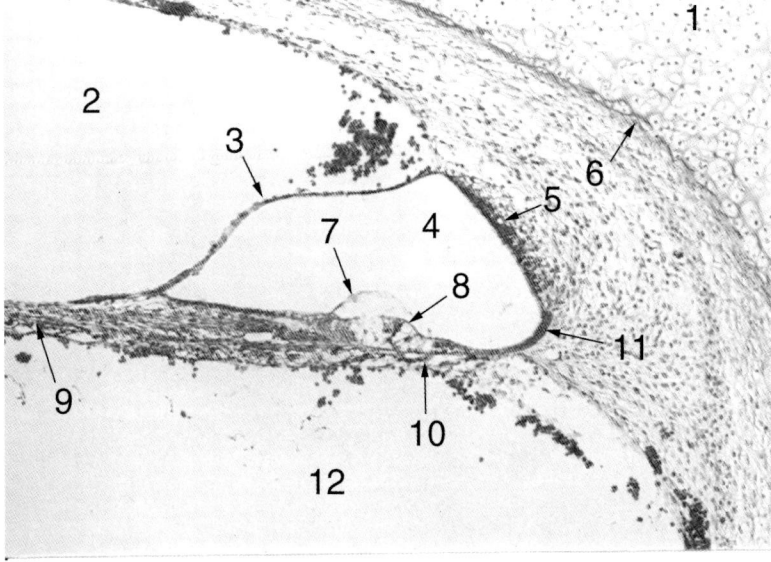

Fig. 1.20

Histological specimen of an 11½-week-old human embryo, depicting the developing organ of Corti. Azan staining.

1. Cartilagineous otic capsule
2. Scala vestibuli
3. Vestibular membrane (of Reissner)
4. Cochlear duct
5. Stria vascularis
6. Perichondrium
7. Tectorial membrane
8. Developing organ of Corti
9. Site of future lamina spiralis ossea (spiral bony lamina)
10. Basilar membrane
11. Sulcus spiralis externus
12. Scala tympani

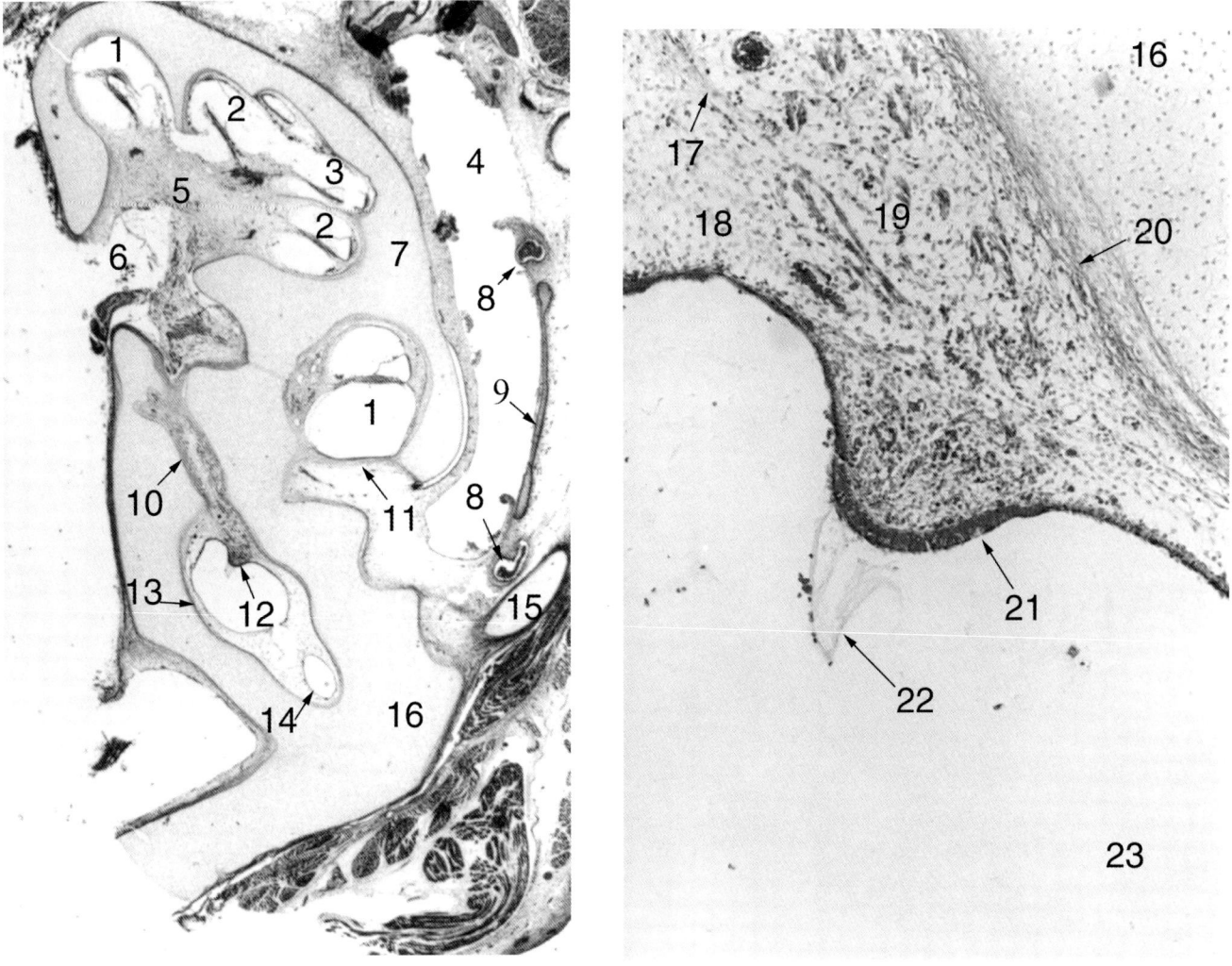

Figs. 1.21 and 1.22

Histological specimens of an 11½-week-old human embryo, demonstrating the development of the temporal bone. Figure 1.22 shows an enlarged field of Fig. 1.21, depicting the developing ampullar crest of the posterior semicircular duct. Azan staining.

1. First turn of the cochlea
2. Second turn of the cochlea
3. Third turn of the cochlea
4. Tympanic cavity
5. Modiolus
6. Internal acoustic meatus
7. Cochlear part of the cartilagineous otic capsule
8. Fibrocartilagineous ring (tympanic annulus)
9. Tympanic membrane
10. Foramen singulare
11. Area of the round window niche
12. Ampullar crest of the posterior semicircular canal
13. Ampullar part of the posterior semicircular canal
14. Posterior semicircular canal (crus simplex)
15. Reichert's cartilage
16. Vestibular part of the cartilagineous otic capsule
17. Border between the vascular layer and future perilymphatic space
18. Future perilymphatic space
19. Vascular layer
20. Perichondrium of otic capsule
21. Neuroepithelium of ampullar crest
22. Cupula
23. Endolymphatic space

ATLAS—EAR

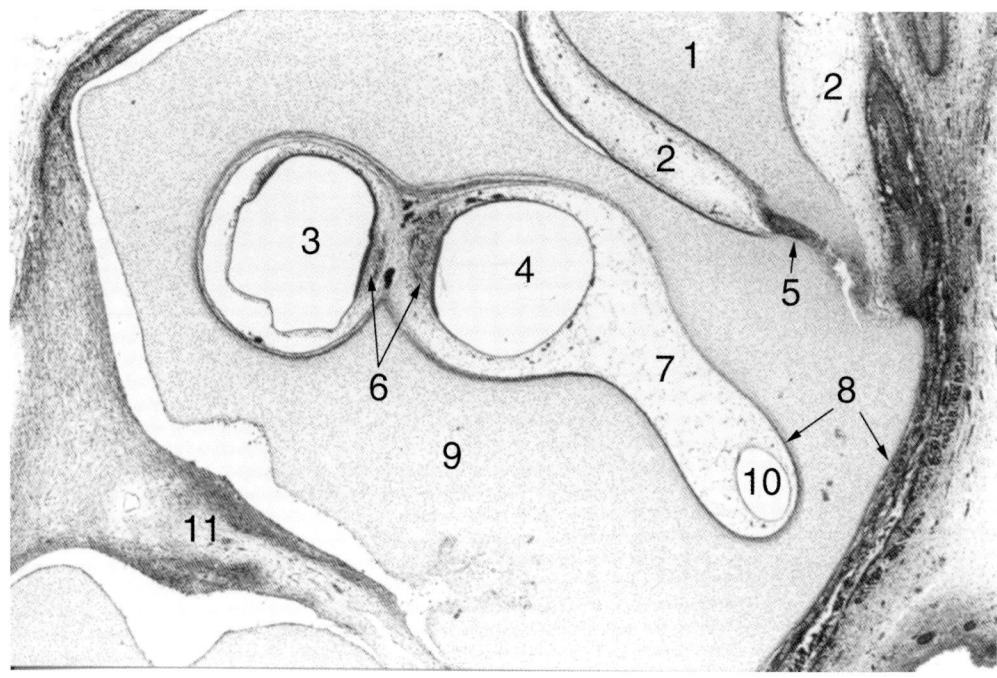

Fig. 1.23

Histological specimen of an 11½-week-old human embryo, demonstrating the development of semicircular canals. Azan staining.

1. Incus
2. Embryonic mesenchyme of the epitympanum
3. Ampullar part of the anterior semicircular canal
4. Ampullar part of the lateral semicircular canal
5. Developing incudial fossa
6. Neuroepithelia of ampullar crests
7. Perilymphatic space
8. Perichondrium
9. Cartilagineous otic capsule
10. Lateral semicircular canal (crus simplex)
11. Connective tissue in the subarcuate fossa

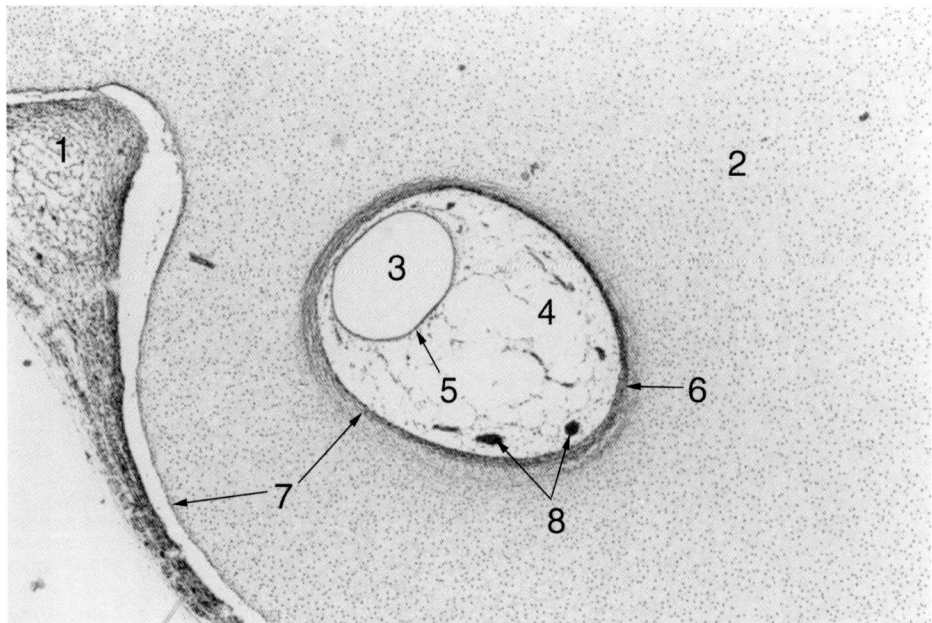

Fig. 1.24

Histological specimen of an 11½-week-old human embryo, demonstrating the development of endolymphatic and perilymphatic spaces. Azan staining.

1. Connective tissue of subarcuate fossa
2. Cartilage of otic capsule
3. Endolymphatic space
4. Perilymphatic space
5. Membranous labyrinth
6. Condensation of perilymphatic mesenchyme
7. Perichondrium
8. Blood vessels in the perilymphatic space

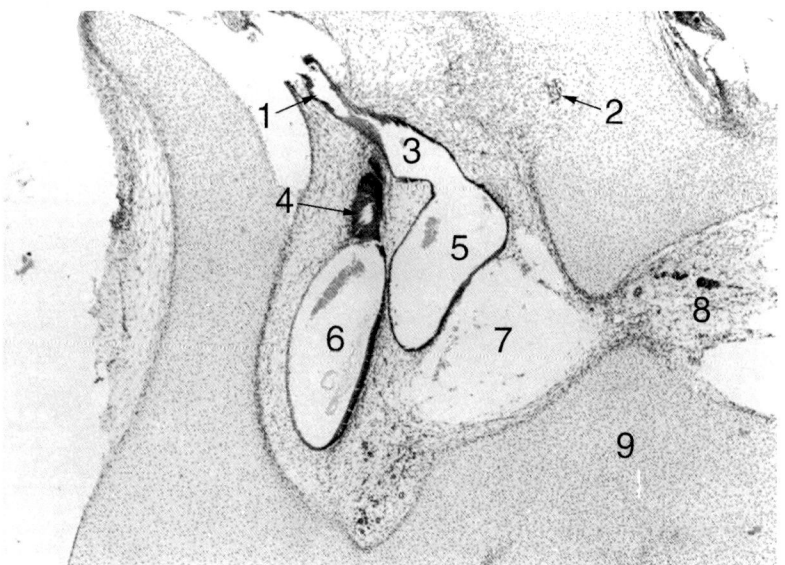

Fig. 1.25

Histological specimen of an 8½-week-old human embryo, demonstrating the developing vestibule. HE staining.

1. Endolymphatic duct
2. Primordium of cochlear canalicle
3. Utriculosaccular duct
4. Utriculosaccular duct (tangential section)
5. Saccule
6. Utricle
7. Vestibule (perilymphatic space)
8. Cochlea
9. Precartilage of the otic capsule

Development of the Middle Ear

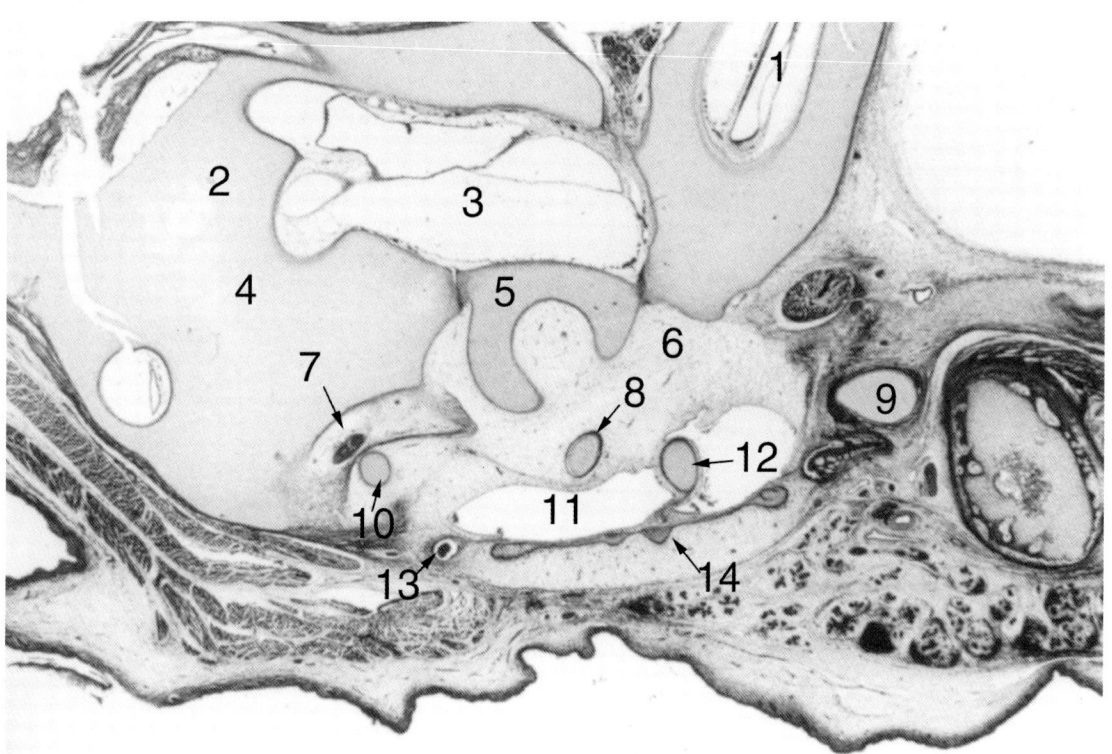

Fig. 1.26

Histological specimen of an 11½-week-old human embryo, demonstrating the development of the middle ear and auditory ossicles. Azan staining.

1. Cochlea
2. Cartilagineous otic capsule
3. Vestibule
4. Posterior semicircular canal
5. Stapes
6. Embryonic mesenchyme in the middle ear
7. Facial nerve
8. Crus longum of incus
9. Meckel's cartilage
10. Reichert's cartilage
11. Tympanic cavity
12. Manubrium of malleus
13. Fibrocartilagineous ring (tympanic annulus)
14. Tympanic membrane

Fig. 1.27 and 1.28

Histological specimens of an 11½-week-old human embryo, demonstrating the development of the tympanic cavity, tympanic membrane and external acoustic meatus. Figure 1.28 shows an enlarged view of the developing tympanic membrane. Azan staining.

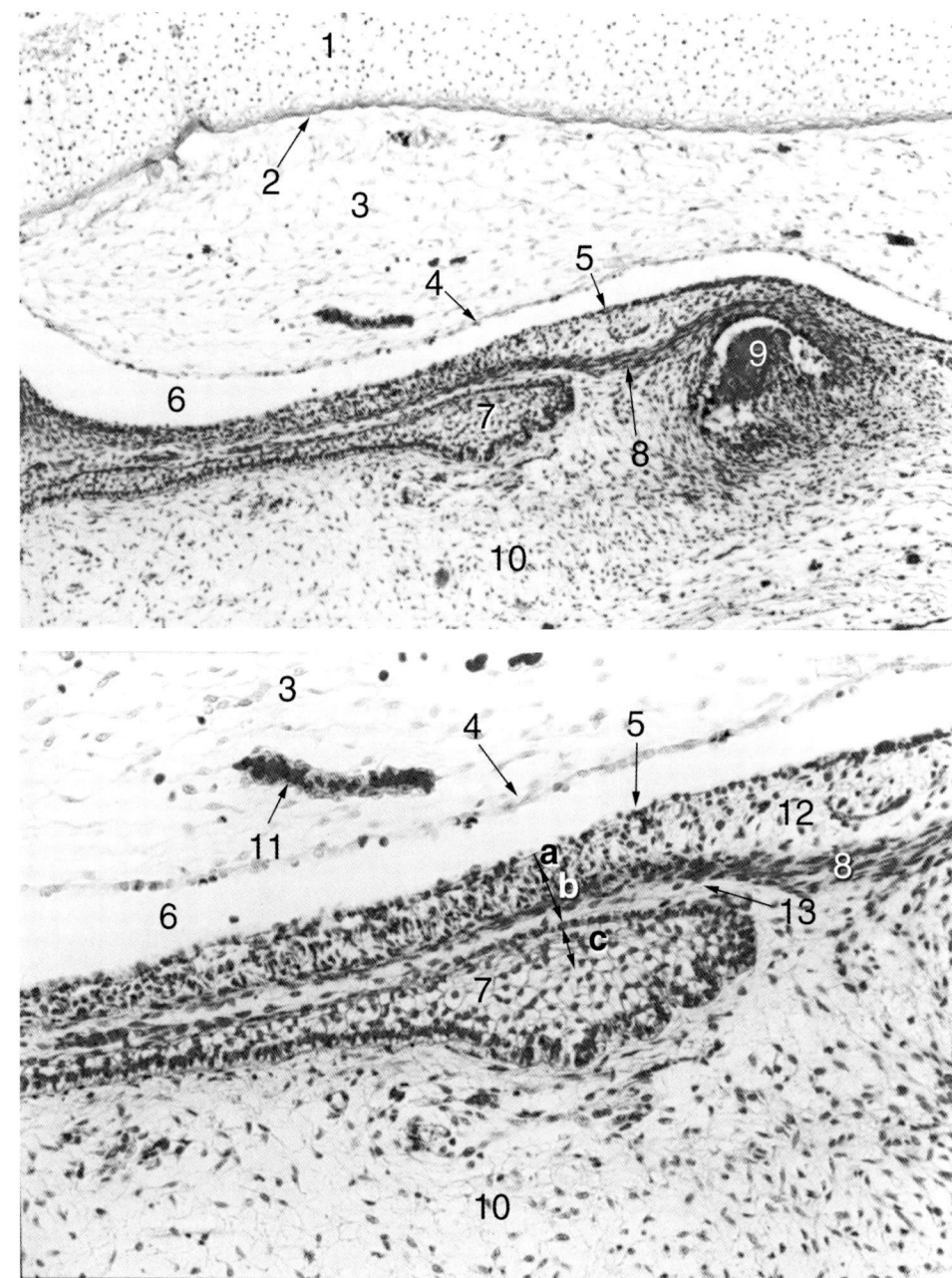

1. Cartilagineous otic capsule
2. Perichondrium
3. Embryonic mesenchyme in the middle ear
4. Internal wall of the tubotympanic mucosa
5. External wall of the tubotympanic mucosa
6. Tympanic cavity
7. Mass of ectodermal cells developing into the external acoustic meatus by means of canalization (meatal plug)
8. Radially oriented connective tissue forming the intermediate layer of the future tympanic membrane
9. Fibrocartilagineous ring (tympanic annulus)
10. Connective tissue
11. Blood vessel in the embryonic mesenchyme
12. External part of the middle layer of future tympanic membrane
13. Internal part of the middle layer of future tympanic membrane

Sublayers of developing tympanic membrane
    a mucous stratum.
    b intermediate fibrous stratum.
    c cuticular stratum

Development of the External Ear

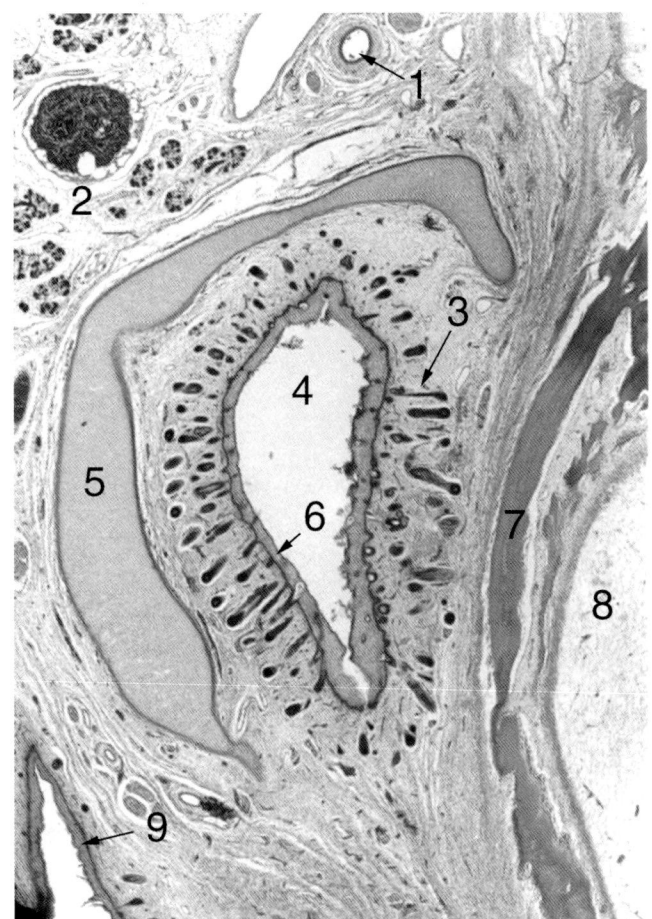

Fig. 1.29

Histological specimen of a 20-week-old human embryo, demonstrating the developing external acoustic meatus. HE staining.

1. Superficial temporal artery
2. Developing parotid gland
3. Developing hair follicles and sebaceous glands of external acoustic meatus
4. Lumen of external acoustic meatus
5. Cartilage of auricle
6. Skin of the external acoustic meatus
7. Squamous temporal bone
8. Embryonic mesenchyme of the tympanic cavity
9. Temporal skin

Fig. 1.30

Histological specimen of a 20-week-old human embryo, demonstrating the developing pinna. HE staining.

1. Temporal skin
2. Cartilage of auricle
3. Developing integument of auricle
4. External acoustic meatus
5. Developing hair follicles of the earlobe

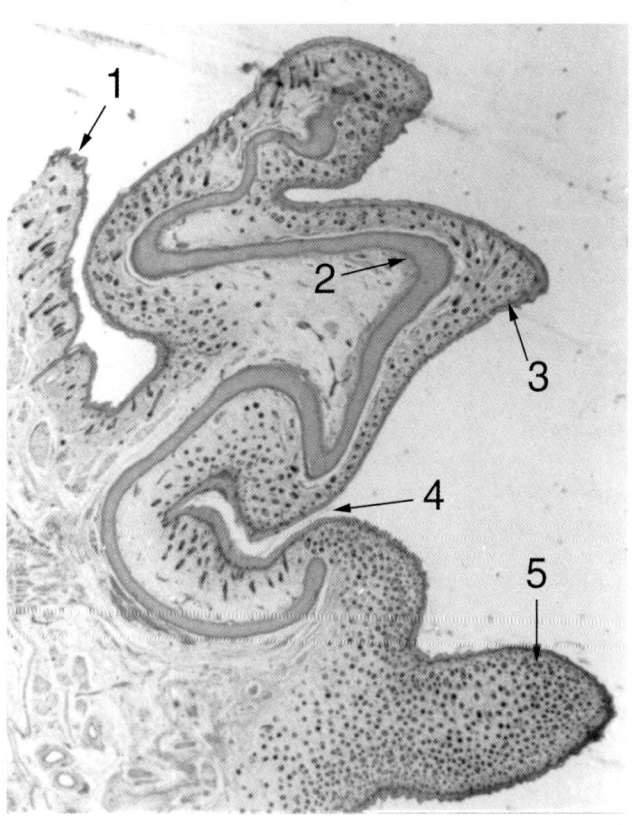

ATLAS—EAR 27

ATLAS PLATES II (MICRODISSECTION AND ENDOSCOPY)

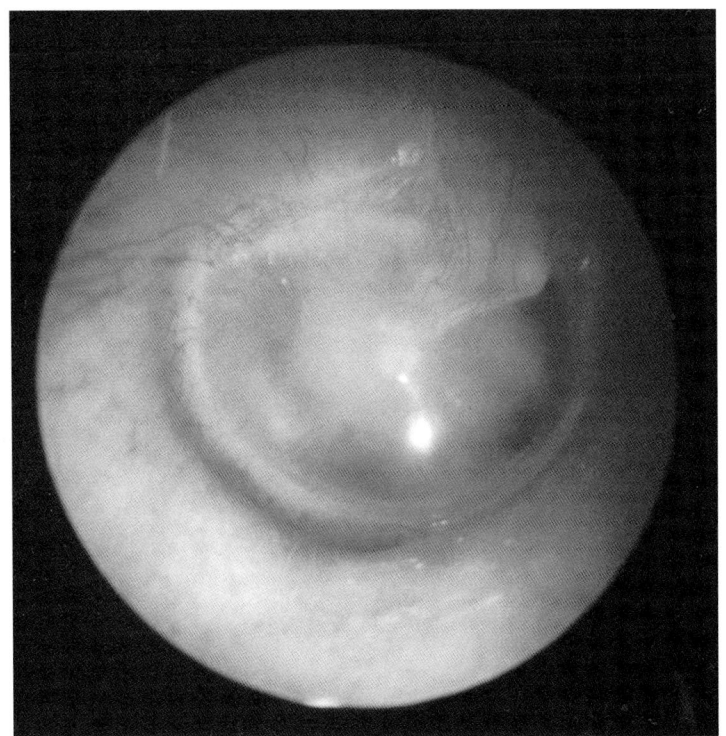

Fig. 1.31

Otoscopic image of the right tympanic membrane. The structures appearing through the translucent tympanic membrane are indicated by asterisk. Lateral view.

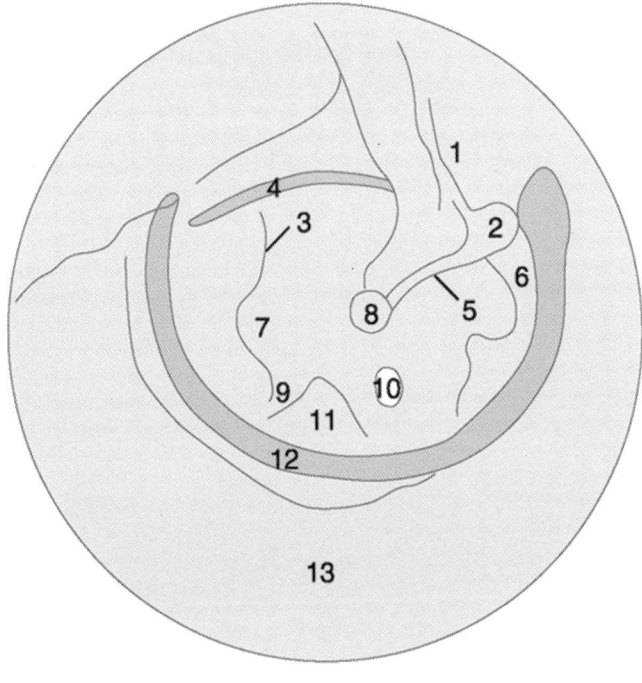

1. Pars flaccida
2. Prominentia mallei (malleolar prominence)
3. Round niche (round window)*
4. Chorda tympani
5. Handle of malleus (manubrium)
6. Tympanic opening of pharyngotympanic tube*
7. Pars tensa
8. Umbo
9. Sustentaculum promontorii*
10. Light reflex triangle
11. Hypotympanic air cells*
12. Fibrocartilagineous ring
13. External acoustic meatus

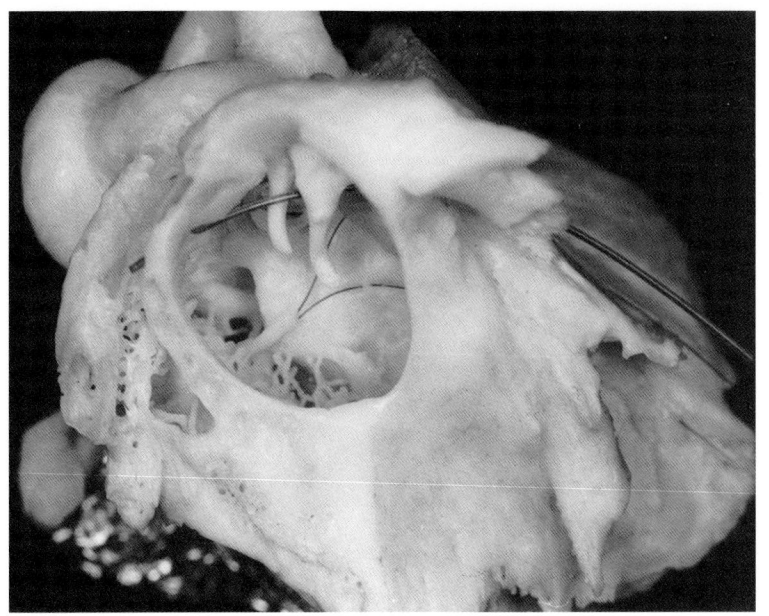

Fig. 1.32.

Microdissection specimen of the right middle ear (tympanic cavity and ossicles). The external acoustic meatus was removed. The grey probe demonstrates the position of the chorda tympani nerve, and the blue probe shows the course of the tympanic nerve on the promontory.

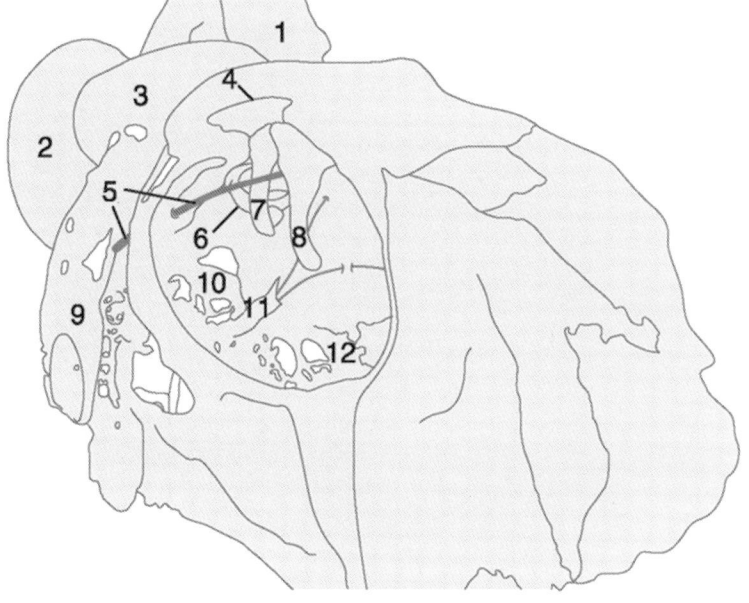

1. Anterior semicircular canal
2. Posterior semicircular canal
3. Lateral semicircular canal
4. Incisura tympanica (Rivini)
5. Chorda tympani nerve
6. Stapes
7. Incus
8. Manubrium of malleus
9. Facial canal
10. Round window
11. Sustentaculum promontorii (canaliculus tympanicus)
12. Hypotympanum

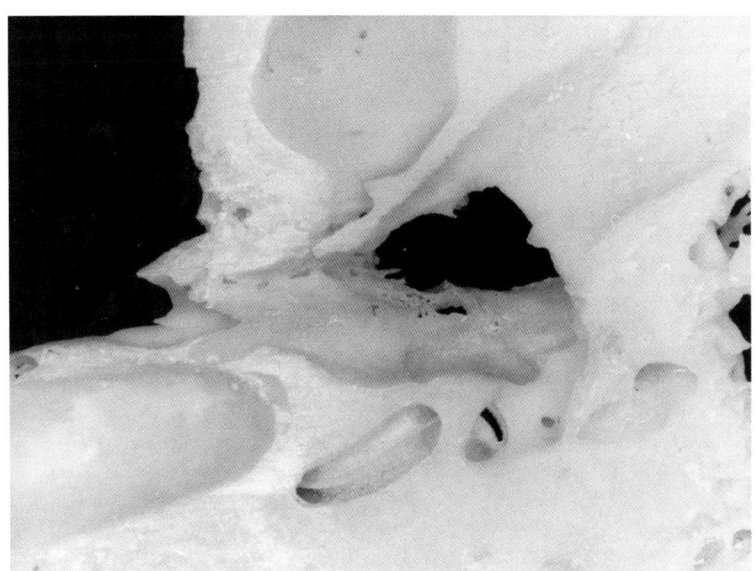

Fig. 1.33.

Horizontal section of the left temporal bone through the external acoustic meatus. Inferior view. Note the proximity of mandibular fossa and external acoustic meatus, and the thin partition between the carotid canal and the first turn of cochlea.

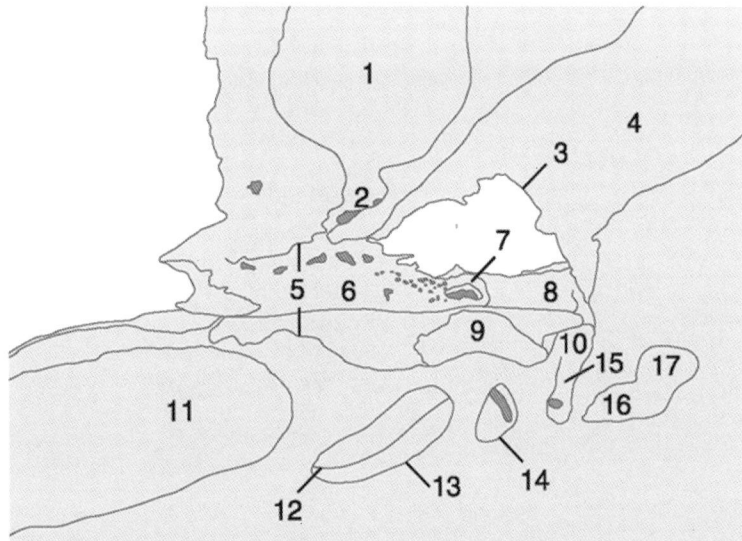

1. Fossa articularis
2. Fissura Glaseri
3. Incisura tympanica (Rivini)
4. External acoustic meatus
5. Pharyngotympanic tube
6. Semicanal of tensor tympani muscle
7. Cochleariform process
8. Facial canal, tympanic part
9. Stapedial fossa
10. Pyramidal eminence
11. Carotid canal
12. Lamina spiralis ossea
13. Cochlea, first turn
14. Round window
15. Tympanic sinus
16. Cavity of pyramidal eminence
17. Facial canal, mastoid part

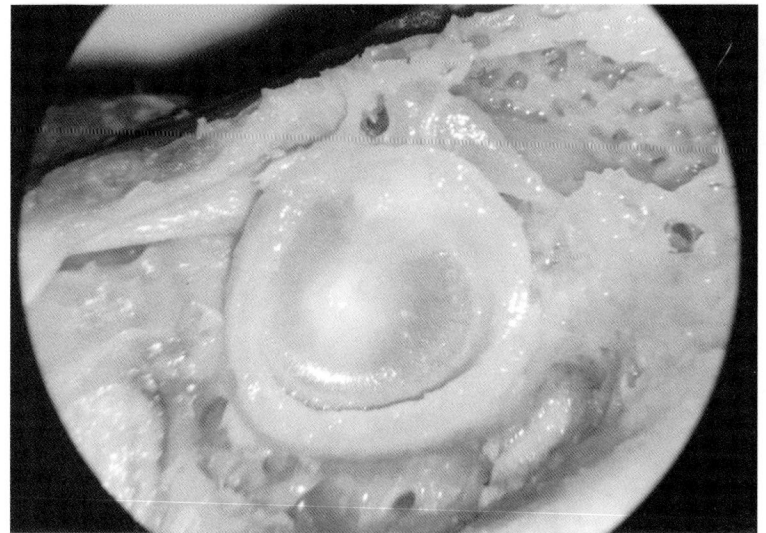

Fig. 1.34.

Microdissection specimen of the lateral wall of middle ear. The tympanic part of temporal bone was cut away by a surgical drill. Lateral view.

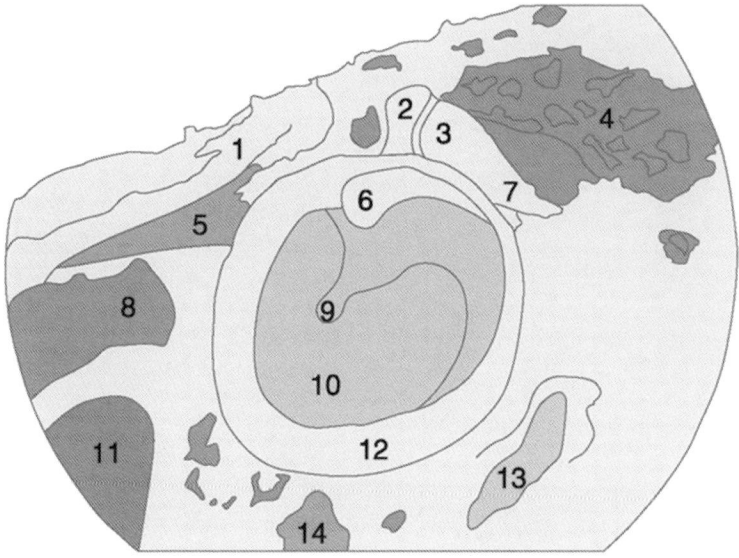

1. Tegmen tympani
2. Head of malleus
3. Body of incus
4. Mastoid antrum
5. Tensor tympani muscle
6. Prominentia mallei
7. Posterior crus of incus
8. Pharyngotympanic tube
9. Stria malleolaris (handle of malleus appearing through the tympanic membrane)
10. Tympanic membrane
11. Internal carotid artery
12. Remnant of tympanic part of temporal bone
13. Facial nerve
14. Hypotympanic air cells

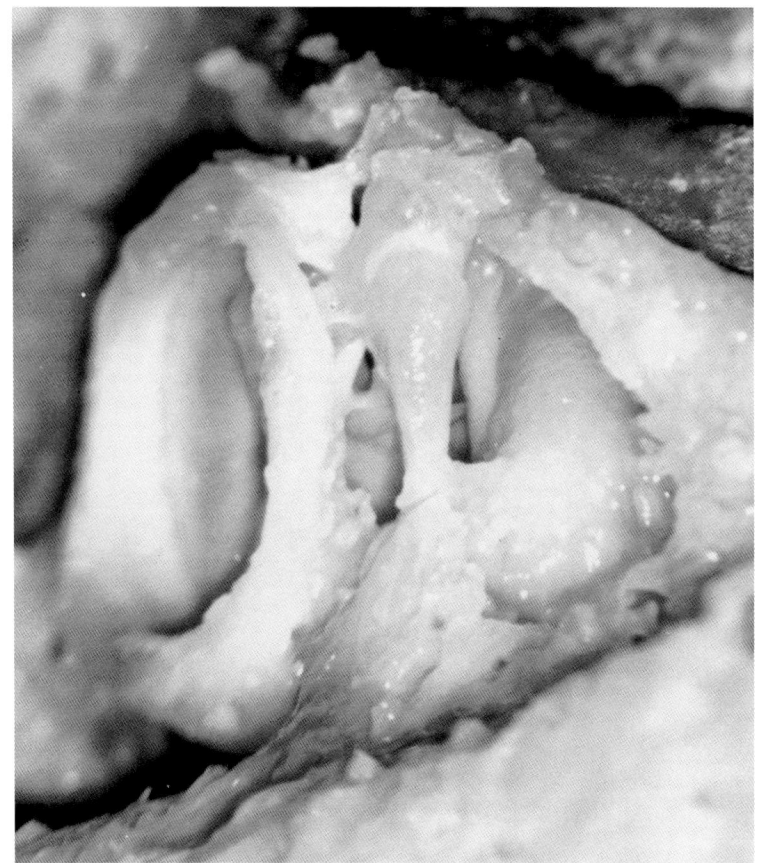

Fig. 1.35

Microdissection specimen of middle ear following removal of tegmen tympani and the posterior wall of tympanic cavity. Major part of external acoustic meatus has been also removed. Postero-superior view.

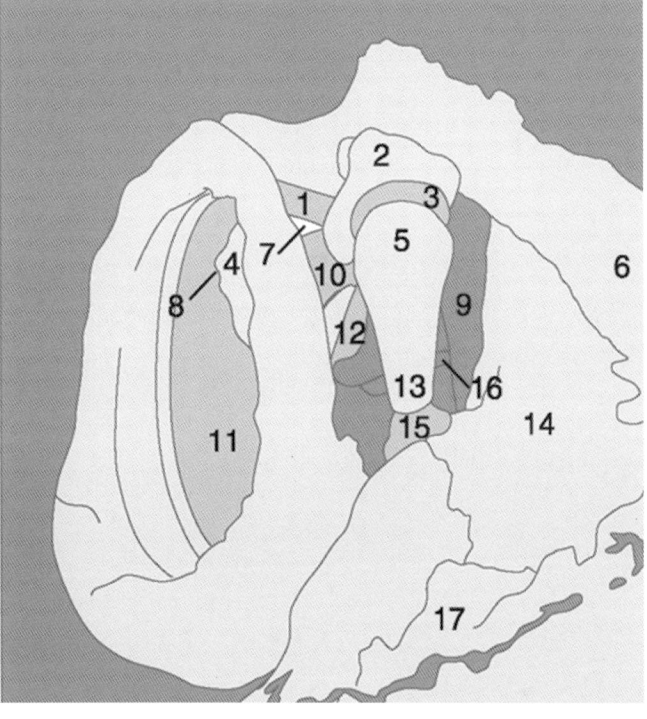

1. Lateral mallear fold
2. Head of malleus
3. Incudomalleolar joint
4. Pars flaccida of tympanic membrane
5. Body of incus
6. Tegmen tympani
7. Prussak's space
8. Prominentia mallei
9. Facial canal
10. Lateral incudal fold (plica membranae tympani posterior)
11. Pars tensa of tympanic membrane
12. Chorda tympani nerve
13. Posterior crus of incus
14. Lateral semicircular canal
15. Posterior ligament of incus
16. Posterior crus of stapes
17. Mastoid air cells

Fig. 1.36

Microdissection specimen of middle ear following removal of tegmen tympani. Superior view.

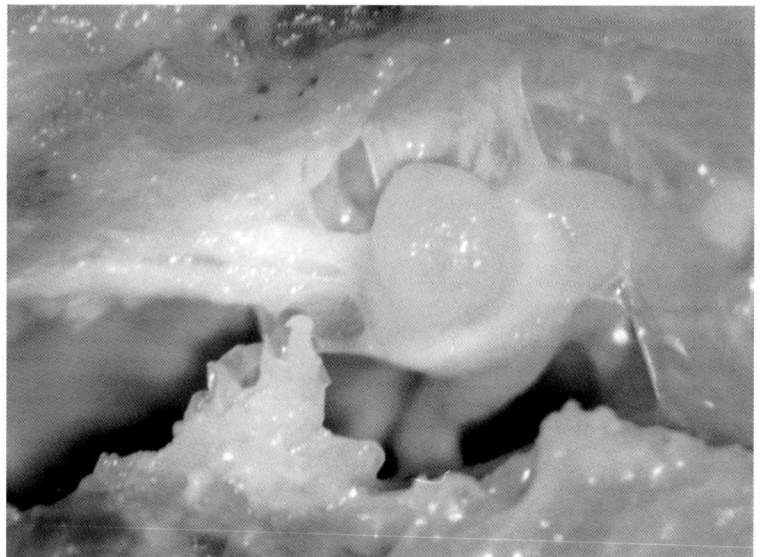

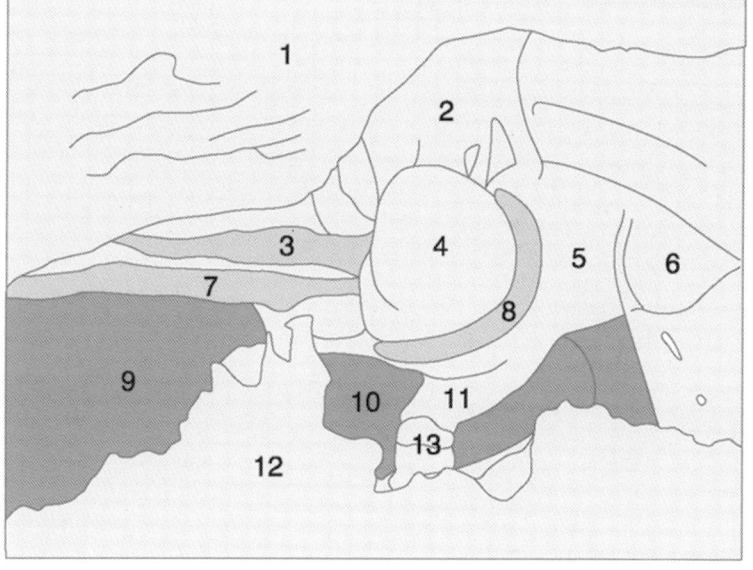

1. Squamous temporal bone
2. Lateral epitympanic space
3. Anterior ligament of malleus
4. Head of malleus
5. Body of incus
6. Posterior crus of incus
7. Chorda tympani nerve
8. Incudomalleolar joint
9. Tympanic opening of pharyngotympanic tube
10. Lateral epitympanic space
11. Long crus of incus
12. Petrous temporal bone
13. Incudostapedial joint

ATLAS—EAR

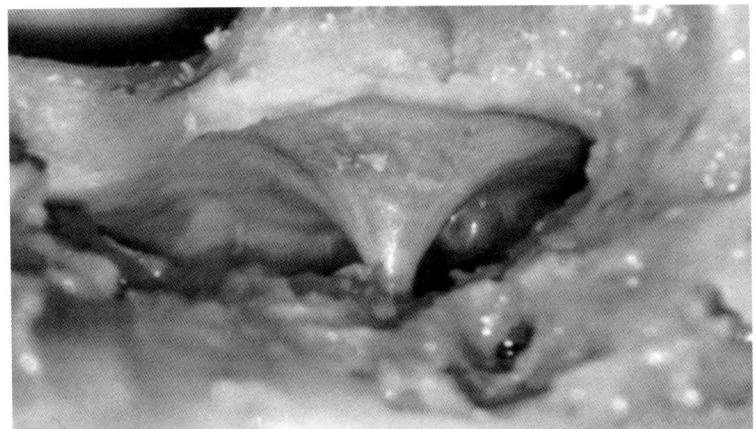

Fig. 1.37

Microdissection specimen of tympanic membrane. The hypotympanic air cells have been removed. Inferior view.

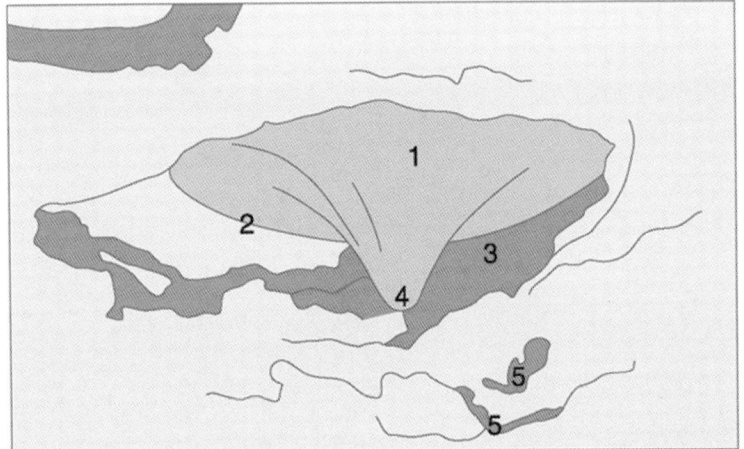

1. Tympanic membrane
2. Fibrocartilagineous ring
3. Tympanic cavity
4. Umbo
5. Air cells

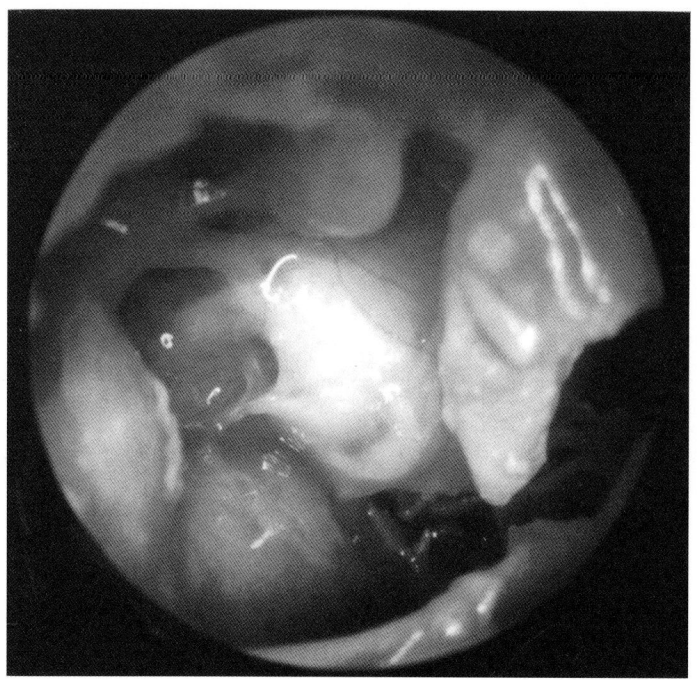

Fig. 1.38

Tympanic cavity. Endoscopic image obtained by fresh cadaver examination following removal of tympanic membrane. Lateral view.

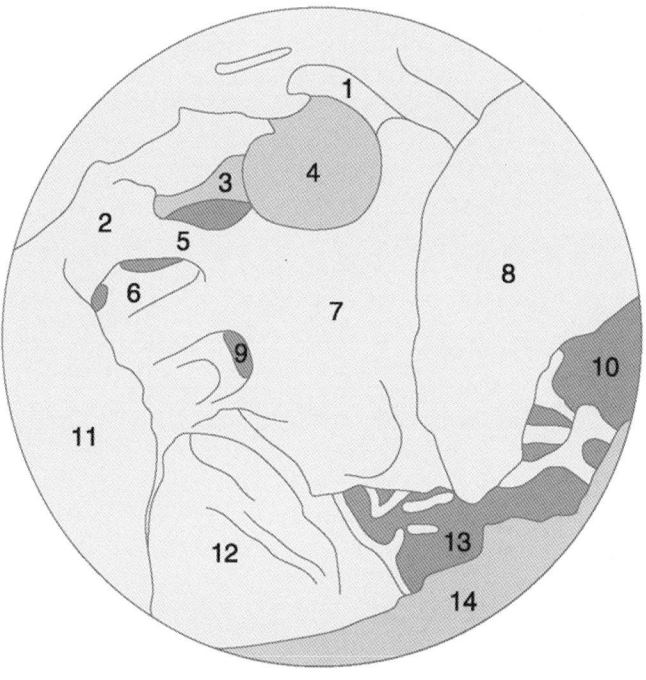

1. Long crus of incus
2. Pyramidal eminence
3. Tendon of stapedius muscle
4. Incudostapedial joint
5. Ponticulus medialis
6. Sinus tympani
7. Promontory
8. Handle of malleus (manubrium)
9. Round window
10. Tympanic opening of pharyngotympanic tube
11. Styloid prominence
12. High bulb (tympanic protrusion of internal jugular vein)
13. Hypotympanic air cells
14. Remnant of tympanic membrane

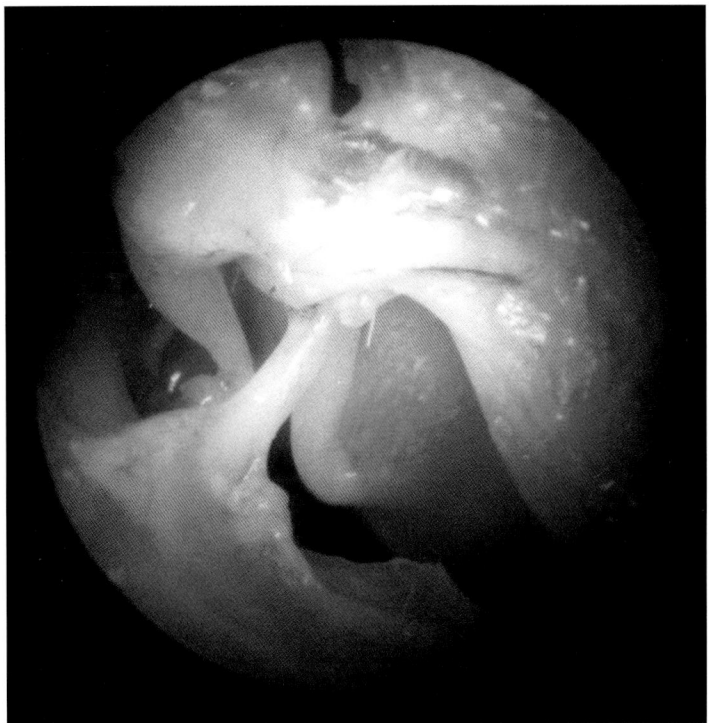

Fig. 1.39

Tympanic cavity. Antero-supero-medial view. Endoscopic image obtained by fresh cadaver examination. Note the characteristic course of the chorda tympani nerve, including two adjoining arches, one behind the manubrium of malleus and another in front of the manubrium. The anterior tympanic artery is also visible.

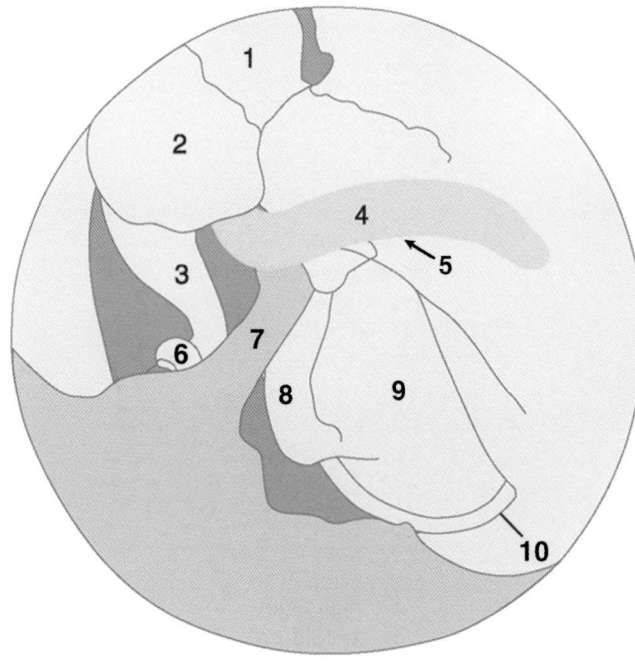

1. Head of malleus
2. Body of incus
3. Crus longum of incus
4. Chorda tympani nerve
5. Anterior tympanic artery
6. Lenticular process
7. Tendon of tensor tympani muscle
8. Manubrium of malleus
9. Tympanic membrane
10. Fibrocartilagineous ring (annulus tympanicus)

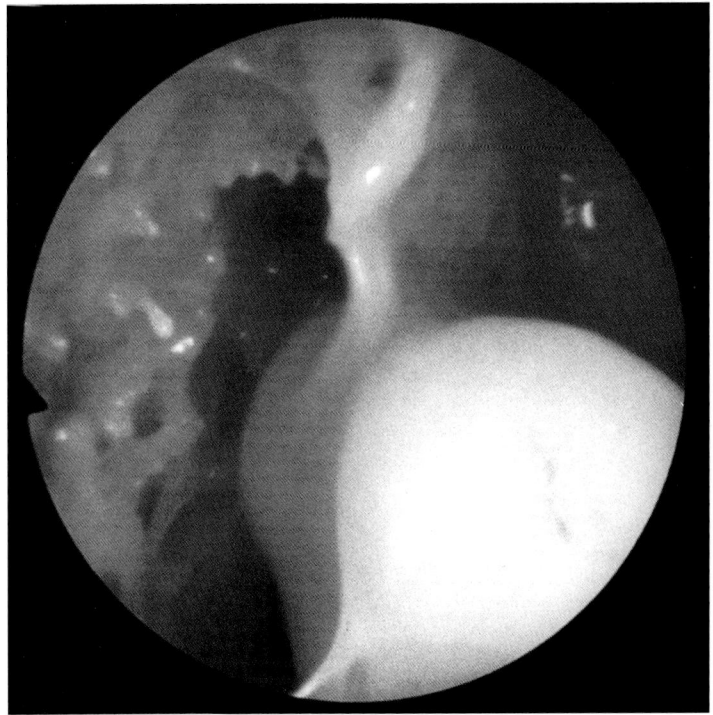

Fig. 1.40

Epitympanic space. Endoscopic image obtained by fresh cadaver examination.
Posterior view.

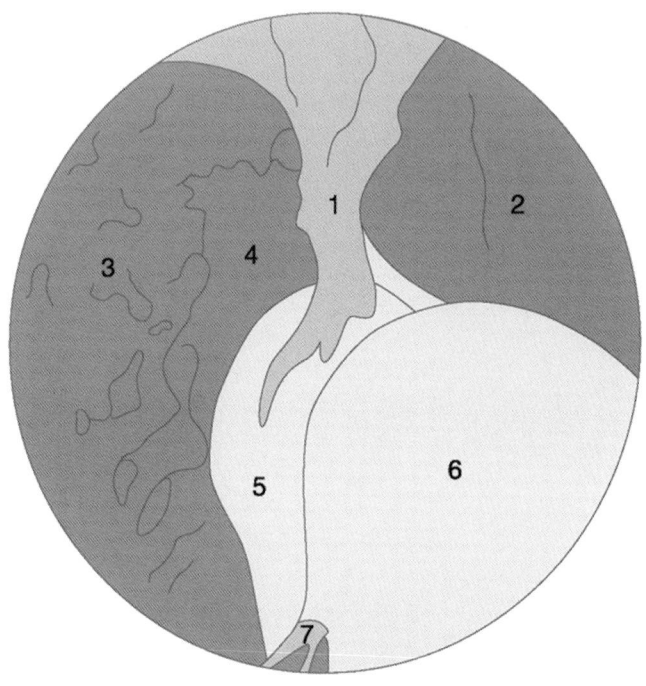

1. Superior ligament of malleus
2. Medial epitympanic space
3. Tegmen tympani
4. Lateral epitympanic space
5. Head of malleus
6. Body of incus
7. Mucosal fold

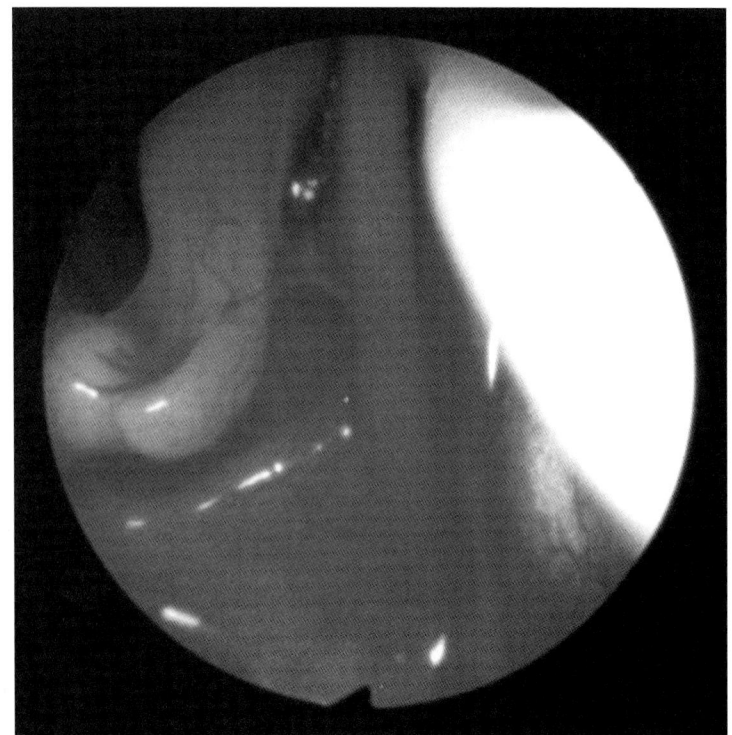

Fig. 1.41

Tympanic cavity, posterolateral part. Anteromedial view. Endoscopic image obtained by fresh cadaver examination.

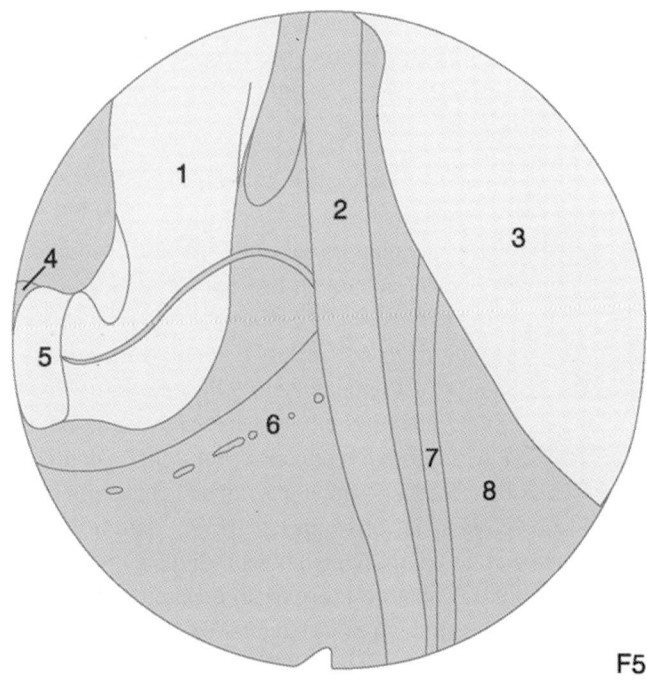

1. Crus longum of incus
2. Chorda tympani nerve
3. Handle (manubrium) of malleus
4. Incudostapedial joint
5. Lenticular process
6. Ponticulus lateralis
7. Fibrocartilagineous ring
8. Tympanic membrane

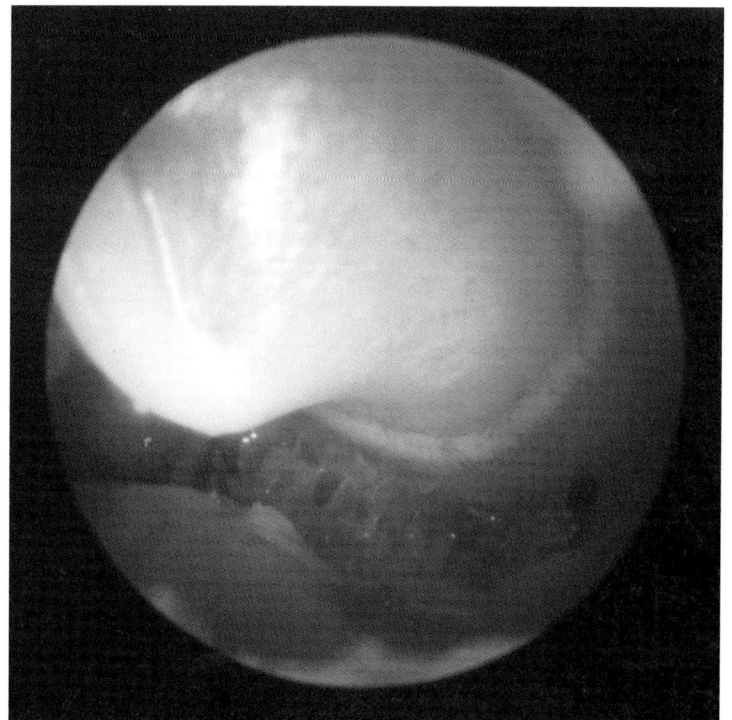

Fig. 1.42

Internal surface of tympanic membrane. Endoscopic image obtained by fresh cadaver examination. Anteromedial view.

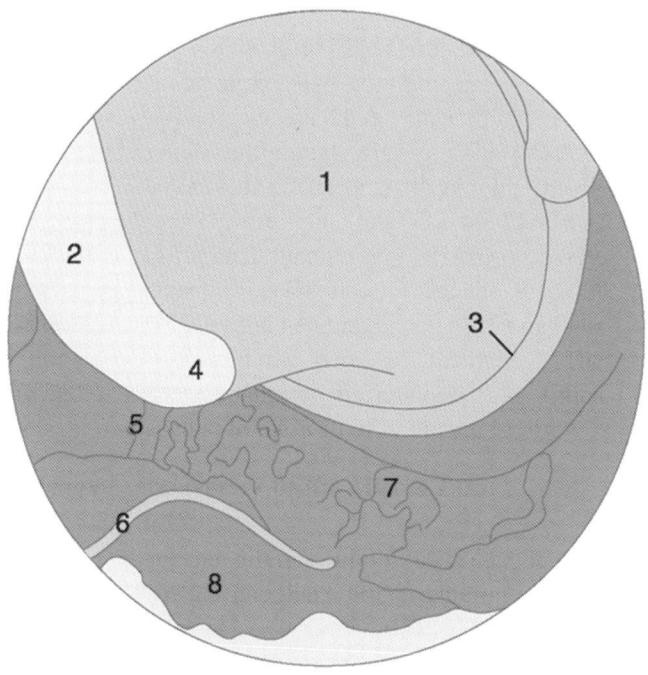

1. Tympanic membrane
2. Handle of malleus (manubrium)
3. Fibrocartilagineous ring
4. Umbo (with spatula of manubrium)
5. Angustia tympanica (hourglass-shaped narrowing between the promontory and the tympanic membrane)
6. Jacobson's canal (tympanic nerve and inferior tympanic artery)
7. Hypotympanic air cells
8. Promontory

ATLAS—EAR

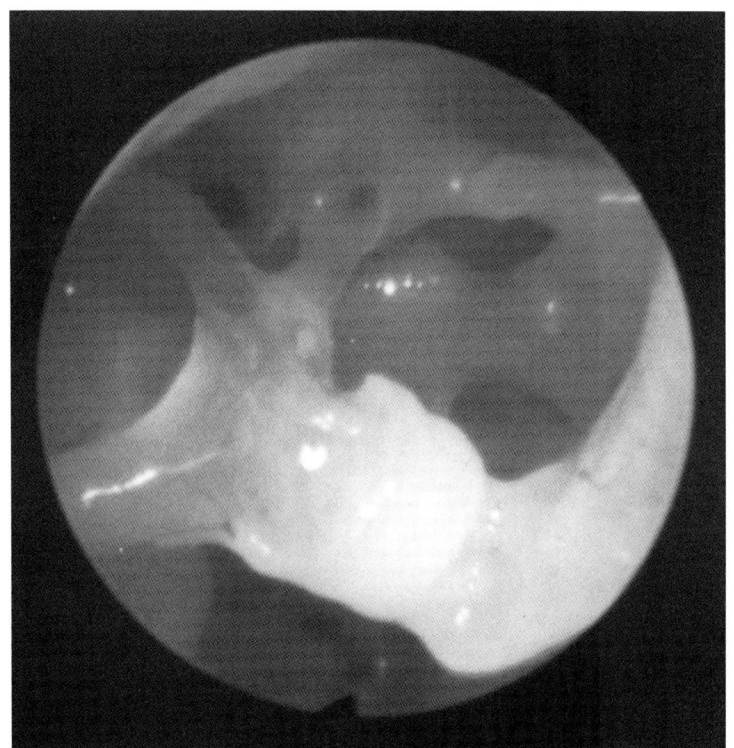

Fig. 1.43

Incudostapedial joint. Endoscopic image obtained by fresh cadaver examination. Note the apposition between the lenticular process of incus and head of stapes. The stapedius muscle is attached at the transition between the posterior crus and head of stapes (muscular process). Antero-supero-lateral view.

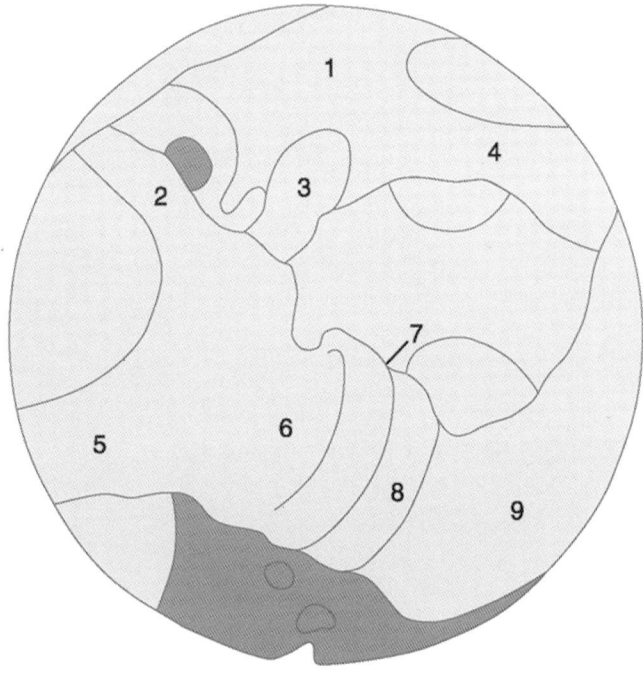

1. Pyramidal eminence
2. Posterior crus of stapes
3. Tendon of stapedius muscle
4. Ponticulus lateralis
5. Anterior crus of stapes
6. Head of stapes
7. Incudostapedial joint
8. Lenticular process
9. Long crus of incus

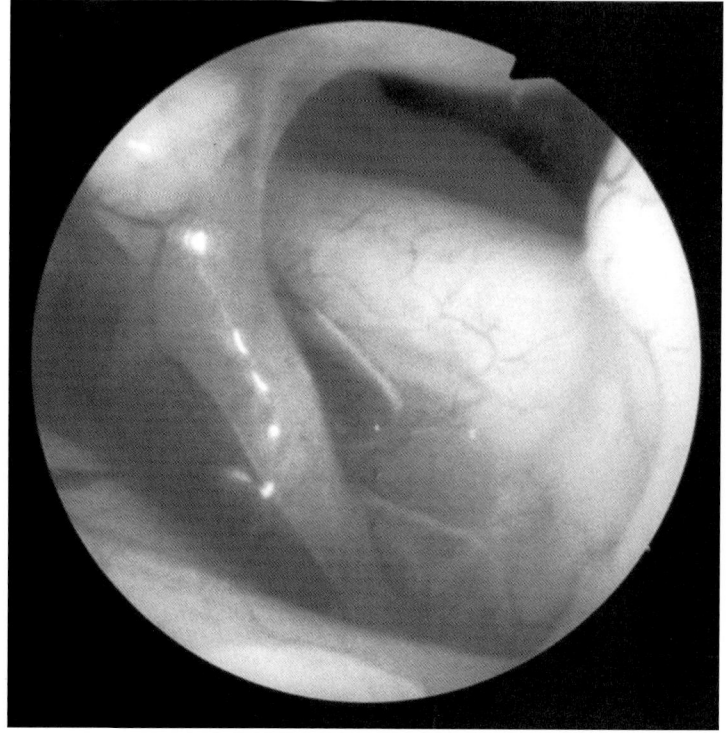

Fig. 1.44
Stapedial fossa. Antero-inferior view. Endoscopic image obtained by fresh cadaver examination.

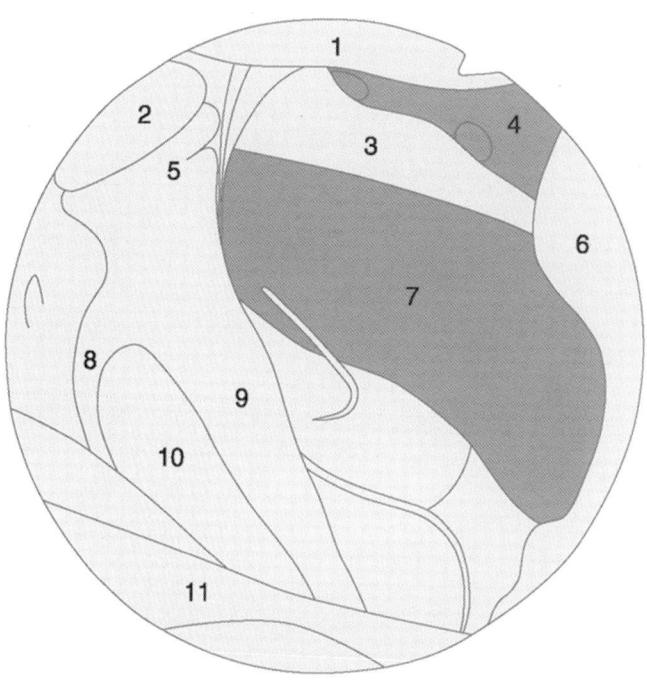

1. Long crus of incus
2. Lenticular process
3. Lateral semicircular canal
4. Aditus ad antrum
5. Head of stapes
6. Cochleariform process
7. Facial canal, tympanic portion
8. Posterior crus of stapes
9. Anterior crus of stapes
10. Stapedial fossa
11. Promontory

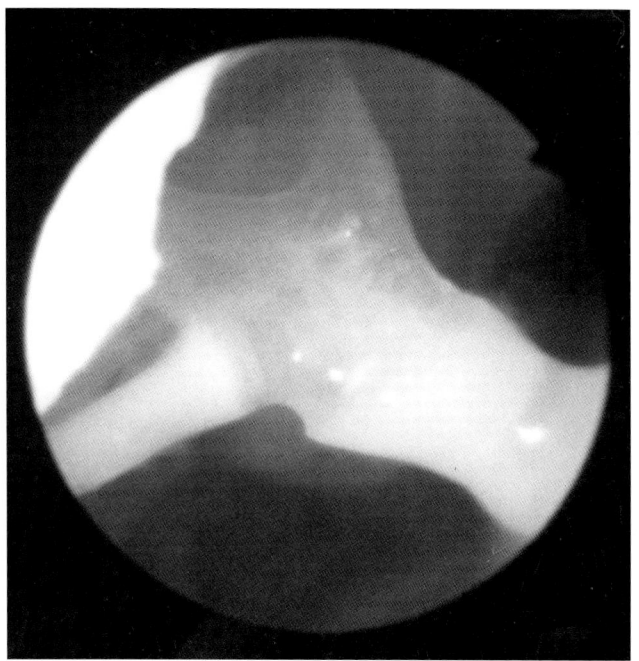

Fig. 1.45

Stapes and related structures. Endoscopic image obtained by fresh cadaver examination. Postero-superior view.

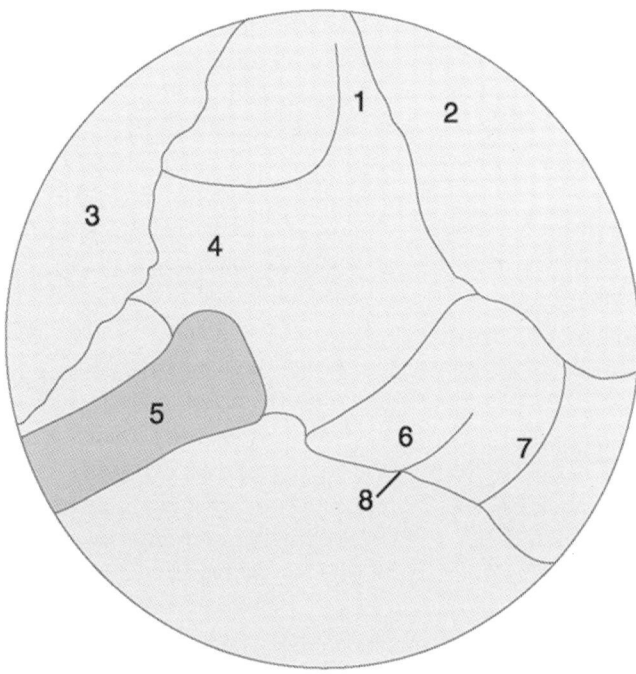

1. Anterior crus of stapes
2. Promontory
3. Facial canal
4. Posterior crus of stapes
5. Tendon of stapedius muscle
6. Head of stapes
7. Lenticular process
8. Incudostapedial joint

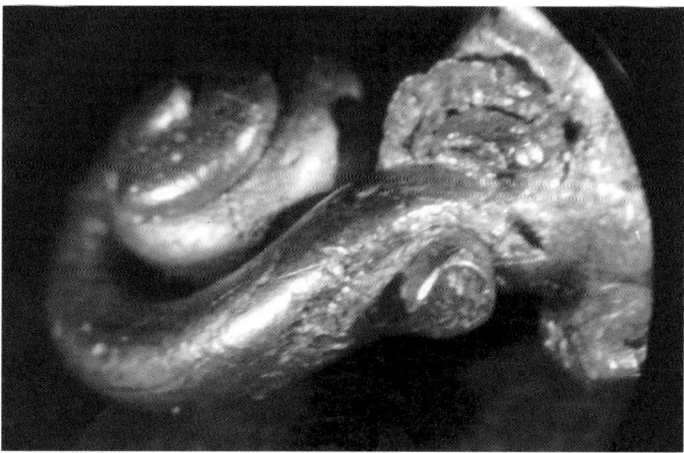
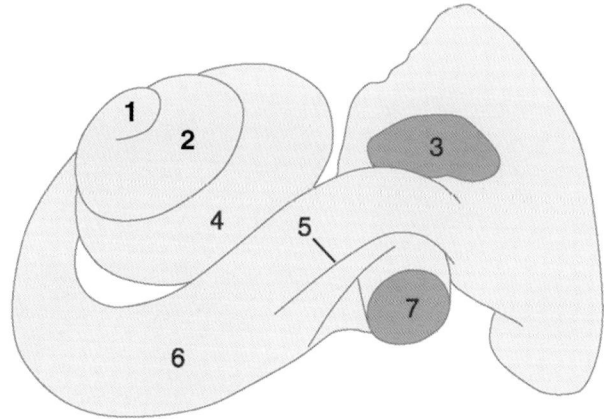

Fig. 1.46

Metal cast of left cochlea, infero-lateral view.

1. Cupola
2. Third turn
3. Oval window
4. Second turn
5. Groove of secondary spiral lamina
6. First turn
7. Round window

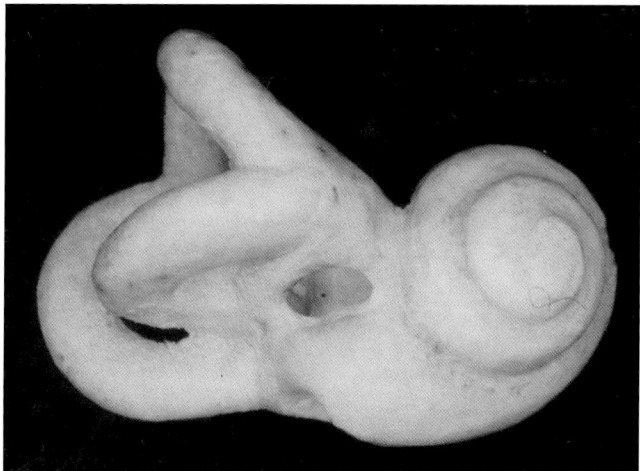
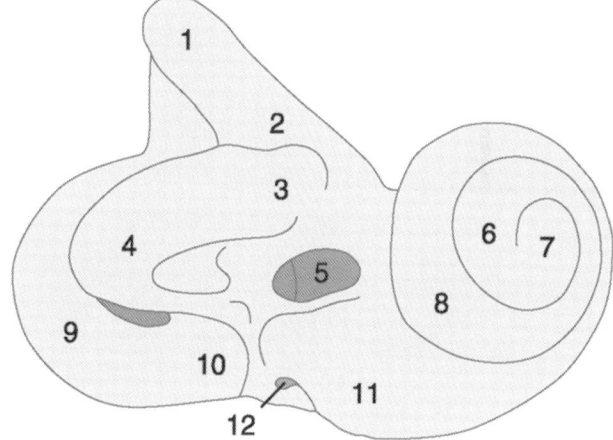

Fig. 1.47

Microdissection specimen of the right osseous internal ear. Lateral view.

1. Anterior semicircular canal
2. Ampulla of anterior semicircular canal
3. Ampulla of lateral semicircular canal
4. Lateral semicircular canal
5. Oval window
6. Third turn of cochlea
7. Cupola
8. Second turn of cochlea
9. Posterior semicircular canal
10. Ampulla of posterior semicircular canal
11. Basal turn of cochlea
12. Round window

ATLAS—EAR

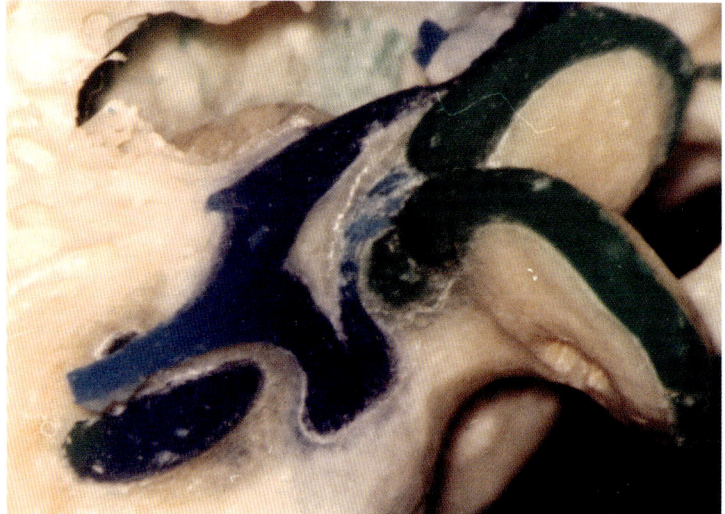

Fig. 1.48

Plastic cast demonstrating the semicircular canals (green) and the facial canal (blue). Superior view.

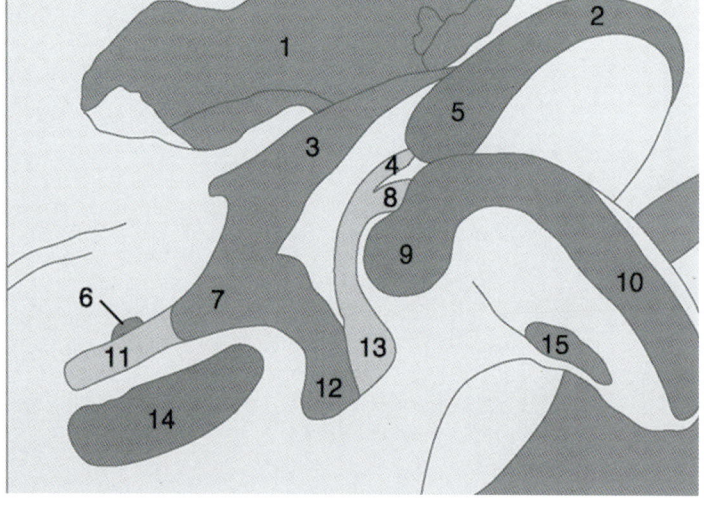

1. Tympanic cavity
2. Lateral semicircular canal
3. Tympanic segment of facial canal
4. Lateral ampullar branch of utriculoampullar nerve
5. Ampulla of lateral semicircular canal
6. Second turn of cochlea
7. Geniculum of facial canal
8. Anterior ampullar branch of utriculoampullar nerve
9. Ampulla of anterior semicircular canal
10. Anterior semicircular canal
11. Greater (superficial) petrosal nerve
12. Labyrinthine segment of facial canal
13. Utriculoampullar nerve
14. First turn of cochlea
15. Petromastoidal canal (continuation of subarcuate fossa toward the mastoid air cells)

ATLAS PLATES III (RADIOLOGY)

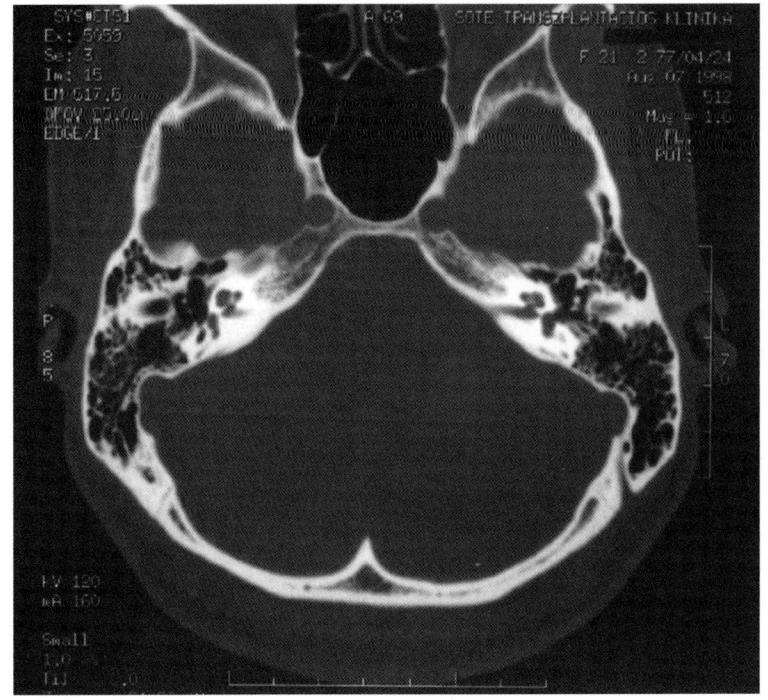

Fig. 1.49

Axial CT image of skull. Courtesy of K. Halbák.

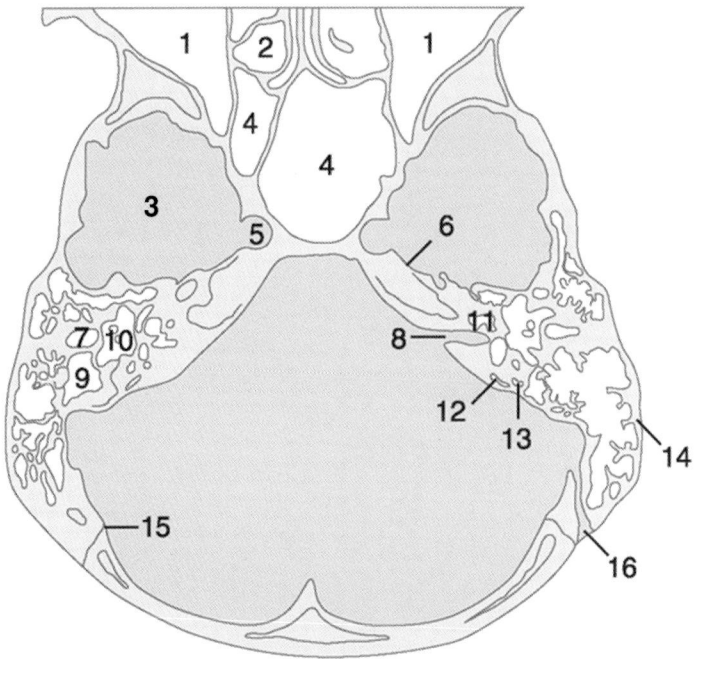

1. Orbit
2. Ethmoidal air cells
3. Middle cranial fossa
4. Sphenoidal sinus
5. Carotid sulcus
6. Meckel's cave (trigeminal impression)
7. External acoustic meatus
8. Internal acoustic meatus
9. Antrum
10. Tympanic cavity with auditory ossicles
11. Cochlea
12. Vestibular aqueduct
13. Posterior semicircular canal
14. Mastoid process
15. Occipitomastoid suture
16. Mastoid foramen with emissary vein

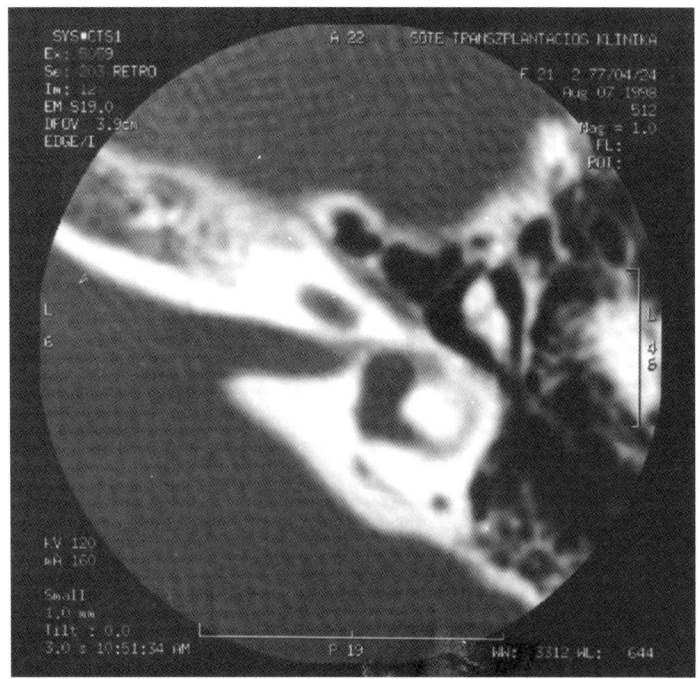

Fig. 1.50

Axial CT image of left temporal bone.
Courtesy of K. Hrabák.

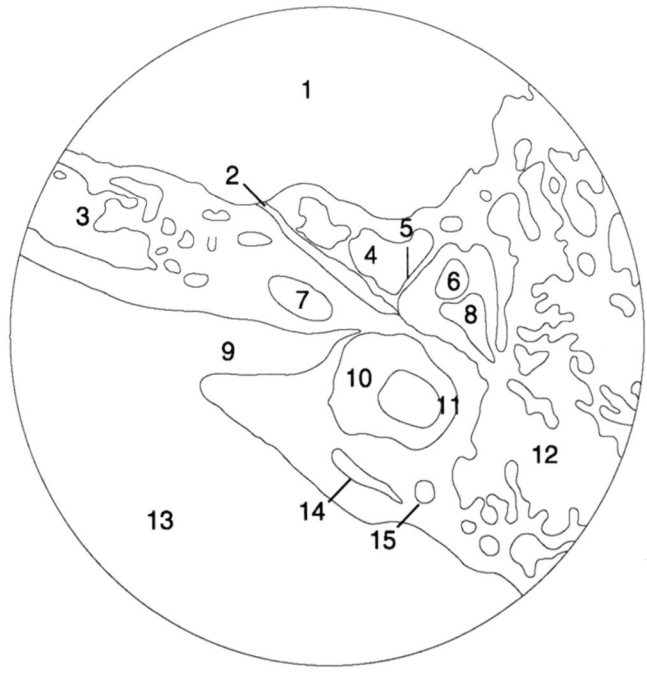

1. Middle cranial fossa
2. Groove for greater petrosal nerve
3. Apex of petrous temporal bone
4. Anterior epitympanic space
5. "Cog" (bony plate hanging from tegmen tympani between the tympanic ring and cochleariform process)
6. Malleus in the posterior epitympanic space
7. Cochlea
8. Incus in the posterior epitympanic space
9. Internal acoustic meatus
10. Vestibule
11. Lateral semicircular canal
12. Antrum
13. Posterior cranial fossa
14. Vestibular aqueduct
15. Posterior semicircular canal

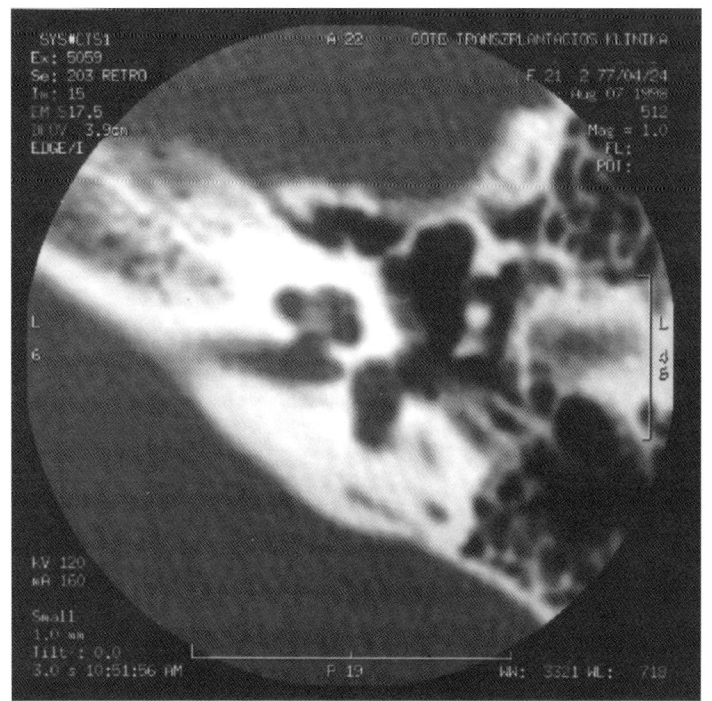

Fig. 1.51

Axial CT image of left temporal bone.
Courtesy of K. Hrabák.

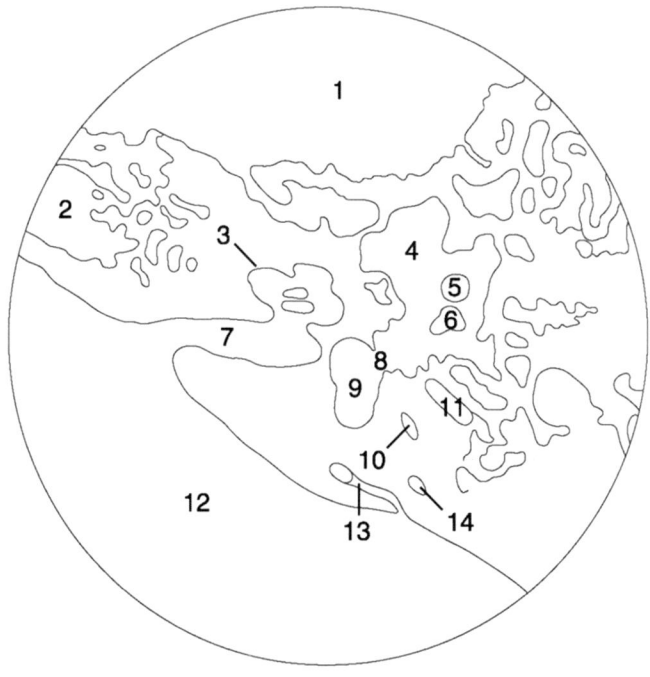

1. Middle cranial fossa
2. Apex of petrous temporal bone
3. Cochlea
4. Tympanic cavity
5. Malleus
6. Incus
7. Internal acoustic meatus
8. Oval window (fenestra ovalis)
9. Vestibule
10. Tympanic sinus (tangential section)
11. Transition between mastoid and tympanic parts of facial canal
12. Posterior cranial fossa
13. Vestibular aqueduct
14. Posterior semicircular canal

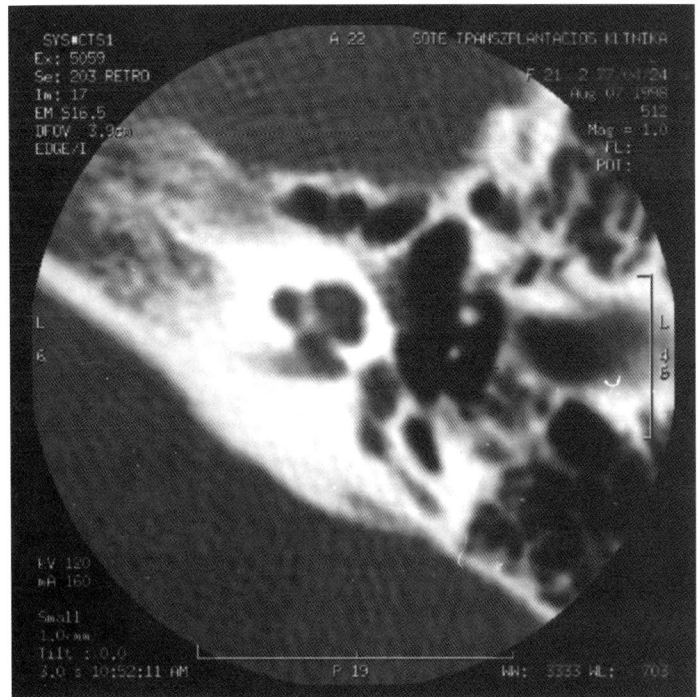

Fig. 1.52

Axial CT image of left temporal bone. Courtesy of K. Hrabák.

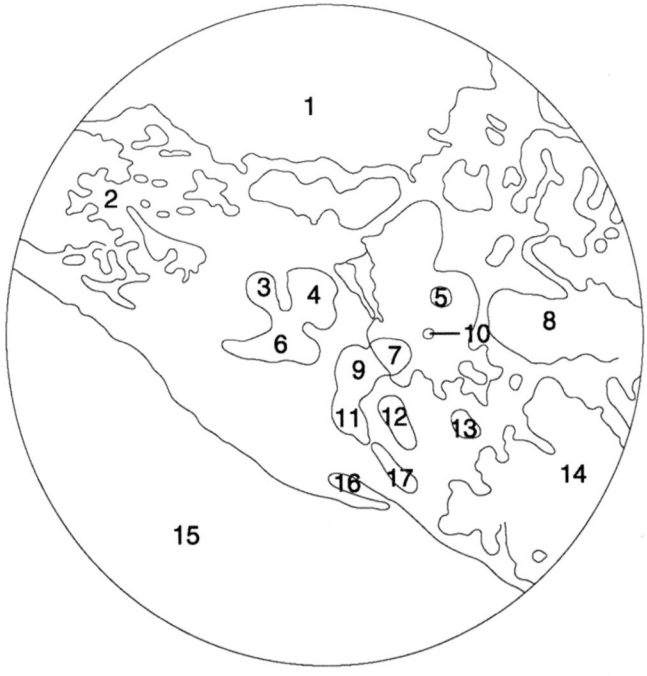

1. Middle cranial fossa
2. Apex of petrous temporal bone
3. First turn of cochlea
4. Second and third (apical) turn of cochlea
5. Manubrium of malleus
6. Internal acoustic meatus
7. Stapes
8. External acoustic meatus (tangential section)
9. Vestibule
10. Crus longum (long limb) of incus
11. Ampulla of posterior semicircular canal
12. Tympanic sinus
13. Mastoid part of facial canal
14. Antrum
15. Posterior cranial fossa
16. Vestibular aqueduct
17. Posterior semicircular canal

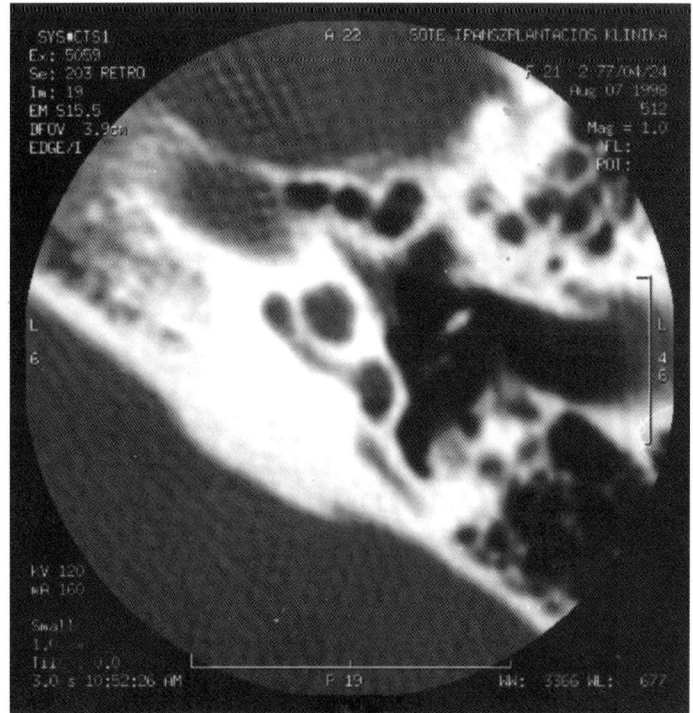

Fig. 1.53

Axial CT image of left temporal bone. Courtesy of K. Hrabák.

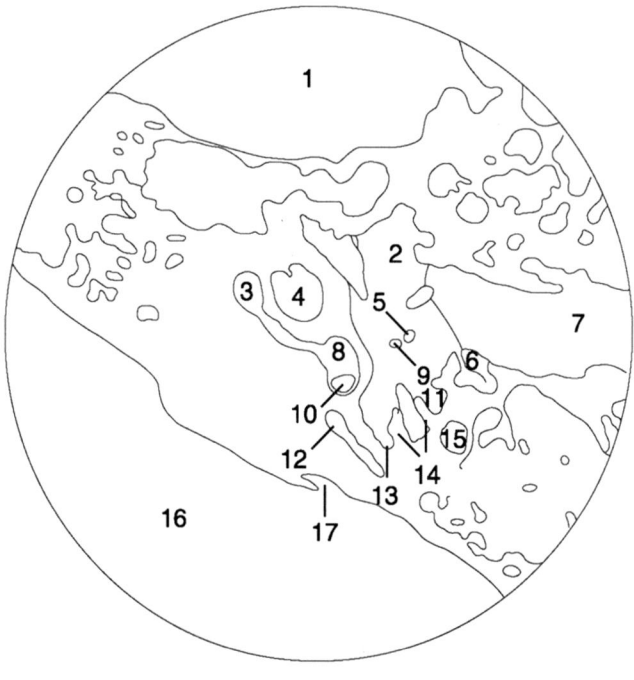

1. Middle cranial fossa
2. Manubrium of malleus
3. First turn of cochlea
4. Second and third (apical) turn of cochlea
5. Crus longum (long limb) of incus
6. Posterior canaliculus for chorda tympani nerve
7. External acoustic meatus
8. Basal turn of cochlea
9. Head of stapes
10. Round window (air-filled niche of tympanic cavity)
11. Facial (posterior) sinus
12. Posterior semicircular canal
13. Tympanic sinus
14. Pyramidal eminence
15. Mastoid part of facial canal
16. Posterior cranial fossa
17. Vestibular aqueduct

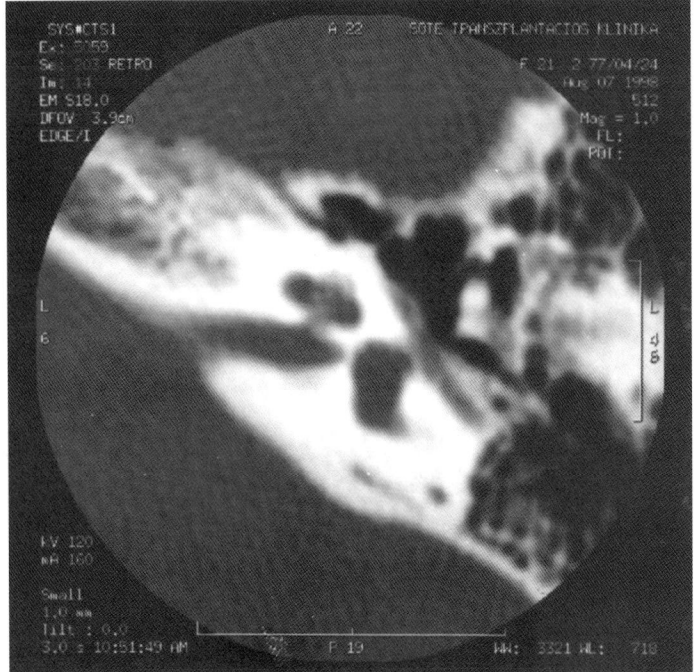

Fig. 1.54

Axial CT image of left temporal bone.
Courtesy of K. Hrabák.

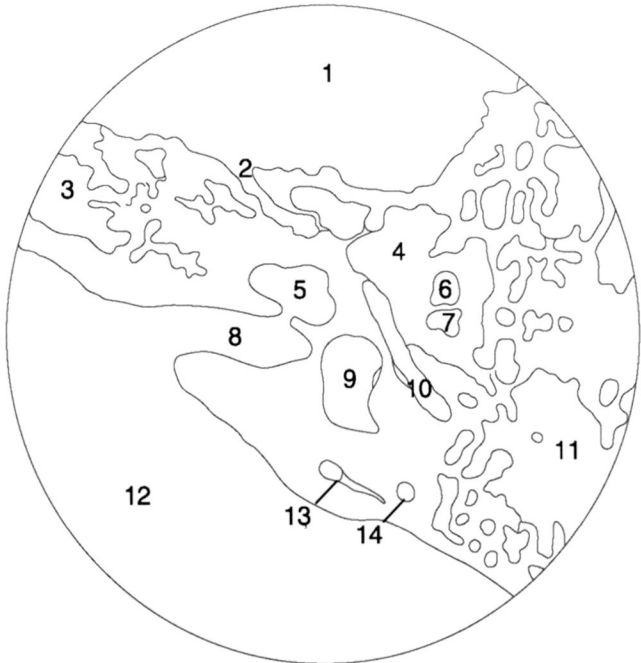

1. Middle cranial fossa
2. Groove for greater petrosal nerve
3. Apex of petrous temporal bone
4. Tympanic cavity
5. Cochlea
6. Malleus
7. Incus
8. Internal acoustic meatus
9. Vestibule
10. Tympanic part of facial canal
11. Antrum
12. Posterior cranial fossa
13. Vestibular aqueduct
14. Posterior semicircular canal

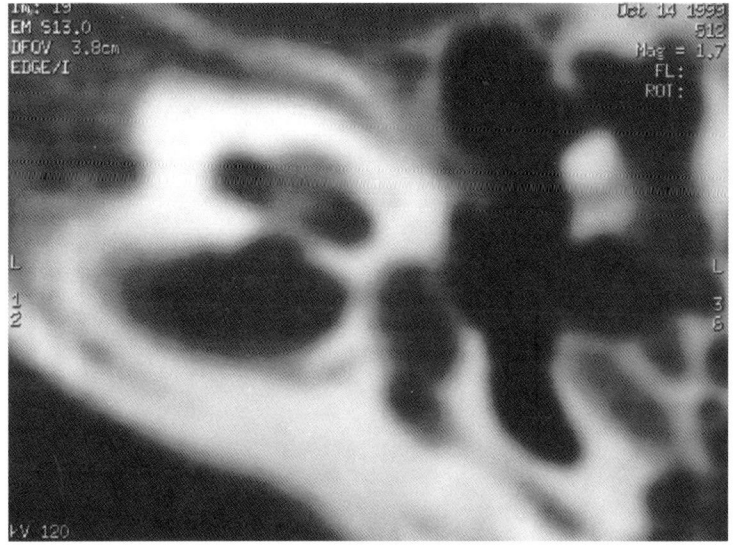

Fig. 1.55

Axial CT image of left temporal bone.
Courtesy of K. Hrabák.

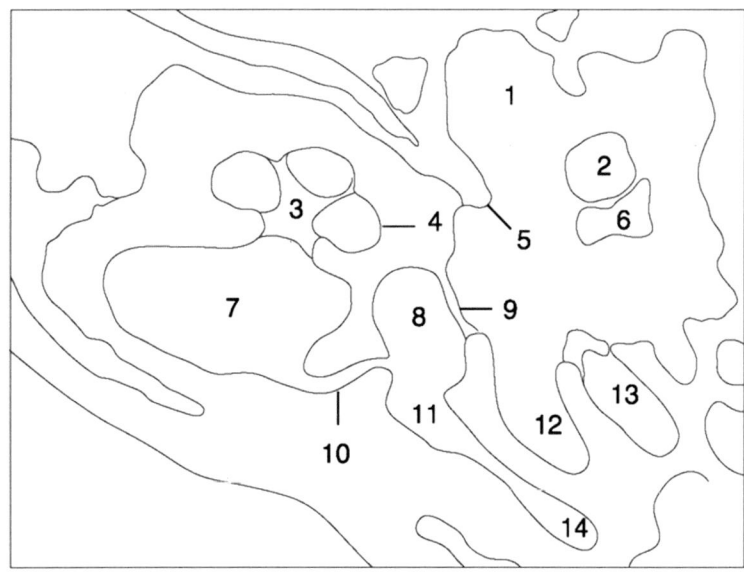

1. Tympanic cavity
2. Head of malleus
3. Modiolus
4. Cochlea
5. Cochleariform process
6. Body of incus
7. Internal acoustic meatus
8. Vestibule
9. Round window
10. Foramen singulare
11. Ampulla of posterior semicircular canal
12. Tympanic sinus
13. Pyramidal eminence
14. Posterior semicircular canal

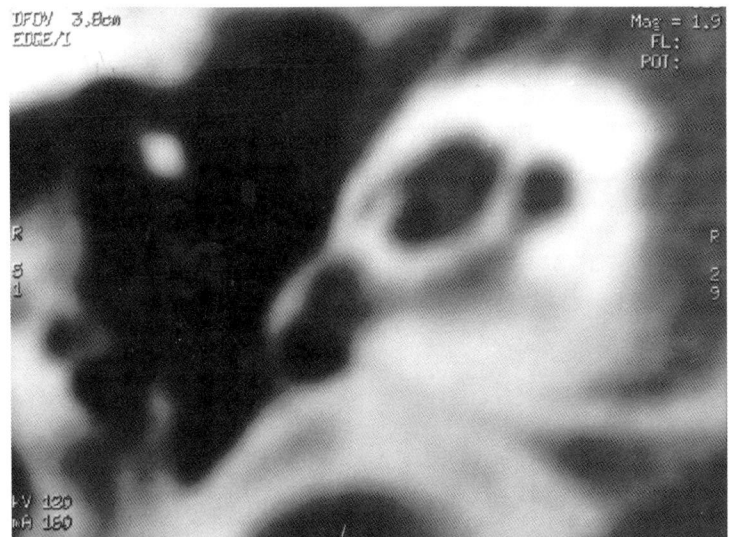

Fig. 1.56

Axial CT image of right temporal bone.
Courtesy of K. Hrabák.

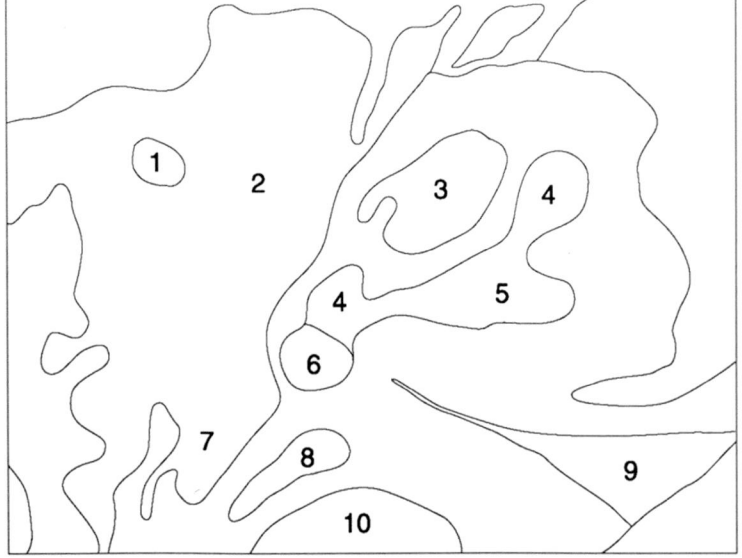

1. Manubrium of malleus
2. Tympanic cavity
3. Second and third turns of cochlea
4. Basal turn of cochlea
5. Internal acoustic meatus
6. Round window niche
7. Tympanic sinus
8. Posterior semicircular canal
9. Cochlear canaliculus
10. Jugular bulb

Fig. 1.57
Longitudinal CT image of temporal bone.
Courtesy of K. Hrabák.

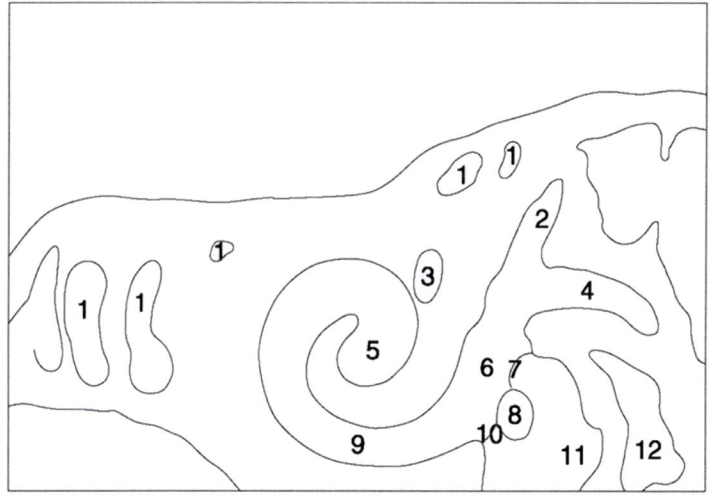

1. Air cells of temporal bone
2. Anterior semicircular canal
3. Labyrinthine part of facial canal
4. Lateral semicircular canal
5. Second turn of cochlea
6. Vestibule
7. Oval window (fenestra ovalis)
8. Subiculum of promontory
9. First turn of cochlea
10. Round window (fenestra rotunda)
11. Tympanic cavity
12. Mastoid part of facial canal

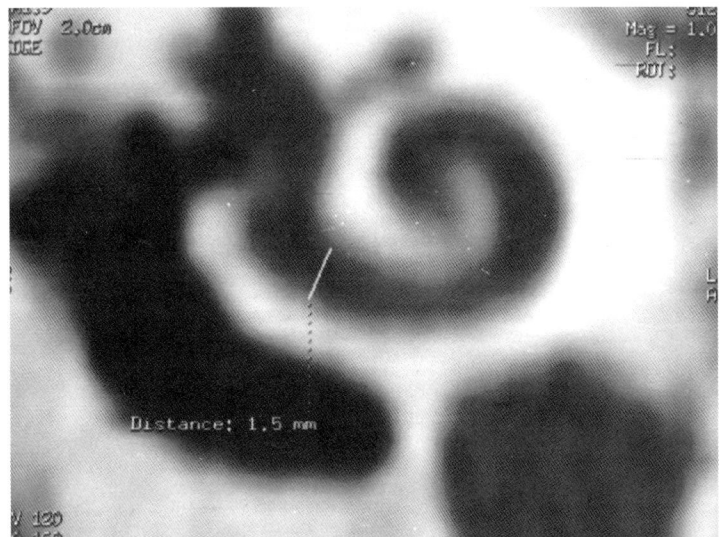

Fig. 1.58

Longitudinal CT image of temporal bone. Courtesy of K. Hrabák.

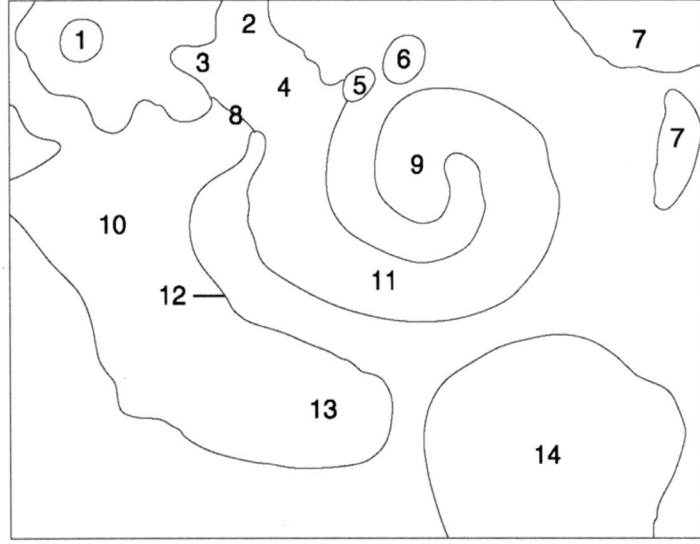

1. Lateral semicircular canal
2. Ampulla of anterior semicircular canal
3. Ampulla of lateral semicircular canal
4. Vestibule
5. Superior vestibular area of internal acoustic meatus
6. Labyrinthine part of facial canal
7. Air cells of temporal bone
8. Oval window (fenestra ovalis)
9. Second turn of cochlea
10. Tympanic cavity
11. First turn of cochlea
12. Promontory
13. Hypotympanum
14. Carotid canal

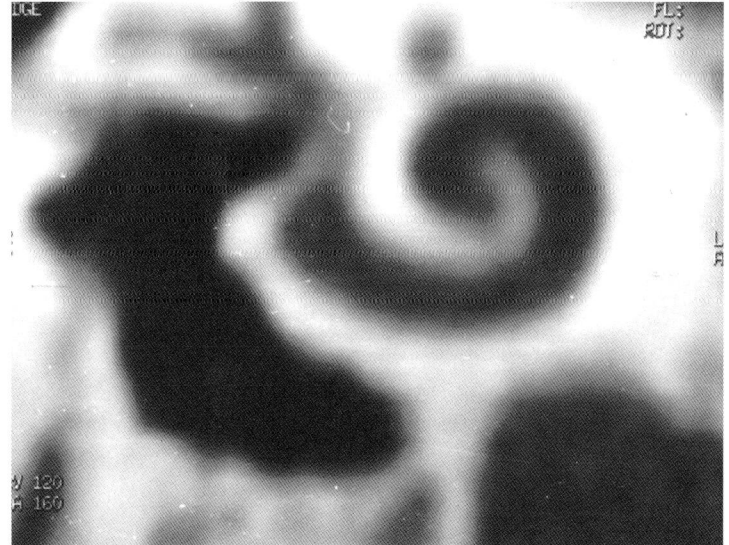

Fig. 1.59

Longitudinal CT image of temporal bone. Courtesy of K. Hrabák.

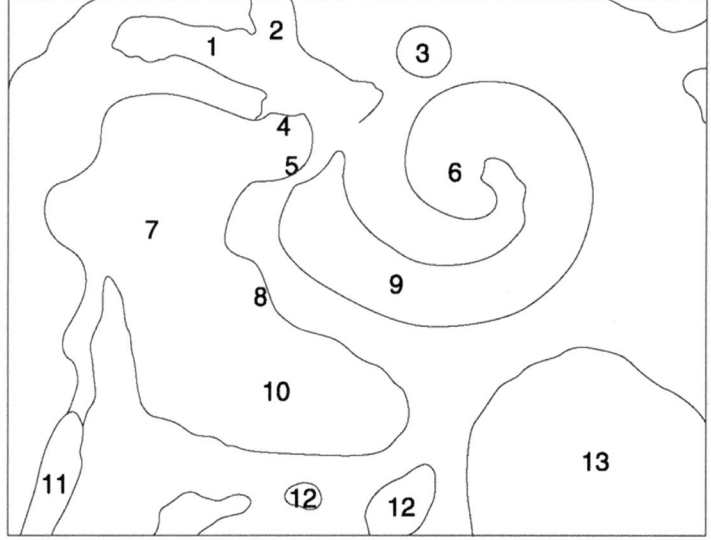

1. Lateral semicircular canal
2. Ampulla of anterior semicircular canal
3. Labyrinthine part of facial canal
4. Oval window (fenestra ovalis)
5. Round window (fenestra rotunda)
6. Second turn of cochlea
7. Tympanic cavity
8. Promontory
9. First turn of cochlea
10. Hypotympanum
11. Mastoid part of facial canal
12. Air cells of temporal bone
13. Carotid canal

ATLAS—EAR

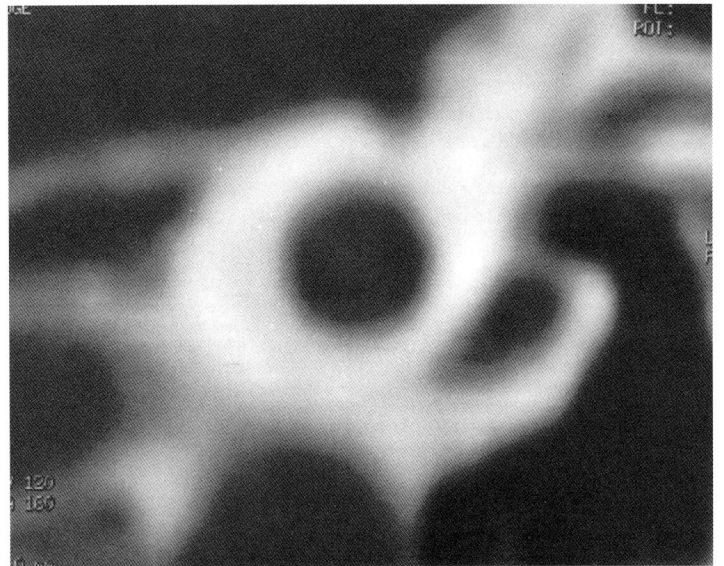

Fig. 1.60

Coronal CT image of temporal bone.
Courtesy of K. Hrabák.

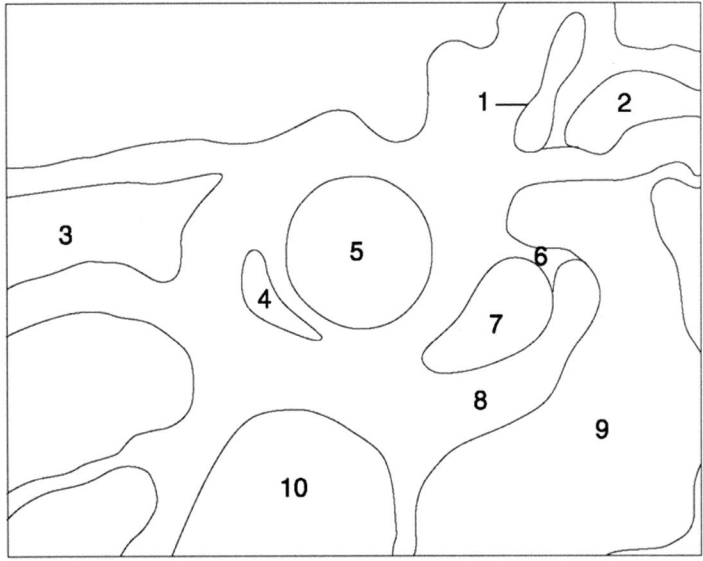

1. Anterior semicircular canal
2. Lateral semicircular canal
3. Air cells of apex
4. First turn of cochlea
5. Cupula of cochlea
6. Round window
7. Basal turn of cochlea
8. Promontory
9. Tympanic cavity
10. Carotid canal

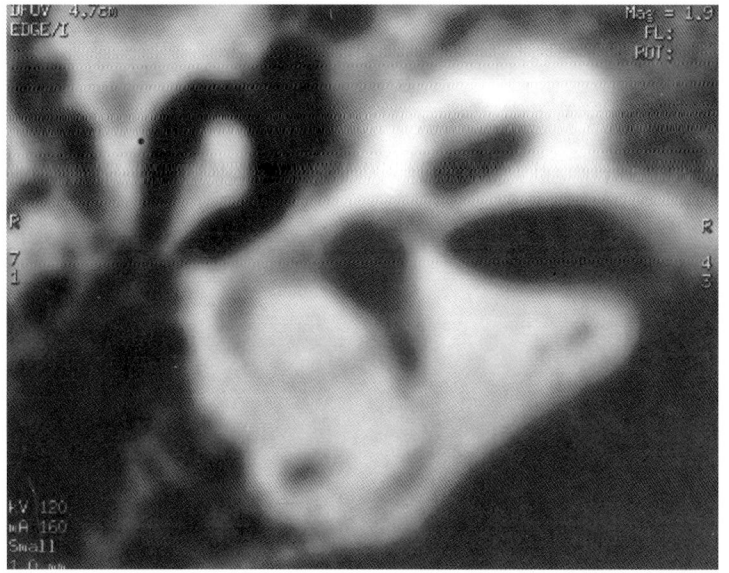

Fig. 1.61

Axial CT image of right temporal bone.
Courtesy of K. Hrabák.

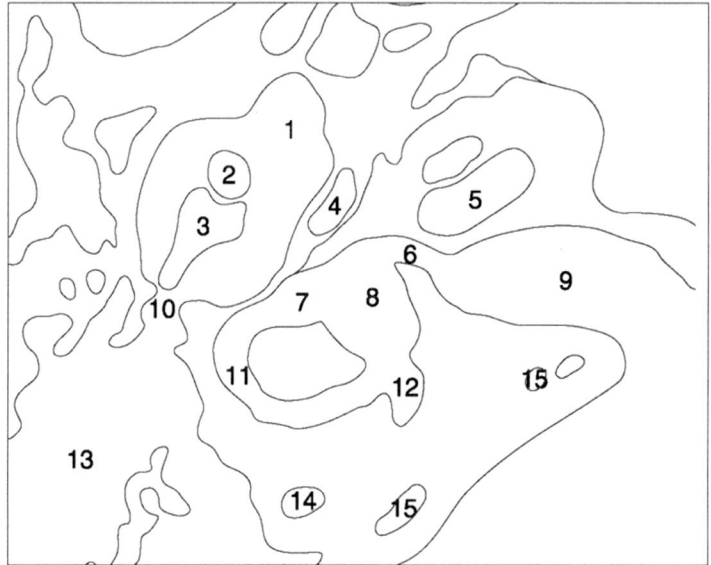

1. Tympanic cavity
2. Head of malleus
3. Body of incus
4. Tensor tympani muscle
5. Cochlea
6. Utriculo-ampullar nerve
7. Ampulla of lateral semicircular canal
8. Vestibule
9. Internal acoustic meatus
10. Crus breve of incus in incudial fossa
11. Lateral semicircular canal
12. Ampulla of posterior semicircular canal
13. Antrum
14. Posterior semicircular canal
15. Air cells behind the labyrinth

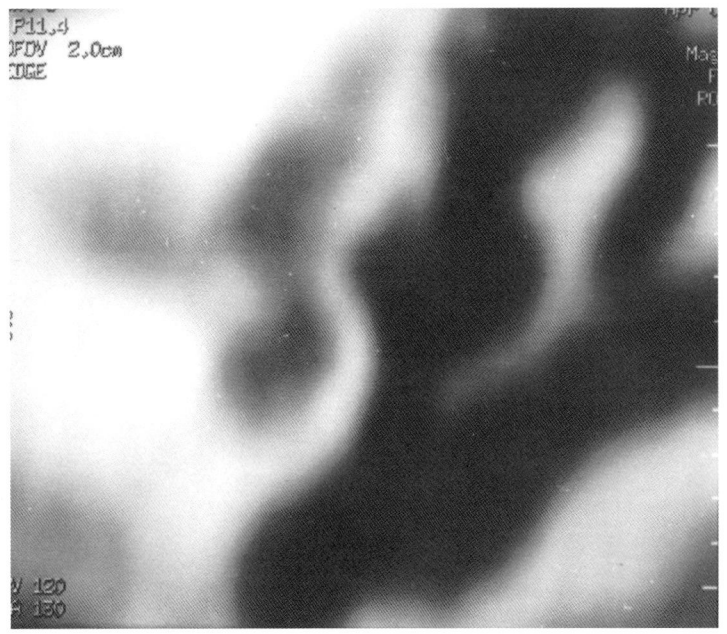

Fig. 1.62

Coronal CT image of temporal bone.
Courtesy of K. Hrabák.

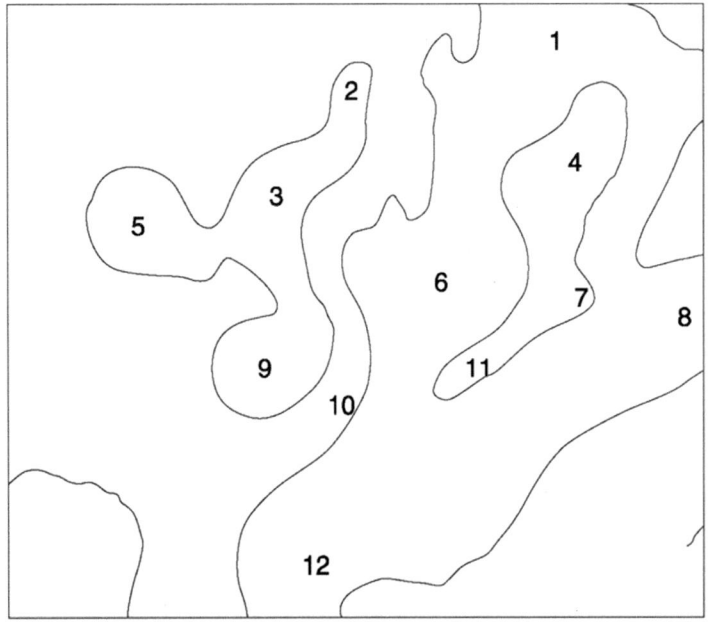

1. Epitympanum
2. Anterior semicircular canal
3. Vestibule
4. Head of malleus
5. Internal acoustic meatus
6. Mesotympanum
7. Lateral process of malleus
8. External acoustic meatus
9. First turn of cochlea
10. Promontory
11. Manubrium of malleus
12. Hypotympanum

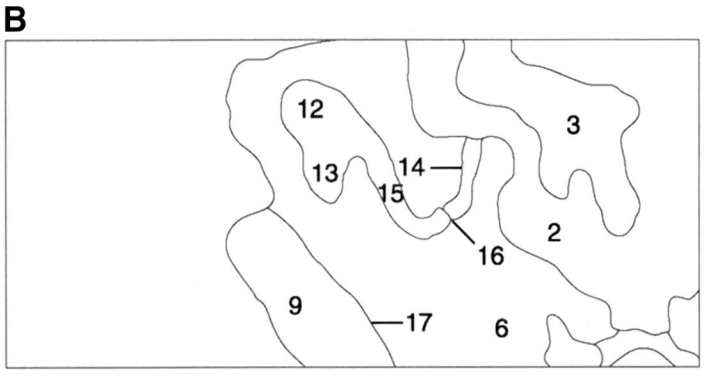

Fig. 1.63

Multiplanar reconstruction CT image (coronal view) of temporal bone. Courtesy of K. Hrabák.

1. Head of malleus
2. Otic capsule
3. Vestibule
4. Oval window
5. Neck of malleus
6. Tympanic cavity
7. Subiculum
8. Lateral process of malleus
9. External acoustic meatus
10. Manubrium of malleus
11. Round window
12. Body of incus
13. Crus breve of incus
14. Stapes
15. Crus longum of incus
16. Incudostapedial joint
17. Tympanic membrane

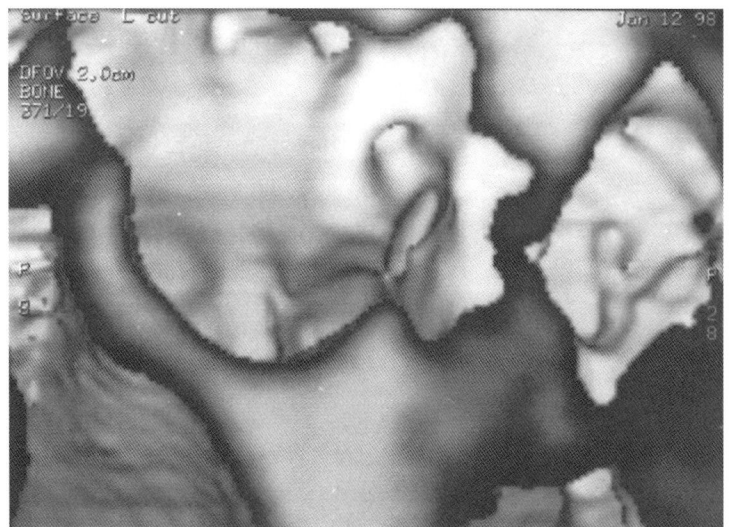

Fig. 1.64

3D CT reconstruction image of tympanic cavity. Courtesy of K. Hrabák.

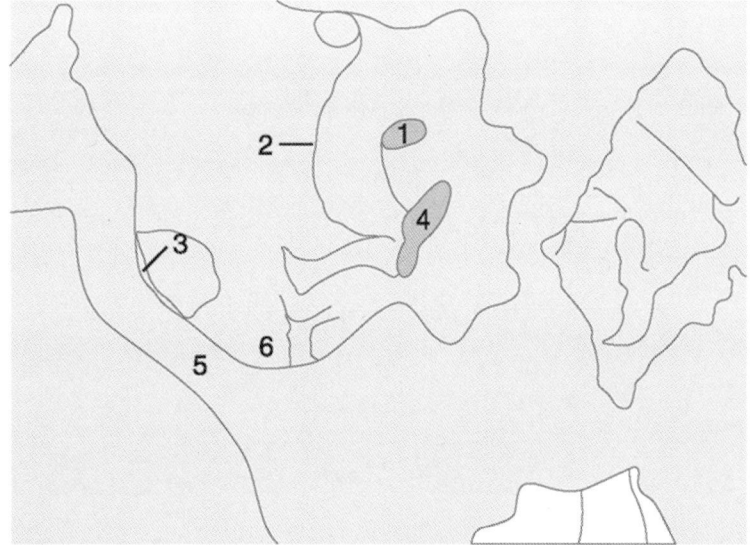

1. Oval window (fenestra ovalis)
2. Canaliculus for the tympanic nerve (of Jacobson)
3. Tympanic sulcus
4. Round window (fenestra rotunda)
5. Tympanic ring (annulus)
6. Hypotympanum

Fig. 1.65

Cranial nerve branches in the internal acoustic meatus. T2-weighted axial MR image highlighting the subarachnoid space and its perineural extension (evagination). Courtesy of K. Hrabák.

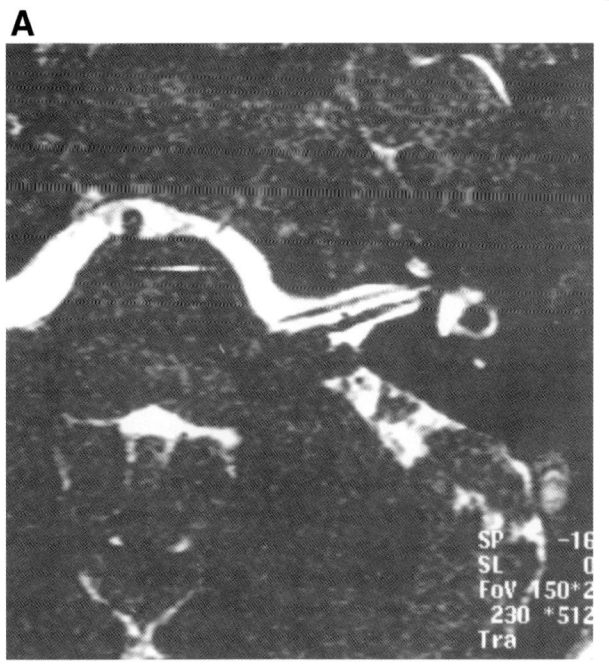
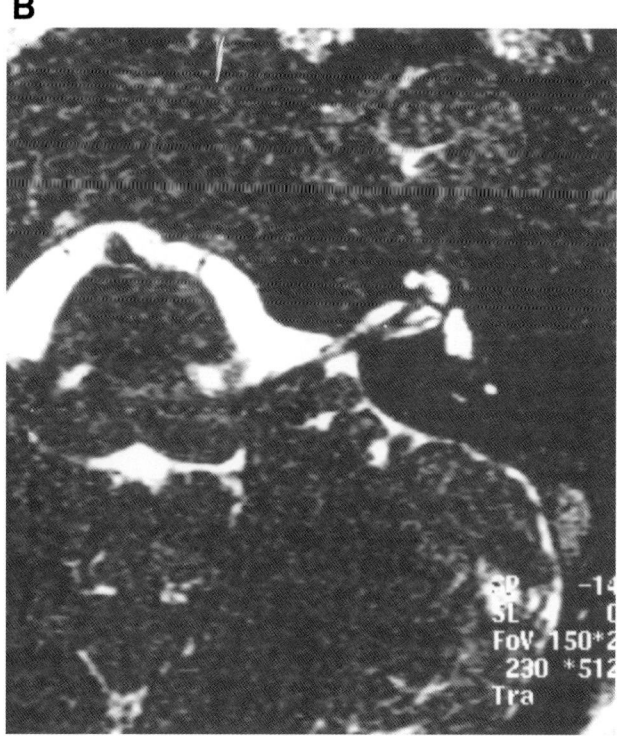
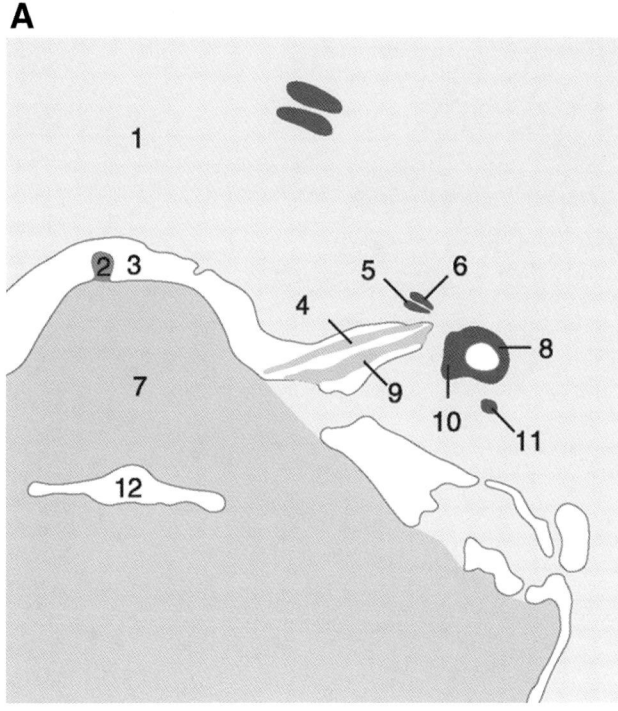
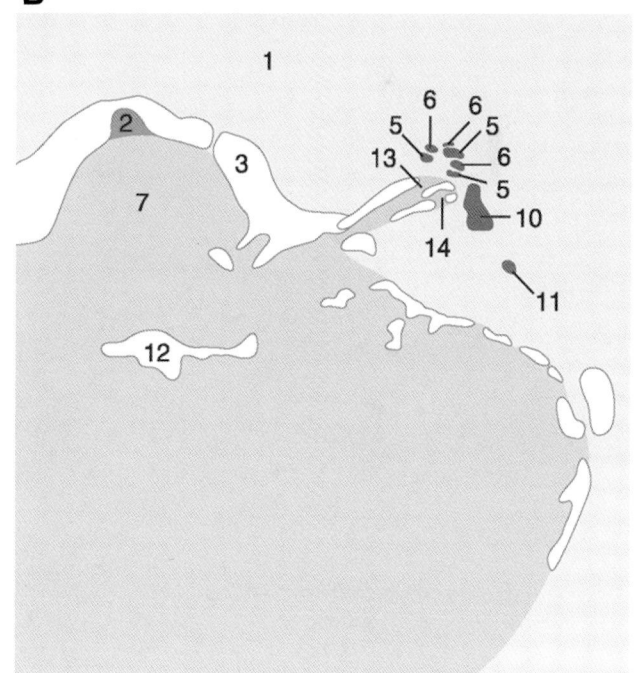

1. Clivus
2. Basilar artery
3. Prepontine cistern
4. Facial nerve
5. Scala tympani
6. Scala vestibuli
7. Pons
8. Lateral semicircular canal
9. Utriculo-ampullar nerve
10. Vestibule
11. Posterior semicircular canal
12. Fourth ventricle
13. Cochlear nerve
14. Saccular nerve

Fig. 1.66

3D-CISS (constructive interference in steady state) reconstruction images from T2-weighted gradient echo MR sequences of the right labyrinth. A – dorsolateral aspect, B – lateral aspect. Courtesy of K. Hrabák.

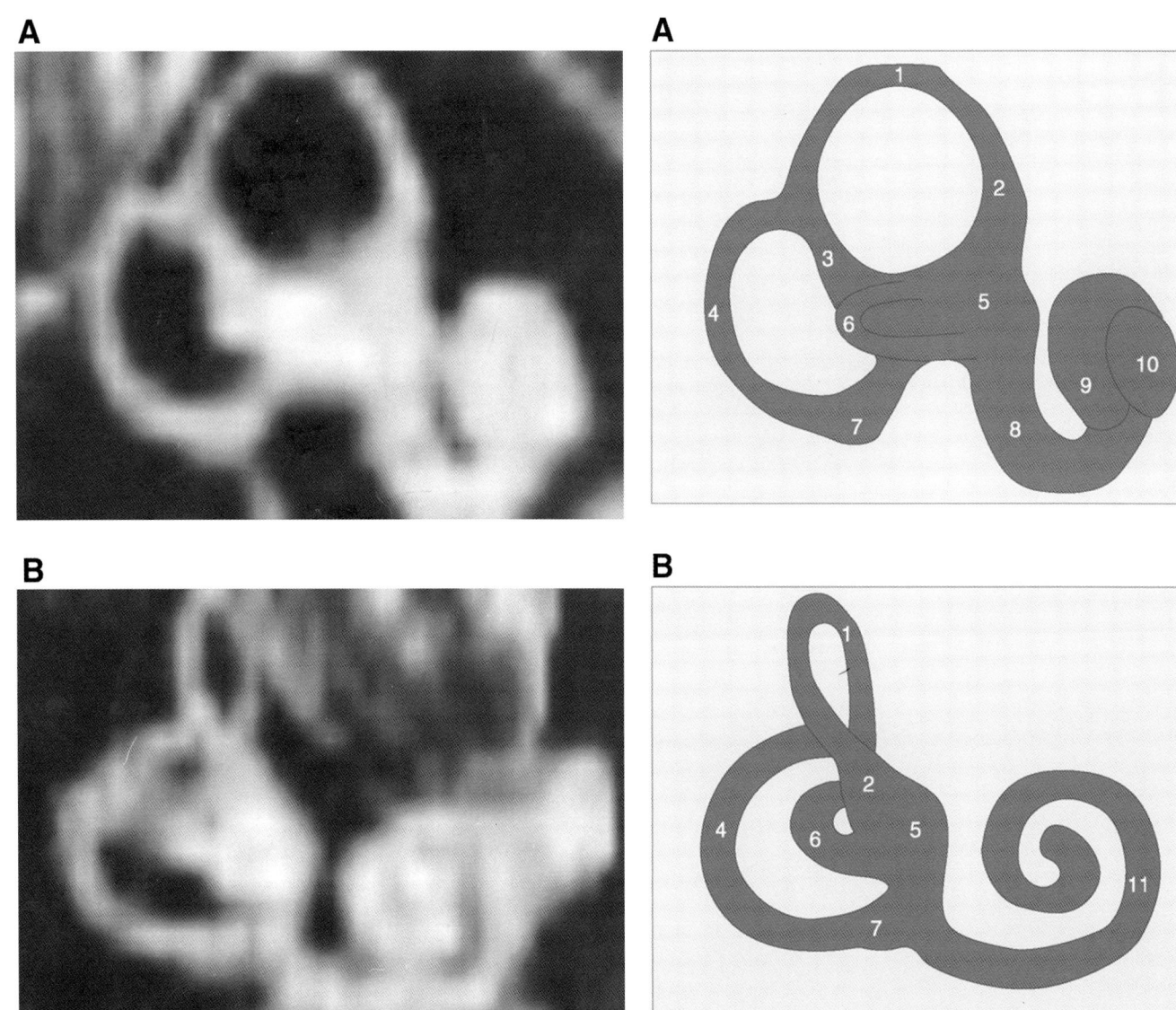

1. Anterior semicircular canal
2. Ampulla of anterior semicircular canal
3. Crus commune
4. Posterior semicircular canal
5. Vestibule
6. Lateral semicircular canal
7. Ampulla of posterior semicircular canal
8. First turn of cochlea
9. Second turn of cochlea
10. Third turn of cochlea
11. Cochlea

ATLAS PLATES IV (HISTOLOGY)

Fig. 1.67

External acoustic meatus and tympanic cavity. Decalcinated histological specimen from newborn, HE. Near-horizontal section. The ossicles are still cartilagineous and the pneumatisation of tympanic cavity is incomplete.

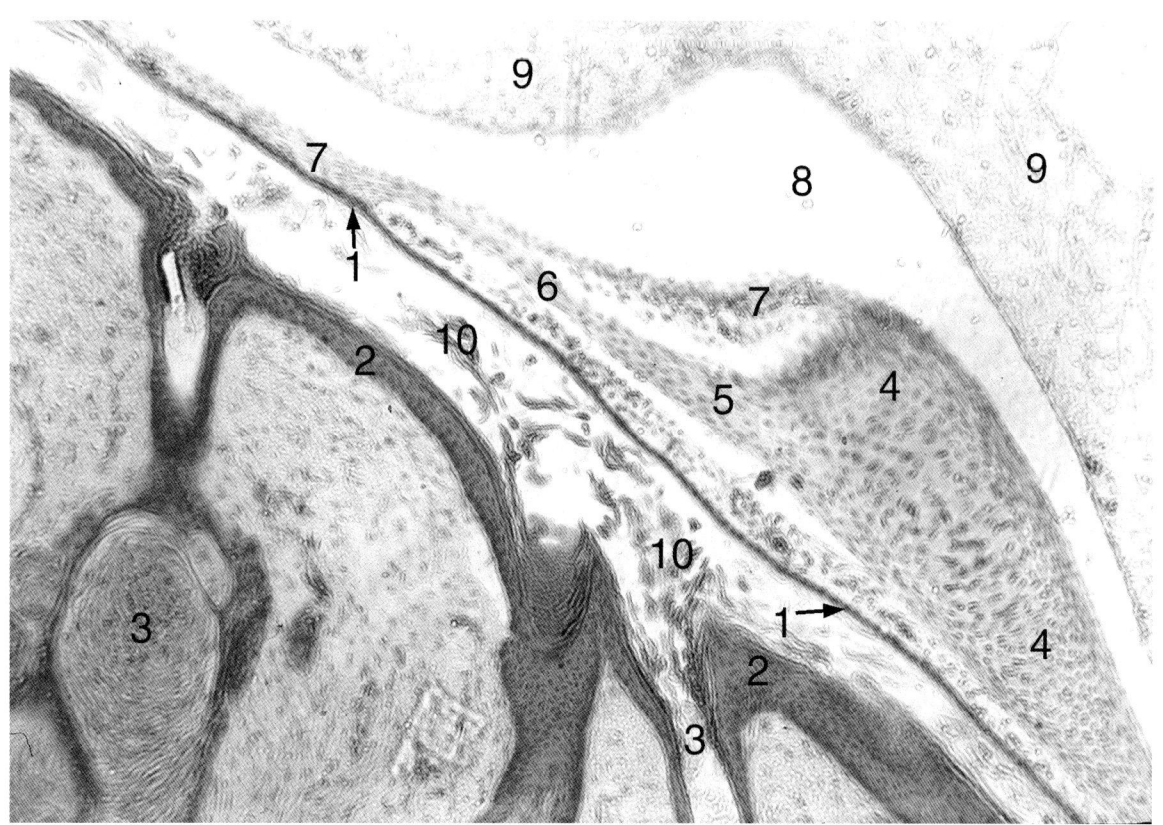

1. Tympanic membrane
2. External aoustic meatus, epithelium of the skin
3. Hair follicles and sebaceous glands
4. Malleus
5. Anterior process of malleus
6. Anterior ligament of malleus
7. Tympanic cavity, mucosa covering the auditory ossicles
8. Tympanic cavity, air-filled space
9. Tympanic cavity, the mucosa covering the inner wall
10. Keratinized and detached epithelial cells in the external acoustic meatus

Fig. 1.68

External acoustic meatus and tympanic cavity. Decalcinated histological specimen from newborn, HE. Near-horizontal section. The ceruminous glands are modified apocrine glands characteristic of the external acoustic meatus. Note the absence of subcutaneous adipose tissue in the external acoustic meatus.

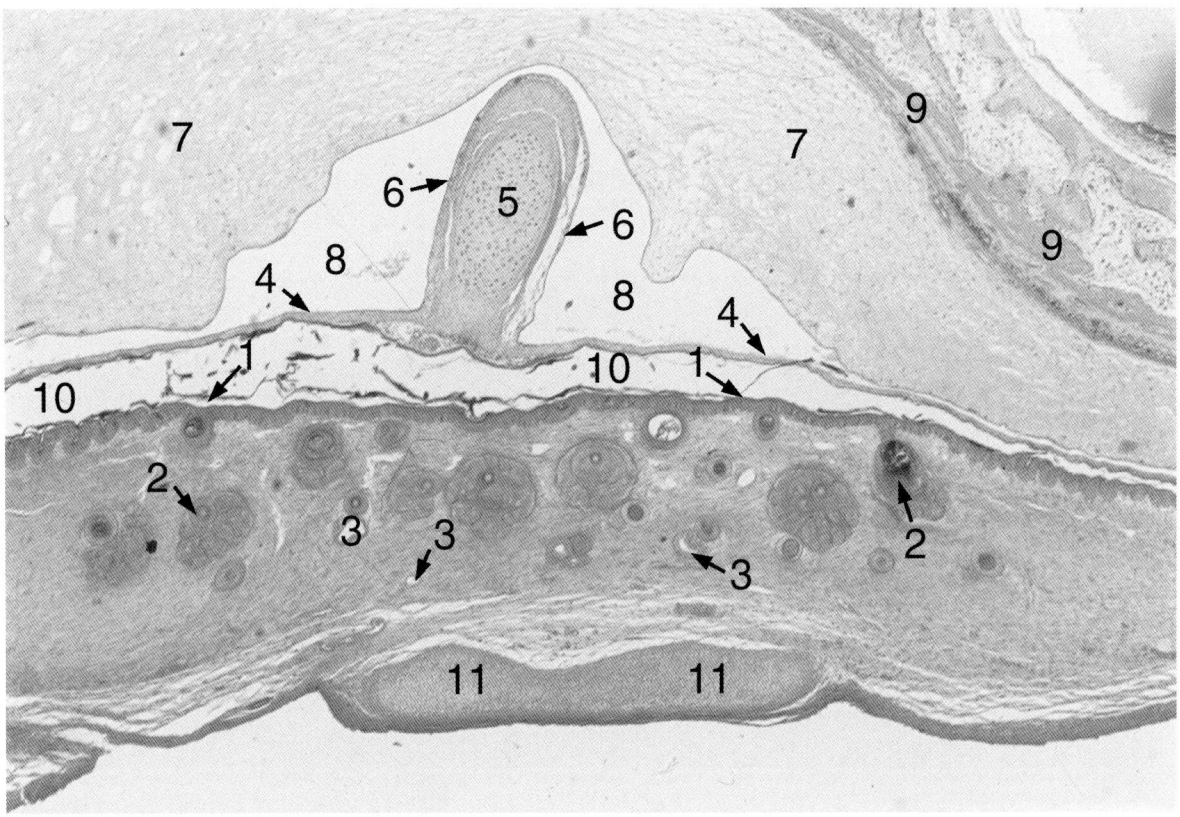

1. External acoustic meatus, arrow points to the skin
2. Hair follicles and sebaceous glands
3. Ceruminous glands
4. Tympanic membrane composed of two epithelial layers and connective tissue between them
5. Malleus, the manubrium is still cartilagineous
6. Tympanic cavity, arrow points to the mucosa covering the ossicles
7. Tympanic cavity, the mucosa covering the inner (medial) wall
8. Tympanic cavity, air-filled space
9. Tympanic cavity, ossification in the inner (medial) wall
10. External acoustic meatus, air-filled space
11. External acoustic meatus, the wall is still cartilagineous

Fig. 1.69

Auditory (Eustachian) tube, general view. Decalcinated histological specimen (HE) taken from the border between the osseous and cartilagineous parts (isthmus). The slit-like lumen lies actually in the vertical plane.

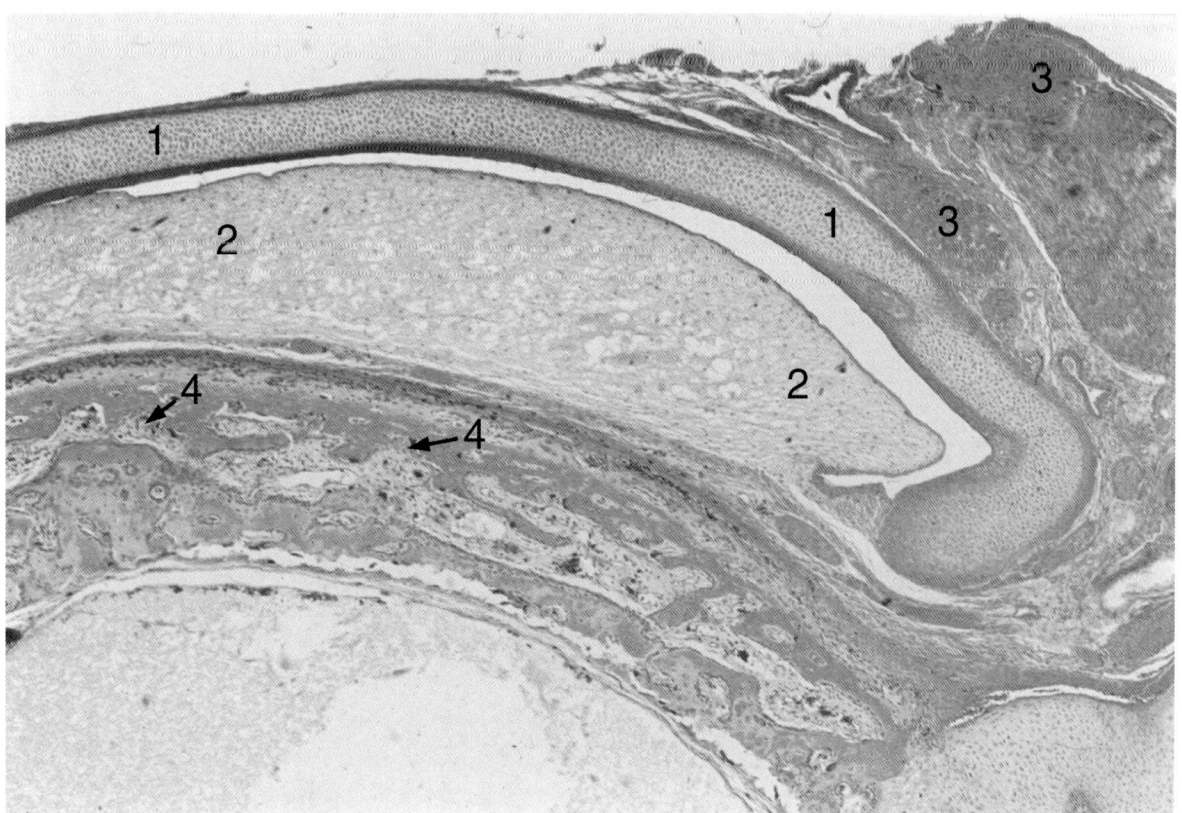

1. Cartilage
2. Mucosa
3. Tensor veli palatini muscle (remnants)
4. Bone, arrow points to an osteoclast

Fig. 1.70

Cochlea, fetal specimen, general view. Semithin section, toluidine blue staining.

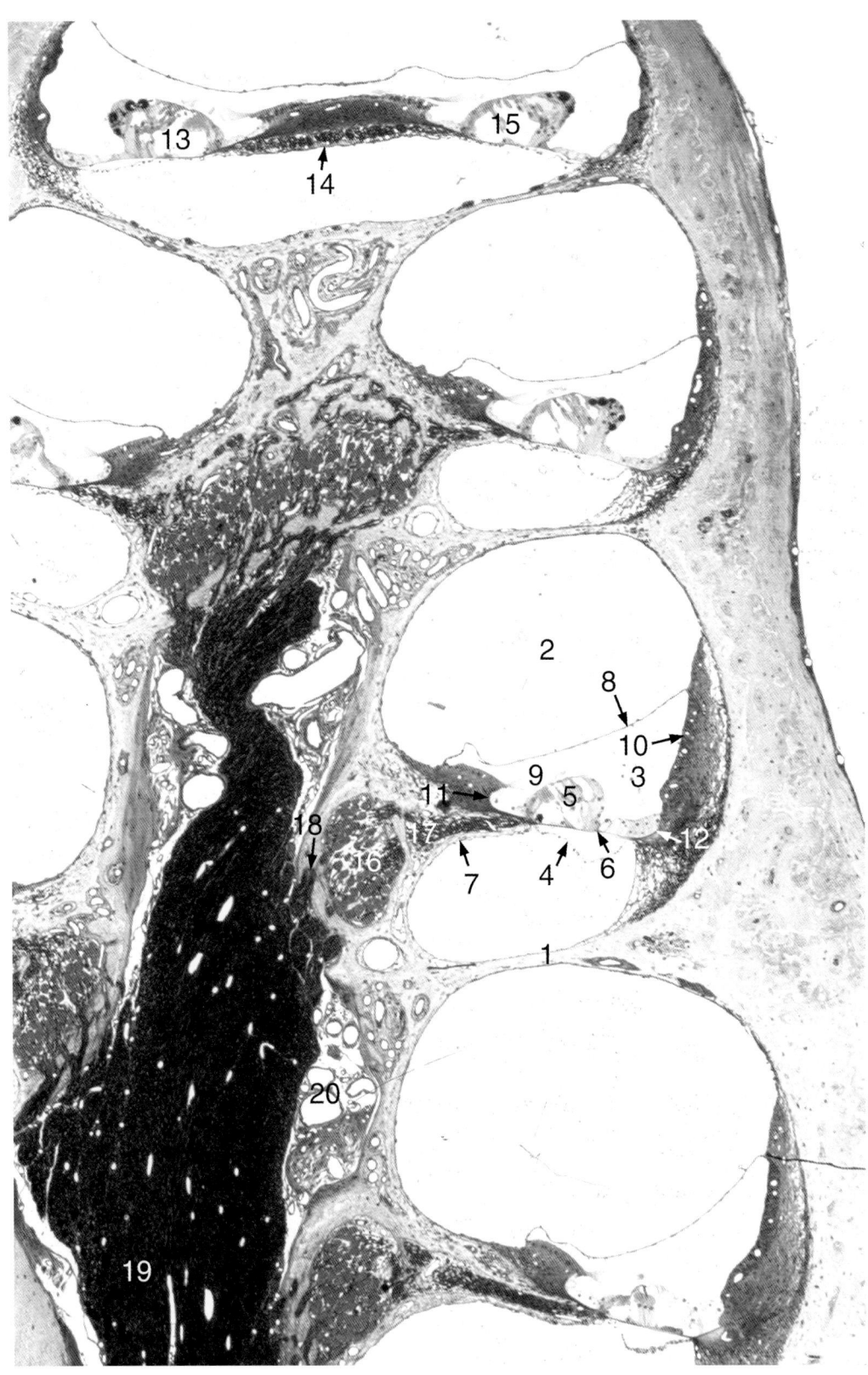

1. Bony septum between the turns
2. Scala vestibuli
3. Scala media (cochlear duct)
4. Scala tympani, arrow point to the detached lining cells (perilymphatic cells)
5. Organ of Corti
6. Basilar membrane
7. Osseous spiral lamina
8. Vestibular membrane (of Reissner)
9. Tectorial membrane
10. Stria vascularis (the only vascularised epithelium in the human)
11. Internal spiral sulcus (sulcus spiralis internus)
12. External spiral sulcus (sulcus spiralis externus)
13. Cochlea, third (apical) turn, a section of the turning organ of Corti on one side
14. Cochlea, third (apical) turn, the osseous spiral lamina lies in the middle in this plane of section
15. Cochlea, third (apical) turn, a section of the turning organ of Corti on the other side
16. Spiral ganglion with bipolar neurons in the spiral canal of the modiolus (of Rosenthal)
17. Spiral ganglion, peripherally directed nerve fibers
18. Longitudinal canal of modiolus with the centrally directed fibers of bipolar neurons
19. Modiolus with the cochlear nerve
20. Blood vessel

Fig. 1.71

Section through a turn of the cochlea. Semithin section, toluidine blue staining. The vestibular membrane (of Reissner) consists of a thin layer of connective tissue covered by epithelium on the side of the cochlear duct, while on the opposite side a layer of modified connective tissue cells can be found. The perilymphatic space is typically lined by a layer of such modified connective tissue cells called perilymphatic cells.

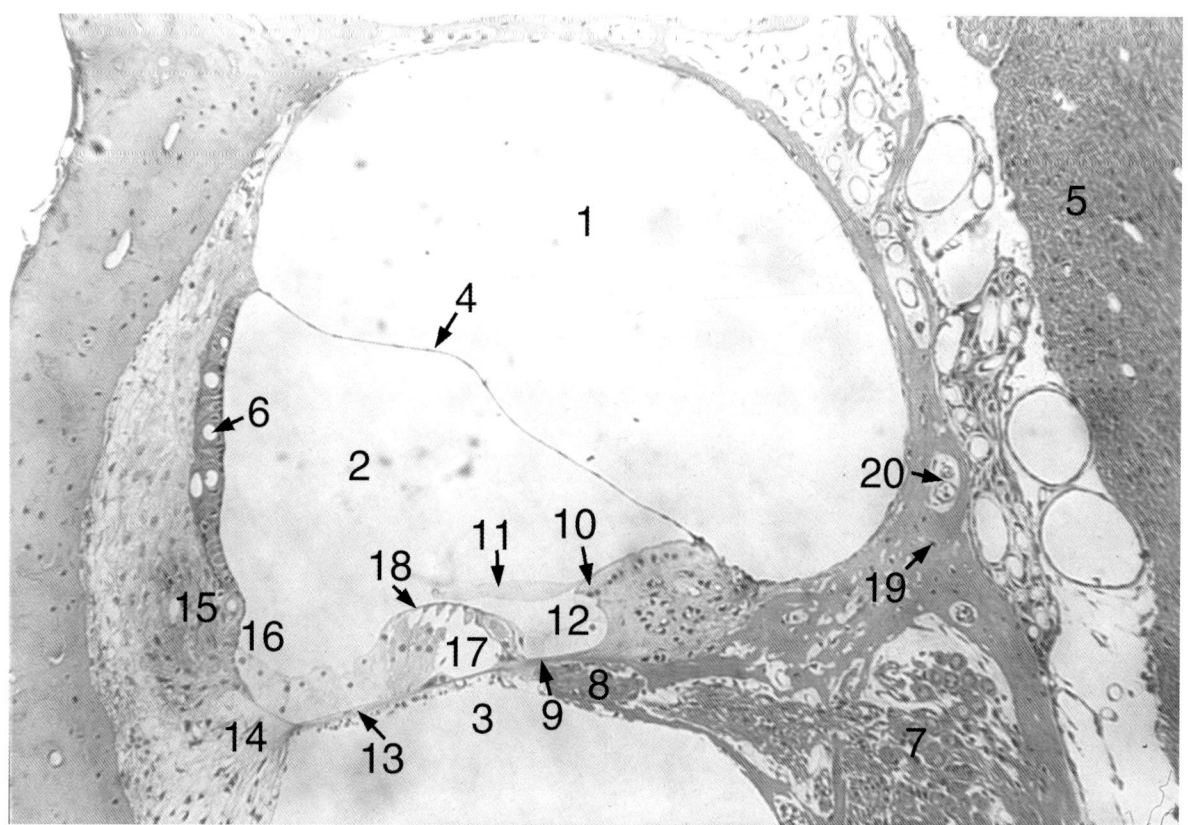

1. Scala vestibuli
2. Scala media (cochlear duct)
3. Scala tympani
4. Vestibular membrane (of Reissner)
5. Modiolus with the cochlear nerve
6. Stria vascularis
7. Spiral ganglion with bipolar neurons in the spiral canal of modiolus
8. Osseous spiral lamina and the peripherally directed neurites of the bipolar neurons
9. Tympanic lip of the spiral limbus
10. Vestibular lip of the spiral limbus, arrow points at the interdental cells
11. Tectorial membrane
12. Internal spiral sulcus
13. Basilar membrane
14. Spiral ligament (here the secondary spiral lamina appears in the basal turn)
15. Spiral prominence
16. External spiral sulcus
17. Tunnel of Corti (cuniculum internum)
18. Reticular lamina
19. Osteocyte
20. Blood vessel

ATLAS—EAR

Fig. 1.72

Cochlea, the organ of Corti. Semithin section, toluidine blue staining.

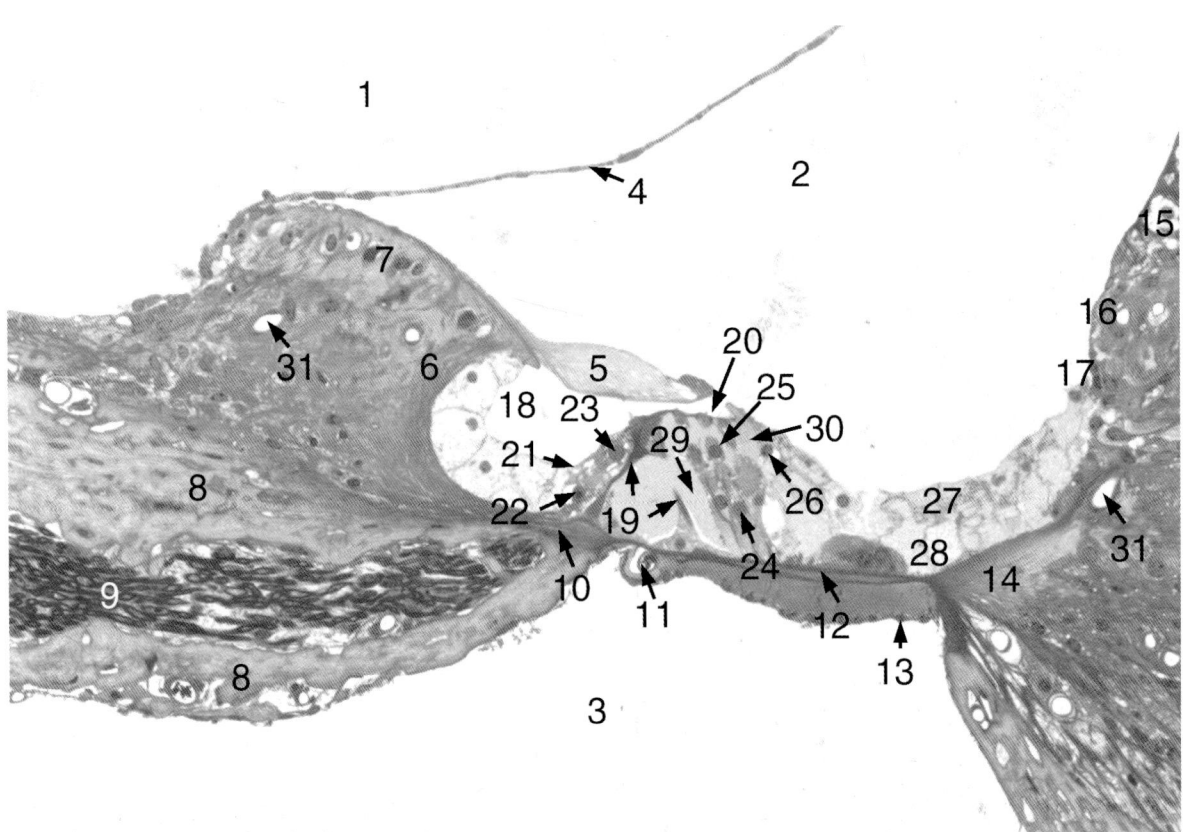

1. Scala vestibuli
2. Scala media (cochlear duct)
3. Scala tympani
4. Vestibular membrane (of Reissner)
5. Tectorial membrane
6. Vestibular lip of spiral limbus
7. Interdental cells
8. Osseous spiral lamina
9. Myelinated nerve fibers, the peripherally directed neurites of the bipolar neurons of spiral ganglion
10. Tympanic lip of spiral limbus
11. Vas spirale (blood vessel running along the basilar membrane)
12. Basilar membrane
13. Scala tympani, arrow points to perilymphatic cells
14. Spiral ligament
15. Stria vascularis
16. Spiral prominence
17. External spiral sulcus
18. Internal spiral sulcus
19. Tunnel of Corti (cuniculum internum), arrow point to pillar cells (rods)
20. Reticular lamina
21. Cell of Held (inner border cell)
22. Inner phalangeal cell, arrow points to the nucleus
23. Inner hair cell, arrow points to the nucleus
24. Outer phalangeal cell of Deiters
25. Outer hair cell
26. Cells of Hensen
27. Cells of Claudius
28. Cells of Boettcher
29. Spaces of Nuel (cuniculum intermedium) between the outer rods and the outer hair cells
30. Outer tunnel (cuniculum externum) between the outermost hair cells and the cells of Hensen
31. Capillary

Fig. 1.73

Scanning electronmicrographs (SEM) of the sensory epithelia in the organ of Corti, and in the utricle and saccule of the rat. Courtesy of A. Seoane and J. Llorens. A – Top view of the endolymphatic surface of cochlear hair cells exposed by the removal of the tectorial membrane. B – Side view SEM of the organ of Corti through a tear in the tissue. C – SEM of the exposed surface of the macula of utricle. D – SEM of the exposed surface of the macula of saccule.

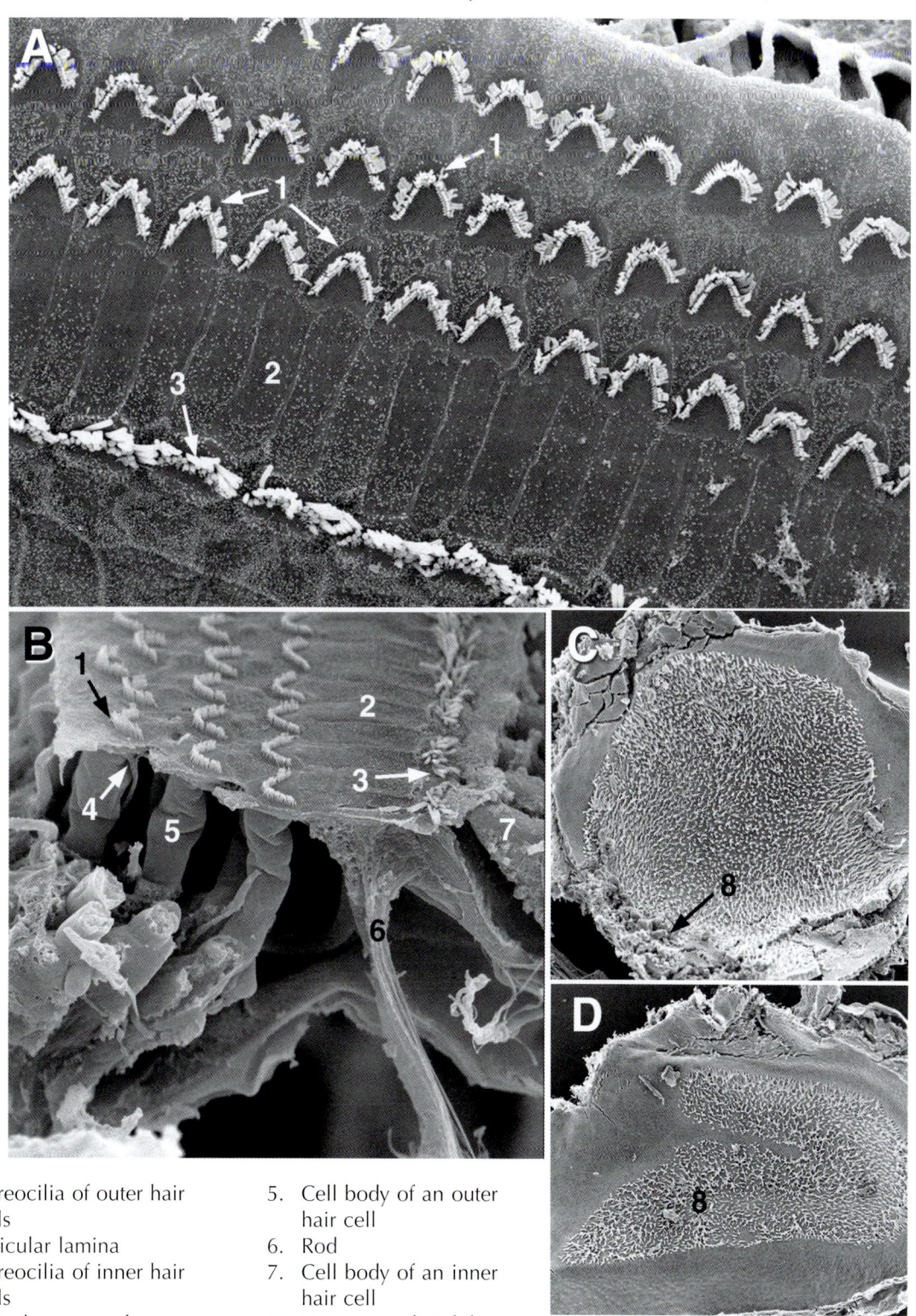

1. Stereocilia of outer hair cells
2. Reticular lamina
3. Stereocilia of inner hair cells
4. Apical process of a phalangeal (Deiters') cell
5. Cell body of an outer hair cell
6. Rod
7. Cell body of an inner hair cell
8. Remnants of otolith membrane after removal of cupula

Fig. 1.74

Macula of the saccule. Semithin section, toluidine blue staining.

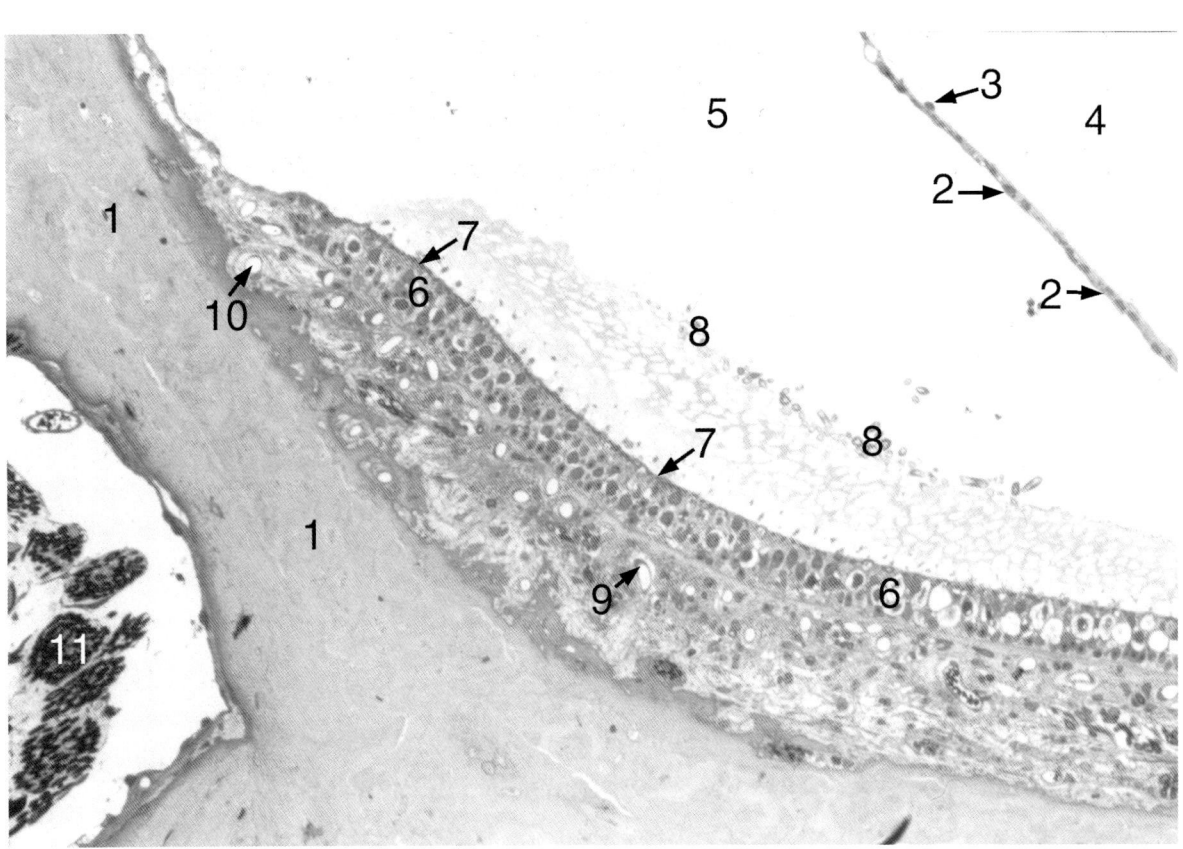

1. Bony wall of labyrinth
2. Membranous wall of labyrinth (corresponds to the vestibular membrane of cochlea)
3. Perilymphatic cells
4. Perilymphatic space
5. Endolymphatic space
6. Sensory epithelium consisting of supporting cells and hair cells
7. Cilia of hair cells embedded in the jelly of the otolithic membrane
8. Otoliths (calcium containing precipitates) embedded in the jelly of the otolithic membrane
9. Capillary, arrow points to the nucleus of the endothelial cell
10. Capillary, arrow points to a pericyte
11. Saccular nerve

Fig. 1.75

Ampulla of a semicircular canal. Histological specimen, HE staining

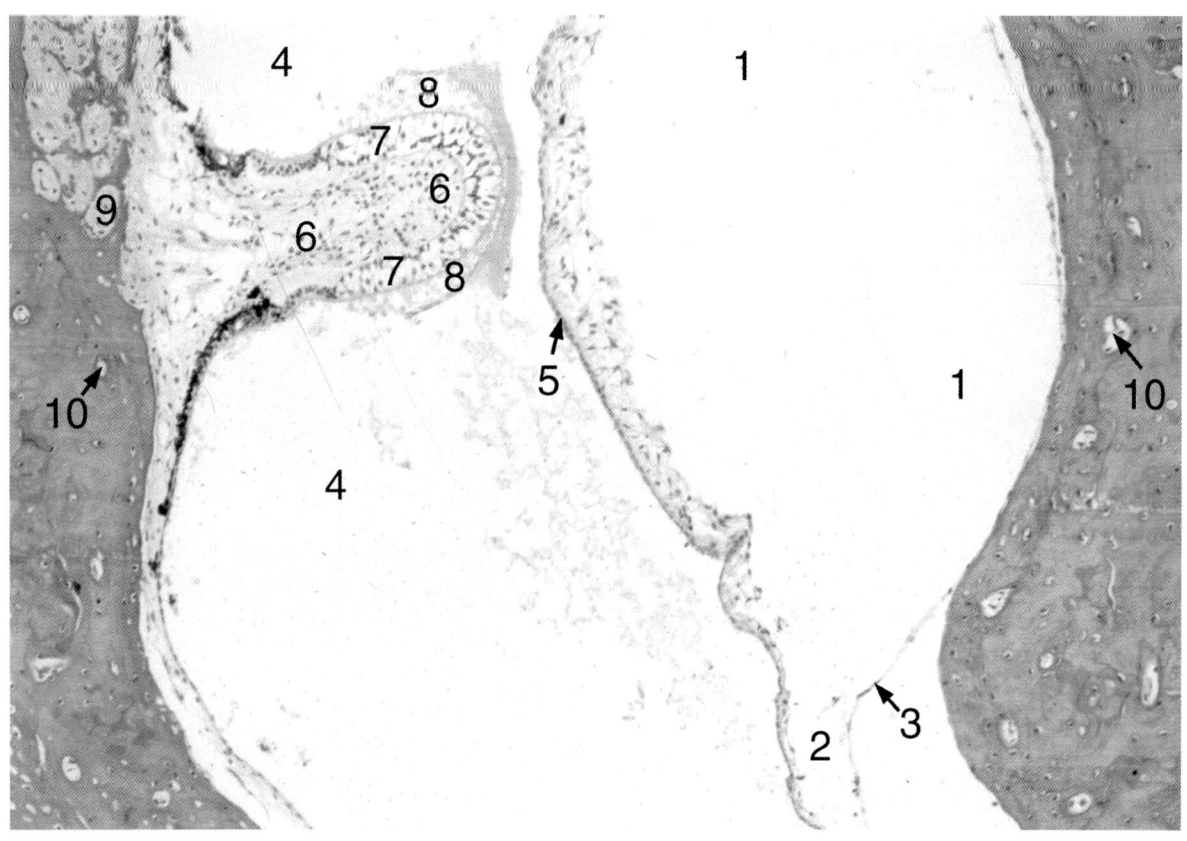

1. Perilymphatic space of the ampulla of semicircular canal
2. Perilymphatic space of the crus of semicircular canal
3. Bony wall of labyrinth, arrow points to detached perilymphatic cells
4. Endolymphatic space
5. Membranous wall of labyrinth
6. Crista ampullaris, inner mass of connective tissue
7. Crista ampullaris, sensory epithelium consisting of supporting cells and hair cells
8. Cupula
9. Branches of the ampullar nerve
10. Blood vessel in the bone

Fig. 1.76

The sensory epithelium in the crista ampullaris of the rat. Courtesy of E. Balbuena, A. Seoane and J. Llorens. A – Light micrograph of the apical part of a crista showing type I (I) and type II (II) hair cells with stereocilia (s) and nerve fibers (n). B – Scanning electronmicrograph of the ampullar surface of a crista after removal of the cupula. The polygonal fields (outlined by arrows) represent the surface of tightly linked supporting cells with microvilli. The hair cells project their bundles of stereocilia into the ampullar space. C – Side view of the hair cells of a crista through a tear in the tissue.

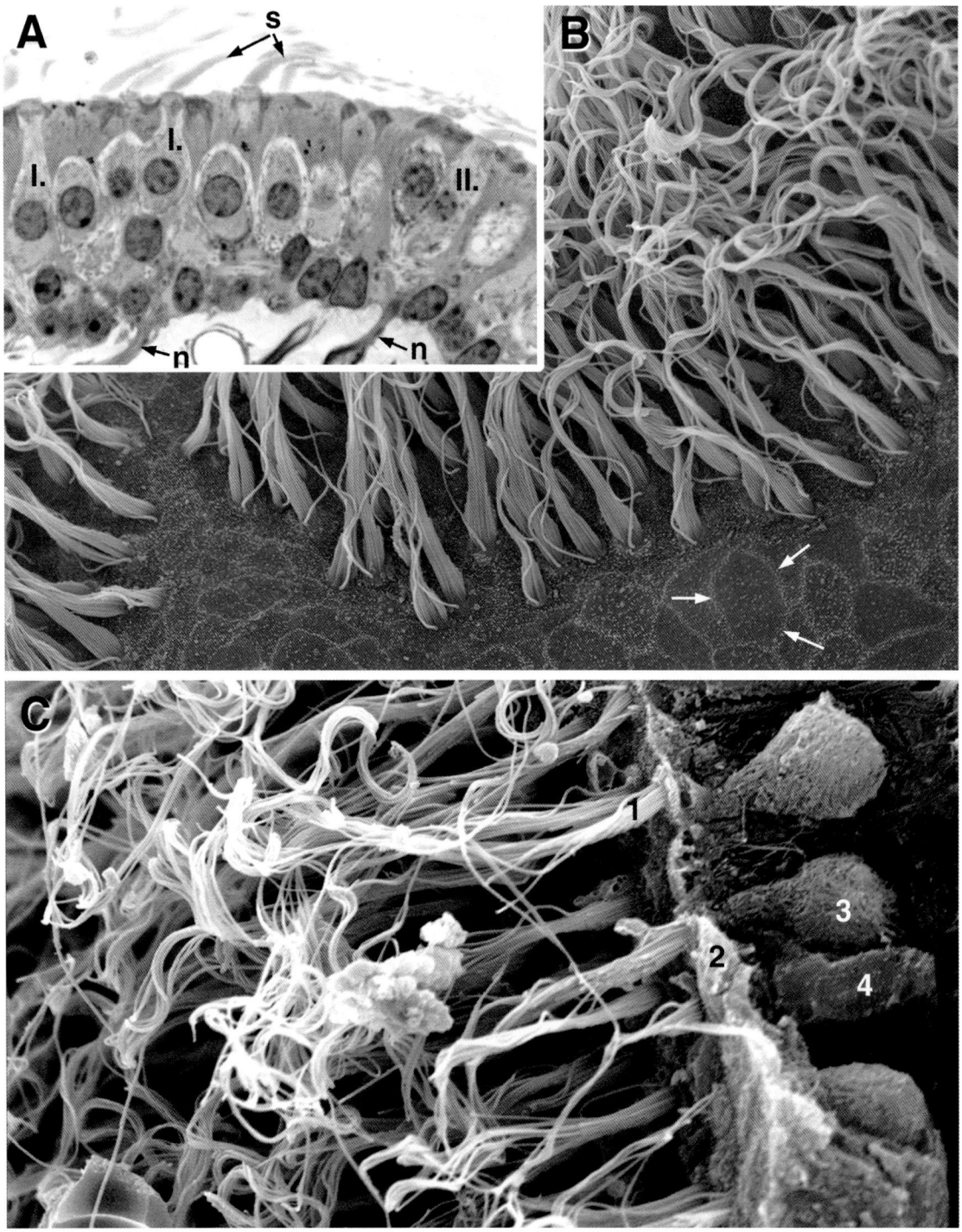

1. Stereocilia of hair cells
2. Lamina reticularis separating the cell bodies from the endolymphatic space
3. Bottle-shaped type I hair cell
4. Cylindric type II hair cell

The Auditory Pathway

I. COCHLEAR NUCLEI (FIGS.1.77; 1.78)

Most of the descriptive and experimental morphological data have been obtained in non-human species, particularly the cat. However, whenever possible, an attempt is made here to discuss the relevant human structures along with their mammalian counterparts.

The signal conveyed by the cochlear nerve (centrally directed neurites of the spiral ganglion) reaches first the cochlear nuclei (CN) located at the pontomedullary border. There are two such nuclei on each side, the ventral cochlear nucleus (VCN) and dorsal cochlear nucleus (DCN), both adjacent and external to the inferior cerebellar peduncle. On entering the medulla, the acoustic nerve cuts across the VCN, subdividing it into anteroventral (AVCN) and posteroventral (PVCN) divisions, and branches into an ascending fascicle, terminating in the AVCN, and a descending fascicle, arborizing in the PVCN and DCN. The incoming fibers are arranged according to a strict tonotopic order, i.e., those representing high-frequency sounds are located dorsally in the nuclei, whereas those representing low-frequency sounds tend to be distributed in the ventral divisions of CN. The main cell types of the VCN are called bushy (spherical), stellate (multipolar), and octopus cells. Bushy cells respond to tones with a sharp "primary-type" excitation, much like the acoustic nerve fibers themselves. They project through the trapezoid body (TB) to the medial superior olivary nucleus bilaterally. Those bushy cells with more extensive dendritic fields project to the medial nucleus of the trapezoid body (MNTB) of the cat and to the dorsal nucleus of the lateral lemniscus (DNLL). The stellate cells of the VCN, which project to the inferior colliculus (IC), respond to a single stimulus by repetitive firing ("chopper" response). Octopus cells typically showing an "on-response" send their projections to the periolivary region and to the contralateral ventral nucleus of the lateral lemniscus (VNLL). Apparently, the human DCN lacks the lamination observed in the cat, its main cell type being giant "pyramidal" neurons. The response pattern of DCN neurons is more complex than that found for VCN and these neurons send parallel projections to higher auditory centers such as the IC. Thus, auditory processing is very prominent already at the level of the cochlear nuclei, and this is the most central site in which a lesion brings about deafness in the ipsilateral ear. Unilateral lesions at higher levels can no longer produce such deafness because of the existence of multiple commissural pathways (see below).

The main efferent pathways arising from the CN are as follows. The TB leaves the dorsomedial aspect of the VCN and passes to the level of the superior olivary complex both ipsi- and contralaterally, thereby representing the caudalmost commissural pathway of the auditory system. The intermediate acoustic stria (IAS), containing mainly octopus cell projections, emerges from the posterior tip of the VCN, whereas the dorsal acoustic stria (DAS) is an efferent pathway from the DNC, crossing over to the contralateral lateral lemniscus, destined for the IC.

II. SUPERIOR OLIVARY COMPLEX (SOC) (FIGS. 1.77; 1.78)

An important relay center of the auditory pathway, this nuclear complex lies in the caudal pontine tegmentum ventromedial to the rostral portion of the facial nucleus and lateral to the trapezoid body. Main parts of the SOC are the lateral superior olivary (LSO) nucleus, medial (accessory) superior olivary (MSO) nucleus, and the medial nucleus of the trapezoid body (MNTB). Unlike the situation found in the cat, the human LSO is a relatively small cluster of neurons, whereas the MSO is compact and well developed. This fact is likely to be associated with a relatively poor sensitivity of humans to high-frequency tones (which are predominantly processed in the LSO). As a result, humans tend to use low-frequency cues (also including speech) for the localization of sound source, with a concomitant enlargement of MSO. The most prominent cell type of MSO is spindle-shaped, bipolar, and transversely oriented with two dendritic arbors projecting in opposite directions. The processes directed medially receive input from the contralateral CN, whereas those directed laterally are innervated from the ipsilateral CN. This anatomical situation renders bipolar cells highly efficient detectors of the sound source because the cells are sensitive to the interaural time difference. The signal from the ear that is closer to the sound source reaches the bipolar cell before the signal from the other ear does, in precise proportion to the azimuth of sound source. Conversely, the binaural cells of the LSO receiving ipsilateral excitation but contralateral inhibition are more sensitive to interaural intensity (rather then timing) differences of high-frequency sounds.

In the cat, the contralateral inhibitory input to the binaural cells of LSO is supplied by the MNTB (whose cells are monaural). The most prominent feature of this nucleus is the presence of unusually large nerve endings termed the calyces of Held. They are terminals of bushy cell projections from VCN forming multiple synapses with the somata of principal MNTB neurons. Although the existence of a true MNTB in humans has been disputed for some time, it is now generally accepted that a similar nucleus lies at the level of the exiting abducent nerve just

Fig. 1.77

Hierarchical illustration of the auditory pathway in relation to the brainstem auditory-evoked potentials (BAEP). The bottom graph represents an original recording from a normal individual.

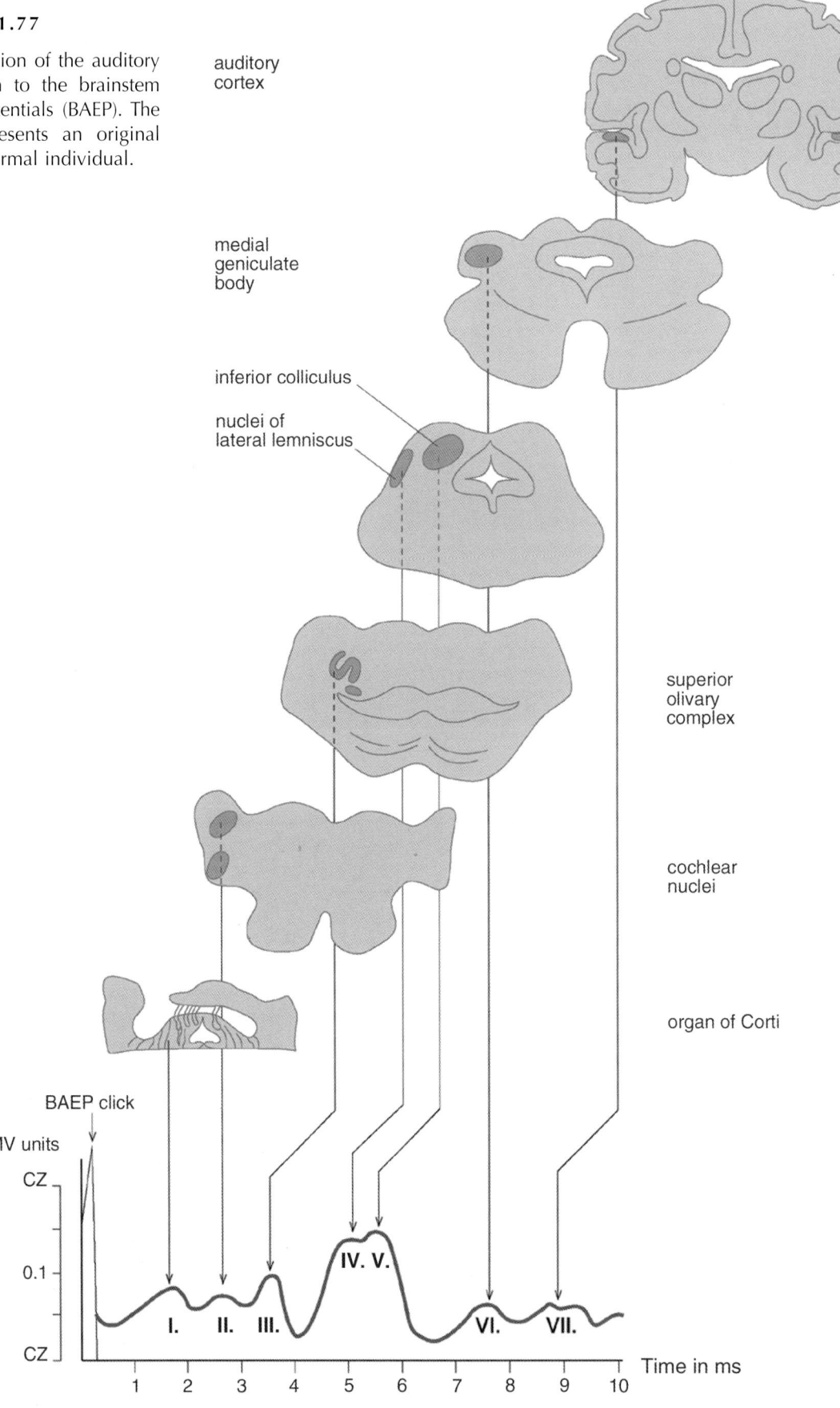

Fig. 1.78

Experimental studies on the auditory pathway of the cat. I. Cochlear nuclei and superior olivary complex. A – Retrogradely labeled projection neurons of DCN arranged as a single layer (dark field image); B – Retrogradely labeled projection neurons of VCN in a multilaminar arrangement (dark field image); C – bipolar neurons of MSO (Golgi-Kopsch impregnation); D – bipolar neurons of MSO retrogradely labeled from IC by horseradish peroxidase (dark field image); E – bipolar neuron of MSO (silver impregnation). Note the surface indentations of the cell body (arrows), corresponding to sites of axosomatic terminals; F – Electronmicrograph showing two branching cochlear terminals (CT1 and CT2) synapsing with the dendrite of a bipolar MSO neuron, similar to in Figs. C–E. Arrow points to the site where the myelin sheath of CT1 axon ends. Synapses are marked by arrowheads. Courtesy of K. Majorossy and Á. Kiss.

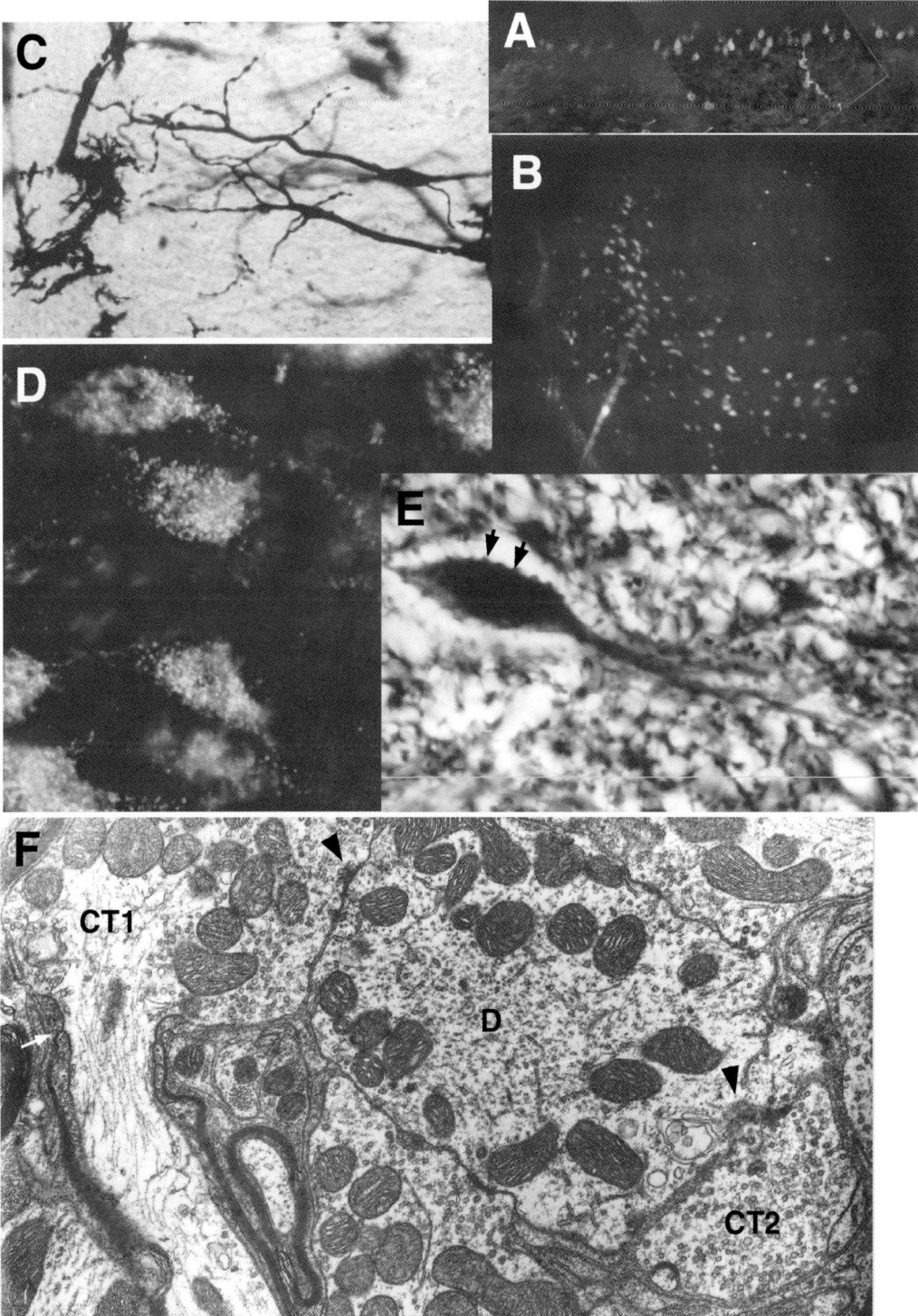

medial to the MSO. Large axosomatic terminals resembling (albeit not identical with) the calyces of Held have been described also in humans.

In addition to ascending pathways, the region surrounding the SOC is the source of an important descending connection called olivocochlear bundle. This pathway originates from a limited number of neurons in the retro-olivary cell group, rather poorly delineated in humans but probably corresponding to the dorsal, medioventral and lateroventral periolivary nuclei described in other mammals. Fibers from here bilaterally innervate the hair cells of the cochlea. More precisely, the outer hair cells receive direct efferent connections, whereas the inner hair cells are affected indirectly, through synapses on their afferent boutons. Presumably, olivocochlear efferents selectively control the sensitivity of the spiral organ to certain acoustic signals.

III. NUCLEI OF THE LATERAL LEMNISCUS (NLL) (FIGS. 1.77; 1.79)

These nuclei are embedded in the lateral lemniscus, a large pathway ascending from the CN and the SOC to the IC. The dorsal, intermediate, and ventral nuclei extend from the rostral pole of SOC to the level of IC. In the cat, the dorsal part receives most of its afferents from the SOC and, as a result, the neurons are mainly binaurally represented. The ventral nucleus has more monaural units. Although the NLL definitely exist in the human brain, their function is poorly understood. An important crossing of the auditory pathway (Probst tract or Probst's commissure) lies at the level of the NLL.

IV. INFERIOR COLLICULUS (IC) (FIGS. 1.77; 1.79)

The main target nucleus of the lateral lemniscus, the IC consists of a central nucleus, an adjacent dorsal cortex, and further smaller nuclei (dorsomedial and lateral). The most characteristic features of the central nucleus are its laminar arrangement and a tonotopic distribution of incoming lemniscal fibers: low or high frequencies being represented in the dorsal or ventral laminae, respectively. Although the dorsal cortex has a layered appearance, a similar arrangement has not been observed in the dorsomedial or lateral nuclei. Most of the cells in the IC are binaurally driven and the IC is likely to participate in the spatial mapping of the auditory receptive field. This is at least suggested by analogy with avian studies, in particular the barn owl, where the IC-homologue tectal region does have a similar spatial map. A precise mapping of the auditory receptive field enables these nocturnal predators to locate their prey (usually small rodents) by the faintest rustle coming from the undergrowth. Contrary to the common belief concerning an extraordinary night vision capability of owls, what really matters in the preycatching of these birds is the acuity of hearing.

Efferents from the IC reach the medial geniculate nucleus (MGN) via the brachium of the inferior colliculus (BIC). This pathway is mainly ipsilateral but a crossed connection (intercollicular commissure) is also present at this level.

V. MEDIAL GENICULATE BODY (MGB) (FIGS. 1.77; 1.80)

This nucleus lies ventromedial to the pulvinar of thalamus and separated from the latter by the brachium of the superior colliculus (superior quadrigeminal brachium). The MGB comprises three divisions: medial (magnocellular), dorsal (posterior), and ventral. Only the ventral division receives a topographic input from the central nucleus of IC via the BIC, with a low- to high-frequency gradient extending from lateral to medial parts. The ventral division contains laminae, which are defined by the dendritic ramification of principal cells. The dorsal division of MGB, with no clear laminar arrangement, is innervated by fibers mainly from the dorsal cortex of IC and also from other auditory nuclei of the brainstem. A tonotopic arrangement of fibers is not evident here. The medial (magnocellular) division is less well studied in humans but it is likely to receive an even wider input, including the superior colliculus. As a result, higher order processing in MGB already involves modalities other than purely acoustic. According to increasing levels of complexity, the ventral division (with purely acoustic and tonotopically arranged fiber input) projects to the primary auditory cortex, the dorsal division sends fibers to cortical regions surrounding the primary auditory cortex, whereas efferents from the medial division reach a broader field including the insular and opercular cortices.

VI. AUDITORY CORTEX (FIGS. 1.77; 1.81; 1.82)

Fibers ascending from the medial geniculate body (geniculocortical tract) pass through the posterior limb (sublenticular part) of the internal capsule and terminate in the temporal operculum of the ipsilateral hemisphere. This region contains the primary auditory cortex (AI), situated in the transverse temporal gyri (Heschl's gyri), coextensive with area 41 and, partly, with area 42 of Brodmann. Owing to a particularly prominent layer IV (internal granular layer), these subregions appear "grainy" under the microscope and are collectively referred to as koniocortex (meaning dusty in Greek). Detailed studies in monkeys have shown that the AI is parcellated according to a strict tonotopy, high frequencies being represented posteriorly and low frequencies anteriorly. Interlaced with this rostro-caudal gradient of frequency distribution is

Fig. 1.79

Experimental studies on the auditory pathway of the cat. II. Commissural pathways of the brainstem. A – Electronmicrograph of the inferior colliculus. The axons of the contralateral IC, anterogradely labelled with the pathway tracer biotinylated dextran amine (BDA) (A1, A2), synapse with the dendrite of a principal neuron (D). B – Electronmicrograph of the dorsal nucleus of the lateral lemniscus (LLD). Two BDA labelled axon terminals (A3, A4) arising from the contralateral ventral cochlear nucleus synapse with a large dendrite (LD). The latter belongs to a projection neuron labelled retrogradely with horseradish peroxidase (HRP) from the contralateral IC via the Probst tract. The HRP deposit is marked by arrows. The cartoon in the top right of the figure shows the site of tracer injections (cannula and wire: BDA iontophoresis; hypodermic syringe: HRP pressure injection). Courtesy of Prof. K. Majorossy and Dr. Á. Kiss.

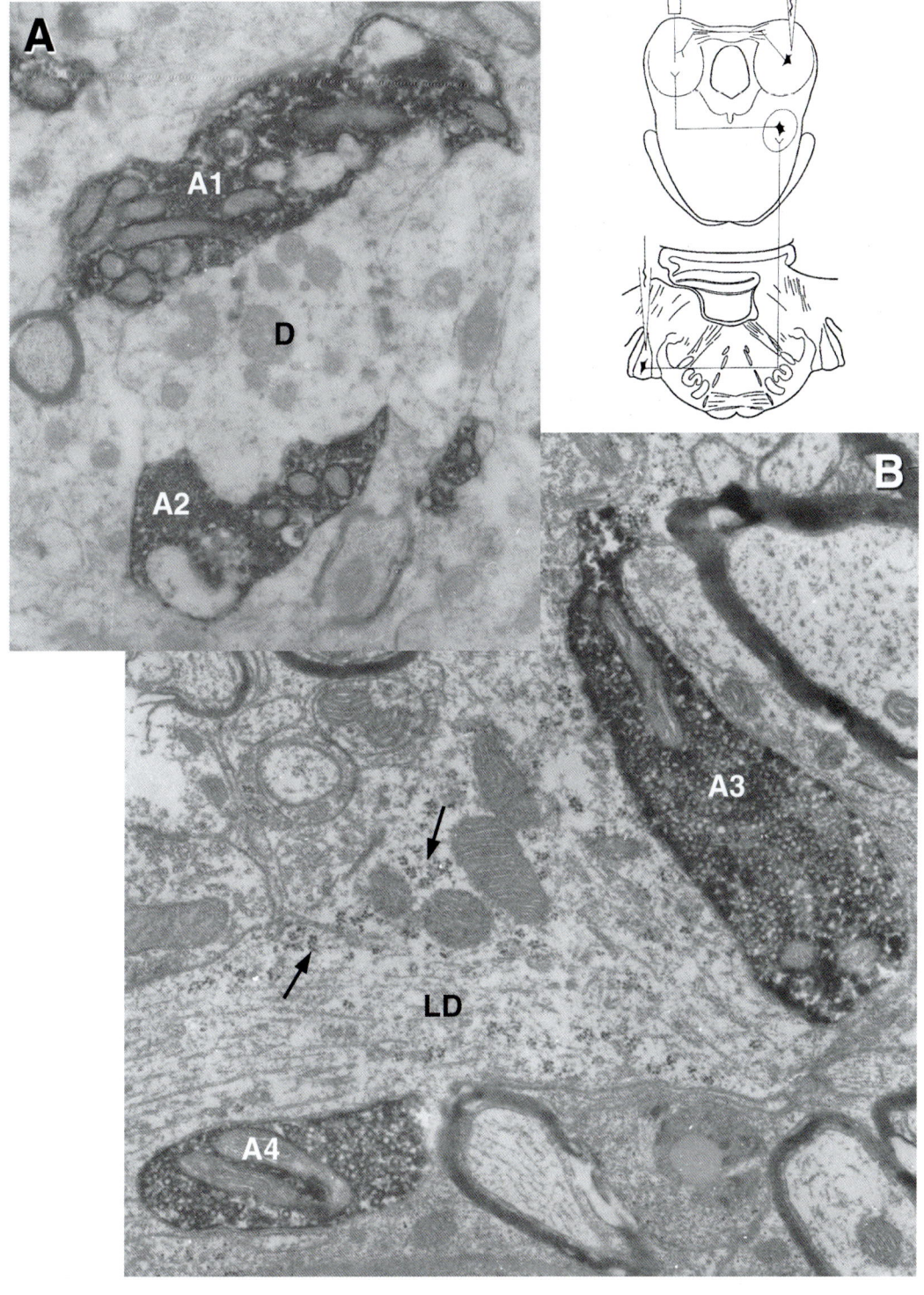

Fig. 1.80

Experimental studies on the auditory pathway of the cat. III. Medial geniculate body. A – Site of HRP injection in the auditory cortex. B – Dark field image showing a band of retrogradely labelled geniculocortical relay neurons (arrow, representing a tonotopic lamina) in the MGB. C – High magnification image of two HRP-filled neurons in the same lamina of MGB. D – Electronmicrograph of a synaptic glomerulus in the MGB. The scheme of the glomerulus is depicted in the cartoon (top right of the figure). SA – specific afferent terminal (from the IC), CA – cortical afferent terminal, RD – relay cell dendrite, GD – Golgi cell dendrite, GA – Golgi axon terminal. Courtesy of Prof. K. Majorossy and Dr. Á. Kiss.

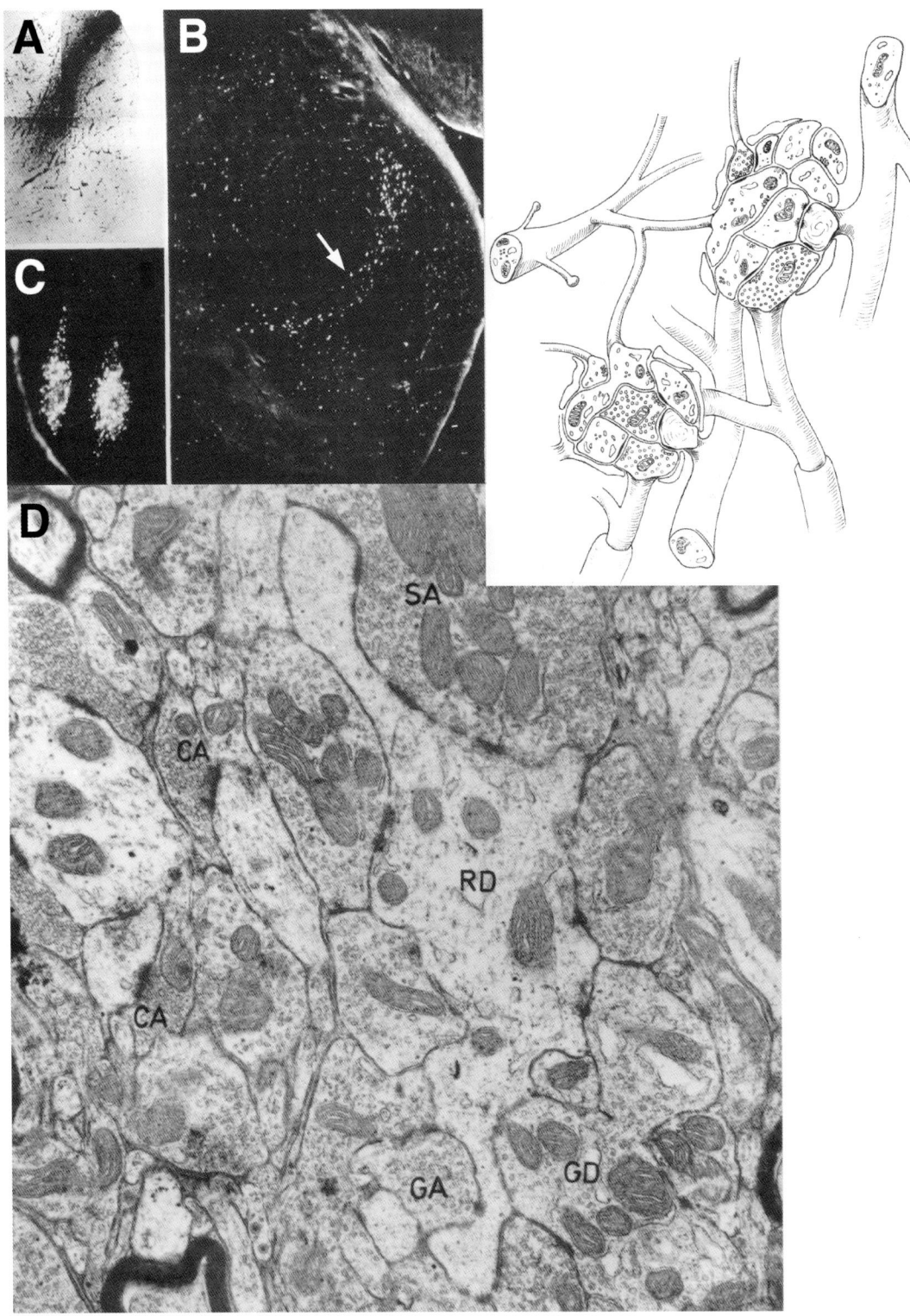

another arrangement of cellular bands (perpendicular to the previous one), representing different tone intensities. A third and even more complex pattern emerges in the cat from the alternation of bands containing two kinds of binaural neurons. Summation bands contain excitation/excitation (E/E) neurons, excited by stimuli to both ears, whereas suppression bands are composed of neurons that can be excited by stimulus from the contralateral ear but inhibited when the ipsilateral ear is stimulated (E/I neurons). The presence of summation and suppression bands is likely to underlie the role of the primary auditory cortex in spatial orientation. However, studies concerning this function in man are still far from being conclusive.

Moving to the areas surrounding AI, the picture gets even more confusing. What seems to be evident is that the tonotopic gradient is repeated (albeit in an inverse direction) in the region lying just rostral to the AI (area R), and again in the adjacent rostrotemporal (RT) area. Even the area posterior (caudal) to AI (area C) has some tonotopic arrangement. Such multiple representations are known from other sensory cortical fields, including the visual cortex. To the "core" regions of auditory cortex (AI, R, and RT), further "belt" regions, situated in the deep temporal operculum, planum temporale, and superior temporal gyrus are attached. These areas are still acoustic in character but their anatomical parcellation and functional significance have not been established with certainty. However, recent functional studies in human subjects and primates have revealed areas of particular significance even beyond those mentioned here. A region selectively responsive to human voice was described bilaterally along the upper bank of the superior temporal sulcus, using functional magnetic resonance imaging (fMRI). The area may be functionally similar to

Fig. 1.81

Coronal histological section of human brain demonstrating the primary auditory cortex. Section thickness: 20 μm. Levanol staining. Courtesy of Prof. M. Palkovits. NC – caudate nucleus, Th – thalamus, F – fornix, IC – internal capsule, Pu – putamen, GP – globus pallidus, V III – third ventricle, TO – optic tract.

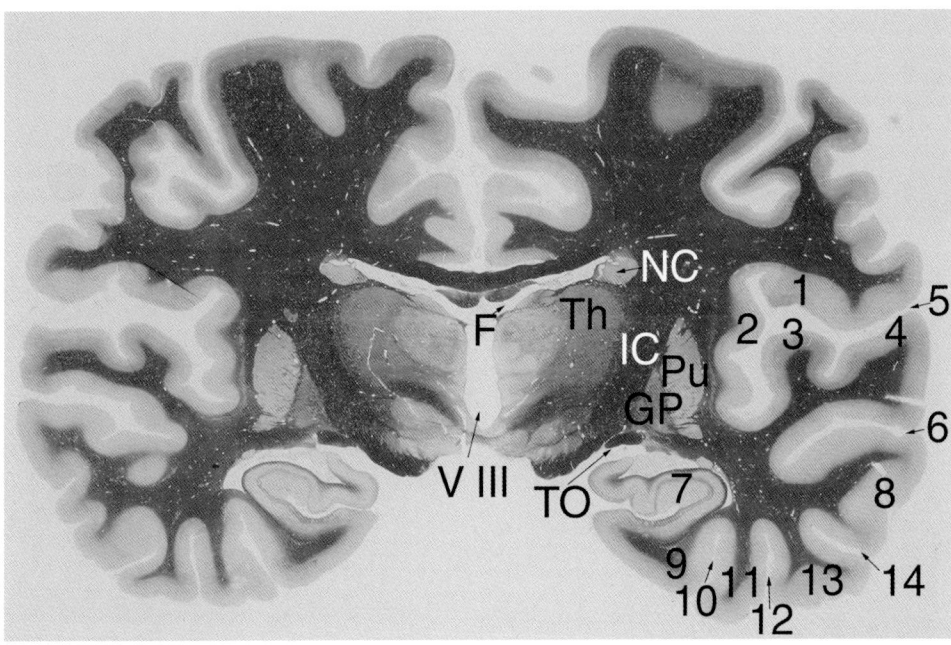

1. Parietal operculum
2. Insula
3. Anterior transverse temporal gyrus (Heschl's gyrus)
4. Planum temporale and superior temporal gyrus, T1
5. Lateral fissure
6. Superior temporal sulcus
7. Hippocampus
8. Middle temporal gyrus, T2
9. Entorhinal cortex and uncus
10. Collateral sulcus
11. Fusiform gyrus, T4
12. Lateral occipitotemporal sulcus
13. Inferior temporal gyrus, T3
14. Inferior temporal sulcus

Fig. 1.82

Experimental studies on the auditory cortex. Descending cortical pathways. A – HRP injection in the MGB of cat, B – Dark field photomontage demonstrating the position of HRP-filled neurons in layer V of the cat auditory cortex. p – pial surface, C – High magnification of a typical pyramidal neuron. D, E – Layer III pyramidal neuron of human auditory cortex, retrogradely labelled post-mortem with the fluorescent tracer DiI deposited in a neighbouring cortical area. The dendritic spines of the association neuron are particularly prominent in Fig. E. F – Golgi-Kopsch impregnated pyramidal neuron of cat auditory cortex. G – Camera lucida drawing of the pyramidal neuron of Fig. F. Note the thick and densely spiny apical and basal dendrites, and the rich network of thin axon collaterals. H – Golgi-Kopsch impregnated non-pyramidal (fusiform) neuron of cat auditory cortex. The cell bodies of the pyramidal and non-pyramidal cell types are shown enlarged in Figs. I and J, respectively. Courtesy of Prof. K. Majorossy and Á. Kiss (Figs. A, B, C, F, G, H, I, J), and Dr. E. Tardif and Prof. S. Clarke (Figs. D, E).

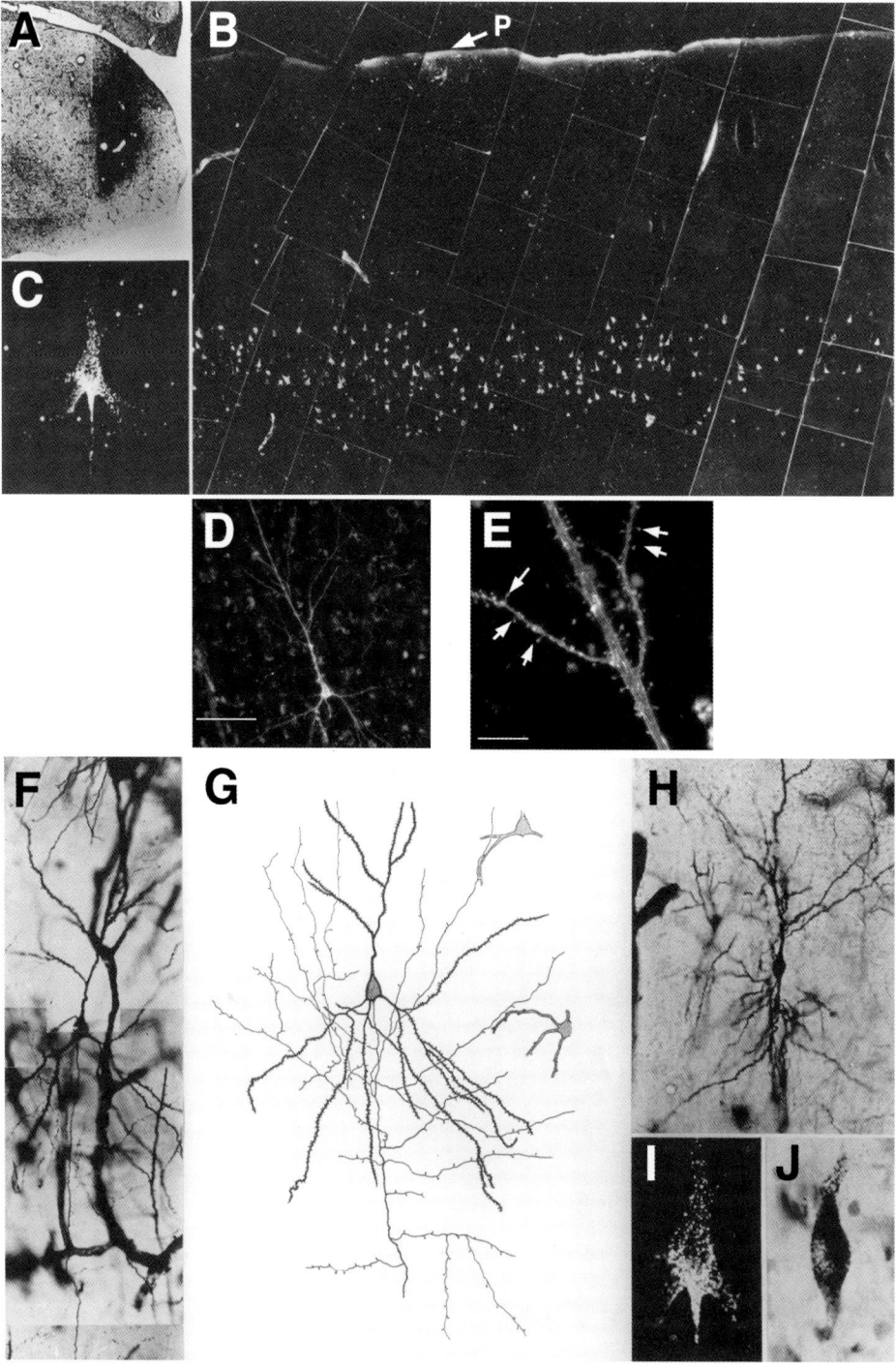

the face recognition area of the visual system. Other regions (T3 of the right planum temporale and the adjacent right parietal cortex) were found to be activated by moving, but not by stationary sound signals. Interestingly, the right hemisphere tends to be dominant also in the motion detection of visual cues.

Both the core and belt regions are reciprocally connected with the MGB, in particular the ventral division of the MGB sends fibers mainly to the core, whereas the medial and dorsal divisions project to both core and belt areas. In turn, the core projects to the central nucleus and the belt to other nuclei of the IC.

The latter pathways belong to the widespread system of descending connections, reaching stepwise all subcortical stations of the auditory tract. Essentially, each level reports back to lower levels either immediately below itself or down to more caudal levels. An example for these "skipping" connections is the cortico-collicular pathway described above. The least well established of such descending connections is the projection from the MGB to IC, although its existence has been suggested in humans. Descending connections are likely to influence the attentional selection of incoming acoustic signals.

Information from the belt areas of the auditory cortex spreads radially into other adjacent and more distant cortices of the same hemisphere and also to identical areas of the contralateral hemisphere. With such irradiation of signal the acoustic modality is combined with other modalities (visual, somatosensory, motivational, etc.). Some of these combinations may be quite surprising if perfectly "logical" on their own account. For example, the ventral premotor cortex (an area of the frontal lobe just anterior to the primary motor cortex) contains neurons that encode visual information only from a space near the body, i.e., no more than approx 30 cm from the tactile receptive field. In the macaque monkey, a large proportion of such neurons were found to be trimodal, that is, they responded to tactile, visual, and acoustic input, but only when the sound source was near to the animal's head (~10 cm). Apparently, the ventral premotor cortex of primates stores multimodal information salient to nearby objects.

Because of the strongly bilateral character of the auditory pathway, unilateral injuries of the auditory cortex cause little if any hearing defect. Even bilateral lesions may only lead to a diminished ability of the patient to recognize and interpret tones in a particular context (various forms of acoustic agnosia). On the other hand, there is a clear hemispheric asymmetry in the extent of Heschl's gyrus and planum temporale in most humans. The most prominent of such asymmetries concerns a region that is obviously connected to the acoustic cortex though is not part of it. The speech area of Wernicke, covering the posterior parts of superior and middle temporal gyri, supramarginal gyrus, and adjacent inferior parietal cortex (including areas 39 and 40 of Brodmann) is active only in the dominant hemisphere. This is the case, at least, in right-handed people but, interestingly, in left-handed individuals the speech center of Wernicke (as well as the motor speech area of Broca) may be bilaterally represented. Injuries to this speech center in the dominant hemisphere lead to serious difficulty of speech comprehension, without loss of speech production (sensory aphasia).

Vestibular Pathways (Fig. 1.83)

These pathways begin with the primary afferents arising from the sensory epithelia of the cristae ampullares and the maculae of the the utricle and saccule. The afferents terminate in the vestibular nuclei of the brainstem. Secondary pathways from here form extensive links with the cerebellum and the oculomotor nuclei, and descending connections descend as far as the spinal cord. Owing to the extremely diffuse nature of vestibular pathways and their intricate connections with the entire locomotor system including postural reflexes and ocular movements, the central representation of the organ of equilibrium can no longer be dealt with as a specific sensory system. We therefore chose to include a scheme with only those principal brainstem areas and connections that are immediately relevant to vestibular input. Further important points, in particular the role of the medial longitudinal fascicle and its bearing on the organization of ocular movement, are discussed in the chapter on the eye.

Fig. 1.83

Schematic illustration of the vestibular pathways in craniocaudal arrangement (A) and at four different coronal sections of the brainstem (B – rostral midbrain, C – caudal midbrain, D – pons, E – medulla).

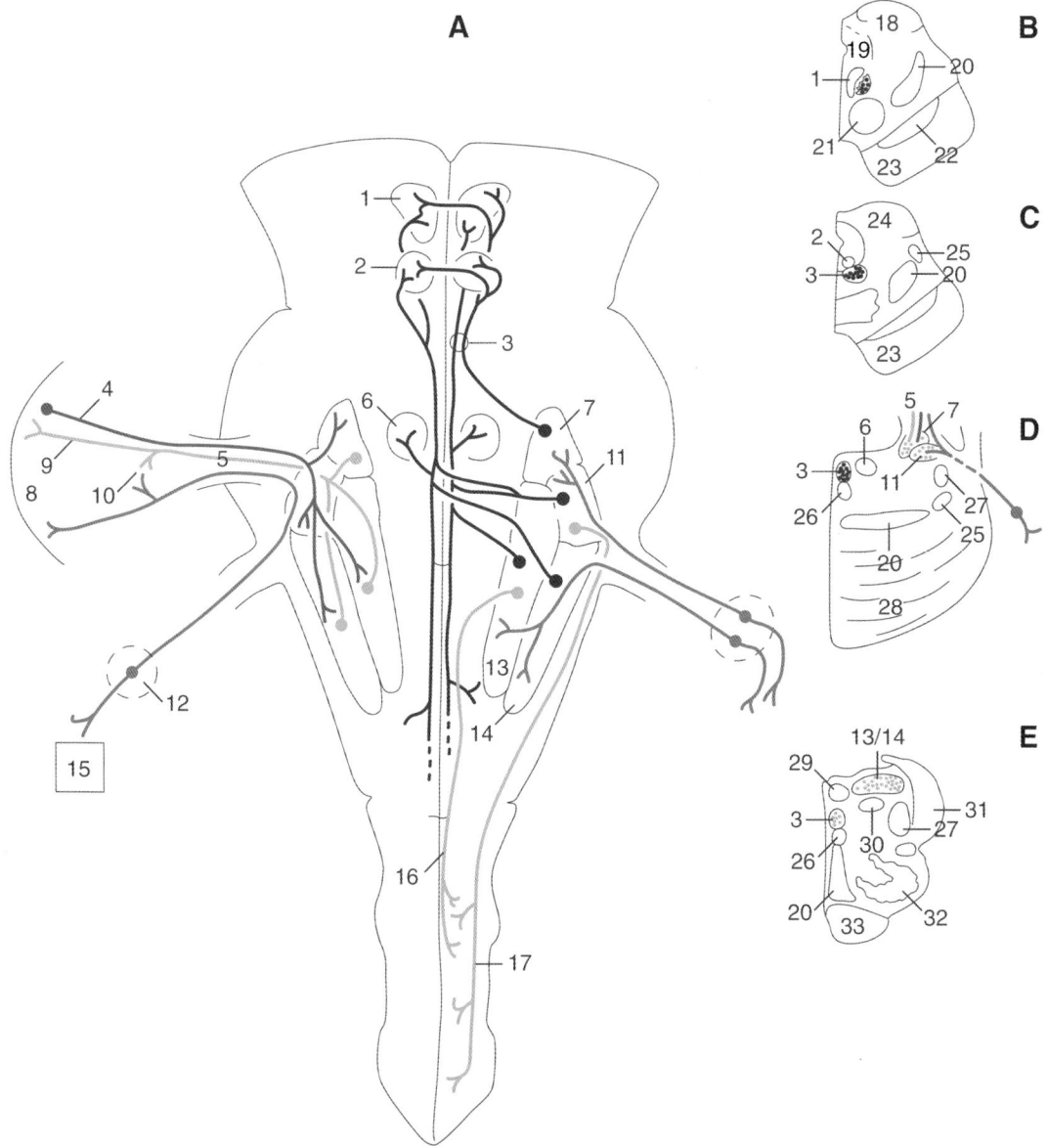

1. Oculomotor nucleus
2. Trochlear nucleus
3. Medial longitudinal fascicle
4. Cerebellar corticovestibular fibers
5. Juxtarestiform body
6. Abducens nucleus
7. Superior vestibular nucleus
8. Cerebellar cortex
9. Vestibulocerebellar fibers, secondary
10. Cerebellar nuclei
11. Lateral vestibular nucleus
12. Vestibular ganglion
13. Medial vestibular nucleus
14. Inferior (spinal) vestibular nucleus
15. Crista ampullaris, maculae of utricle and saccule
16. Medial vestibulospinal tract
17. Lateral vestibulospinal tract
18. Superior colliculus
19. Periaqueductal gray
20. Medial lemniscus
21. Red nucleus
22. Substantia nigra
23. Crus cerebri
24. Inferior colliculus
25. Anterolateral system
26. Tectospinal tract
27. Spinal trigeminal tract
28. Basilar pons
29. Hypoglossal nucleus
30. Solitary tract and nucleus
31. Restiform body (inferior cerebellar peduncle)
32. Inferior olivary nucleus
33. Pyramid

RECOMMENDED READINGS

Textbooks and Handbooks

1. Bannister LH, Berry MM, Collins P, Dyson M, Dussek JE, Ferguson MWJ. Gray's Anatomy, 38th ed., Churchill Livingstone, Edinburgh, 1999.
2. Brodal P. The Central Nervous System. Structure and Function, Oxford University Press, New York, NY, Oxford, 1992.
3. Drews U. Color Atlas of Embryology, Thieme, Stuttgart, New York, NY, 1995.
4. Hinrichsen KV (ed.). Humanembryologie. Lehrbuch und Atlas der vorgeburtlichen Entwicklung des Menschen, Springer, Berlin, 1990.
5. Smith C.U.M. Biology of Sensory Systems, John Wiley & Sons, Chichester, 2000.

Reviews and Research Reports

1. Baumgart F, Gaschler-Markefski B, Woldorff MG, Heinze H-J, Scheich H. A movement-sensitive area in auditory cortex. Nature 1999;400:724-726
2. Belin P, Zatorre RJ, Lafaille P, Ahad P, Pike B. Voice selective areas in human auditory cortex. Nature 2000;403:309-312.
3. Dallos P, Evans BN. High-frequency motility of outer hair cells and the cochlear amplifier. Science 1995; 267:2006-2009.
4. Graziano MSA, Reiss LAJ, Gross CG. A neuronal representation of the location of nearby sounds. Nature 1999;397:428-430.
5. Huffman RF, Henson OW Jr. The descending auditory pathway and acousticomotor systems: connections with the inferior colliculus. Brain Res Rev 1990;15:295-323.
6. Moore JK. Organization of the human superior olivary complex. Microsc Res Techn 2000;51:403-412.
7. Moore JK, Simmons DD, Guan Y. The human olivocochlear system: Organization and development. Audiol Neuro-Otol 1999;4:311-325.
8. Rivier F, Clarke S. Cytochrome oxidase, acetylcholinesterase and NADPH-diaphorase staining in human supratemporal and insular cortex: Evidence for multiple auditory areas. Neuroimage 1997;6:288-304.
9. Tardif E, Clarke S. Intrinsic connectivity of human auditory areas: A tracing study with DiI. European J Neurosci 2001;13:1045-1050.
10. Zwislocki JJ, Nguyen M. Place code for pitch: A necessary revision. Acta Otolaryngol (Stockh) 1999;119: 140-145.

2 The Organ of Vision

András Csillag

ANATOMICAL OVERVIEW OF THE EYE

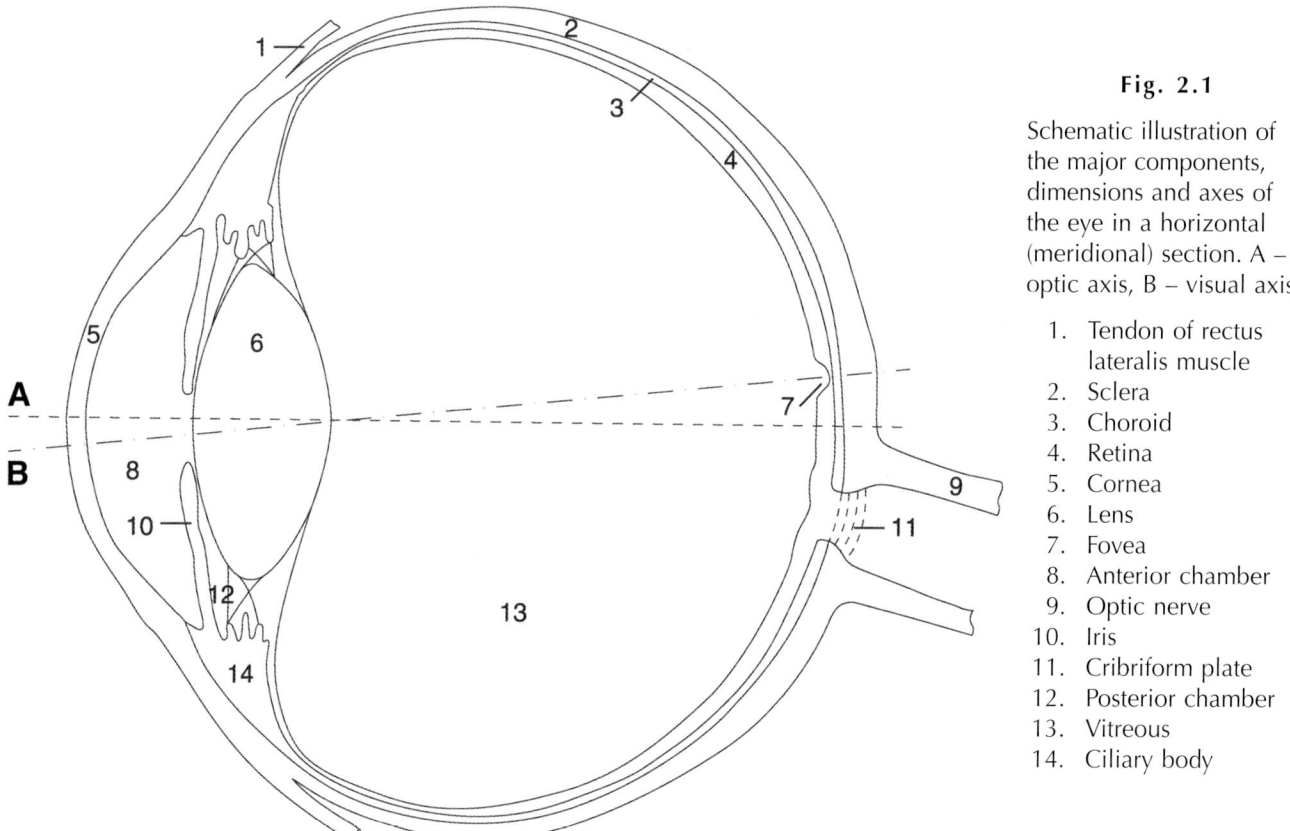

Fig. 2.1

Schematic illustration of the major components, dimensions and axes of the eye in a horizontal (meridional) section. A – optic axis, B – visual axis

1. Tendon of rectus lateralis muscle
2. Sclera
3. Choroid
4. Retina
5. Cornea
6. Lens
7. Fovea
8. Anterior chamber
9. Optic nerve
10. Iris
11. Cribriform plate
12. Posterior chamber
13. Vitreous
14. Ciliary body

Light sensitivity is an ancient faculty of living organisms. Visible light falls in a well-defined range of the electromagnetic spectrum and it consitutes a large component of the rays emitted by our parent star, the Sun. The Earth and its living inhabitants have been exposed to this type of radiation for billions of years, enough time for highly sensitive detecting mechanisms to evolve. Yet the receptors themselves may not be all that sophisticated. The capturing of photons by bioorganic molecules looks easy when compared to the daunting task: converting this primary physical effect to a meaningful bioelectric signal that is lasting enough to have an impact, yet volatile enough to allow for repeated stimulation and plasticity. Those simple light-sensitive cutaneous cells of ancient animal species have evolved into the highly specialized retina of complex vertebrates, including man. Most importantly, this process was accompanied by an equally impressive development of the visual pathways, whereas other structures together with the retina formed an elaborate organ, the eye.

Remarkably, the basic structure of the eye has remained largely unchanged throughout the vertebrate classes including mammals. Although early mammals probably evolved in the shadow of dinosaurs as small nocturnal predators or rodents, heavily dependent on smell rather than sight (this problem will be discussed in more detail in Chapter 3), primates and their descendent humans adopted a lifestyle that is clearly visually orientated. Good vision in terms of both distance judgment and a precise determination of the shape and color of objects had its obvious evolutionary payoff for our immediate ancestors and, ultimately, shaped our mind into something we, humans, are today.

We begin the anatomical description of the human eye with the eyeball (globe) and its coats or tunics, followed by the ocular refractive media, the extraocular muscles, orbit and the protective apparatus.

THE GLOBE

The paired eyeballs or globes are situated in the bony sockets of the skull called orbits. The globes can be directed toward the object of interest and allow free passage of light through specialized refractive media (cornea, lens) to the retina where the primary imaging takes place.

The shape of the human eye is near-spherical and, as with the globe of the Earth, one can distinguish poles and equator. The line connecting the anterior pole with the posterior pole is the geometric axis. However, because the site of acute vision (fovea) is laterally displaced from the posterior pole, the true axis of vision can be generated by an inward rotation of the geometrical axis. The equatorial plane transects the globe in its widest part at a right angle with the geometric axis. Those planes hitting both of the poles perpendicular to the equator are called meridians. A typical meridional section of the globe is shown in Fig. 2.1. Here the position of the optic disc (the site where the optic nerve fibers leave the eyeball) and the fovea can also be observed.

Light from an illuminated object passes through the refractive media of the eye and falls on the retina as an inverted and diminished image. In the normal eye, the light rays from a fixated object converge in the fovea, the site of acute vision. When the point of convergence lies behind or in front of the plane of the retina the resulting conditions are known as hypermetropia (farsightedness) or myopia (nearsightedness), respectively. Both can be corrected by the use of eyeglasses, convex lens for hypermetropia and concave lens for myopia. Unlike the case with many vertebrate animals, the forward-directed eyes of humans are situated in front of the skull. Thus, most of the visual field is covered by both eyes (binocular), only small temporal crescent-shaped fields remaining visible to either eye only (uniocular). Binocular vision is especially important in space perception (stereopsis, see also later). When the visual axes of the two eyes converge on a point of fixation the image falls on corresponding retinal sites. The locus of all those points in the visual space, which are perceived singly, and in the same frontal plane as the fixation point, to the right or left of the fixation point is termed the horopter (Vieth-Müller circle).

The globe consists of three concentric tissue coats called ocular tunics. The outermost fibrous tunic comprises the cornea in front and the sclera in other parts of the globe. The intermediate vascular tunic (uveal layer or uveal tract) consists of the iris, ciliary body, and choroid. The innermost coat (tunica nervosa) corresponds to the retina.

Cornea (Figs. 2.29; 2.30; 2.38; 2.39)

Contrary to common belief, the cornea, rather than the lens, is the principal refractive medium of the eye.

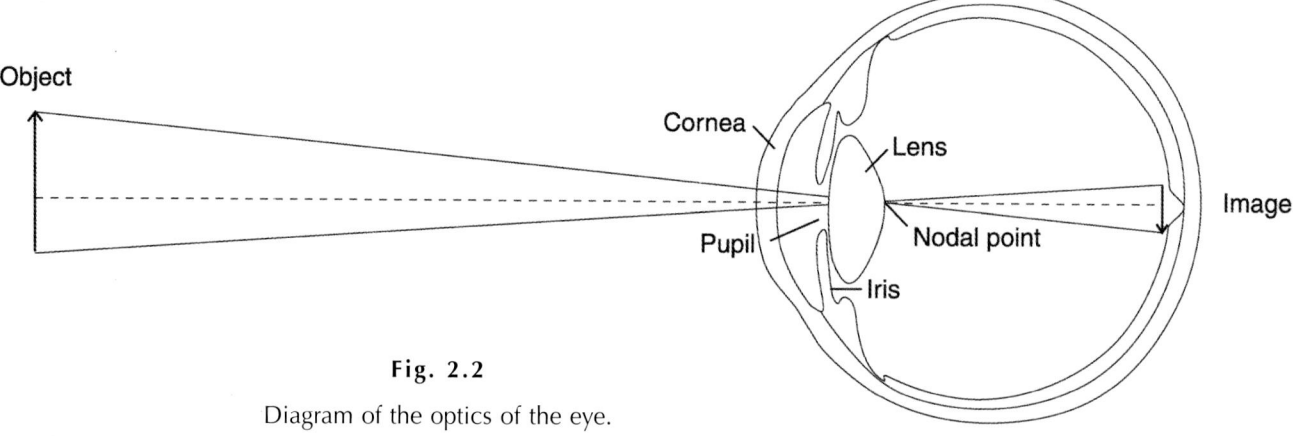

Fig. 2.2

Diagram of the optics of the eye.

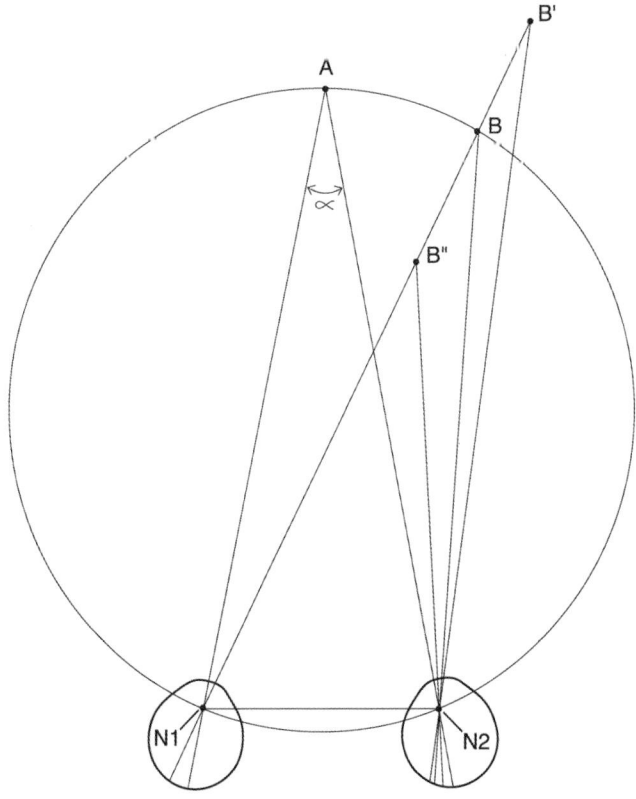

Fig. 2.3

Diagram of the horopter (Vieth-Müller circle). A – fixation point at the intercept of the visual axes, B – point on the horopter projecting on corresponding retinal points of the eyes, B' – point beyond the horopter projecting on non-corresponding retinal points (generating divergent disparity signal), B" – point within the horopter projecting on non-corresponding retinal points (generating convergent disparity signal), N_1, N_2 – nodal points, α – angle of convergence

The shape of a watchmaker's glass continuous with the sclera, the cornea is a transparent membrane in the center of the visible eye. The circular edge separating it from the "white of the eye" (sclera) is called the limbus. The pivotal role of cornea is demonstrated by frequent pathological alterations: an irregular radius of curvature results in astigmatism, cloudiness (opacity) of the cornea owing to scarring or inflammation may necessitate a transplant. The transparency of the organ is maintained through continuous lubrication by the lacrimal fluid (blinking).

The cornea consists of the following tissue layers: the anterior epithelium, Bowman's layer, substantia propria, Descemet's membrane, and endothelium. The anterior epithelium represents a five to six-layered stratified non-keratinizing epithelium. As with similar epithelia found elsewhere in the body, the cells of the basal (germinal) layer divide continually and migrate to the surface only to be lost to the lacrimal fluid. Such renewal process is essential for the integrity of the cornea. Apart from common epithelial cells (known here as keratocytes), the cornea also contains Langerhans (dendritic) cells and, occasionally, lymphocytes, and macrophages. The intermediate layer is composed of "wing" cells, with long interdigitating processes. The cells in the superficial layer are flattened, with few or none internal organelles apart from numerous microvilli.

A product of the underlying propria, the Bowman's layer is a network of fine collagen fibres embedded in a mucopolysaccharide matrix.

The thickest of the corneal layers, the substantia propria is a highly organized array of about 200 collagen lamellae. Within each lamella, the parallel running collagen fibers span the entire width of the cornea from limbus to limbus. Each lamella, however, is positioned at a large angle to those above or below it. The collagen fibers are produced by fibroblasts (also termed here corneal corpuscles) scattered among the fibers. The latter are embedded in a matrix composed of proteoglycans, glycoproteins, inorganic salts, and water.

The Descemet's membrane is clearly visible under the light microscope and it corresponds to a true basement membrane of the endothelium.

The endothelium is a single layer of flattened cells facing the anterior chamber. These cells are not capable of proliferating, yet they are highly active metabolically, which is signified by the abundance of organelles. The transparency of the cornea critically depends on an incessant supply of water and salts from the lacrimal fluid, as well as an active outward transport of HCO_3^- and Na^+ together with water from the matrix toward the aqueous humour. This seems to be the main function of corneal endothelium and it also requires the continual presence of oxygen and glucose. Otherwise, excess water would be trapped by the glycosaminoglycan residues of the matrix, inevitably resulting in swelling, perturbance of the collagen lamellae and, ultimately, corneal opacity.

One important factor of corneal transparency is the lack of blood vessels within the boundary of the limbus. However, nerve fibers from the ophthalmic division of the trigeminal nerve do occur in the outer layer of the propria (see also the chapter on sensory nerve endings). These nerves (essentially pain fibers) subserve corneal sensitivity and the nociceptive corneal reflex (including the blinking reflex).

Sclera (Figs. 2.39; 2.50)

This structure is a continuation of the propria layer of the cornea. A tough connective tissue coating, the sclera forms a protective capsule around the softer and more delicate tissues of the globe. The visible part of the sclera in front (covered by the conjunctiva) is known as the white of the eye. Posteriorly, the sclera surrounds the optic disc and continues in the outer sheath of the optic nerve. The outermost layer of sclera (episclera) contains blood vessels, in particular the large vortex veins (venae vorticosae), which perforate the sclera in the equatorial plane. Further perforations are found in the optic disc area for the optic fibres (cribriform plate), near the limbus (for the anterior choroidal vessels) and near the posterior pole (for the posterior choroidal vessels).

The histological structure of the sclera resembles that of the substantia propria of the cornea. However, the collagen fibers of the sclera are less uniform in diameter and less regularly arranged than those of the cornea. The sclera also contains a small amount of elastic fibers and, of course, blood vessels, just like any other dense connective tissue.

Iris (Figs. 2.39; 2.40)

The iris is one of the most characteristic features of the visible eye and thereby the facial expression. The limitless variety of its color and the individually different markings have long been noted by scientists, artists, and laymen alike. Interestingly, the very word "iris" is derived by its original Greek etymology as well as in many modern languages from the word "rainbow." A similar association with the multicolor character of iris is reflected in the English word "iridescent."

A strongly vascularized circular disc overhanging the anterior surface of the lens, the iris has much of the characteristics of an optical diaphragm, including a central opening called the pupil. The size of the latter is significant in controlling the quantity of light reaching the retina under varying ambient illumination. Pupil size can also determine the depth of focus: the smaller the pupil diameter the greater the depth of focus. This effect is particularly important when near objects are viewed with a high precision.

The tissue layers of the iris are as follows. The anterior surface is uneven, with invaginations between collagen fibers, covered by fibroblasts and occasional melanocytes. This surface is exposed to the aqueous humor of the anterior eye chamber, however, an endothelial cell lining typical of the cornea is not observed, at least in the adult iris.

The next layer, called stroma, which is continuous with the stroma of the ciliary body, contains loose connective tissue, with numerous blood vessels (arterioles, venules, and capillaries), mast cells, and melanocytes. The latter are of mesenchymal origin and their number and distribution largely determine the appearance of the iris. Individuals with many and heavily pigmented melanocytes have dark eyes, whereas those with fewer and less pigmented melanocytes tend to have blue eyes. Furthermore, if less variable, pigment is present also in the posterior layer of the iris (see below). A genetically determined deficiency of melanin pigment (albinism) leads to a complete lack of iris pigment, too. As a result, the iris of albinoes appears red owing to the vasculature as well as light reflection from the retina. The stroma of the iris, in particular in the vicinity of the sphincter pupillae, contains a peculiar cell type called clump cell. This is a large (more than 100 µm) ovoidal structure packed with pigment inclusions. Most clump cells correspond to macrophages but a subpopulation is likely to represent neuroectodermal cells having a common origin with the iris sphincter.

Beneath the stroma there is a layer of smooth muscle with circular fibers (sphincter pupillae). The muscle is particularly prominent near the pupillary margin and is innervated by parasympathetic fibers from the ciliary ganglion. Contraction of the sphincter muscle makes the aperture of the pupil smaller.

External to the sphincter and continuous with the pigment epithelium of the retina and ciliary body is a single layer of myoepithelium with long processes extending into the stroma, collectively called dilator pupillae muscle. It is innervated by sympathetic fibers and its contraction widens the aperture of the pupil. Although it is more obvious in the case of the dilator pupillae, both pupillary muscles are derivatives of the ectoderm.

A continuation of the blind part of retina (pars iridica retinae), the posterior pigmented epithelium is a single row of epithelial cells packed with melanin pigment granules. This pigment layer protrudes through the pupillary aperture to form a visible dark border on the anterior side around the pupil (pupillary ruff).

Ciliary Body (Figs. 2.39; 2.41)

A wide ring-shaped structure adjoining the iris laterally, the ciliary body is the middle part of the uveal tract, projecting towards the axis of the eye. It consists of two subregions: posteriorly the flat pars plana (orbiculus ciliaris) and anteriorly the elevated pars plicata (corona ciliaris). The latter is characterised by about 80 fringe-like ciliary processes, from which thin fibers (suspensory ligaments) pass to the lens, forming the zonule of Zinn. The richly vascularized ciliary processes secrete the aqueous humor and, through the action of the ciliary muscle, are instrumental in the accomodation (focusing the lens).

The stroma of the ciliary body is continuous with that of the choroid including its profuse vascular supply. The bulk of the ciliary body is composed of smooth muscle (ciliary muscle) whose fibers run in three planes: the meridional fibers (muscle of Brücke) are anchored in the sclera near the ora serrata (see below), whereas the circular (muscle of Müller) and radial fibers are attached to an inward projection of the limbal sclera termed scleral spur. The ciliary muscle is innervated by postganglionic parasympathetic nerve fibres from the ciliary ganglion. By accomodation (focusing at near objects) collective action of the ciliary muscle fibres moves the corona ciliaris inward, i.e. closer to the lens (partly by making the ciliary body more bulky and elevated and partly by narrowing the ring formed by the ciliary body). As a result, the tension on the zonule fibers is relaxed and the lens becomes more spherical by virtue of its own elasticity. Once this inherent elasticity is lost (which is normally the case around the age of 40 and older), the lens is no longer capable of regaining the spherical shape and we begin to have difficulty seeing objects at short distance (a condition called presbyopia). Typically, this is the time for wearing reading glasses.

On the inner (vitreous) side the ciliary processes are covered by a double epithelial border comprising the outer pigmented and inner nonpigmented epithelial cells. Both layers are continuations of the blind retina anterior to the ora serrata. Apparently, these epithelial cells play an important role in the secretion and composition of the aqueous humor. Although the outer pigmented cells are only loosely connected to each other, the inner nonpigmented cells possess genuine tight junctions. Therefore, the primary ultrafiltrate seeping through the wall of fenestrated capillaries of the stroma can easily find its way in the intercellular spaces between the pigmented epithelial cells. However, the nonpigmented epithelial cells represent a real barrier, reminiscent of the blood–cerebrospinal fluid barrier, where active transport processes including the extrusion of Na^+ and bicarbonate together with water and organic molecules such as glucose, glutathione, or lactic acid are necessary for a definitive composition of aqueous humor to be adjusted. Further details of the circulation and drainage of the aqueuos fluid are discussed below.

Choroid (Fig. 2.44)

This constitutes the main part of the uveal tract and is directly apposed to the neural part of the retina (retina vera) behind the ora serrata. Accordingly, the blood vessels of the choroid supply the pigment epithelium of the retina. The thickness of the choroid varies between 0.1 and 0.2 mm (the higher value applying for the posterior part).

The tissue layers of the choroid (starting from the scleral border) are as follows. The suprachoroid layer with melanocytes and a rich innervation used to be interpreted as part of the sclera itself (lamina fusca sclerae) because it often remains attached to the latter in dissection. In nocturnal animals this layer forms a well-developed light-reflecting surface (tapetum lucidum) that causes the eye of the animal light up in the dark (hence the word "cat's eye" used for the light reflectors on vehicles or motorways).

The vascular layer corresponds to the stroma and contains a dense meshwork of arterioles supplied by 15–20 short posterior ciliary arteries and the long posterior ciliary artery together with the accompanying venules.

The next layer is called choriocapillaris because it contains the terminal branchings of unusually dilated capillaries supplied by the overlying vascular layer. The capillaries are separated from the retinal pigment epithelium by a thin (1–4 µm) and structureless "glassy" membrane (lamina vitrea or Bruch's membrane). The sublayers of the latter are the basement membrane of the capillaries in the choriocapillaris, outer collagenous layer, elastic layer, inner collagenous layer and the basement membrane of retinal pigment cells. Experiments have shown that the Bruch's membrane is permeable even to large macromolecules such as plasma proteins.

Circulation of Aqueous Humor and Filtration Angle (Figs. 2.4.; 2.5; 2.42)

A summary of the circulation of the aqueous humor along with further histological details of the corneoscleral junction is justified by the preeminent clinical importance of this question. As already mentioned, aqueous humor is secreted from the ciliary processes posterior to the iris. From here, the fluid flows through the posterior chamber between the iris and the lens, then it passes through the pupil into the anterior chamber. Apparently the aperture of the pupil is no limiting factor for the flow of the fluid.

Fig. 2.4

Diagram of the circulation of aqueous humor.

1. Vitreous
2. Sclera
3. Posterior chamber
4. Lens
5. Ciliary body
6. Iris
7. Pupillary aperture
8. Drainage (filtration) angle
9. Schlemm's canal
10. Conjunctiva
11. Cornea

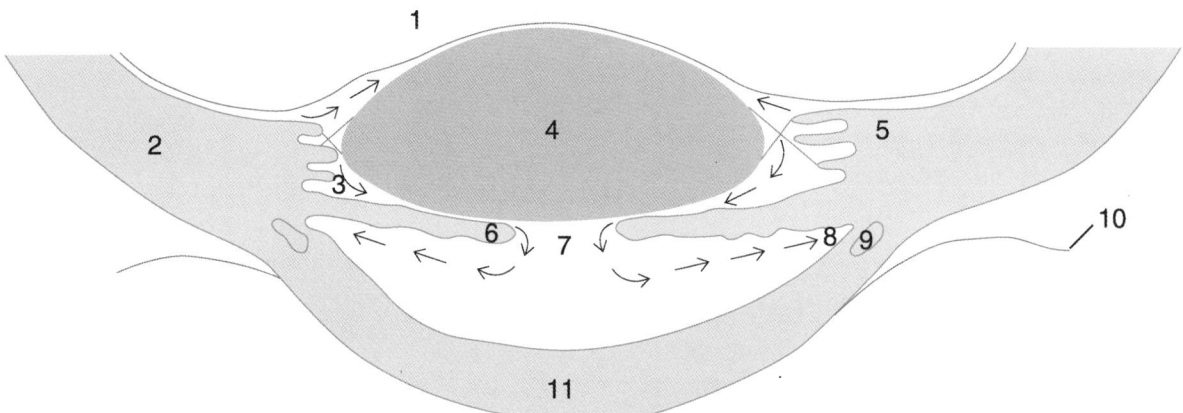

The critical element is its absorption at the "filtration angle" (drainage angle), the site where the sclera meets the base of the iris at an acute angle (in other words, the iris is "suspended: at the corneoscleral border). At this site, both the iris and the overlying sclera are remarkably loosened (pectinate ligaments of iris, spongiosa sclerae). Here, aqueous fluid percolates first through the uveal trabeculae bordering the spaces of Fontana and then it drains through the corneoscleral trabeculae and the internal channels of Sondermann into Schlemm's canal (sinus venosus sclerae). The latter is a dilated vessel that encircles the limbus of the cornea. From here, the fluid passes through the external channels of Maggiore to aqueous veins and thereafter to the episcleral veins leading to the anterior ciliary veins and back into the central circulation.

A smaller contingent of the aqueous humor follows a more direct alternative (uveoscleral) route through the ciliary muscle into the suprachoroid layer and then on to the scleral veins. This way of drainage is permitted mainly in the relaxed state of the ciliary muscle.

The flow of aqueous humor is alleviated by contraction of the pupil (myosis), when the tissues in the filtration angle are less congested, and hindered by dilation of the pupil (mydriasis) due to an overall compression of the filtration angle. The balance between the secretion and elimination of aqueous fluid determines the intraocular pressure (IOP) with normal values between 10 and 20 mmHg. Smooth flow of the fluid is essential for the nutrition of the lens and (as we have seen before) the maintenance of corneal transparency. Should the drainage process fail, the IOP is bound to increase and may ultimately lead to severe retinal damage, even blindness. Such condition is called glaucoma, which may be caused by an increased resistance of the trabecular meshwork (open angle glaucoma), or an obstruction of drainage by the peripheral iris coming to contact with the lateral cornea and narrowing the angle recess (primary angle closure glaucoma).

Lens (Figs. 2.6; 2.7; 2.8; 2.43)

A biconvex disc suspended from the ciliary processes via the suspensory ligaments of the ciliary zonule, the lens is the only adjustable component of the ocular refractive media. The latter comprise (from front to back) the cornea, aqueous humor, lens and vitreous body. The cornea has already been discussed above.

The lens has an equator and anterior and posterior poles. Its anterior surface lies adjacent to the iris and is related to the anterior chamber via the pupillary aperture. The lateral part faces the posterior chamber and, through the zonule, is related to the ciliary processes. The posterior surface, more curved than the anterior one, is apposed to the vitreous body behind (through a narrow retrolenticular space and the hyaloid membrane), accomodated by the patellar fossa. The lens is embedded in a tough, yet elastic capsule, essentially a thick basement membrane produced by the lens epithelium. The zonular fibers are attached to the equatorial part of this capsule as the zonular lamella, keeping the lens under constant radial tension. As described earlier, passive relaxation of the suspensory ligaments by an active contraction of the ciliary muscle makes the lens through the inherent elasticity of the capsule more spherical (focusing to close objects, accomodation).

Inside the capsule the lens consists of an outer epithelium and a system of highly specialized and elongated cells termed lens fibers. The epithelium is a single layer of polyhedral cells of normal composition of organelles. The presence of tight junctions was not verified by recent studies. The epithelial cells are found only on the anterior surface of the lens, and the lateral (pre-equatorial) zone of lens epithelium (well away from the potentially harmful ultraviolet rays bombarding the central zone) is the main site for mitotic divisions.

The mature lens fibers, however, are devoid of nuclei or other organelles but for numerous gap junctions allowing an exchange of metabolites between the neighboring fibers. Remarkably, lens fibers are never replaced throughout life, they just get overgrown by new layers as time goes by. Thus, a centrally located nucleus of lens (actually, several nuclei from embryonic through adult, encapsulating each other like a Russian doll) is surrounded by a cortex made of the most recently formed fibers. The new fibers are generated by elongation and migration of the parent epithelial cell. The apical part of the fiber grows in an anterior direction under the lens epithelium, whereas the basal end grows in a posterior direction, keeping in touch with the capsule. The deposition of successive layers forms the lens bow. In the process the once nucleated lens fibers lose their organelles fairly abruptly and develop into strap-like structures arching over the lens in concentric layers. The fiber tips meet those of other fibers along dividing lines called sutures. In the fetal lens there is an anterior Y and a posterior inverted Y suture but in the adult lens more and more sutures are added to this simple pattern, ultimately to form a nine-point star of the mature cortex.

The lens owes its transparency to the presence of proteins known as crystallins within its fibers. Related to the heat-shock proteins, a family of "molecular chaperons," α-crystallin is likely to prevent the formation of light-scattering protein aggregates. Once the regular arrangement of lens fibers is perturbed or an aggregation of crystallins takes place, the result will be opacity of the lens. The commonest manifestation of this disease is known as cataract with its many different forms (age-related cataracts being only one of them).

The suspending apparatus (zonule of Zinn) consists of anterior zonule from the pars plana to the pre-equatorial lens (supplemented from the pars plicata), posterior zonule from the pars plicata to the postequatorial lens (supplemented from the pars plana), and equatorial zonule from pars plicata to the equator. Further zonular fibers deep to the main system connect the pars plana with the edge of the patellar fossa. Anchorage of the zonular plexuses to the ciliary processes is further strengthened by tension fibers that help transform the angular traction of zonular fibres into an equatorial force.

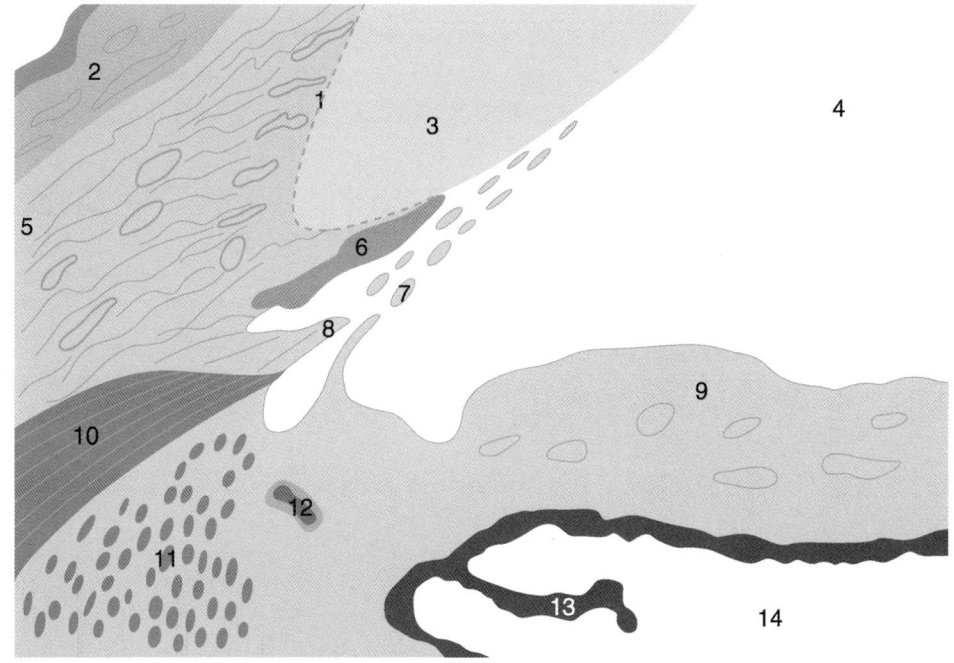

Fig. 2.5

Schematic illustration of the corneoscleral limbus and drainage angle.

1. Corneoscleral limbus
2. Conjunctiva
3. Cornea
4. Anterior chamber
5. Sclera
6. Schlemm's canal
7. Trabecular meshwork
8. Scleral spur
9. Iris
10. Longitudinal portion of ciliary muscle
11. Circular and radial portions of ciliary muscle
12. Major arterial circle
13. Ciliary process
14. Posterior chamber

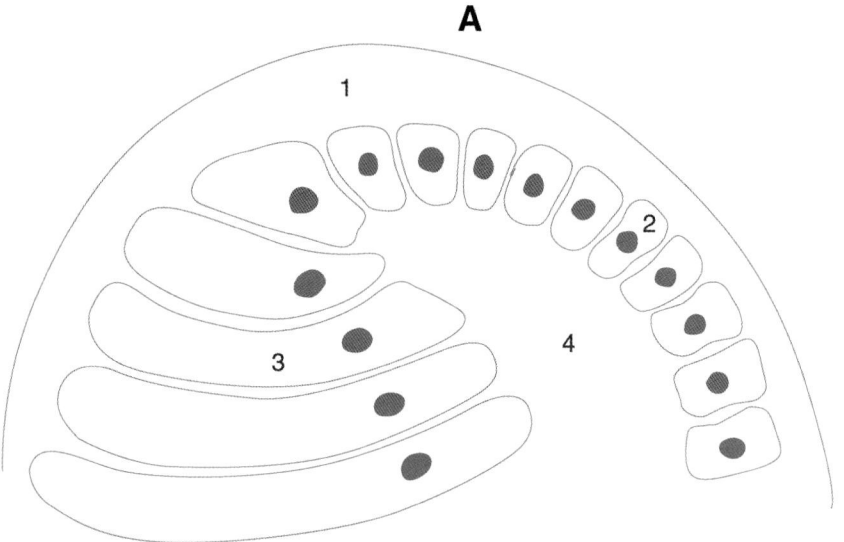

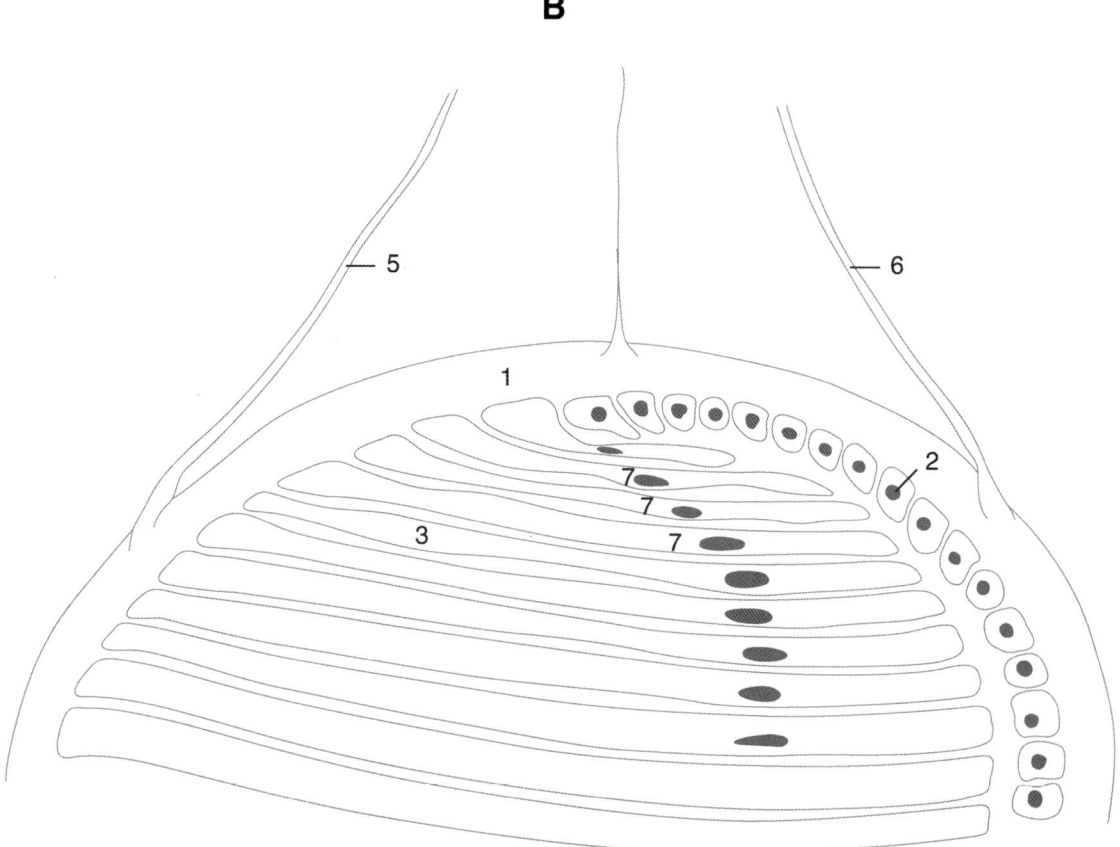

Fig. 2.6

Schematic illustration of the development of lens fibers. A – early phase, B – late phase.

1. Lens capsule corresponding to the basement membrane of lens epithelium
2. Anterior lens epithelium
3. Developing lens fibers
4. Lens vesicle
5. Posterior zonule
6. Anterior zonule
7. Lens bow

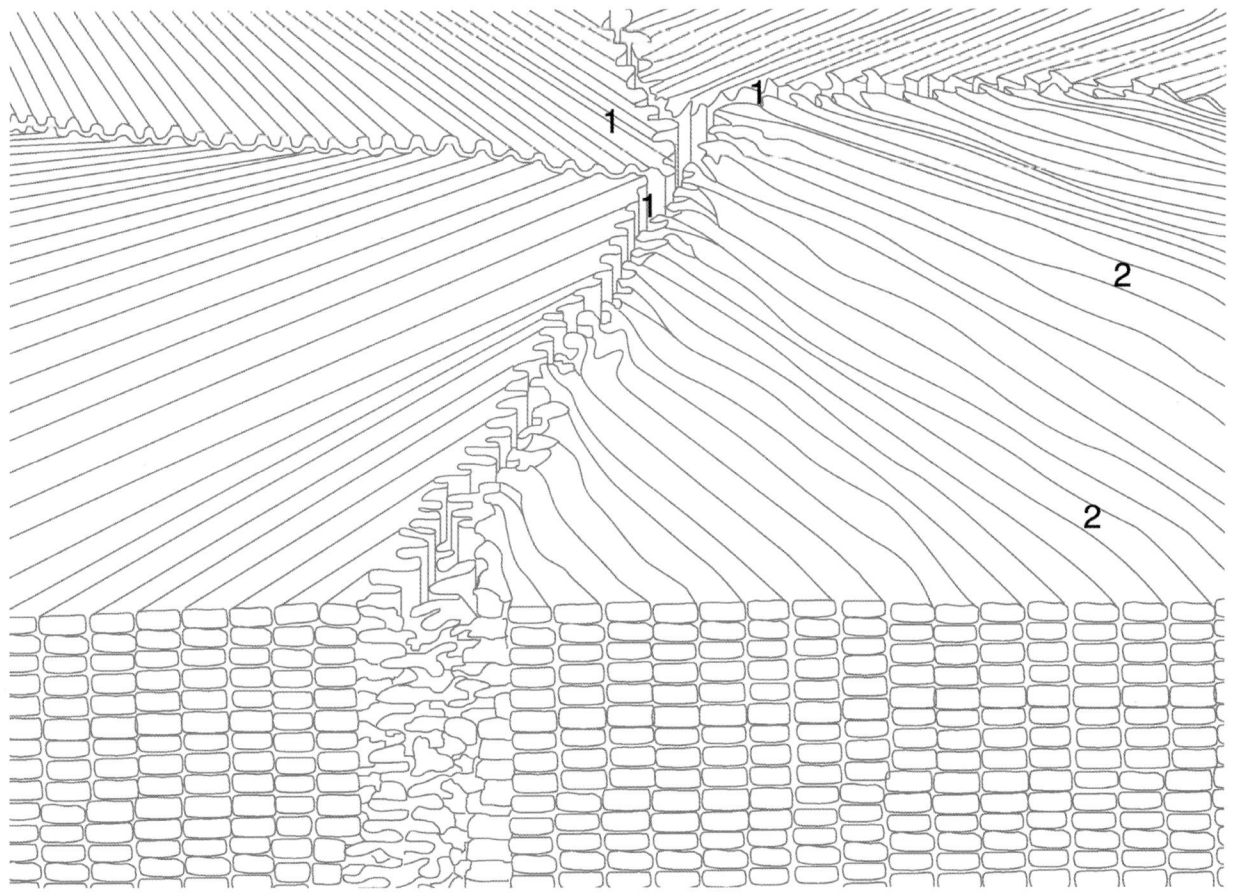

Fig. 2.7

Schematic illustration of the arrangement of lens fibers (after Hogan MJ, Alvarado JA and Weddell JE, 1971).

1. Sutures   2. Lens fibers

The Chambers of the Eye (Figs. 2.4; 2.5)

The borders of the anterior chamber are anteriorly the internal surface of the cornea, and posteriorly the anterior surface of ciliary body (peripherally), the anterior surface of the iris and the lens (within the pupillary aperture). The drainage angle is formed by the convergence of the anterior and posterior boundaries. The anterior chamber communicates with the extracellular spaces of the ciliary body, iris and the trabecular meshwork, as well as with the posterior chamber (via the pupil).

The posterior chamber is a space in front of the anterior vitreous face and posteromedial to the iris and ciliary body. It is subdivided into three compartments: prezonular, zonular and retrozonular with repect to the suspensory ligament. The prezonular compartment is formed by the iris, lens and zonules. The apex of this triangular space (in meridional section) points at the junction between the pupillary margin of iris and the lens. The base is formed by the ciliary processes and the valleys between these (recesses of Kuhnt). The zonular compartment lies essentially within the zonular apparatus (described above). The retrozonular compartment corresponds to the narrow space of Petit, separated from the retrolental space (of Berger) by the ligament of Wieger (lig. hyaloideocapsulare).

Vitreous Humor

This jelly-like substance (of a consistency similar to egg white) fills the remaining part of the globe. It is not entirely spherical because the anterior surface is depressed to accommodate the lens (patellar fossa). At the periphery of this fossa, the vitreous is adherent to the capsule of the lens over a ring-shaped zone (ligament of Wieger) and, lateral to this ligament, it is apposed to the ciliary processes and zonule. A remnant of the embryonic hyaloid artery, the canal of Cloquet passes within the vitreous from the optic disc to the retrolental space.

Despite its uniform watery appearance the vitreous has some internal structure (e.g., channels and tracts) as a

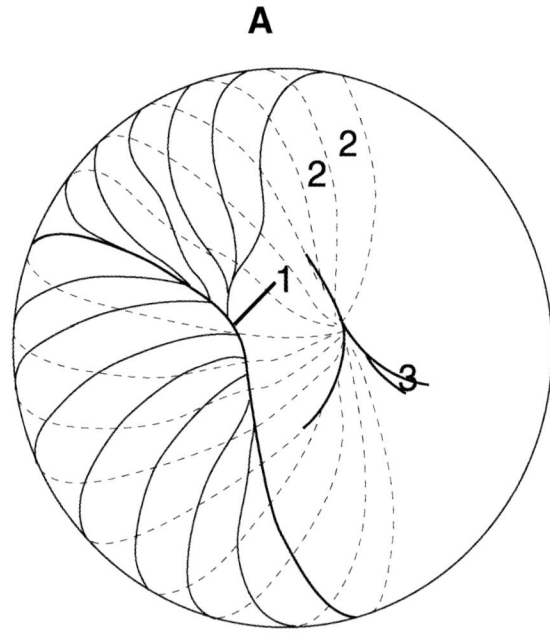 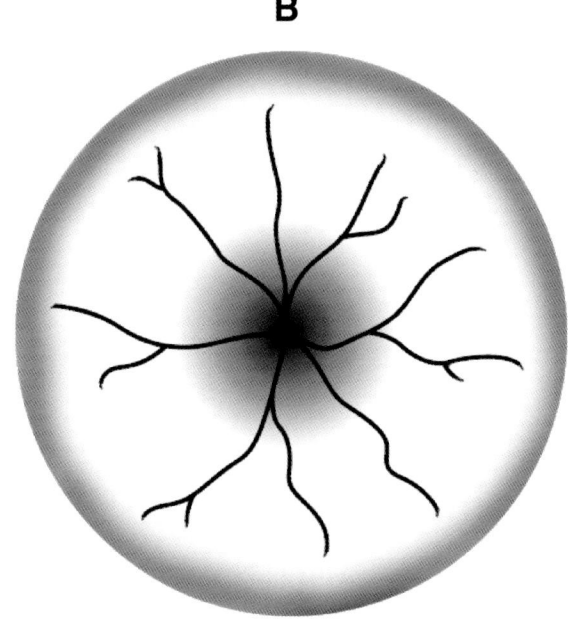

Fig. 2.8

Sutures of the lens. A – fetal pattern, oblique view; B – adult pattern, surface view (after Hogan MJ, Alvarado JA and Weddell JE, 1971).

1. Anterior Y suture
2. Bundles of lens fibers
3. Posterior inverted Y suture

result of postnatal development, and subdivisions can also be distinguished. Most of the vitreous is cell-free apart from a few phagocytic cells (hyalocytes) occurring in the outer region (cortex) adjacent to the retina and ciliary body. The cells produce hyaluronic acid molecules, which form a network that is capable of binding water in enormous quantity (99% of the vitreous is constituted by water, the highest value in the entire body). The vitreous is held in place by collagen fibers attached to the internal limiting membrane of the retina, especially around the optic disc. These connections are important also for the firm attachment of the retina although the most critical factor is intravitreous pressure itself. In old age (or in the myopic eye), alteration of the structure of the vitreous may cause veil-like disorders of light scattering and, more dangerously still, vitreous detachment as a forerunner of retinal detachment.

Retina (Figs. 2.9; 2.10; 2.31; 2.32; 2.33; 2.34; 2.44; 2.45; 2.46; 2.47; 2.48; 2.49)

A piece of "ectopic brain" with all the features characteristic of the central nervous system, the retina forms the internal coating of the globe. During the development of the eye cup the outer layer forms the pars pigmentosa or retinal pigment epithelium (RPE), whereas the inner layer is transformed into a stratified complex structure, the pars nervosa. Such a transition occurs, however, only in the posterior two thirds of the inner ocular tunic (retina vera or pars optica). Anterior to the ora serrata (the circular line behind the ciliary body) the retina is reduced to a double layer of epithelium corresponding to the pars ciliaris and pars iridica (collectively termed pars caeca—blind retina).

Anatomically, the retina can be subdivided into central and peripheral retinae. The central retina occupies the posterior fundus temporal to the optic disc and is demarcated by the upper and lower arcuate and temporal vessels. The central area has a diameter of approx 5.5 mm and it can be further parcellated into concentrically arranged areas of clinical importance. Approximately 1.8 mm across, the fovea lies in the centre of the central retina. In turn, this spot also has a central depression (foveola) of about 0.35 mm diameter, surrounded by sloping walls (clivus). On the other hand, the fovea is surrounded by belt regions: the internal belt called parafovea (~0.5 mm wide), and the external belt called perifovea (~1.5 mm wide). The often used classical term macula lutea ("yellow spot") or macula has been a source of some confusion. This is visible in the freshly dissected eye but cannot easily be discerned on ophthalmoscopic examination. Although it appears smaller than the central retina, the presence of the yellowish carotenoid pigment is detectable almost as far as the outer border of central retina (~5 mm).

Another conspicuous feature of the posterior fundus, the optic disc (also called papilla) is a near-circular spot of approx 1.5 mm diameter, with a central depression known as the physiological cup. The optic disc is the site where the axons of retinal ganglion cells leave the retina to form the optic nerve. Furthermore, it passes the retinal vessels. The photoreceptors are absent from the optic disc, therefore this site corresponds to the blind spot of Mariotte.

The pars optica consists of the following layers starting from the Bruch's membrane (Figs. 2.44; 2.45).

1. Stratum pigmentosum (retinal pigment epithelium, RPE)
2. Stratum neuroepitheliale (rods and cones, outer and inner segments, OS, IS)
3. Membrana limitans externa (external limiting membrane, ELM)
4. Stratum granulosum externum (outer nuclear layer, ONL)
5. Stratum plexiforme externum (outer plexiform layer, OPL)
6. Stratum granulosum internum (inner nuclear layer, INL)
7. Stratum plexiforme internum (inner plexiform layer, IPL)
8. Stratum ganglionare (ganglion cell layer, GCL)
9. Stratum fibrosum nervi optici (optic fiber layer, OFL)
10. Membrana limitans interna (internal limiting membrane, ILM)

Layer 1: The Pigment Epithelium

The pigment epithelium (Fig. 2.46) is a single layer of closely packed cells, which are roughly hexagonal in shape when viewed from the surface. Their base is apposed to the Bruch's membrane whereas the apical part is thrown into long cytoplasmic processes engulfing the outer segments of rods and cones. The pigment cells are connected by tight junctions and contain melanin deposits in the cytoplasm. These cells are not merely supporting elements or light insulators, instead they play a dynamic role in the process of photoreception. One important function of the pigment epithelium is elimination of the detached lamellar units of rod and cone outer segments by phagocytosis. Another is capturing and recycling of the "used" retinal molecules released from the outer segments after photogenic *cis-trans* conversion. Moreover, the RPE allows and controls passage of nutrients and other substances between the choriocapillaris and the neural retina.

Layer 2: The Neuroepithelial Layer

The neuroepithelial layer contains the rods and cones (Figs. 2.10; 2.47; 2.48). These are highly specialized processes, essentially modified cilia, of the photoreceptor neurons situated in the ONL. Rods are long and slender, whereas cones have a more bulky shape. Both rods and cones consist of inner and outer segments. The inner segment has an outer ellipsoid region (packed with mitochondria) and an inner myoid zone (with endoplasmic reticulum and Golgi apparatus). The outer segment is connected through a cytoplasmic isthmus containing a connecting cilium or basal body. The rest of the outer segment is packed with stacks of flattened double lamellae (discs), which are generated by invagination of the plasma membrane. In cones, the spaces between the discs are open to the extracellular environment, in rods, however, the discs are fully detached from the outer membrane, thus forming closed cytoplasmic enclosures. In this sense, rods show a higher degree of specialization than cones. Nevertheless, the suggestion that cones preceded rods in evolution is still disputed.

The discs of outer segments are highly effective photon traps. In rods, they contain a protein known as visual purple (rhodopsin) composed of large opsin molecules and a small chromophore unit called 11-*cis*-retinal. The energy of photons of the appropriate wavelength (visible light falls between 400 and 700 nm) is transformed into the activation energy required for an intramolecular rearrangement to produce the lower energy all-*trans*-retinal. The latter can no longer bind to opsin and it diffuses out of the membrane only to be captured by the RPE for re-isomerization. The alcohol form of all-*trans*-retinal (all-*trans*-retinol, vitamin A) is resupplied partly from the blood via the choriocapillaris. This explains why vitamin A deficiency leads to visual defect, in particular in the dark-adapted retina (poor night vision is a dangerous condition, affecting e.g., drivers, especially because their eyesight appears normal in daytime, when cones are more active). The abandoned opsin molecules undergo a conformation change that leads to the generation of a bioelectric signal through coupling with special proteins (G proteins, transducins) and the action of the enzyme cGMP-phosphodiesterase (for further details of this process the reader is referred to textbooks of biochemistry and physiology). Notably, in resting (in the dark) the rod cells are slightly depolarised and the light-induced activity brings about hyperpolarization, that is, the direction of membrane potential change is opposite to what is normally seen in stimulated nerve cells.

The cones are quite similar to rods in their basic structure and function but, instead of the visual purple of rods, they contain one of three different color pigments called iodopsin or cone opsin, whose absorption maxima fall around 440 nm (blue), 540 nm (green) and 577 nm

Fig. 2.9

Schematic diagram of the retinal layers and main synaptic connections. The layers are numbered by roman numerals other structures by arabic numerals.

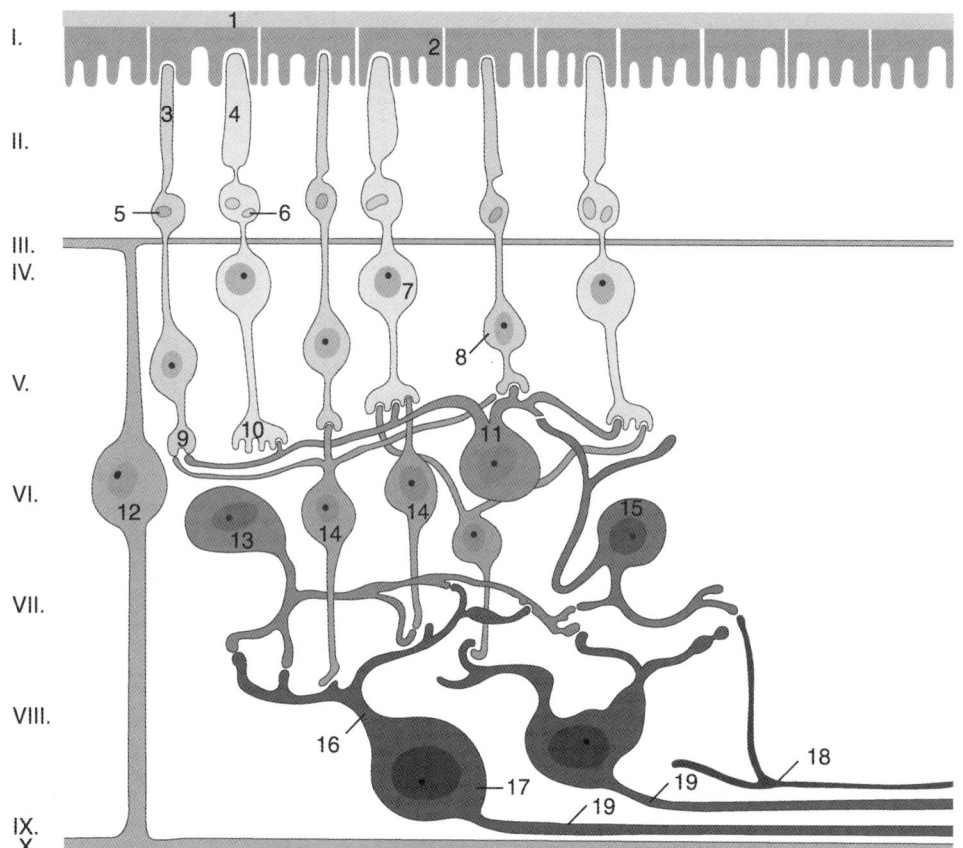

1. Bruch's membrane
2. Pigment epithelial cell
3. Rod outer segment
4. Cone outer segment
5. Rod inner segment
6. Cone inner segment
7. Cone cell
8. Rod cell
9. Rod spherule
10. Cone pedicle
11. Horizontal cell
12. Müller cell
13. Amacrine cell
14. Bipolar cells
15. Interplexiform cell
16. Ganglion cell dendrite
17. Cell body of ganglion cell
18. Efferent axon
19. Afferent axons (optic fibers

I. Stratum pigmentosum (retinal pigment epithelium, RPE)
II. Stratum neuroepitheliale (rods and cones, outer and inner segments, OS, IS)
III. Membrana limitans externa (external limiting membrane, ELM)
IV. Stratum granulosum externum (outer nuclear layer, ONL)
V. Stratum plexiforme externum (outer plexiform layer, OPL)
VI. Stratum granulosum internum (inner nuclear layer, INL)
VII. Stratum plexiforme internum (inner plexiform layer, IPL)
VIII. Stratum ganglionare (ganglion cell layer, GCL)
IX. Stratum fibrosum nervi optici (optic fiber layer, OFL)
X. Membrana limitans interna (internal limiting membrane, ILM)

(orange-red), respectively. This constitutes the stuctural basis for trichromatic vision. The differences in absorption spectra are due to differences in the amino acid sequence of the opsin component, rather than to the chromophore (which is the same retinal, as with rods). The evolution of opsins has gradually led to a series of gene duplications and mutations, the last decisive splitting for the lineage of primates occurring between the red and green opsins. However, by no means does the trichromacy of humans represent some kind of "evolutionary peak." Birds, for example, have tetrachromatic vision with an extra retinal pigment in the ultraviolet range. The color selectivity of birds and some other species is further enhanced by the presence of light filtering oil droplets in the inner segments of cones.

The density of cones is highest (between 100,000 and 300,000/mm^2) in the foveola, the site of maximal visual acuity. Conversely, rods are absent from the center of the fovea, otherwise their density is greater in the central than in the peripheral retina, the peak being around 140,000 to 160,000/mm^2. Altogether there are about 90 million rods and 4.5 million cones in the human eye (i.e., rods outnumber cones by 20:1). The photoreceptors are arranged in a mosaic form. By using antibodies to visual pigments the different subtypes of cones can now be separated histologically in the primate retina (Fig. 2.49).

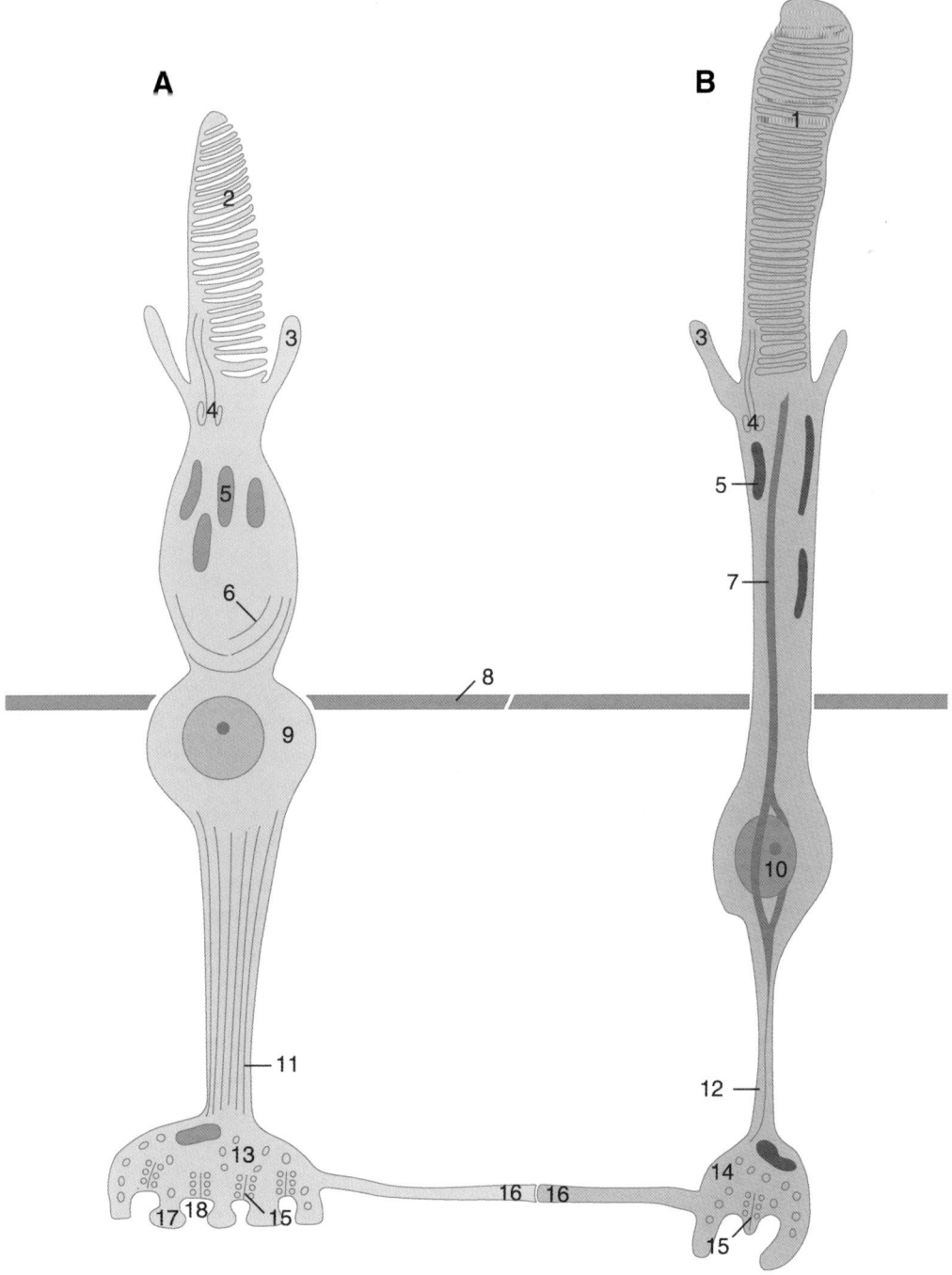

Fig. 2.10

Schematic illustration of the structure of cones (A) and rods (B).

1. Discs in rod outer segment segregated from the outer plasma membrane
2. Discs in cone outer segment continuous with the outer plasma membrane
3. Cytoplasmic processes
4. Cilia in connecting stalk
5. Ellipsoid area of inner segment with mitochondria
6. Myoid area of inner segment
7. Microtubular array
8. External limiting membrane
9. Cell body of cone with nucleus
10. Cell body of rod with nucleus
11. Inner fiber of cone with microtubules
12. Inner fiber of rod with microtubules
13. Pedicle
14. Spherule
15. Ribbon synapse
16. Lateral connection with gap junction
17. Basal junction area
18. Arciform density area

LAYER 3: THE EXTERNAL LIMITING MEMBRANE

The external limiting membrane stretches from the optic disc to the ora serrata and separates the RPE and neuroepithelial layers (outer chamber) from the overlying compartments (inner chamber). This separation is required for holding the photoreceptors firmly in place as well as a barrier for nutrients and other components. The ELM is composed of the end-feet of Müller cells connected by zonulae adherentes to each other and to villus-like processes of the myoids of photoreceptor cells.

LAYER 4: THE OUTER NUCLEAR LAYER

The outer nuclear layer contains the cell bodies of the photoreceptor cells. It is densely packed with nuclei arranged in eight to nine rows. The outer fibers from these cells form the rods or cones, whereas the inner fibers (axons) project to the EPL to form synapses. The cone nuclei lie close to the ELM, therefore their inner fibers are long. The opposite is true for rods, whose inner fibers traverse a shorter distance to reach their synaptic connections.

LAYER 5: THE OUTER PLEXIFORM LAYER

Both rod and cone inner fibers terminate in specialized end bulbs within the outer plexiform layer (Fig. 2.48). The synaptic complex of the rod consists of an end bulb termed spherule, ribbon synapses and connecting processes of horizontal and bipolar cells (Fig. 2.10). The ribbon synapse has a synaptic membrane specialization that is perpendicular to the apposition between the pre- and post-synaptic structures. It is surrounded by synaptic vesicles and the neurotransmitter is thought to be glutamate. Rod spherules make contact with horizontal cells (which pass information in both directions, hence it is not appropriate to call their processes either dendrites or axons) in deeply invaginated parts, and also with bipolar cells in less indented parts. On average, one spherule is contacted by four processes, and each bipolar cell contacts ca. 50 spherules in the perifovea and several hundred at the periphery of retina. A horizontal cell contacts a spherule only once but any spherule may synapse with several horizontal cells. In addition, rod spherules and cone pedicles also form lateral connections via gap junctions.

The terminal end-feet of cones, termed pedicles, are larger than the spherules and accomodate more synaptic sites (Fig. 2.10). Typically, they form synaptic triads in the deep presynaptic invaginations (arciform density), with three connecting processes: two horizontal cells and one bipolar cell or three horizontal cells together. Here, too, ribbon synapses occur. Of the bipolar cells, midget bipolars participate in the synaptic triads, whereas the flat bipolars tend to contact the superficial shallow indentations (basal junctions) of cone pedicles. The midget bipolar cells are connected to one cone only. Flat bipolars are arranged more diffusely: they may contact as many as six cones, and each repeatedly, without the presence of ribbons. Lateral connections are abundant between the pedicles (but not between the spherules).

LAYER 6: THE INNER NUCLEAR LAYER

The inner nuclear layer contains the somata of the following neuronal types: bipolar, horizontal, interplexiform and amacrine. In addition, the cell bodies of radial gliocytes called Müller cells are also situated in this layer. The cells in the INL are generally larger than those of the ENL and less densely packed. The nuclei are arranged in 8–12 rows, forming sublayers that consist of horizontal cells (the outermost sublayer), bipolars, Müller cells (intermediate sublayer), amacrine, and interplexiform cells (innermost sublayer).

Bipolars are the main relay cells between the photoreceptors and the ganglion cells. Although rod bipolars are fairly uniform, cone bipolars belong to different subtypes. Because this question is still somewhat disputed, and the cone subtypes seem to vary across the different mammalian species, a simplified description is given here. The dendrites of rod bipolars contact the spherules of the EPL in a synaptic complex that also involves the dendrites of horizontal cells. The axons of rod bipolars synapse mainly with the processes of amacrine cells and also with the dendrites and somata of diffuse ganglion cells in the IPL. In the dark adapted retina, the lateral gap junctions between spherules and pedicles are closed due to a slow depolarisation of rods. Therefore, any light-triggered rod activity (hyperpolarization) is directed exclusively toward the rod bipolar, which in turn is depolarized ("sign-inverting" response).

Apart from various subtypes, the cone bipolars belong to two main categories: midget or diffuse. The midget bipolars represent the most direct route of information transfer because they contact one single cone, at least at the fovea. The invaginating type of midget bipolars forms synaptic triads in the arciform densities of pedicles as mentioned above, whereas the flat type is involved in more superficial contacts. The diffuse cone bipolars contact many (up to 40) cone pedicles. These, too, have invaginating and flat types and their stuctural arrangement implies an involvement in the convergence of cone input on ganglion cells. Physiologically, cone bipolars belong to "On" or "Off" response categories. In the former, the bipolar cell is activated (depolarized) when the light falling on the center of its receptive field is switched on ("sign-inverting" response). Conversely, the other type of bipolar is depolarized when the light is switched off ("sign-conserving" response). Owing to the synaptic interactions defining the receptive field (this is mainly the function of horizontal cells) those "On" bipolars that are active when the center of their receptive field is illuminated tend to give an "Off" response when the periphery of their receptive field is illuminated (and vice versa). This is one of the bases for the center-surround and contrast phenomena, of preeminent importance in retinal imaging.

Horizontal cells are instrumental in the spreading of information in a lateral direction. They all have many dendrites which contact cone pedicles at the triads (see above) and a long neurite that can travel a long distance (as much as 1 mm) within the EPL to synapse with rod spherules (also with the dendrites of rod bipolars) or other cones. The former (HI) type of horizontal cell is likely to be associated with luminosity, whereas those contacting cones only (HII and HIII) are probably related to chromatic vision. Owing to a large number of gap junctions, horizontal cells constitute a functional network.

Unlike horizontal cells, which arborize in the EPL, amacrine cells exert their effect in the IPL. They have a flask-shaped perikaryon and numerous processes of a uniform type which cannot really be categorized either as a dendrite or as an axon. A wide variety of morphological types of amacrines have been described, nearly 30 in the primate retina and 24 in humans alone. Ultimately, however, all these types belong to two main categories: diffuse and stratified. Diffuse amacrines form a dense plexus on the inner side of IPL, those associated with rod bipolars having wider fields. Stratified amacrines arborize in one or more sublayers, usually outer or intermediate, of the IPL. Gap junctions are abundant between amacrine cells and a variety of neurotransmitters (e.g., dopamine, serotonin) and neuromodulators (e.g., enkephalin, substance P, neurotensin, somatostatin) have been demonstrated in these cells. Most of the cells contain γ-aminobutyric acid (GABA) or glycine, which have an inhibitory effect on retinal ganglion cells.

Interplexiform cells are sometimes regarded as a variety of amacrins. However, their distinctive feature is that they pass information from the IPL back to the EPL. The interplexiform cell receives its input from amacrine cell processes and also from efferent (retinopetal) fibres from the brain. Its axon synapses mainly with cone horizontal cells in the EPL. Dopaminergic and glycinergic forms have been reported.

Müller cells represent the most characteristic glial element of the retina. Radially orientated, these cells span most of the width of retina between the external and internal limiting membranes. The latter are formed by the contiguous terminal processes of neighbouring Müller cells. More specifically, the ILM contains tight junctions, forming an impassable barrier between the retina and the vitreous, whereas in the ELM, only zonulae adherentes are found. The cells put out many microvilli projecting into the photoreceptor layer and other processes form a delicate meshwork around the neural elements in the more superficial retinal layers. Apart from their role as migratory guides in the embryonic retina, Müller cells are active in controlling the metabolic activity of neurons (e.g., glucose uptake, capturing and decomposition of neurotransmitters such as GABA) and they also serve as CO_2 and K^+ buffers.

LAYER 7: THE INNER PLEXIFORM LAYER

The inner plexiform layer is the site for the second synaptic junction in the line. It is wider and has more synapses than the EPL. The main pathway of information is from the bipolars to the retinal ganglion cells, however this pattern is modified by the participation of amacrin and, to a lesser extent, interplexiform cells. The most common form of synapse found in this layer is known as the dyad, in which a bipolar axon contacts an amacrine cell process and a ganglion cell dendrite side by side, with a ribbon synapse. Less frequently, two amacrine cells or ganglion cell dendrites may also participate in synaptic dyads. In reciprocal synapses, an amacrine cell process postsynaptic to a dyad forms a nearby synapse back to the same bipolar axon. Two consecutive synapses between adjacent amacrine processes and a third synapse with a ganglion cell dendrite or bipolar axon is termed serial synapse. All of these synaptic contacts (which can only be described as promiscuous) subserve an elaborate processing (spreading, narrowing, summation, or averaging) of the information to be passed on to the ganglion cells. There is one important stratification within the IPL. The inner sublayer (lamina b) contains mainly the dendrites of "On" ganglion cells and the corresponding bipolar cell terminals (see above), whereas the outer sublayer (lamina a) is the site for the dendrites of the "Off" ganglion cells and the connecting bipolar cell axons.

LAYER 8: THE GANGLION CELL LAYER

Retinal ganglion cells constitute the definitive output elements that are capable of generating action potentials traveling along the optic fibers. They are considerably larger and wider apart than the other retinal neurons. Categorization of the ganglion cells is difficult because the morphological data available in humans do not completely match those obtained by physiological methods in various animal species. Therefore, one has to rely on several criteria, according to which a reasonable degree of overlap between the classes can be attained. The first argument is the terminal projection area of retinal ganglion cells in the lateral geniculate body (see later). Those projecting to the magnocellular layers are categorized as M cells and those terminating in the parvocellular layers are termed P cells. In the ensuing description, the ganglion cells are sorted by this first argument, and then further categorized by their morphology and physiology.

The P cells (probably corresponding to the β cells of the cat, although this is disputed) have medium sized perikarya and comprise the midget ganglionic (P1) neurons and another (P2) subtype. The P1 cell has a small and focused, if densely ramifying, dendritic tree and, in most cases, it receives input from a single cone via a single midget bipolar cell. This type of ganglion cell constitutes approx 90% of all ganglionic neurons in the fovea. As mentioned earlier, the P1 cells can be further subdivided into "Off-center" (a-type) with dendrites ramifying in the outer sublamina of the IPL, and "On-center" (b-type) with dendrites in the inner sublamina of the IPL. P1 cells exhibit color-opponent responses (i.e., they respond to light stimuli of different wavelength in an opposite fashion). Furthermore, these cells tend to have a small center-surround field and show a sustained response to illumination. According to the electrophysiologist, such ganglion cells are termed X-cells, implicated in the precise detection of objects. Larger than the P1 type and with a wider dendritic field, P2 cells are more numerous at the periphery of the retina and they show a "blue-on" type response.

The M ganglion cells (probably corresponding to the α cells of the cat) are considerably larger than the P cells, with several main dendrites ramifying over a wide area. Basically, all polysynaptic ganglion cells fall into this category, including those contacted by rod bipolars. These cells exhibit nonopponent responses that is, they are probably not involved in colour vision. Instead, the M cells give quick transient responses even on continuous illumination, signalling the gross features and movements of objects. By the categorisation of the electrophysiologist, such cells are called Y cells.

Layer 9: The Optic Fiber Layer

The optic nerve fiber layer contains the axons of the approx 1.2 million retinal ganglion cells. The fibers are unmyelinated as far as the optic disc (hence the excellent transparency of the OFL), the myelin sheaths appear where the fibers enter the optic disc to pass through the cribriform plate. The axons arising from the nasal retina converge on the optic disc in a simple fashion, whereas those arising from the temporal retina have to bypass the papillomacular fasciculus above or below the dividing line called raphe. The papillomacular bundle containing those axons originating from the fovea is the first to develop and its axons keep their separate position within the optic nerve and tract. In addition to retinofugal afferents, the OFL also contains efferent (retinopetal) fibers, astrocytes, numerous capillaries, and microglia.

Layer 10: The Internal Limiting Membrane

The internal limiting membrane is composed of the plasma membranes of tightly linked Müller cells and other glial elements together with their basement membrane, as well as collagen fibrils and proteoglycans of the vitreous. The vitreous components of the ILM constitute the hyaloid membrane.

The Retina of the Macular Region (Fig. 2.31)

Apart from an overall structural similarity, the site of acute vision shows modifications to the general pattern, which promote the unattenuated access of light to the photoreceptors. Both the central pit (foveola) and the sloping walls of the fovea are devoid of rods. Instead, this is the site of the densest accumulation of cones, which are elongated here, resembling rods. Apart from gliocytic processes, only the inner and outer segments of cones are found inside the foveola. Thus, light can access these photoreceptors without any other interposed layer. The cells and inner fibres belonging to the foveolar cones are stretched out along the sloping part of the fovea, and the other layers of the retina commence only beyond the edge of the fovea. Given its unparalleled transparency, the high density of cones connecting via the "direct route" (midget bipolars to midget ganglion cells) and the absence of blood vessels, the foveola and fovea are set for high resolution photopic (daytime) vision. The retina is thickest in the parafoveal region due to the abundance of ENL and INL displaced from the fovea. In the perifoveal region the number of cones drops sharply, with a concomitant increase in rods.

Blood Vessels of the Eye (Figs. 2.11; 2.12; 2.31; 2.32; 2.33)

A branch of the internal carotid artery, the ophthalmic artery enters the orbit via the optic canal. Apart from the orbital branches, the arteries supplying the globe itself belong to either of two systems: the vessels of the uveal tract or those supplying the retina.

The arteries of the uveal tract supply the choroid, ciliary body, and iris. The short posterior ciliary arteries pass through perforations of the sclera near the posterior fun-

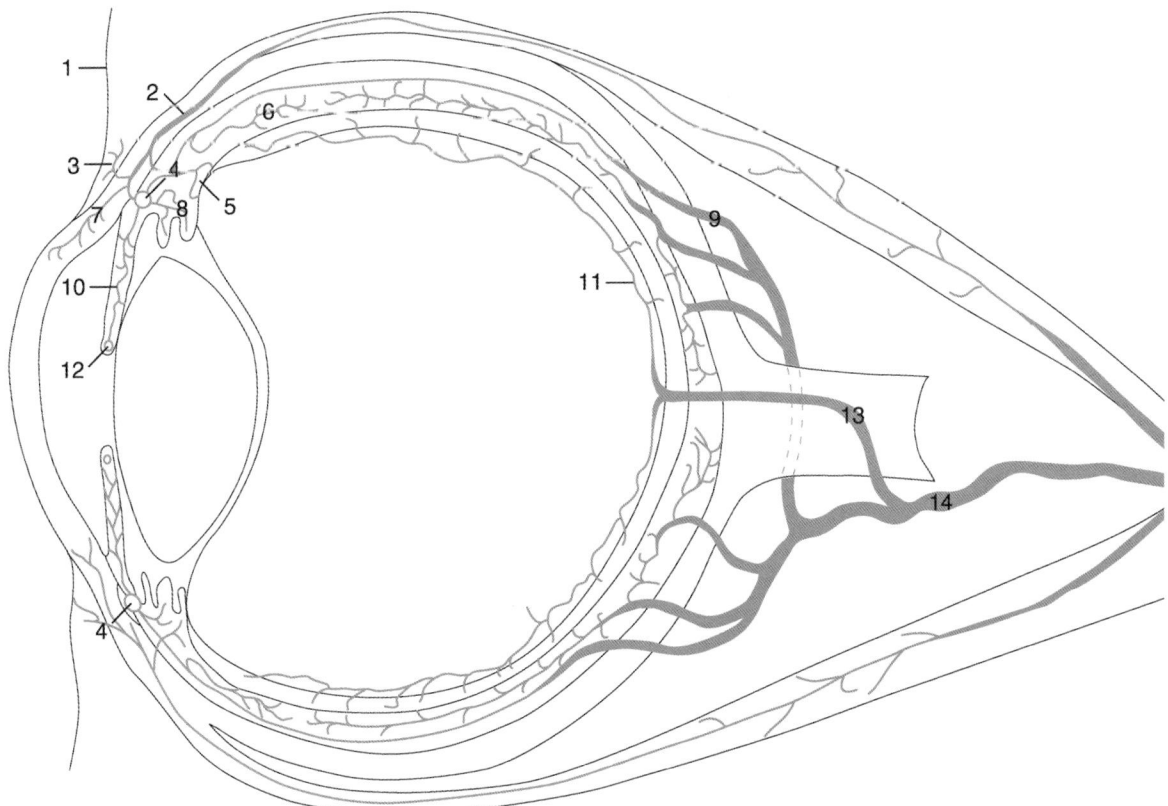

Fig. 2.11

Scheme of the arterial supply of the globe.

1. Conjunctiva
2. Tendon of rectus muscle
3. Episcleral and conjuctival branches of the anterior ciliary artery
4. Major arterial circle of the iris
5. Branches of the arterior ciliary artery for the pars plana of the ciliary body
6. Recurrent branches of the long posterior ciliary artery for the anterior part of choroid
7. Branches of the anterior ciliary artery for the corneoscleral border
8. Branches of the major arterial circle for the pars plicata of the ciliary body
9. Long posterior ciliary artery
10. Branches for the iris
11. Branch of the central retinal artery for the internal layers of the retina
12. Minor arterial circle of the iris
13. Central retinal artery in the optic nerve
14. Ophthalmic artery

dus. The paraoptic group of these vessels forms a proximal and a distal ring around the optic nerve head (circles of Haller and Zinn, respectively). More lateral branches of the posterior ciliary arteries supply the posterior choroid (behind the equator) and a part of the anterior choroid. The vertical dividing line between the supply zones of the temporal and nasal posterior ciliary arteries runs mostly across the optic disc. The remaining part of the choroid is supplied by the long posterior ciliary arteries, which travel a long way in the suprachoroid layer on both sides of the globe before giving off the recurrent branches for the anterior choroid and then the final branches to form the major arterial cicle of the iris. The terminal branches of the choroidal vessels form the choriocapillaris. This rich capillary bed is no longer envisaged as a continuous sheet but one consisting of territorial units. Whether such units are formed around centrilobular arteries or veins is as yet a matter of debate.

The anterior ciliary arteries derive from the muscular rami of the rectus muscles (usually two for each). After emerging from the tendons, these arteries divide into deep (scleral) branches, which perforate the sclera and supply the ciliary mucle, and episcleral branches for the sclera, limbus, conjunctiva, and iris.

The major arterial circle of the iris (circulus arteriosus iridis major) actually lies in the anterior part of the ciliary

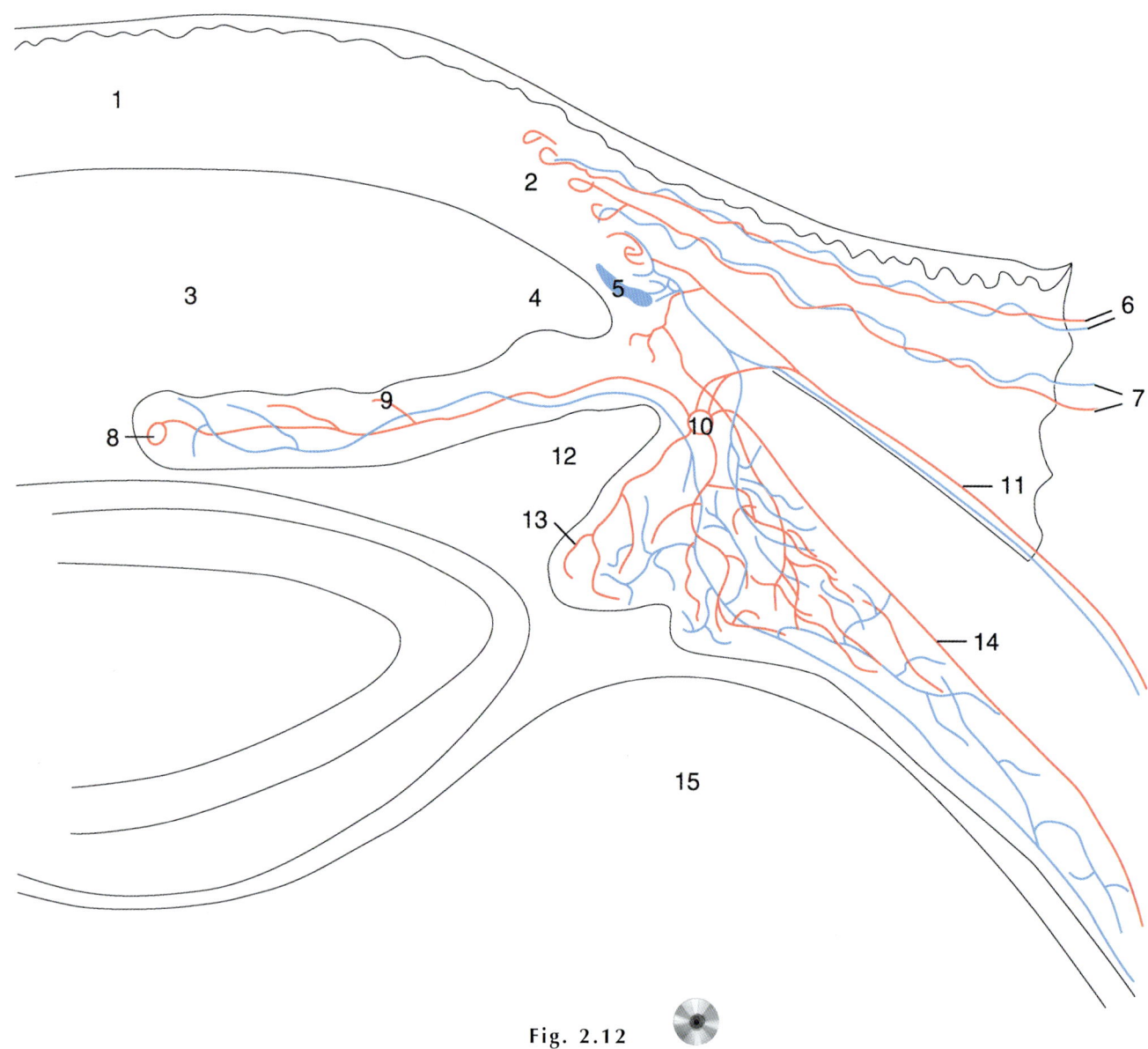

Fig. 2.12

The vessels of the anterior segment.

1. Cornea
2. Corneoscleral border
3. Anterior chamber
4. Drainage angle
5. Schlemm's canal
6. Conjunctival vessels
7. Vascular plexus in Tenon
8. Minor arterial circle of the iris
9. Vascular plexus of the iris
10. Major arterial circle of the iris
11. Anterior ciliary vessels
12. Posterior chamber
13. Ciliary process
14. Long posterior ciliary artery
15. Vitreous

body. Branches from here supply the pars plicata of the ciliary body (forming three distinct vascular territories) together with the inner and anterior parts of the ciliary muscle. Further radially orientated branches enter the iris (these are better seen in blue irides) and converge on the pupillary edge to form the (usually incomplete) minor arterial circle (circulus arteriosus iridis minor).

External to the major arterial circle, the vessels of the outer and more superficial parts of the ciliary muscle (termed intramuscular circle) arise from the anterior ciliary arteries.

Venous blood is returned from the choroid and other parts of the vascular tunic by two routes: the vortex veins (venae vorticosae) and the anterior ciliary veins. The latter drain only part of the ciliary muscle, thus the main outflow of the uveal tract is via the vortex veins. Usually there are four such veins on each side of the superior and inferior rectus muscles, piercing the sclera in an oblique plane behind the equator, the inferior lateral vein being most anterior and the superior lateral vein most posterior. In the myopic eye they are situated further back, near the pole. The two superior vortex veins join the superior ophthalmic vein, whereas the two inferior vortex veins open into the inferior ophthalmic or its anastomosis with the superior ophthalmic vein.

The internal layers of the retina are supplied by the central retinal artery, a branch of the ophthalmic. This artery enters the optic nerve about 12 mm behind the globe and passes through the lamina cribrosa of the optic nerve head. Emerging on the medial side of the optic cup, superonasally displaced, it has two primary branches (superior and inferior), each branching dichotomically into nasal and temporal rami. The nasal branches are more straight, whereas the temporal branches arch around the central retina. The macular area does not contain vessels of visible size, however, its capillary bed is very rich apart from the fovea, which is devoid of blood vessels. The veins follow a similar course as the arteries and also converge on the optic disc as the central retinal vein. In an ophthalmoscopic picture, the veins can be clearly discerned from the arteries by their darker color. Furthermore, the arteries, which are brighter red, also show a streak of light reflexion. The vascular system of the retina, in particular the circulation dynamics, can be well observed by fluorescent angiography (FLAG). The retinal system of circulation is largely separated from the other vascular components of the eye. The terminal branches of the central retinal artery penetrate as far as the INL and do not have anastomoses. Thus, a blockage in one branch leads to a lesion in the given territory. However, there are minor connections between the anastomotic ring of Zinn (around the optic disc) and the adjacent retina via the cilioretinal arteries.

ACCESSORY VISUAL APPARATUS

The term "accessory visual apparatus," also known as the "ocular adnexa" comprises those structures that do not participate directly in the process of vision. The protective apparatus (eyelids, conjunctiva, and lacrimal system) and the extraocular muscles belong to this group.

Eyelids (Palpebrae) (Figs. 2.13, 2.51; 2.52)

These skin folds protect the eye from mechanical injuries and dazzling light. Repeated automatic closure of the lids (blinking) helps in the distribution of lacrimal fluid and has a pumping effect on the lacrimal sac. The upper lid is continuous with the eyebrow over the orbital margin, whereas the lower lid adjoins the skin of the cheek. When the lids are open, the anterior part of the eye is exposed through the elliptical palpebral fissure, whose upper and lower margins meet at the lateral and medial angles (canthi). The lateral canthus is acute, in close contact with the globe, whereas the medial canthus is more obtuse and is separated from the globe by the tear lake (lacus lacrimalis), where the lacrimal caruncle and plica semilunaris are found. In the Eurasian (Mongolian) races, the medial canthus is covered by a dermal fold called epicanthus.

Both eyelids contain a dense fibro-elastic tarsal plate, which is continuous with the orbital septum and serves as a skeleton for the soft tissues. The tissue layers of the lids from the front to back are as follows: skin, subcutaneous areolar tissue, striated skeletal muscle (palpebral part of the orbicularis oculi), submuscular areolar tissue, tarsal plates, septum orbitale, nonstriated muscle, conjunctiva. The skin is loose and thin and it contains special hairs (eyelashes, cilia) lined up in two rows along the palpebral margin. These have no erector muscles but their follicles are richly innervated, corresponding to a sensory function. A group of modified sweat glands, the ciliary glands of Moll discharge their secretion into the ciliary follicles or the ducts of sebaceous glands (of Zeiss). The tarsal glands (of Meibom) are embedded in the tarsal plates as long vertical columns, their ducts opening just in front of the mucocutaneous border. The oily secretion of Meibomian glands prevents an overflow of tears along the palpebral rim and it also traps moisture by forming a thin film. An extension of the levator palpebrae, the nonstriated palpebral (tarsal) muscle of Müller lies posterior to the septum orbitale. Innervated by sympathethic nerves, the muscle is attached to the orbital margin of the superior tarsal plate and its main action is widening the palpebral fissure. The superior palpebral muscle is the most prominent member of the group called capsulo-palpebral muscles, including the retractor muscle of the lower lid and fibres surrounding the palpebral lobe of the lacrimal gland. It is also related to the orbital muscle of Müller.

The mucosal layer contains the accessory lacrimal glands of Krause (in the fornix) and Wolfring (in the palpebral conjunctiva).

Conjunctiva (Figs. 2.13; 2.14; 2.39)

The name, meaning "connecting membrane," is appropriate because the conjunctiva connects the globe to the lids. In so doing, it forms a thin and mucous protective sac in front of the eyeball, which allows a certain degree of movement of the globe and the lids. The epithelium of

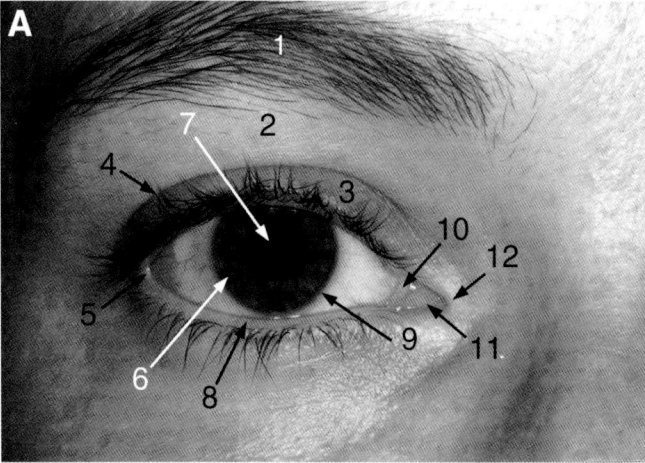
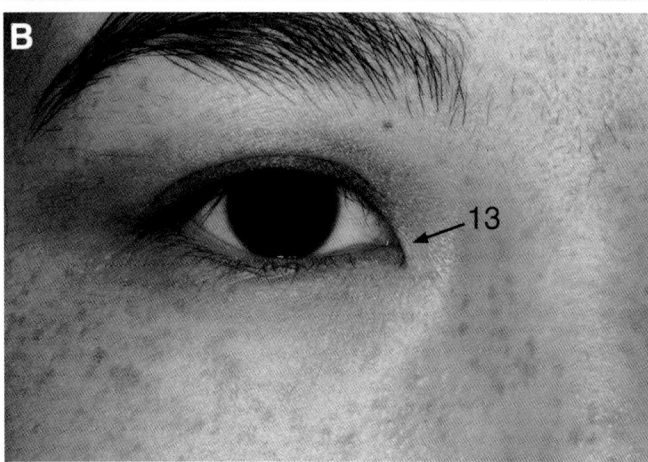

Fig. 2.13

Frontal view of the eye and surroundings of a young Caucasian female (A) and a young Asian (Mongolian) female (B). Note that the eyelids are more inverted in the Mongolian type, in particular the upper lids are retracted under the eyebrow. A prominent epicanthus can also be observed.

1. Superciliary hairs
2. Eyebrow
3. Upper eyelid
4. Eyelashes (cilia)
5. Lateral canthus
6. Iris
7. Pupil
8. Lower eyelid
9. Limbus
10. Plica semilunaris
11. Caruncula lacrimalis
12. Medial canthus
13. Epicanthus (dermal fold covering the medial canthus)

conjunctiva is continuous with that of the cornea over the limbus, after which the conjunctiva covers the anterior surface of the sclera and is reflected as the fornix to the posterior surface of the lids. Accordingly, three parts of the conjunctiva can be distinguished: bulbar, fornical, and palpebral. The bulbar part has a slight elevation near the limbus (limbal conjunctiva). Most of the bulbar conjunctiva is separated from the sclera by loose episcleral tissue, accomodating the anterior ciliary vessels and the rectus tendons. The fornical conjunctiva extends as a ring-like pouch above, below, laterally, and medially, the medial evagination being the smallest. The palpebral conjunctiva is subdivided into marginal, tarsal, and orbital parts. The marginal conjunctiva lies by the mucocutaneous junction of the lids and accomodates the puncta (see below). The tarsal conjunctiva is largely adherent and richly vascularized, whereas the orbital conjunctiva (between the upper tarsal border and fornix) is loosely attached.

The epithelium of the conjunctiva varies according to its location. The palpebral margin is covered by keratinized squamous epithelium, which continues in nonkeratinized statified squamous epithelium of the tarsal conjunctiva. Moving toward the fornices, this is replaced by columnar epithelium, first bilaminar and then trilaminar or stratified in the fornix proper. The latter epithelium continues over most of the bulbar conjunctiva and then it switches fairly abruptly to the stratified squamous nonkeratinized epithelium, characteristic for the cornea. Mucous goblet cells are present mainly in the palpebral and fornical parts and absent near the limbus. However, the surface of the entire conjunctiva contains a glycocalyx component that shows a characteristic staining similar (although not identical) to the glycoproteins of mucin.

The limbal conjunctival epithelium, which is about twice as thick as in the cornea, rests on the tall and radially arranged papillae known as the limbal palisades of Vogt, containing vascular loops reflected at the corneoscleral border. Apart from epithelial cells and goblet cells, the conjunctiva also contains melanocytes and Langerhans (dendritic) cells. The latter belong to the monocyte–macrophage–histiocyte lineage and their main function is antigen presentation and the stimulation of T-lymphocyte response.

Both the propria and the submucosa of the conjunctiva are rich in lymphocytes, in accordance with a heavy exposure to infection.

The arteries of the conjunctiva arise from four main sources: the peripheral and marginal tarsal arcades (connected to the arteries of the face), the anterior ciliary/episcleral arteries, and the deep ciliary system (major arterial circle of the iris). Venous blood from the tarsal, fornical, and posterior bulbar conjunctiva is returned via the palpebral veins, whereas the limbal region is drained by the episcleral veins via the rectus muscle tendons. The lymph vessels of the conjunctiva are collected near the canthi and join the lymphatics of the lids passing toward the parotid nodes laterally and the submandibular nodes medially.

Lacrimal System (Figs. 2.14; 2.53; 2.54; 2.55)

Tears (lacrimal fluid) are necessary for the maintenance of corneal transparency and the protection of the conjunctival sac from foreign bodies and bacteria. The precorneal tear film consitutes the first major refractive interface of the eye. The lacrimal fluid is produced by the lacrimal gland, distributed through blinking over the cornea and drained via the puncta, canalicles, lacrimal sac, and nasolacrimal duct to the nasal cavity. The electrolyte composition of the fluid is very similar to that of the plasma except for a higher content of K^+ and Cl^-. It also contains some enzymes, most importantly lysozyme, which has a bactericidal effect.

The lacrimal gland has two contiguous parts partially separated by the muscle fibers of the levator palpebrae superioris: large orbital and small palpebral. The orbital part lies in the lacrimal fossa situated in the anterolateral part of orbital roof. The gland rests on the levator and extends laterally as far as the lateral rectus and posteriorly as far as the pole of the gobe. Its anterior border is related to the septum orbitale. The palpebral portion of the gland (approximately one-third the size of the orbital part) is adjacent to the lateral border of the upper fornix, palpebral conjunctiva and levator muscle. As many as 12 small excretory ducts (ductuli lacrimales) can be observed to open into the upper fornix.

The microscopic structure of the lacrimal gland resembles that of salivary glands, primarily the parotid, with some notable differences. The lobules of the tubulo-acinar gland are embedded in fatty connective tissue. The acini consist of pyramidal secretory cells with apical microvilli, Golgi apparatus, rough endoplasmic reticulum, and abundant secretory granules. The basal parts of the acini are occupied by myoepithelial cells. Despite an overall serous appearance, mucous cells have also been described in the lacrimal gland, some reports associating the observed differences in the ultrastructure of secretory granules (dense or light) with a respective serous or mucous nature of the acinar cells. Although there is no general agreement over this issue, it is worth noting that the acini of the lacrimal gland have wider lumens than those of the parotid gland, suggesting a "thicker" (partially mucous) type of secretion. Even the possibility of apocrine secretion has been raised because of the reported presence of intact secretory granules in the lumen of some ducts. Another important feature of the lacrimal gland is the abundance of interstitial lymphocytes and plasmocytes. The antibodies

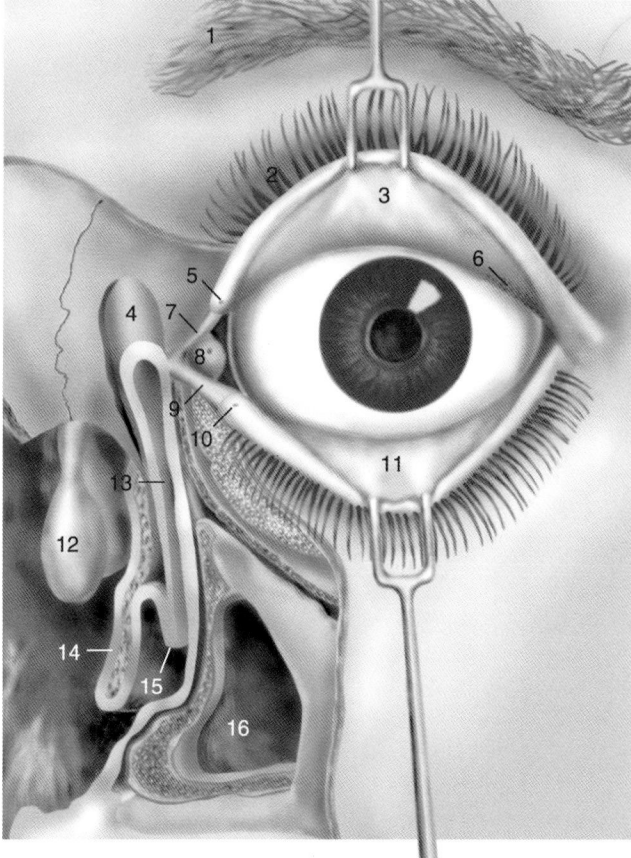

Fig. 2.14

The system of lacrimal drainage.

1. Eyebrow
2. Superior cilia (eyelashes)
3. Upper eyelid (palpebra superior)
4. Lacrimal sac
5. Punctum lacrimale superius
6. Openings of the excretory ducts of lacrimal gland
7. Superior lacrimal canalicle
8. Caruncula lacrimalis
9. Inferior lacrimal canalicle
10. Punctum lacrimale inferius
11. Lower eyelid (palpebra inferior)
12. Middle nasal concha
13. Nasolacrimal duct
14. Inferior nasal concha
15. Plica Hasneri
16. Maxillary sinus

(predominantly immunoglobulin A) generated by the plasma cells pass through the secretory epithelium by a mechanism called adsorptive pinocytosis, to appear in the tears. Thus, apart from lysozyme, the lacrimal fluid has another weapon of antibacterial defense: humoral immune response.

Blood supply for the lacrimal gland is from the lacrimal artery, a branch of the ophthalmic in most cases but it may arise also from one of its anastomotic partners, the middle meningeal or transverse facial arteries. The main route of lymphatic drainage is via the preauricular nodes. The gland is innervated by the lacrimal nerve, a branch of the ophthalmic nerve (V/1) with sensory fibers. The origin of secretory innervation is more complicated and somewhat dubious. The preganglionic parasympathetic fibers arising from the superior salivary nucleus of the pons travel first in the nervus intermedius (together with the facial nerve), then its branch, the greater petrosal nerve, after which they reach the pterygopalatine ganglion via the nerve of the pterygoid canal (Vidian nerve). The latter also contains sympathetic postganglionic fibres. Only the parasympathetic fibres are relayed in the ganglion and the precise fate of the postganglionic fibers is still a matter of debate. According to classical description, they join the zygomatic nerve (V/2) and then are passed through anastomotic branches on to the lacrimal nerve. However, there is strong indication for another pathway, whereby the postganglionic fibres from the pterygopalatine ganglion would enter the orbit in a retro-orbital plexus with direct rami lacrimales for the gland. Sympathetic innervation may reach the lacrimal gland also by the latter route or along the lacrimal artery and nerve.

Reflex secretion from the lacrimal gland (reflex weeping) can be triggered by irritation of the cornea, conjunctiva, nasal mucosa, and even by hot and spicy foods or dazzling light. The neural pathways of psychogenic weeping (a uniquely human phenomenon caused by emotional upset) are poorly understood.

Drainage of the lacrimal fluid begins with the paired puncta lacrimalia, situated in the palpebral border of the upper and lower lids just behind the ciliated part. Each punctum opens on a small elevation called papilla lacrimalis. From the puncta the lacrimal fluid is passed to the lacrimal canaliculi. These are bent at right angle between their vertical and horizontal parts (ampullae) and pierce the periorbita separately before joining in the lacrimal sinus of Maier, an evagination of the lacrimal sac. Both the puncta and canaliculi are surrounded by fibres of the orbicularis oculi. The canaliculi are lined by stratified squamous epithelium.

The lacrimal sac is accomodated by the lacrimal fossa formed by the anterior lacrimal crest of the maxilla and the posterior lacrimal crest of the lacrimal bone. The sac is invested by the lacrimal fascia, a derivative of the periorbita, and is related anteriorly to the medial palpebral ligament and angular vein, laterally to the orbicularis oculi muscle, medially to the bone (and the ethmoid air cells behind it), and posteriorly to the pars lacrimalis of the orbicularis oculi, known as Horner's muscle. The latter has a pumping effect on the sac during blinking.

Downward, the lacrimal sac continues in the nasolacrimal duct with no distinct boundary. The duct lies within the bony canal formed by the maxilla, inferior process of lacrimal bone and lacrimal process of the inferior nasal concha. About 15 mm in length, the duct opens in the inferior nasal meatus under a valve-like mucosal fold (plica lacrimalis, valve of Hasner). A plethora of other "valves" (all known by author's names) have been described in the lacrimal drainage system but hardly any deserve their name because they are mere mucosal elevations or slight constrictions that cannot really block the flow of fluid. The mucosa of the lacrimal sac and nasolacrimal duct contains bilaminar columnar epithelium, ciliated in places, with occasional goblet cells or mucous glands, overlying a fibroelastic lamina propria. The duct is surrounded by a dense vascular plexus similar to the "erectile" tissue of the nasal mucosa. The outermost layer is continuous with the periosteum. The lacrimal passages, in particular their patency, can be investigated by a contrast study known as dacryocystography.

Extraocular Muscles (Figs. 2.15; 2.16; 2.17; 2.19; 2.36; 2.37; 2.50)

These striated skeletal mucles turn the globe in the required direction voluntarily or by involuntary automatism. The movements of the globe are reminiscent of those of a ball and socket joint. The near-spherical eyeball whose firmness is ensured by the intraocular pressure and the taut external coat (sclera) rotates in the matching socket formed by the Tenon's capsule of the orbit. The latter is continuous with the fascial sheaths of the extracoular muscles.

There are six extraocular mucles around the globe, four of these are roughly parallel with the axis of the optic nerve and are termed rectus muscles, whereas the tendons of the two oblique muscles run diagonally. The rectus muscles originate from the common tendinous ring (annulus tendineus of Zinn), an ovoidal fibrous thickening of the orbital periosteum that covers the optic canal and the medial part of the superior orbital fissure. The superior, medial, inferior, and lateral recti pass forward, forming a muscular cone, and their tendons insert in the sclera of the respective quadrant of the globe in front of the equator.

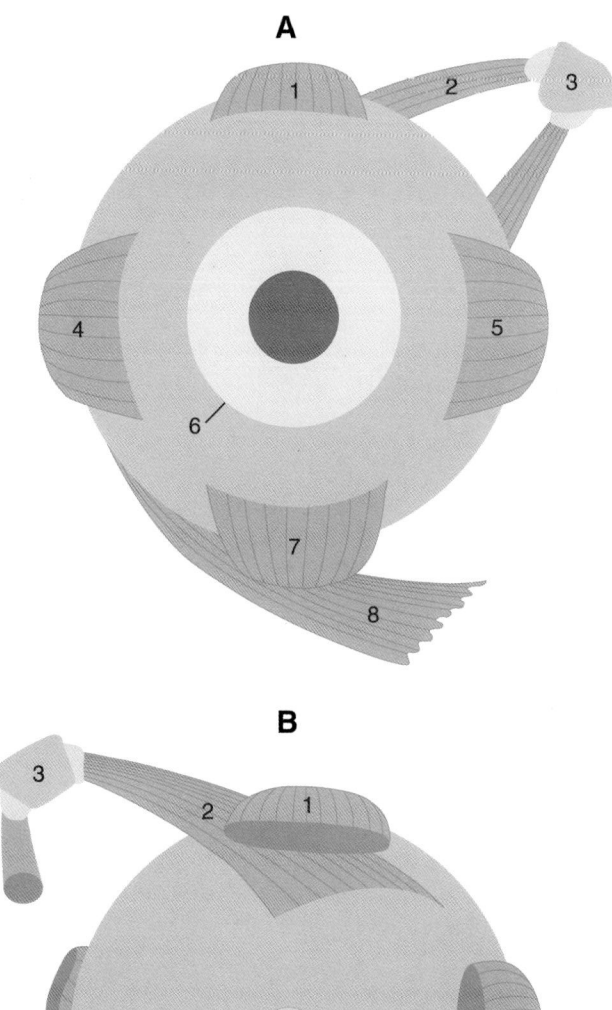

Fig. 2.15

Schematic illustration of the extraocular muscles. A – anterior view, B – posterior view.

1. Superior rectus muscle
2. Superior oblique muscle
3. Trochlea
4. Lateral rectus muscle
5. Medial rectus muscle
6. Corneoscleral limbus
7. Inferior rectus muscle
8. Inferior oblique muscle
9. Optic nerve

The superior rectus is related superiorly to the levator palpebrae superioris, which arises from the sphenoid above the common tendinous ring and, without directly affecting the globe, inserts as a wide tendon in the superior tarsal plate. Both the levator and the superior rectus have a fascial connection with the superior fornix and, as a result, can raise the conjunctival sac.

The inferior rectus muscle is related to the fascial sheath of the inferior oblique, the suspensory ligament of the eye and the lower eyelid. A few detached (nonstriated) fibers of the muscle reach the inferior tarsal plate as the retractor.

The lateral rectus muscle arises from the lateral part of the ring of Zinn but it also has another origin outside the common tendinous ring, from the ala major of sphenoid. The vertical line of insertion is found in the lateral sclera of the globe. The fascial sheath of the muscle is attached to the lateral wall of the orbit as the lateral check ligament.

The largest of the extraocular muscles, the medial rectus muscle lies close to the medial orbital wall. It arises from the medial part of the common tendinous ring and the dural sheath of the optic nerve, and passes to the medial part of the sclera. Similar to the lateral rectus, the fascia of medial rectus also sends a connecting band, known as the medial check ligament, to the wall of the orbit.

The superior oblique muscle arises from the body of the sphenoid bone slightly outside and superomedial to the common fibrous ring. The course of the muscle is rather peculiar because it first passes forward between the medial wall and roof of the orbit but then its round and slender tendon bends around the trochlea (a pulley-like fibrous sling attached to the trochlear fossa near the orbital margin) and passes diagonally backward and downward to the superolateral quadrant of the sclera, where its wide tendon, covered by the superior rectus, inserts behind the equator. The traction of the muscle is determined by the course of its post-trochlear component.

The inferior oblique muscle arises from the floor of the anteromedial part of the orbit, lateral to the nasolacrimal canal. The fibers pass diagonally backward deep to the inferior rectus tendon to insert in the inferolateral quadrant of sclera behind the equator.

With respect to the corneoscleral limbus, nearest is the attachment of the medial rectus, further away the inferior rectus and lateral rectus, and farthest is the superior rectus.

The topography and innervation of the extraocular muscles incuding the levator palpebrae are summarized in the Table 1.

Fig. 2.16

Schematic illustration of the extraocular muscles, lateral view.

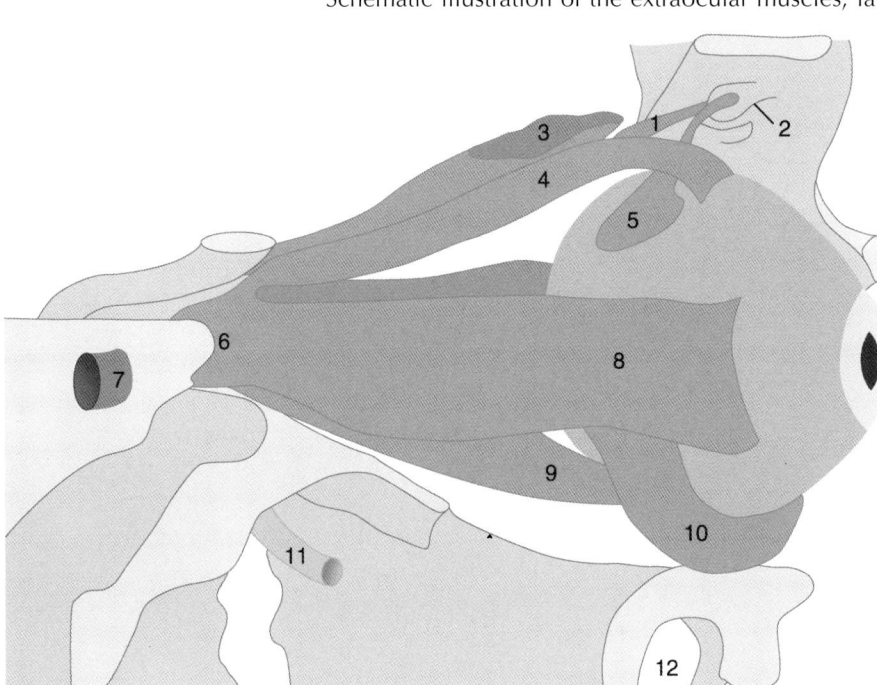

1. Pretrochlear part of the tendon of superior oblique muscle
2. Trochlea
3. Levator palpebrae superioris muscle
4. Superior rectus muscle
5. Insertion of the posttrochear part of the tendon of superior oblique muscle
6. Common tendinous ring of Zinn
7. Internal carotid artery
8. Lateral rectus muscle
9. Inferior rectus muscle
10. Inferior oblique muscle
11. Maxillary nerve
12. Maxillary sinus

Fig. 2.17

Schematic illustration of the common tendinous ring in relation to the extraocular muscles and the structures passing through the optic canal and the superior orbital fissure.

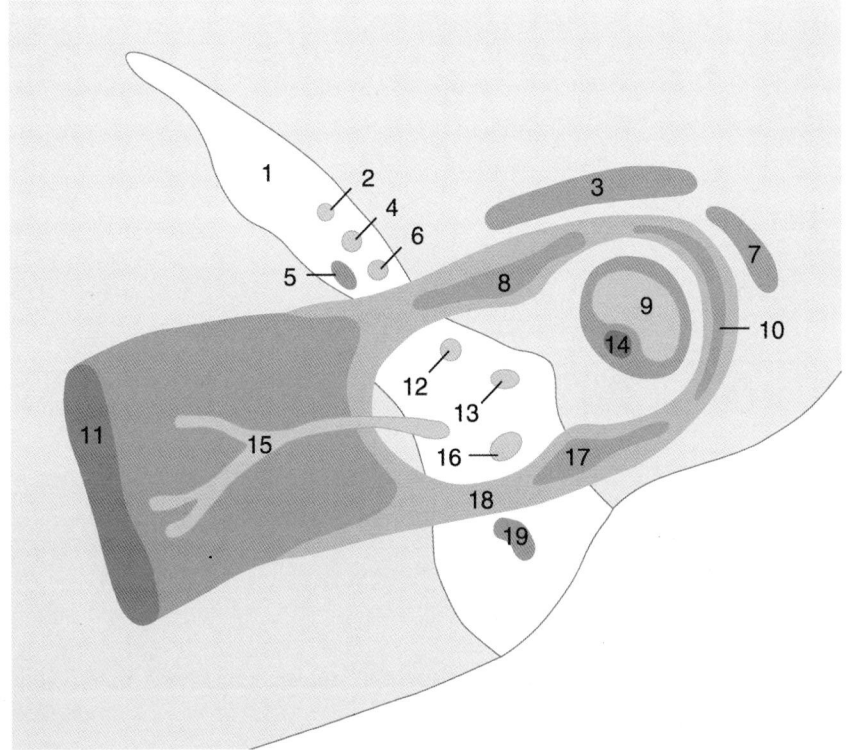

1. Superior orbital fissure
2. Lacrimal nerve
3. Levator palpebrae superioris muscle
4. Frontal nerve
5. Superior ophthalmic vein
6. Trochlear nerve
7. Superior oblique muscle
8. Superior rectus muscle
9. Optic nerve
10. Medial rectus muscle
11. Lateral rectus muscle
12. Oculomotor nerve (superior division)
13. Nasociliary nerve
14. Ophthalmic artery
15. Abducent nerve
16. Oculomotor nerve (inferior division)
17. Inferior rectus muscle
18. Common tendinous ring
19. Inferior ophthalmic vein

Table 1
The Topography and Innervation of the Extraocular Muscles

Muscle	Origin	Insertion	Innervation
Superior rectus	Common tendinous ring	Superior part of sclera, anterior to equator	Oculomotor n. sup. division
Inferior rectus	Common tendinous ring	Inferior part of sclera, anterior to equator	Oculomotor n. inf. division
Lateral rectus	Common tendinous ring ala major of sphenoid	Lateral part of sclera, anterior to equator	Abducent n.
Medial rectus	Common tendinous ring	Medial part of sclera anterior to equator	Oculomotor n. inf. division
Superior oblique	Sphenoid bone superomed. to the optic canal	Superolateral part of sclera posterior to equator	Trochlear n.
Inferior oblique	Floor of orbit anteromedial part	Inferolateral part of sclera posterior to equator	Oculomotor n. sup. division
Levator palpebrae superioris	Sphenoid bone superior to the optic canal	Superior tarsal plate	Oculomotor n. sup. division

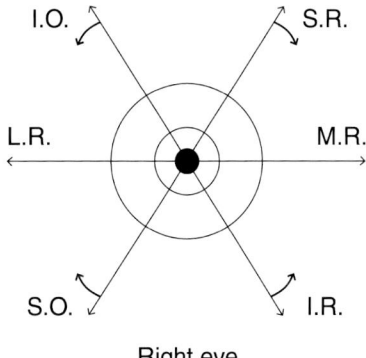
Right eye

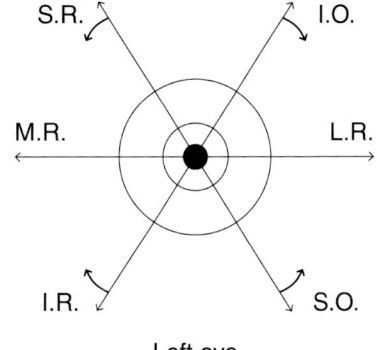
Left eye

Fig. 2.18

Diagram of the extraocular muscles. Frontal view (after Snell and Lemp). Abbreviations: I.O. – inferior oblique, I.R. – inferior rectus, L.R. – lateral rectus, M.R. – medial rectus, S.O. – superior oblique, S.R. – superior rectus.

The movements of the globe result from a concerted action of the extraocular muscles. Around the vertical axis the eyes can turn inward (adduction) or outward (abduction). Around the transverse axis the eyes can turn upward (elevation) or downward (depression). Finally, around the sagittal axis the top of the cornea can turn in a nasal direction (intorsion or inward rotation) or in a temporal direction (extorsion or outward rotation). The latter type of movement is not easily detected in humans because the cornea and pupil are round but in some animals (e.g., cattle, deer) with ovoidal pupils such rotation of the globe is conspicuous.

Table 2 lists actions that the extraocular muscles have on the eyeball when analyzed individually.

From a simplified schematic diagram (of Hering) the different combinations of muscle activities can be read (Fig. 2.19). For example, a straight depression of the globe in the midline requires the collaboration between the inferior rectus and superior oblique, because the secondary and tertiary actions cancel each other out, whilst the primary actions are synergistic. However, for a precise analysis, further points must also be taken into account. For example, the muscle insertions relative to their origins are slightly displaced with any movement of the globe. Therefore, the length and tension of each muscle must be carefully monitored by continual sensory feedback in order for a precise fixation of objects to be attained. Conjugated eye movements in binocular vision belong either to the "version" category (when both eyes move in the same direction) or to the "vergence" category (when the eyes move in opposite directions). As for the latter, the most common form is convergence of the eyes in near fixation.

Innervation of the extraocular muscles is twofold: somatomotor nerves reach the skeletal muscle fibers, whereas the same nerves also carry sensory (proprioceptive) fibers from the muscle spindles, Golgi tendon organs and other specialized nerve endings of the voluntary eye muscles. These sensory afferents are essential for an ade-

Table 2
Actions of Extraocular Muscles on the Eyeball

Muscle	Primary Action	Secondary Action	Tertiary Action
Superior rectus	Elevation	Adduction	Intorsion
Inferior rectus	Depression	Adduction	Extorsion
Lateral rectus	Abduction	—	—
Medial rectus	Adduction	—	—
Superior oblique	Depression	Abduction	Intorsion
Inferior oblique	Elevation	Abduction	Extorsion

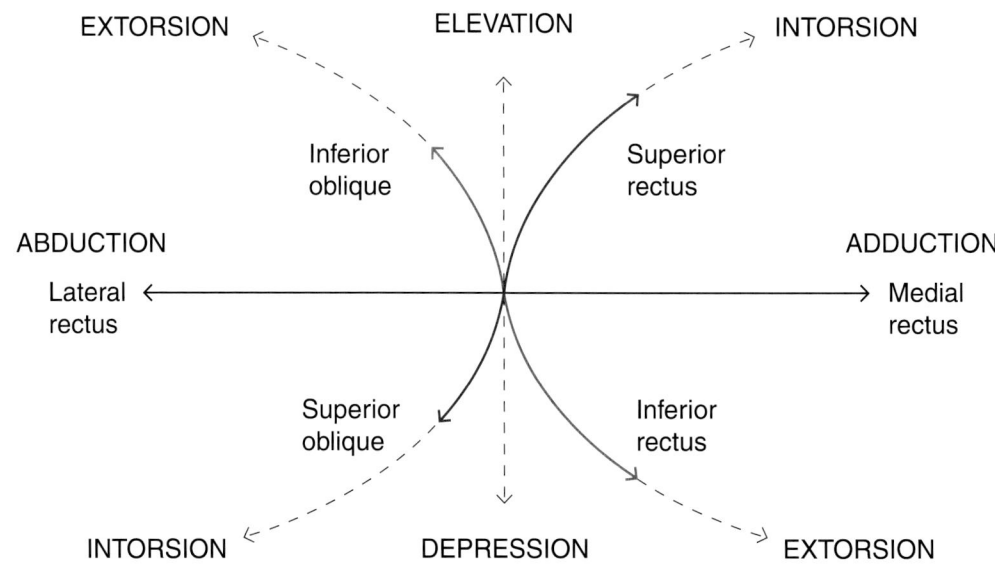

Fig. 2.19

Hering's diagram explaining the actions of extraocular muscles.

quate gauging of the length and tension of extraocular muscles at any time. Apart from the lateral rectus (innervated by the abducent nerve) and superior oblique (innervated by the trochlear nerve), all other extraocular muscles including the levator palpebrae superioris receive their innervation from the oculomotor nerve.

The motor nerve terminals comprise typical myoneural end plates ("en plaque" terminals) on twitch fibers (a fast-acting muscle fiber type), multiple "en grappe" terminals on tonic fibers, and, finally, spiral nerve endings. The latter used to be taken for sensory endings, similar to annulospiral receptors, but it is now increasingly clear that those single or multiple spiral nerve fibers coiling around extrafusal, rather than intrafusal, muscle fibers represent motor innervation, probably subserving overstretch protection.

Sensory terminals of the extraocular muscles occur in three forms: muscle spindles, Golgi tendon organs, and palisade endings. Muscle spindles are encapsulated fusiform end-organs lodged in between extrafusal striated muscle fibers. They contain two types of intrafusal muscle fibers, one with a fusiform accumulation of nuclei (nuclear bag fiber) and another with an even distribution of nuclei (nuclear chain fiber). Both are innervated by γ-motor fibers either with end plates or with finely distributed "flower spray" endings, and also by sensory nerves that are wrapped around the intrafusal fiber (group I afferents belong to both types, whereas group II afferents tend to supply mainly the nuclear chain fibers). Stretching of the intrafusal fibers relative to the muscle in which they are embedded sends a signal down the afferent nerves to the brainstem. Contraction of the intrafusal fibers driven by the activity of γ-motor fibers sets the sensitivity of the receptor organ, enabling the fine tuning required for the coordinated action of extraocular muscles.

Another type of encapsulated receptor, Golgi tendon organs lie in the musculotendinous junctions. They contain collagen fibers interspersed with fine branchings of free nerve endings. Tension building up between the collagen fibers puts pressure on the nerve terminals and elicits an afferent signal. Thus, Golgi tendon organs are sensitive to muscle tension rather than passive stretching. Although widely distributed in various skeletal muscles, classical Golgi tendon organs were rarely observed in the extraocular muscles. Instead, another type of receptor called palisade ending (myotendinous cylinder) appears to play the role of the main tension receptor. Palisade endings are encapsulated nerve endings adjacent to single extrafusal muscle fibres near the myotendinous junction. As with Golgi tendon organs, they are sensitive to compression (see also in Chapter 5).

The proprioceptive fibers arising from the extraocular muscles reach the brainstem via branches communicating with the trigeminal nerve. As with other proprioceptive input carried by the trigeminal nerve (e.g., that of the muscles of mastication), these afferents bypass the Gasserian ganglion and enter the mesencephalic trigeminal nucleus.

Orbital Cavity (Bone, Fasciae, Vessels, Nerves) (Figs. 2.17; 2.20; 2.21; 2.22; 2.35; 2.36; 2.37)

Having discussed the globe and the structures related to it, we conclude the anatomical overview with the orbital cavity including its bony walls, fasciae, blood vessels, and nerves.

An important part of the facial skeleton, the paired orbits have the shape of a truncated pyramid whose forward directed base is open to the exterior. The opening (aditus orbitae) is surrounded by the orbital margin consisting of the following parts: supraorbital (frontal bone), lateral (zygomatic process of frontal bone above and frontal process of zygomatic bone below), infraorbital (zygomatic bone laterally and maxilla medially), and medial (maxillary process of frontal bone above and frontal process of maxilla below). The apex of the orbit is at the medial end of the superior orbital fissure. The orbital cavity has four walls: roof, floor, medial, and lateral.

The roof is formed by the orbital plate of the frontal bone (anteriorly) and the lesser wing of sphenoid posteriorly. Two depressions are situated in this region, anterolaterally the lacrimal fossa and anteromedially the trochlear fossa. The floor is constituted mainly by the orbital surface of the maxilla, supplemented by the zygomatic bone (anterolaterally) and the orbital process of the palatine bone (posteriorly). The inferior orbital fissure runs along the border with the lateral wall but it does not reach the orbital margin. One characteristic structure of the floor is the infraorbital fissure continuing in the infra-

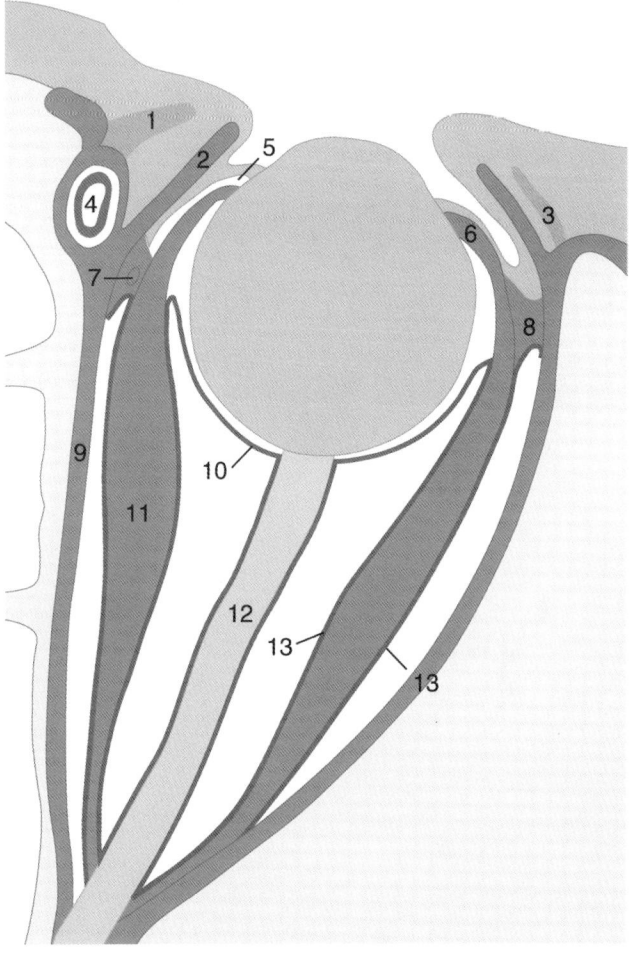

Fig. 2.20

Schematic illustration of the fascial sheaths of the globe in relation to the extraocular muscles. Horizontal section (after Snell and Lemp).

1. Medial palpebral ligament
2. Orbital septum
3. Lateral palpebral ligament
4. Lacrimal sac
5. Tendon of medial rectus muscle
6. Tendon of lateral rectus muscle
7. Medial check ligament
8. Lateral check ligament
9. Periorbita
10. Tenon's capsule
11. Fascial sheath of medial rectus muscle
12. Optic nerve
13. Fascial sheath of lateral rectus muscle

orbital canal and foramen. The lateral wall is formed anteriorly by the zygomatic bone (border with the temporal fossa) and posteriorly by the greater wing of sphenoid (border with the middle cranial fossa). A communication with the middle cranial fossa, the superior orbital fissure

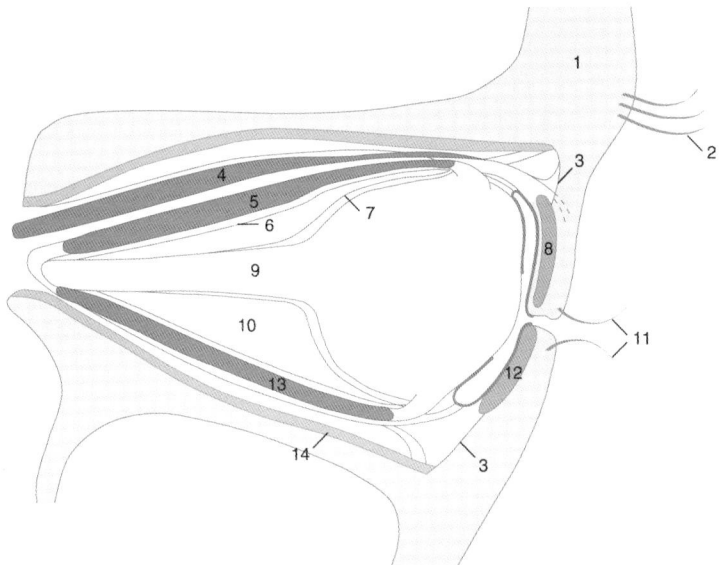

Fig. 2.21

Schematic illustration of the fascial sheaths of the orbit. Sagittal section.

1. Frontal bone
2. Eyebrow (supercilia)
3. Orbital septum
4. Levator palpebrae superioris muscle
5. Superior rectus muscle
6. Fascial sheath of the superior rectus
7. Tenon's capsule
8. Superior tarsal plate
9. Optic nerve
10. Orbital fat
11. Eyelashes (cilia)
12. Inferior tarsal plate
13. Inferior rectus muscle
14. Periorbita

runs along the border between the roof and the medial orbital wall. A small elevation behind the orbital margin, the marginal tubercle is the site of attachment for the tendon of the levator palpebrae superioris, the lateral check ligament and the lateral palpebral ligament. The medial wall, which is largely wafer thin and particularly vulnerable, has the following parts: anteriorly the frontal process of maxilla, then the lacrimal bone, the orbital plate (lamina papyracea) of ethmoid bone and finally the body of sphenoid bone. The groove for the lacrimal sac (fossa sacci lacrimalis) lies between the anterior and posterior lacrimal crests.

The orbital cavity is lined by a periosteal layer called periorbita, which is tightly adherent to bone near the foramina and sutures but not elsewhere in the orbit. An anterior extension of the periorbita, the orbital septum is connected to the eyelids. Further structures attached to the periorbita are the trochlea of the superior oblique muscle, the lacrimal sac and gland and the common tendinous ring of Zinn.

The orbital muscle (of Müller) is a smooth muscle overlying the inferior orbital fissure. Although vestigial in humans, its sympathetic innervation suggests that the muscle may play a role in the stress-related protrusion of the globe.

The connective tissue septa of the orbit comprise the fascia bulbi (Tenon's capsule), the suspensory ligaments of the fornices and the orbital septa. The Tenon's capsule forming the immediate socket for the globe is a derivative of perimysial connective tissue. An important thickening of this envelope, essentially a sling-like support under the eyeball is known as the suspensory ligament (of Lockwood). Further structures connected to the Tenon's capsule are the recently identified rectus muscle pulleys, i.e. con-

Fig. 2.22

Schematic illustration of the fascial sheaths of the globe in relation to the extraorbital muscles. Coronal section of the right orbit (after Snell and Lemp).

1. Levator palpebrae superioris muscle
2. Superior rectus muscle
3. Tenon's capsule
4. Frontozygomatic suture
5. Lateral check ligament
6. Lateral rectus muscle
7. Iris
8. Medial rectus muscle
9. Medial check ligament
10. Medial wall of orbit
11. Suspensory ligament (of Lockwood)
12. Inferior oblique muscle
13. Inferior rectus muscle

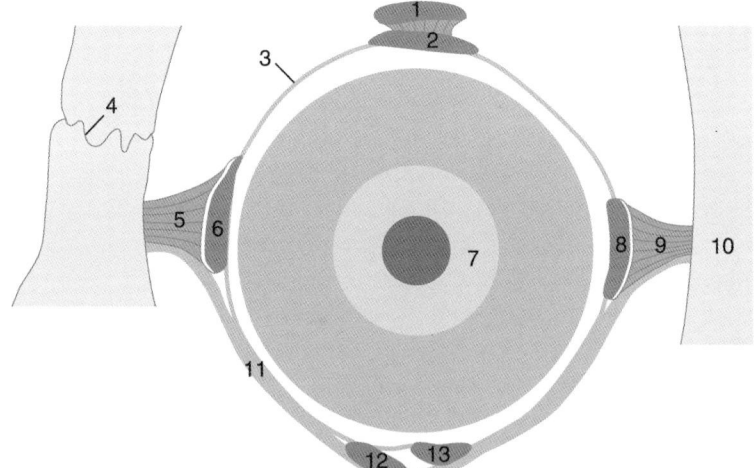

nective tissue sleeves that stabilize the rectus muscle tendons. The suspensory ligaments of the fornices are expansions of the fibrous envelope of the levator and superior rectus, and that of the inferior rectus (reaching the upper and lower fornices, respectively). The orbital septa are condensations of connective tissue and fat for the separation and support of muscles and orbital vessels (mainly veins).

The arteries of the orbital cavity are summarized in Table 3.

VEINS OF THE ORBIT

Three major veins are associated with the orbit, the superior ophthalmic, inferior ophthalmic, and infraorbital veins. The superior ophthalmic vein is collected from two main branches, the supraorbital and angular (terminal branch of facial) veins superomedially near the orbital margin. The supraorbital vein passes through the supraorbital notch (or foramen), whereas the angular vein pierces the orbital septum. They return venous blood from the frontal sinus, the skin and conjunctiva of the upper lid and the skin of the nose. The communication between the angular vein and the superior ophthalmic vein is a potentially dangerous portal of pyogenic infection spreading from the facial area to the orbit and, ultimately, as far as the cavernous sinus. The superior ophthalmic vein also receives the two upper vortex veins, the central vein of the retina and further tributaries corresponding to the similar arteries (see above). The vein crosses the optic nerve and leaves the orbital cavity above the lateral rectus passing through the superior orbital fissure to open into the cavernous sinus.

The inferior ophthalmic vein communicates with three main tributaries, the facial vein, pterygoid plexus and the lower vortex veins, apart from muscular branches. The vein passes near the floor of the orbit and opens either into the cavernous sinus (by joining the superior ophthalmic) or it dissipates in the pterygoid plexus via the inferior orbital fissure.

The infraorbital vein accompanies the infraorbital artery passing backwards in the infraorbital foramen, canal and groove. Its main tributaries are collected in the facial region and in the lower part of the orbit. The vein drains into the pterygoid plexus.

NERVES OF THE ORBIT (FIG. 2.23)

The nerves associated with the orbital cavity are summarized in Table 4.

Fig. 2.23

Half-schematic illustration of the nerves of the orbit. Lateral view following removal of the lateral wall (drawn after an anatomical dissection).

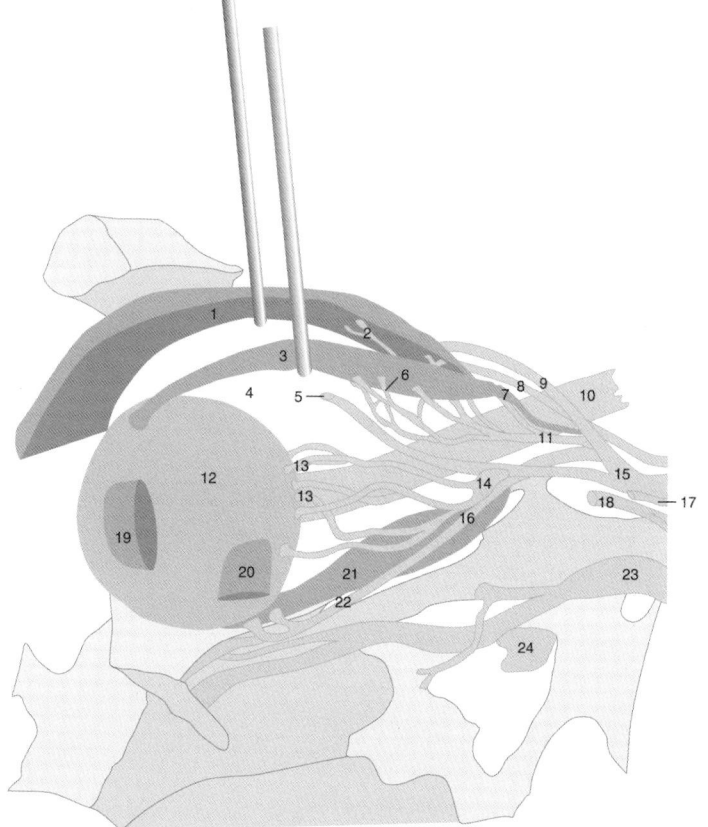

1. Levator palpebrae superioris muscle
2. Branches of oculomotor nerve innervating the levator
3. Superior rectus muscle
4. Orbital fat
5. Nasociliary nerve (cut)
6. Branches of oculomotor nerve innervating the superior rectus
7. Common tendinous ring
8. Trochlear nerve
9. Frontal nerve
10. Optic nerve
11. Superior division of oculomotor nerve
12. Globe
13. Short ciliary nerves
14. Ciliary ganglion
15. Ophthalmic nerve
16. Inferior division of oculomotor nerve
17. Lacrimal nerve (cut)
18. Abducent nerve (cut)
19. Insertion of lateral rectus muscle
20. Insertion of inferior oblique muscle
21. Inferior rectus muscle
22. Branches of oculomotor nerve innervating the inferior rectus
23. Maxillary nerve
24. Pterygopalatine ganglion

Table 3
The Arteries of the Orbital Cavity

Artery[a]	Origin	Major Branches	Supply Region
Central artery of the retina	Ophthalmic	Sup., inf. temporal sup., inf. nasal meningeal central collateral	Retina optic nerve pia mater
Lacrimal artery	Ophthalmic	Lateral palpebral zygomatic muscular recurrent meningeal	Eyelids, conjuctiva temporal fossa lateral rectus muscle dura mater
Rami musculares	Ophthalmic	Muscular anterior ciliary (from rectus tendons)	Extraocular muscles (see anterior ciliary artery)
Long posterior ciliary artery (2)	Ophthalmic	Recurrent communicating	Anterior choroid major arterial circle of iris, pars plicata of ciliary body
Short posterior ciliary artery (7)	Ophthalmic	Choroidal communicating cilioretinal	Posterior choroid ring of Zinn central retina
Anterior ciliary artery (7)	Muscular branches of rectus muscles	Scleral episcleral penetrating	Sclera conjunctiva major arterial circle of iris, pars plana of ciliary body
Supraorbital artery	Ophthalmic	Muscular deep frontal superior palpebral	Levator palpebrae sup. diploë and frontal sinus upper eyelid, scalp, forehead
Posterior ethmoidal artery	Ophthalmic	Dural nasal ethmoidal	Dura of ant. cranial fossa nasal mucosa post. ethmoid air cells
Anterior ethmoidal artery	Ophthalmic	Ethmoidal deep frontal meningeal nasal	Middle and anterior ethmoid air cells frontal sinus dura mater nasal mucosa and skin
Meningeal branch	Ophthalmic	Communicating with the middle meningeal artery	Bone of middle cranial fossa (via sup. orbital fissure)
Medial palpebral artery (2)	Ophthalmic	Sup., inf. peripheral sup., inf. marginal	Eyelids, conjunctiva, nasolacrimal duct
Supratrochlear artery	Ophthalmic (terminal branch)	Communicating with supraorbital	Skin and scalp of forehead, epicranius m.
Dorsal nasal artery	Ophthalmic (terminal branch)	Lacrimal communicating with facial artery	Lacrimal sac dorsum and side of nose
Infraorbital artery	Maxillary	Orbital (others not shown)	Inferior rectus, inferior oblique, lacrimal sac

[a]Numbers in parentheses refer to number of arterial branches bearing the same name.

Table 4
Nerves Associated With the Eye and Orbit

Nerve	Origin	Main Orbital Branches	Supply Region
Oculomotor (third cranial nerve)	Midbrain Main motor nucleus Accessory nucleus (Edinger-Westphal)	Superior ramus Inferior ramus Radix oculomotoria of ciliary ggl. Short ciliary nerves (via ciliary ggl.)	Levator palpebrae m. (bilateral), S.R (C) M.R., I.R., I.O. (I) Preganglionic parasymp. innervation of ciliary ggl. Postganglionic parasymp. innervation of pupillary sphincter and ciliary m.
Trochlear (fourth cranial nerve)	Midbrain Trochlear nucleus decussating fibers	—	S.O. (C)
Abducent (sixth cranial nerve)	Pontomedullary junction	—	L.R. (I)
Nasociliary	Ophthalmic nerve (first division of trigeminal nerve)	Radix sensoria of ciliary ggl. Long ciliary nerves Posterior ethmoidal nerve Anterior ethmoidal nerve Infratrochlear nerve • superior palpebral • inferior palpebral	Sensory fibers from choroid and sclera passing without relay in the ganglion Sensory innervation of ciliary body, iris, cornea Sympathetic fibers for pupillary dilator Posterior ethmoidal air cells, sphenoidal sinus Upper part of nasal cavity, skin of nose (dorsum, vestibule) Upper lid Lower lid, conjunctiva, caruncle, lacrimal sac, skin of nose
Frontal	Ophthalmic nerve (first division of trigeminal nerve)	Supratrochlear nerve Supraorbital nerve • medial ramus • lateral ramus	Upper lid (skin and conjunctiva), glabella, nose Skin of forehead and scalp Upper lid (skin and conjunctiva), frontal sinus
Lacrimal	Ophthalmic nerve (first division of trigeminal nerve)	—	Lacrimal gland, sensory and parasympathetic (disputed!) innervation Lateral part of upper lid (skin and conjunctiva)
Infraorbital	Maxillary nerve (second division of trigeminal nerve)	Inferior palpebral r. (further branches not related to the orbit)	Lower lid (skin and conjunctiva)
Zygomatic	Maxillary	R. communicans Zygomaticofacial n.	Connects to lacrimal nerve Lateral canthus

I, ipsilateral; C, contralateral; S. R., superior rectus; M. R., medial rectus; I. R., inferior rectus; L. R., lateral rectus; I. O., inferior oblique; S. O., superior oblique.

Development of the Eye

Fig. 2.24

Development of the eye. Schematic drawings based on original embryonic specimens published by K.V. Hinrichsen (Human-Embryologie, Springer, 1993). A – 16 somite stage, B – 5 mm embryo, C – 6 mm embryo, D – 6.7 mm embryo, E – 8 mm embryo, F – 13.4 mm embryo.

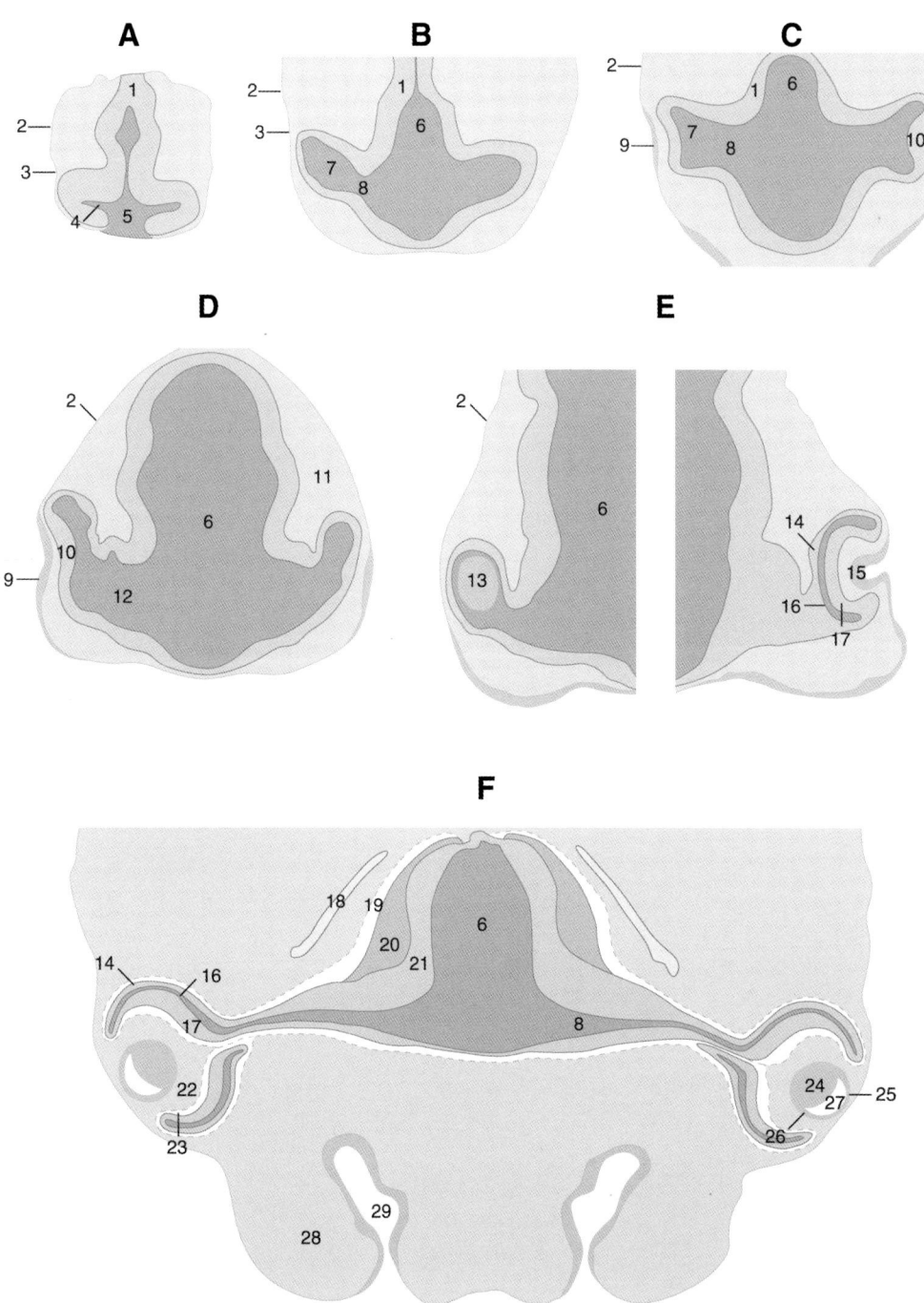

1. Neural tube
2. Ectoderm
3. Optic pit
4. Optic sulcus
5. Anterior neuropore
6. Diencephalon (third ventricle)
7. Optic vesicle
8. Optic stalk
9. Lens placod
10. Indentation of optic vesicle to form the optic cup
11. Condensation of mesenchyme
12. Optic ventricle
13. Tangential section of optic cup
14. Outer (future pigment) layer of optic cup
15. Lens vesicle
16. Cavity of optic cup (future intraretinal space)
17. Inner (future neural) layer of optic cup
18. Developing cranium
19. Marginal layer
20. Intermediate (mantle) layer
21. Ventricular layer
22. Primordium of vitreous body
23. Innermost zone of neural layer, corresponding to the marginal layer
24. Lens hillock
25. Anterior epithelium of lens
26. Equatorial zone
27. Cavity of lens vesicle
28. Lateral nasal prominence
29. Olfactory pit

Fig. 2.25

Development of the eye. Closure of the choroidal fissure and development of the optic nerve (after Snell and Lemp).

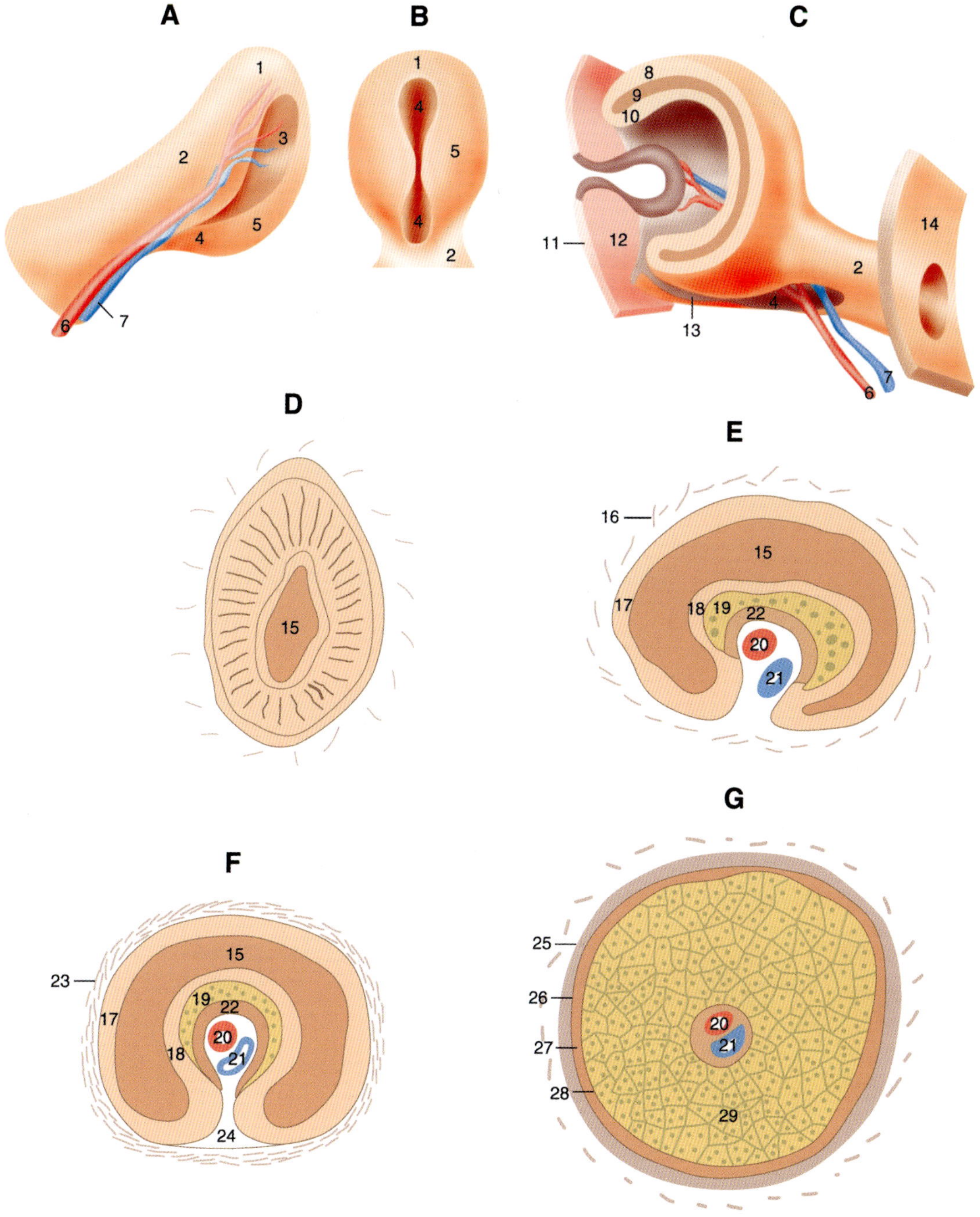

1. Optic cup
2. Optic stalk
3. Optic vesicle
4. Choroidal fissure
5. Fusing lip of choroidal fissure
6. Hyaloid artery
7. Hyaloid vein
8. Outer layer of optic cup
9. Optic ventricle (future intraretinal space)
10. Inner layer of optic cup
11. Lens vesicle
12. Ectoderm
13. Optic cup fissure
14. Diencephalon
15. Cavity of optic stalk
16. Mesenchyme
17. Outer layer of optic stalk
18. Inner layer of optic stalk
19. Optic axons from retinal ganglion cells
20. Central artery of retina
21. Central vein of retina
22. Neuroglial layer
23. Condensation of mesenchyme
24. Choroidal fissure
25. Dura mater
26. Arachnoid
27. Extension of subarachnoidal space
28. Pia mater
29. Bundles of optic axons undergoing myelination

Fig. 2.26

Development of the eye. Early (A, B) and advanced (C, D) stages of differentiation of the ocular tunics (after Snell and Lemp).

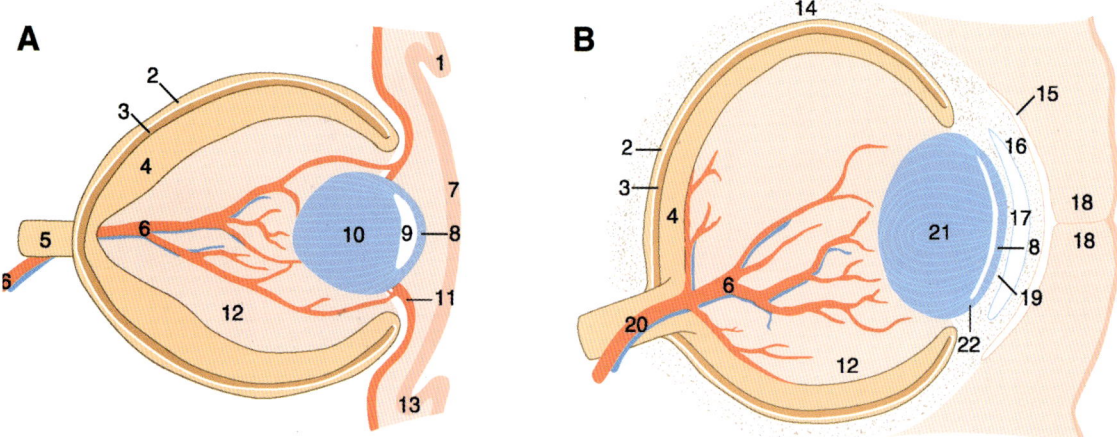
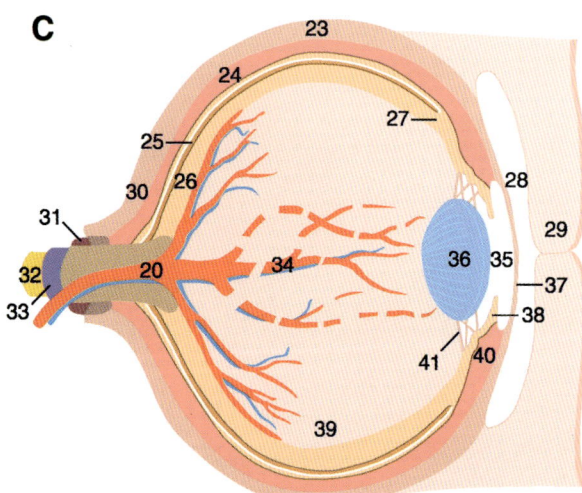
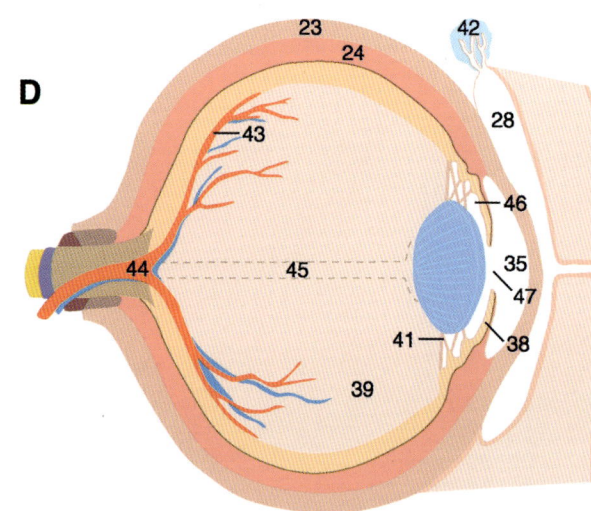

1. Developing upper eyelid
2. Outer layer of optic cup
3. Cavity of optic vesicle (optic ventricle)
4. Inner layer of optic cup
5. Developing optic nerve
6. Hyaloid artery
7. Corneal ectoderm
8. Anterior epithelium of lens
9. Cavity of lens vesicle
10. Posterior layer of lens cells
11. Angular artery
12. Vitreous humour
13. Developing lower eyelid
14. Mesenchyme
15. Conjunctival sac
16. Developing cornea
17. Developing anterior eye chamber
18. Fused eyelids
19. Pupillary membrane
20. Central artery of retina
21. Lens
22. Equatorial zone
23. Sclera
24. Choroid and pigment layer of retina
25. Intraretinal space
26. Neural layer of retina
27. Ora serrata
28. Conjunctival sac
29. Upper eyelid
30. Dura mater
31. Arachnoid
32. Optic nerve
33. Pia mater
34. Degenerating hyaloid artery
35. Anterior chamber
36. Lens
37. Cornea
38. Iris
39. Vitreous body
40. Ciliary body
41. Suspensory ligaments (zonula)
42. Lacrimal gland
43. Branches of retinal vessels
44. Central artery and vein of retina
45. Hyaloid canal
46. Posterior chamber
47. Pupil

ATLAS PLATES I (EMBRYOLOGY)

Fig. 2.27

Development of the eye. Semithin sections from rat embryo, toluidine blue staining.
A – Stage I.; B – Stage II.; C – Stage III.

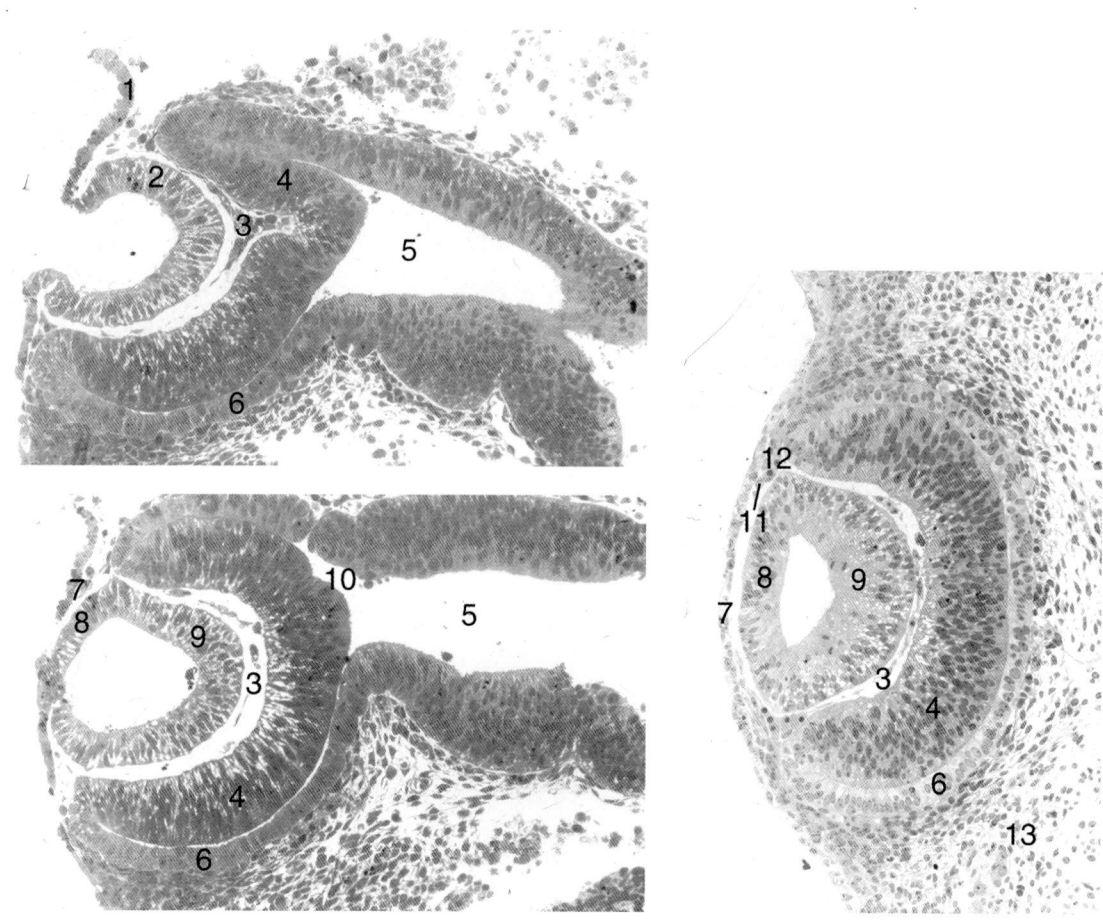

1. Ectoderm
2. Lens vesicle
3. Developing vitreous body
4. Inner layer of optic cup, future pars nervosa of retina
5. Lumen of optic stalk (protrusion of diencephalic ventricle)
6. Outer layer of optic cup, future retinal pigment epithelium
7. Developing cornea
8. Anterior lens epithelium
9. Posterior lens epithelium
10. Intraretinal space
11. Developing iris
12. Developing ciliary body
13. Primordia of sclera and meninges

Fig. 2.28

Development of the eye. Semithin section from rat embryo, toluidine blue staining. Stage IV.

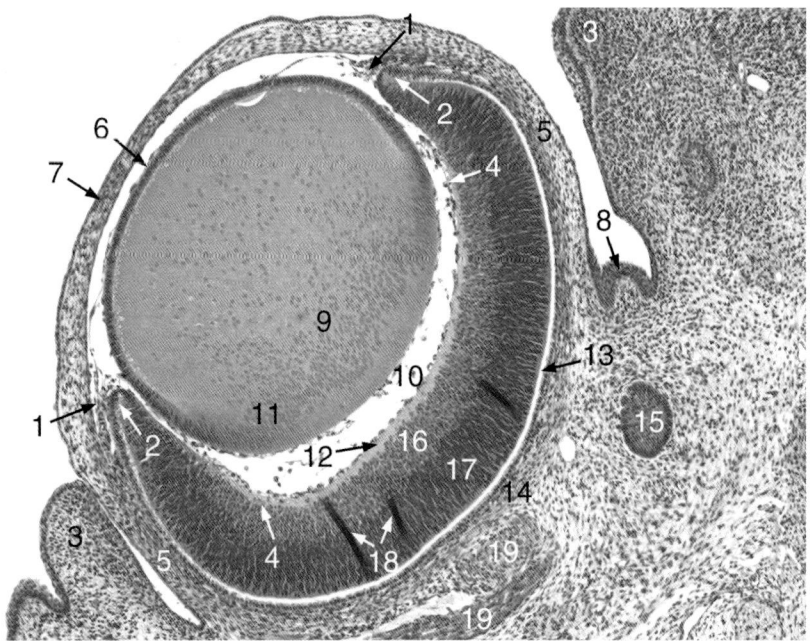

1. Mesodermal anlage of iris, future pupillary membrane
2. Marginal part of optic cup with a transition between the two layers, future retinal layers of iris
3. Primordial eyelid
4. Posterior border of the blind retina, future orra serrata
5. Future conjunctiva apposed to the scleral primordium
6. Simple cuboidal epithelium (anterior lens epithelium) with the primordium of lens capsule (thick modified basement membrane) anterior to it.
7. Developing cornea
8. Fornix of conjunctiva
9. Elongated epithelial cells forming the lens fibers. The nuclei are already sparse and the fibers have filled the cavity of the lens vesicle at this stage
10. Anlage of vitreous body occupied by mesenchyme whose few remaining cells are apposed to the lens and the retina. The vitreous body develops from the persisting ground substance of this mesenchyme.
11. Proliferating epithelium along the equator of the lens with the nuclear bow
12. Optic fiber layer
13. Optic ventricle, a virtual space between the two layers of the optic cup bordered behind by the retinal pigment epithelium.
14. Connecting tissue corresponding to the future choroid and sclera
15. Optic nerve
16. Inner neuroblast layer, primordium of future retinal ganglion cells
17. Outer neuroblast layer, primordium of other retinal neuroepithelial cells
18. Artifacts (section creases)
19. Eye muscles

ATLAS PLATES II (CLINICAL INVESTIGATIONS)

Fig. 2.29

Ultrasonograms of the eye. A – meridional scan, B – oblique scan. Courtesy of E. Márton and ATL Ultrasound Systems.

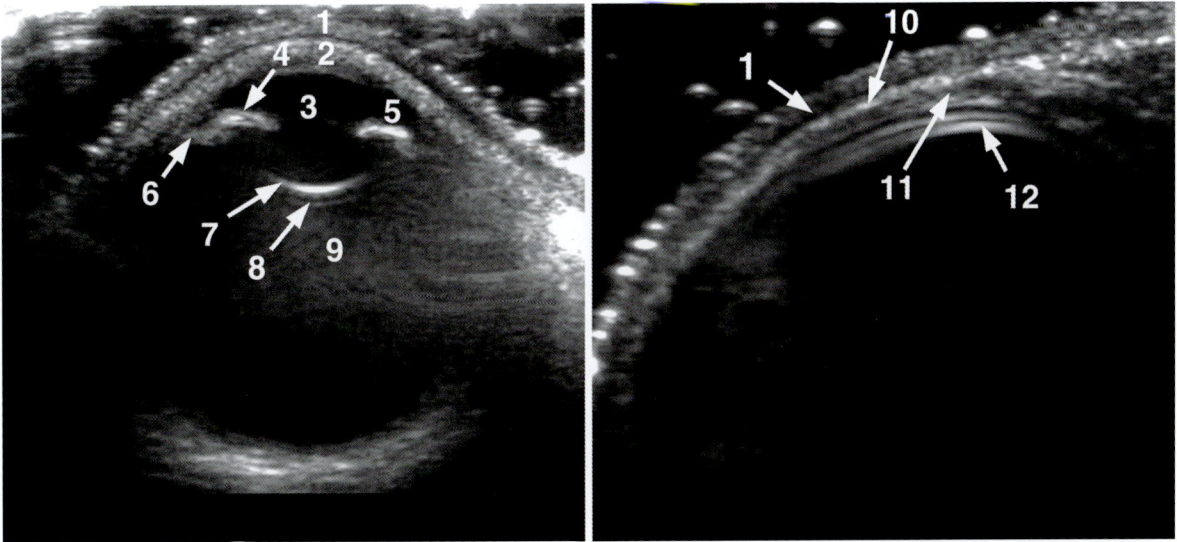

1. Eyelid
2. Cornea
3. Anterior chamber
4. Iris
5. Drainage angle
6. Ciliary body
7. Posterior surface of lens
8. Retrolental space
9. Vitreous
10. Orbital septum
11. Tarsal plate and conjunctiva
12. Sclera with extraocular muscle tendon

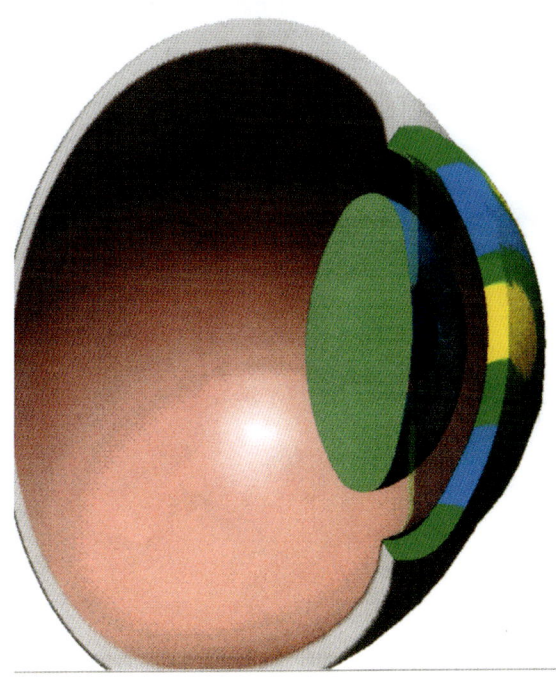

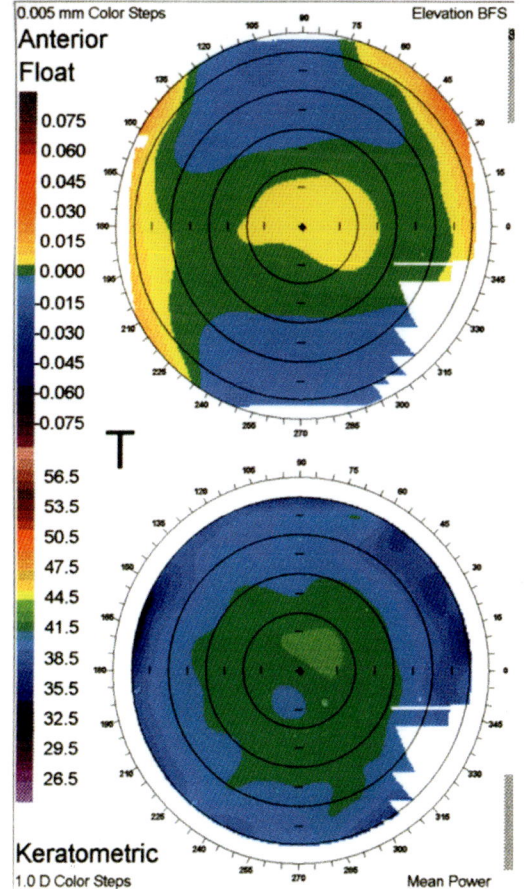

Keratometric image (Orbscan) of the cornea used for establishing the degree of curvature of the different corneal segments, expressed either as surface elevation in mm (top diagram) or in dioptres (bottom diagram). The highest values are found in the central region (corneal cap or apex). The three-dimensional image on the left was generated from the keratometry data by computer. Such examination is necessary for the fitting of contact lenses or for the planning of corneal surgery. Courtesy of J. Györy.

Fig. 2.30

Specular microphotograph (inset) and color-coded image of the corneal endothelium. The red or yellow patches of the histogram indicate a decrease in average cell density and an increase in cell area, corresponding to a compensatory spreading of endothelial cells following an injury. Courtesy of Á. Farkas.

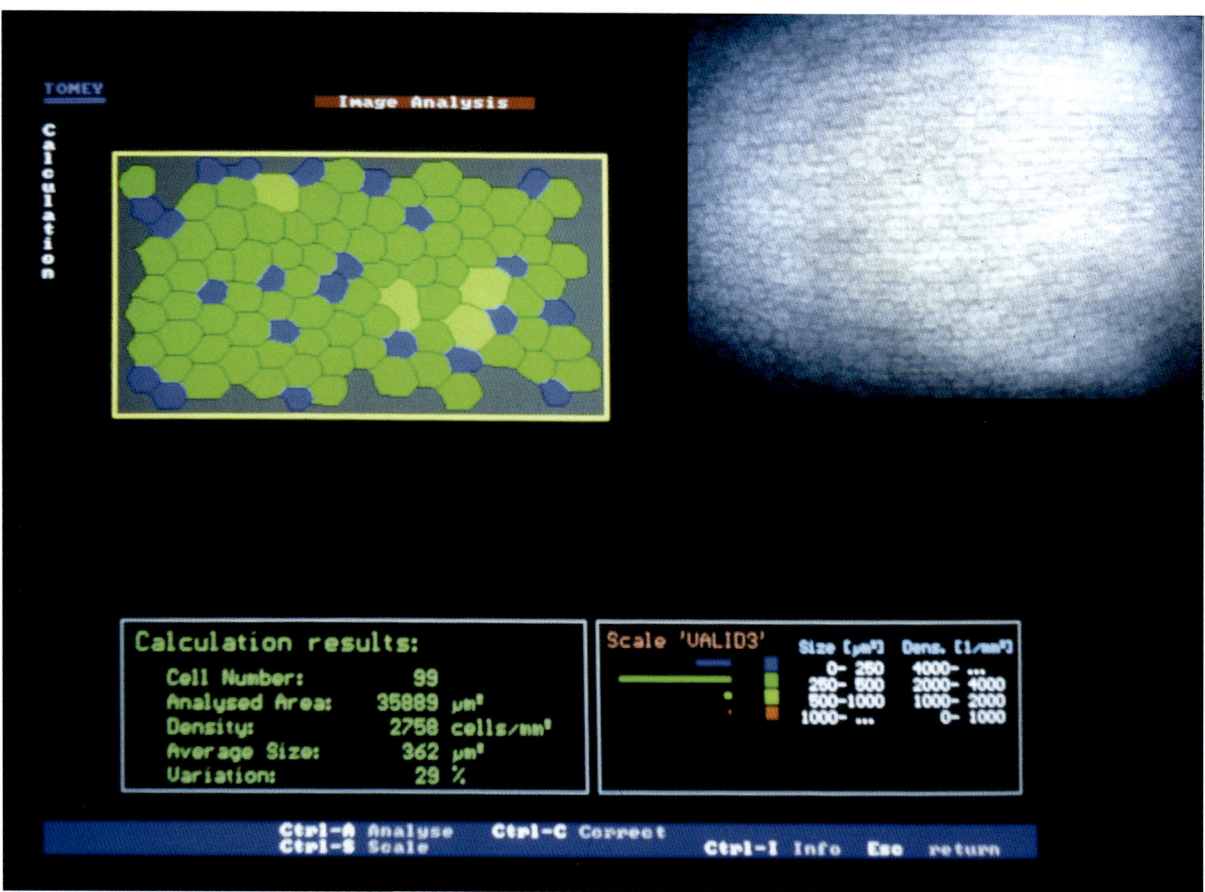

Fig. 2.31

Typical fundus camera images of the right eye. The arteries appear brighter, with a prominent light reflex (longitudinal pale streak). A – Unfiltered image, B – Red-free image (for better visualization of blood vessels), Insert: Enlarged image of the optic cup. Courtesy of Dr. J. Győry.

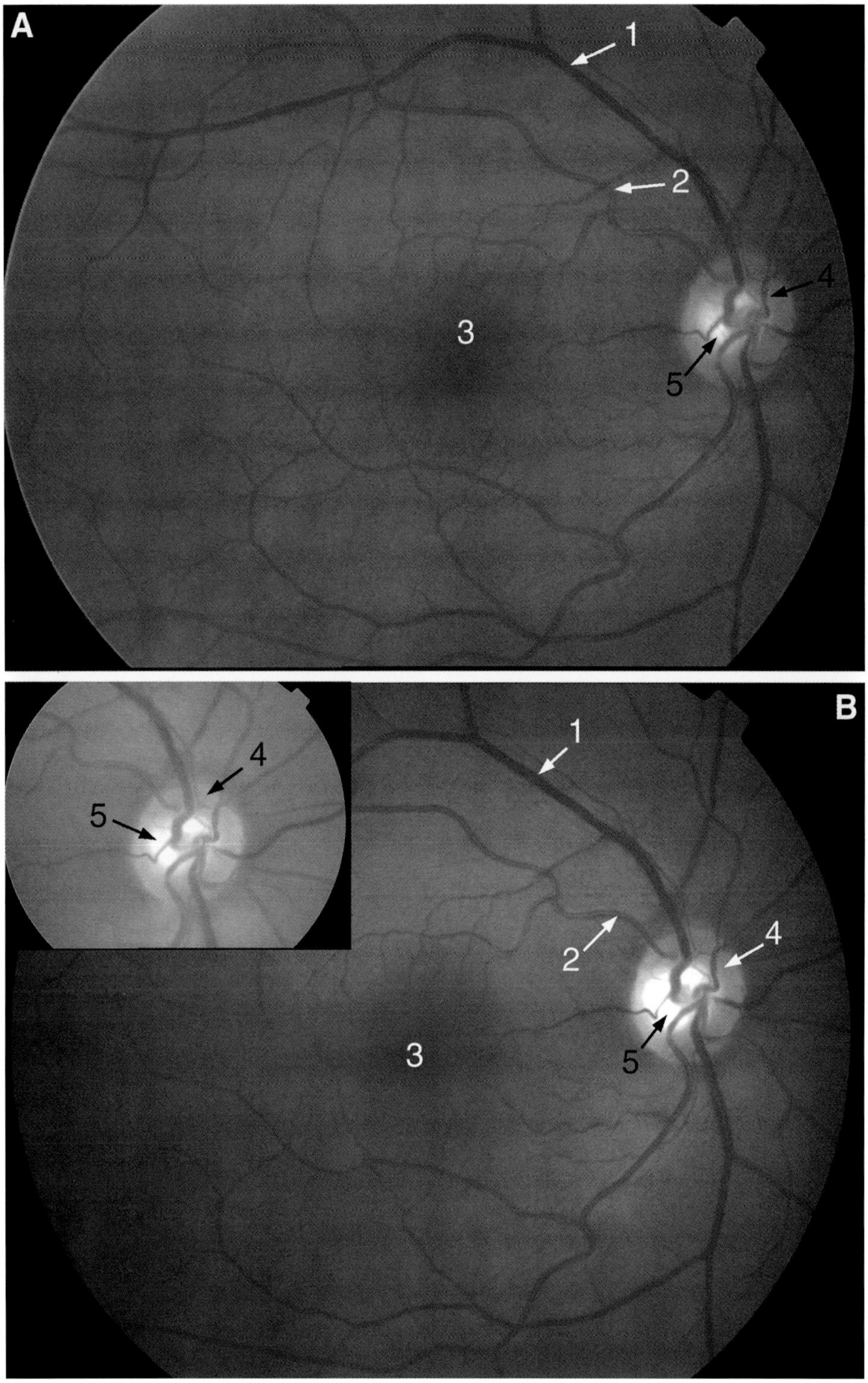

1. Central retinal artery (superior temporal branch)
2. Central retinal vein (superior temporal branch)
3. Fovea
4. Optic cup
5. Lamina cribrosa

Fig. 2.32

Fluorescence angiographic (FLAG) study of a healthy middle-aged individual, using 10% Fluorescein – Na. This low-molecular-weight dye readily passes into the extravascular space and the retinal capillaries. A – Early phase. The main retinal arteries are visible but the small arterial branches, the retinal tissue, and veins are incompletely labeled. B – Mid-venous phase. Tissue filling is near complete. The initial labeling of veins appears on the outer borders owing to the laminar flow of blood. C – Mid-phase. The filling of tissue is now complete. Note the gradual narrowing of the foveal avascular zone (FAZ) owing to the filling of terminal branches. D – Late phase. The dye predominantly labels veins, now completely filled. Courtesy of Dr. J. Györy.

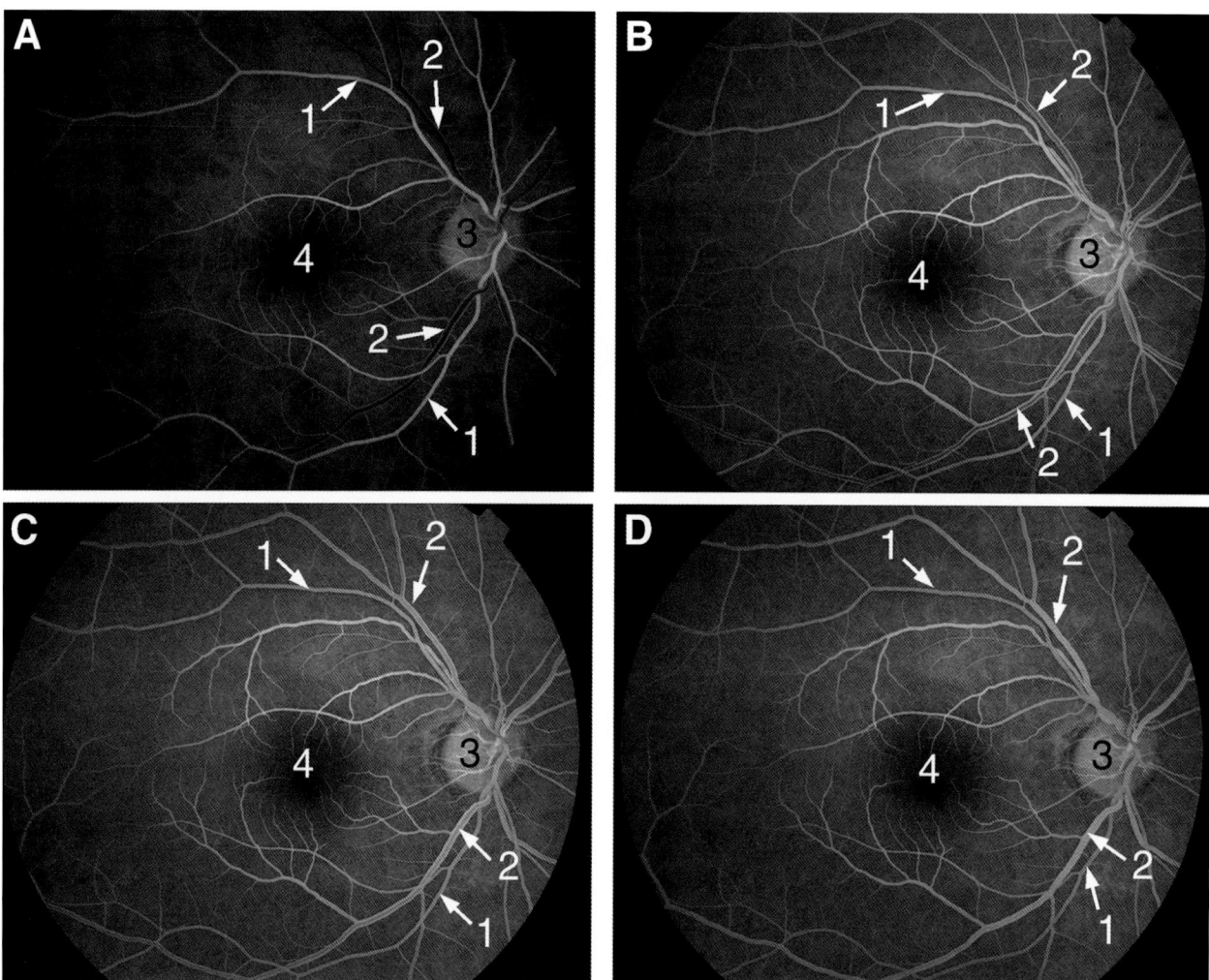

1. Central retinal artery
2. Central retinal vein
3. Optic cup
4. Foveal avascular zone

Fig. 2.33

Fluorescence angiographic (FLAG) studies using Fluorescein – Na (A, B), or Indocyanine Green (ICG) (C, D). The latter is a high-molecular-weight albumin-bound dye, which does not penetrate the retinal capillaries, instead it highlights the larger arterioles of the choroid.

A – Foveal avascular zone limited by capillaries of the central retinal artery, B – In the early phase, the vascular lobules of choroid are visible, C – Marked ICG staining of the choroidal vessels. The branches of the central retinal artery are weakly stained, D – Vortex vein in the peripheral retina. Courtesy of Dr. J. Györy.

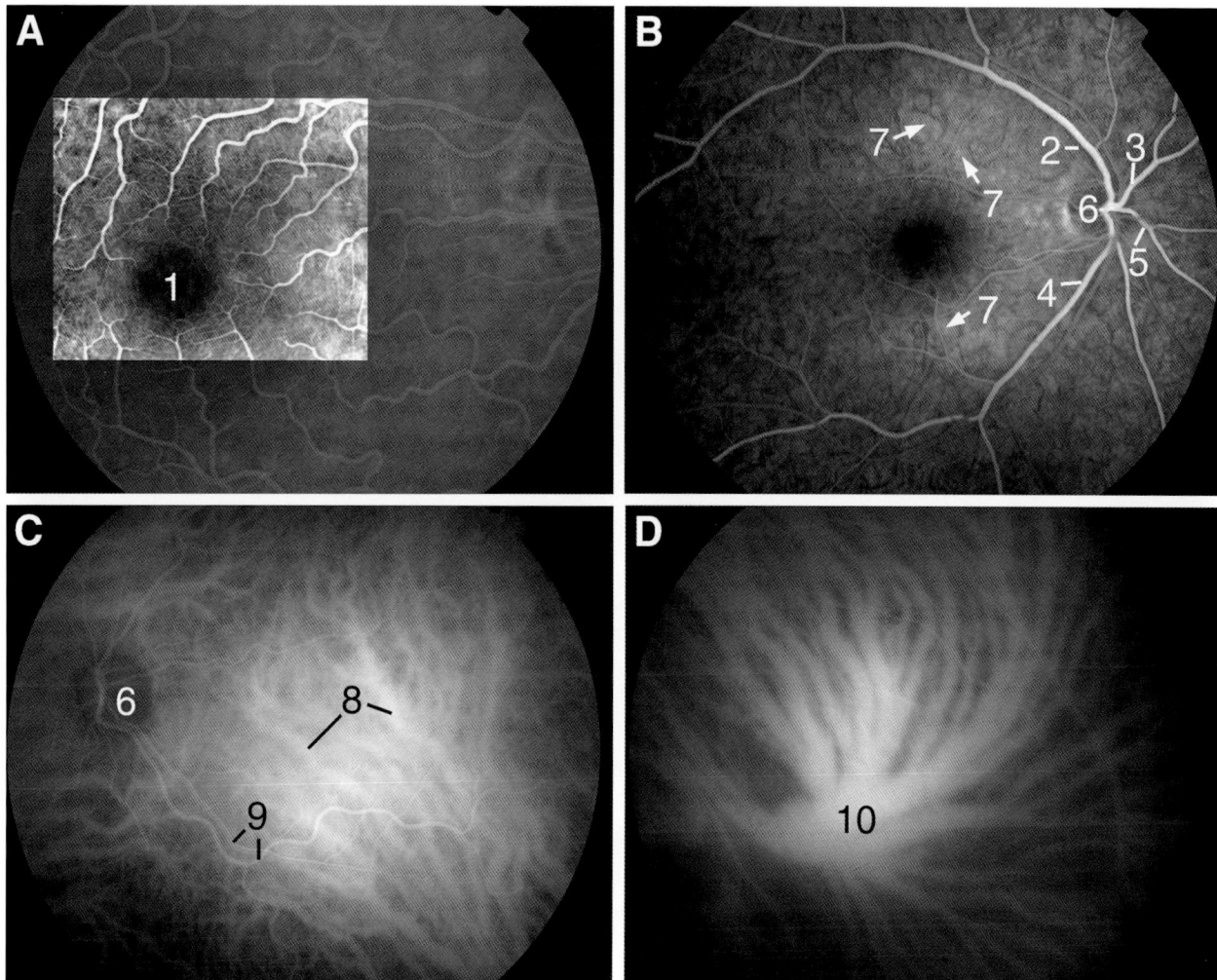

1. Foveal avascular zone
2. Superior temporal branch of central retinal artery
3. Superior nasal branch of central retinal artery
4. Inferior temporal branch of central retinal artery
5. Inferior nasal branch of central retinal artery
6. Optic cup
7. Vascular lobules of choroid
8. Choroidal vessels
9. Central retinal vessels
10. Vortex vein

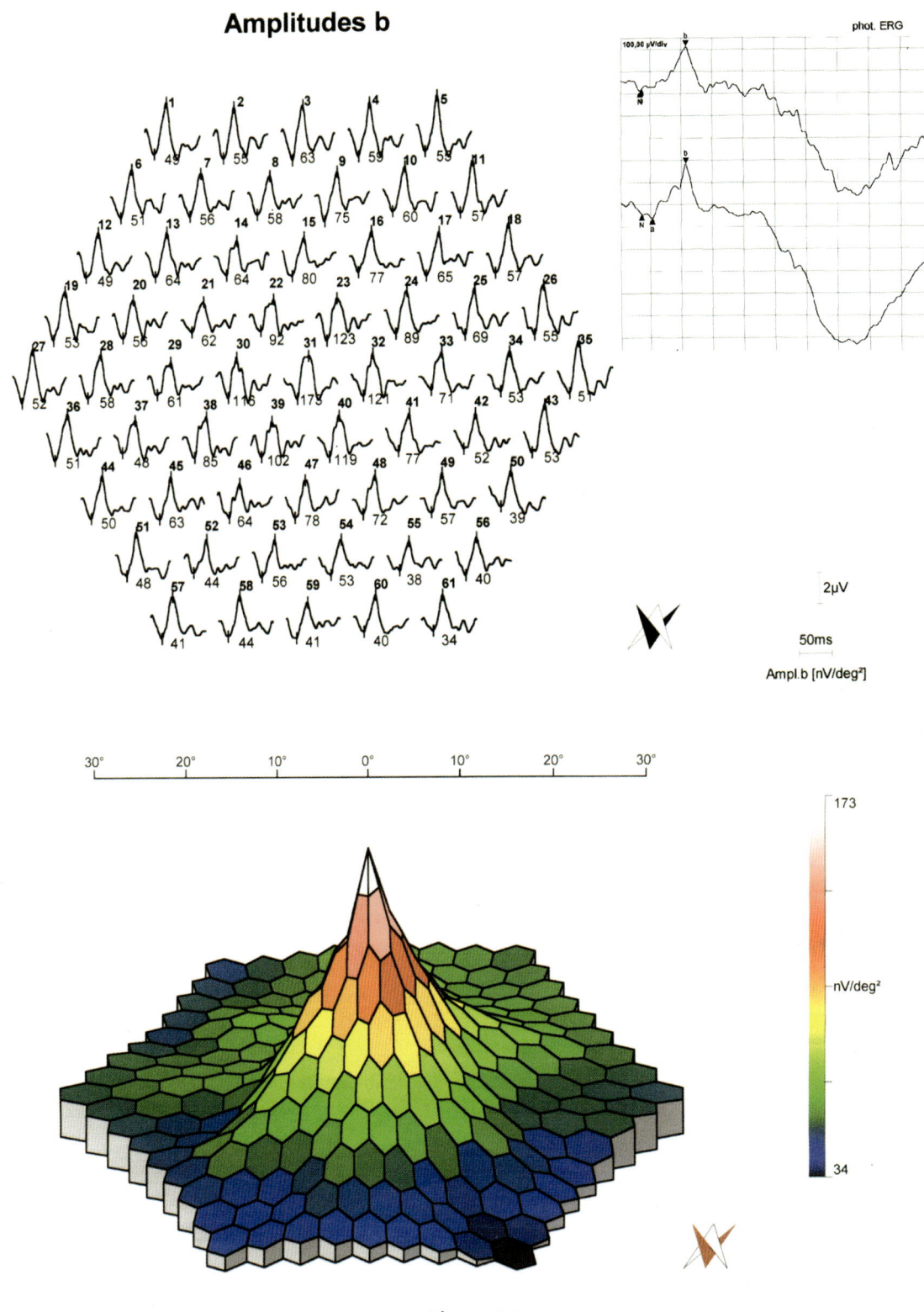

Fig. 2.34

Multifocal electroretinography (ERG). In this examination the summation potentials of retinal photoreceptors, bipolars, horizontal and amacrine cells (but not ganglion cells) are recorded from the posterior retina in response to a systematic random array of flashing lights. Each flash elicits a signal only from the appropriate retinal field. One typical pair of ERG signals is shown in the top right panel. The amplitude of signals (i.e., the intensity of bioelectric response to the light stimulus) is greatest in or around the fovea, as shown by the grid of local response curves (top, see the amplitude values under each curve) and also by the color-coded three-dimensional diagram (below). Courtesy of Á. Farkas.

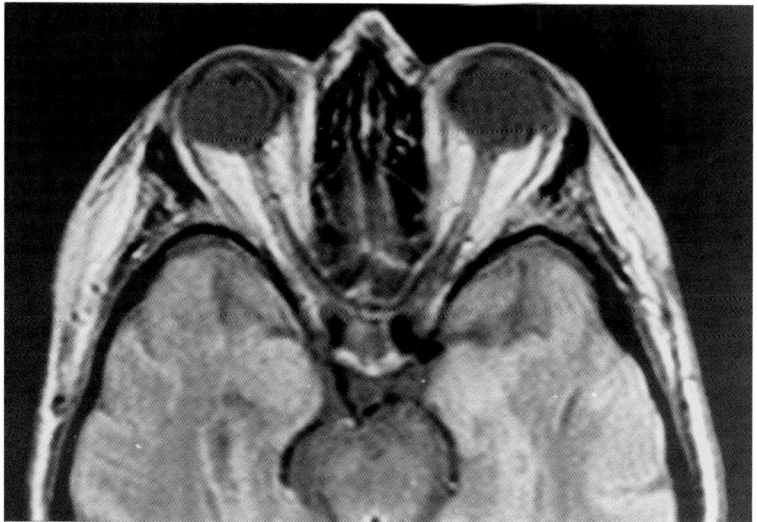

Fig. 2.35

Axial MR image of the orbital cavity.

1. Dorsum of nose
2. Superior tarsal plate
3. Medial palpebral ligament
4. Lateral palpebral ligament
5. Nasal septum
6. Medial check ligament
7. Tendon of medial rectus
8. Globe
9. Tendon of lateral rectus
10. Lateral check ligament
11. Zygomatic arch
12. Temporal fossa
13. Optic nerve
14. Ethmoidal air cells
15. Orbital fat
16. Temporal lobe of brain
17. Optic chiasma
18. Anterior clinoid process
19. Sella turcica (hypophyseal fossa)
20. Dorsum sellae
21. Parahippocampal gyrus
22. Interpeduncular fossa
23. Cerebral aqueduct

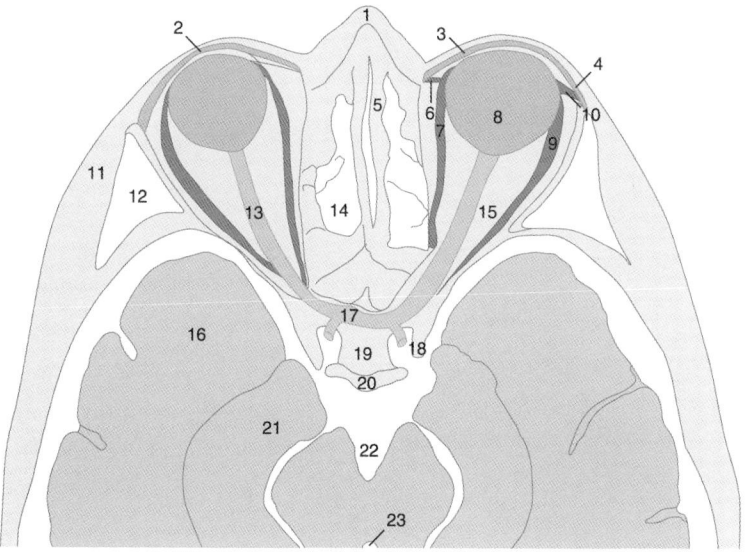

ATLAS—EYE

Fig. 2.36
Parasagittal MR images of the orbital cavity.

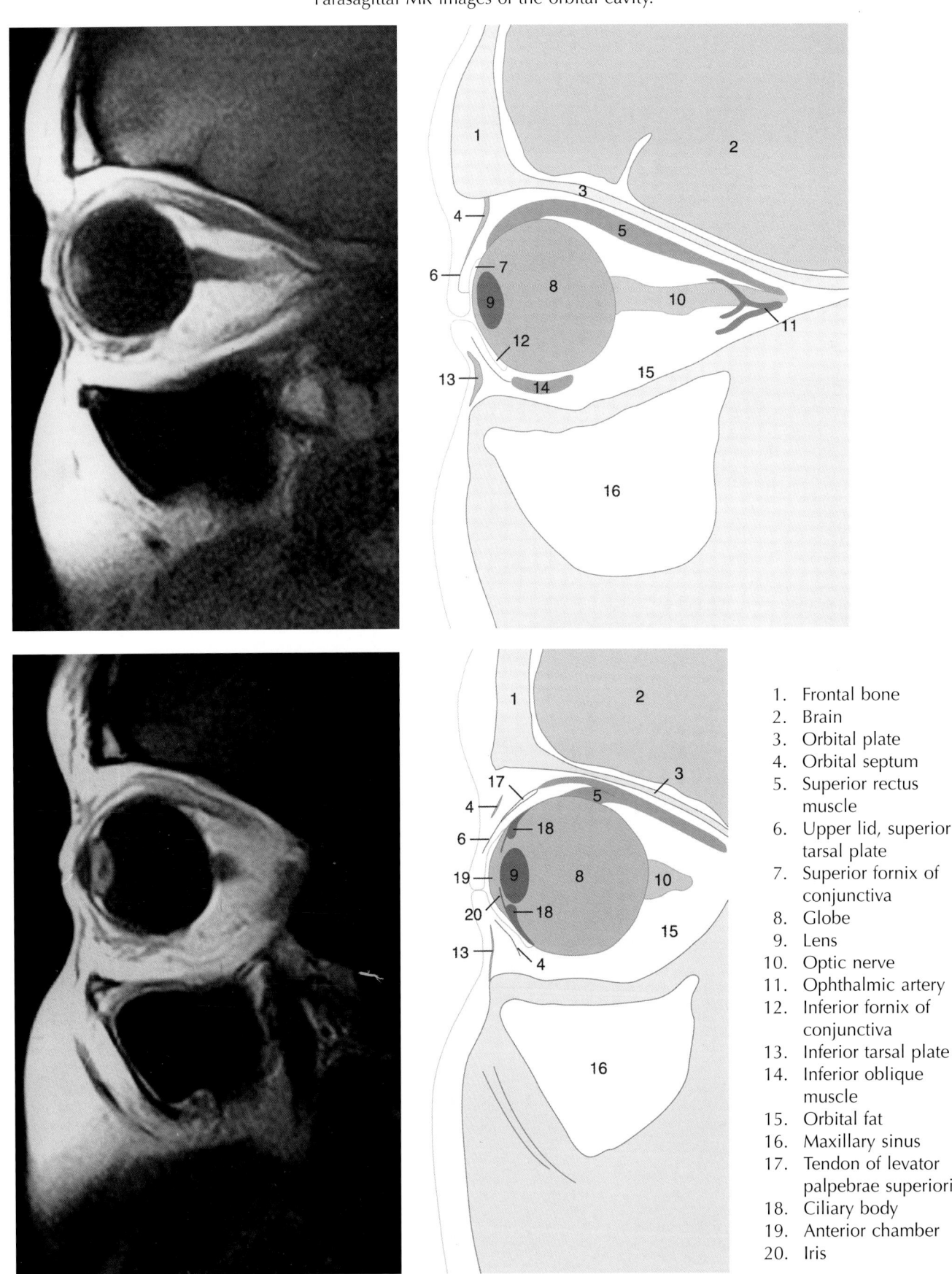

1. Frontal bone
2. Brain
3. Orbital plate
4. Orbital septum
5. Superior rectus muscle
6. Upper lid, superior tarsal plate
7. Superior fornix of conjunctiva
8. Globe
9. Lens
10. Optic nerve
11. Ophthalmic artery
12. Inferior fornix of conjunctiva
13. Inferior tarsal plate
14. Inferior oblique muscle
15. Orbital fat
16. Maxillary sinus
17. Tendon of levator palpebrae superioris
18. Ciliary body
19. Anterior chamber
20. Iris

Fig. 2.37

Coronal MR image of the right orbital cavity.

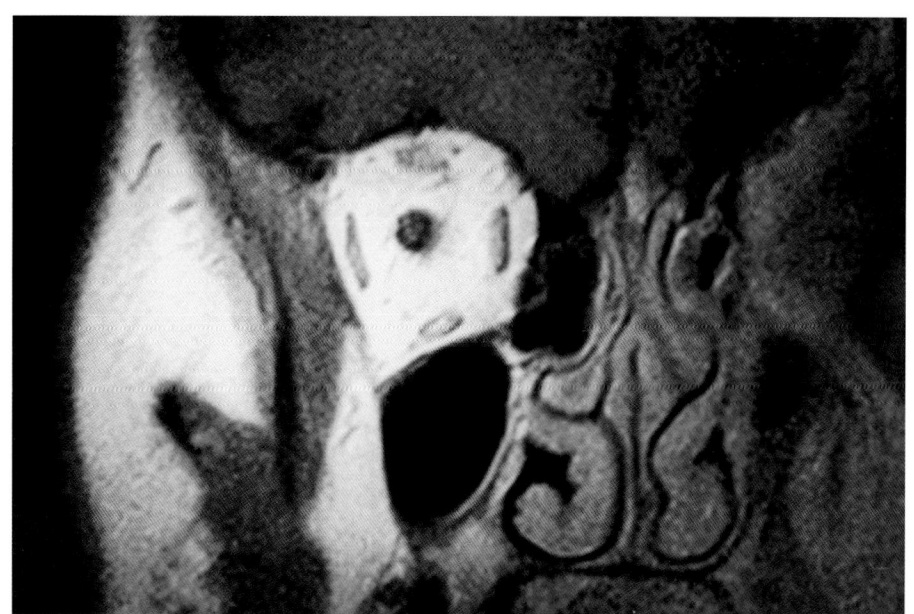

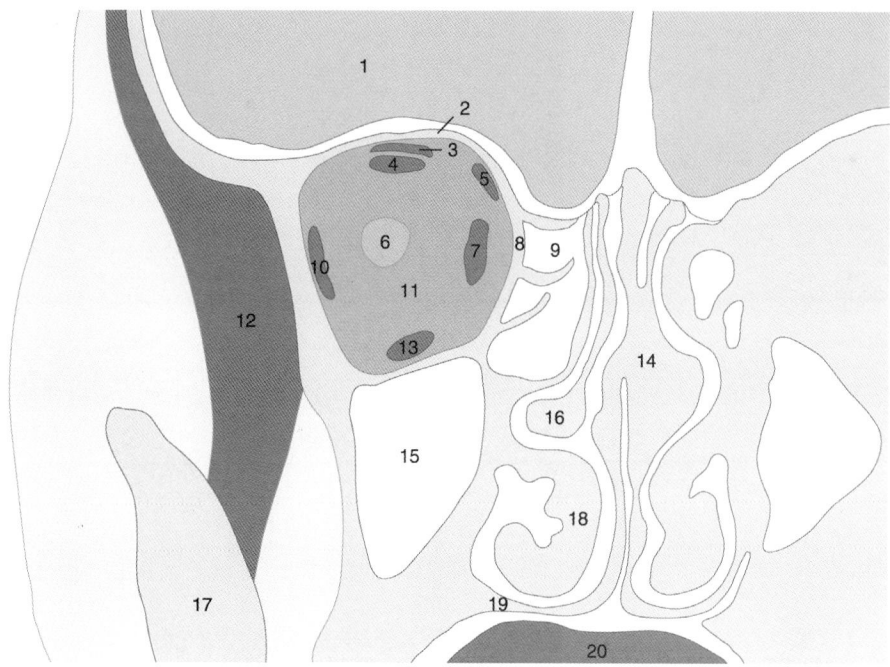

1. Frontal lobe of brain
2. Orbital plate of frontal bone
3. Levator palpebrae superioris muscle
4. Superior rectus muscle
5. Superior oblique muscle
6. Optic nerve
7. Medial rectus muscle
8. Orbital lamina of ethmoid
9. Ethmoidal air cells
10. Lateral rectus muscle
11. Orbital fat
12. Temporalis muscle
13. Inferior rectus muscle
14. Nasal septum
15. Maxillary sinus
16. Middle nasal concha
17. Mandible
18. Inferior nasal concha
19. Palate
20. Tongue

ATLAS—EYE

ATLAS PLATES III (HISTOLOGY)

Fig. 2.38

Histology of cornea. A–C light micrographs, Azan staining, D – electronmicrograph. Courtesy of I. Süveges and J. Tóth (D).

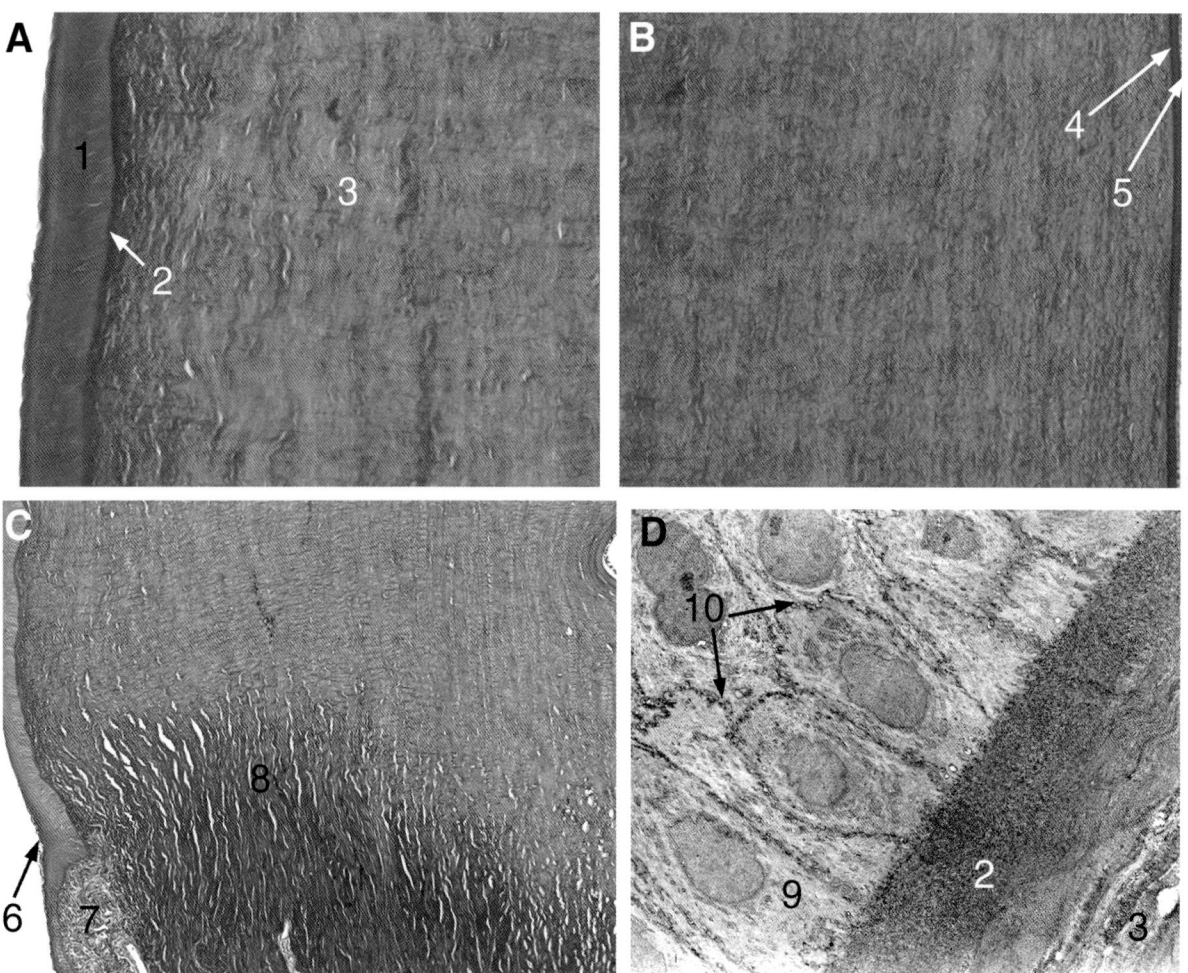

1. Stratified squamous non-keratinizing epithelium
2. Bowman's layer
3. Substantia propria
4. Descemet's membrane
5. Endothelium
6. Corneal limbus
7. Conjunctiva
8. Blood vessels terminating in the sclerolimbal junction zone
9. Epithelial cell of the basal layer
10. Intercellular junctions

Fig. 2.39

The anterior segment of the eye. Histological section, HE staining.

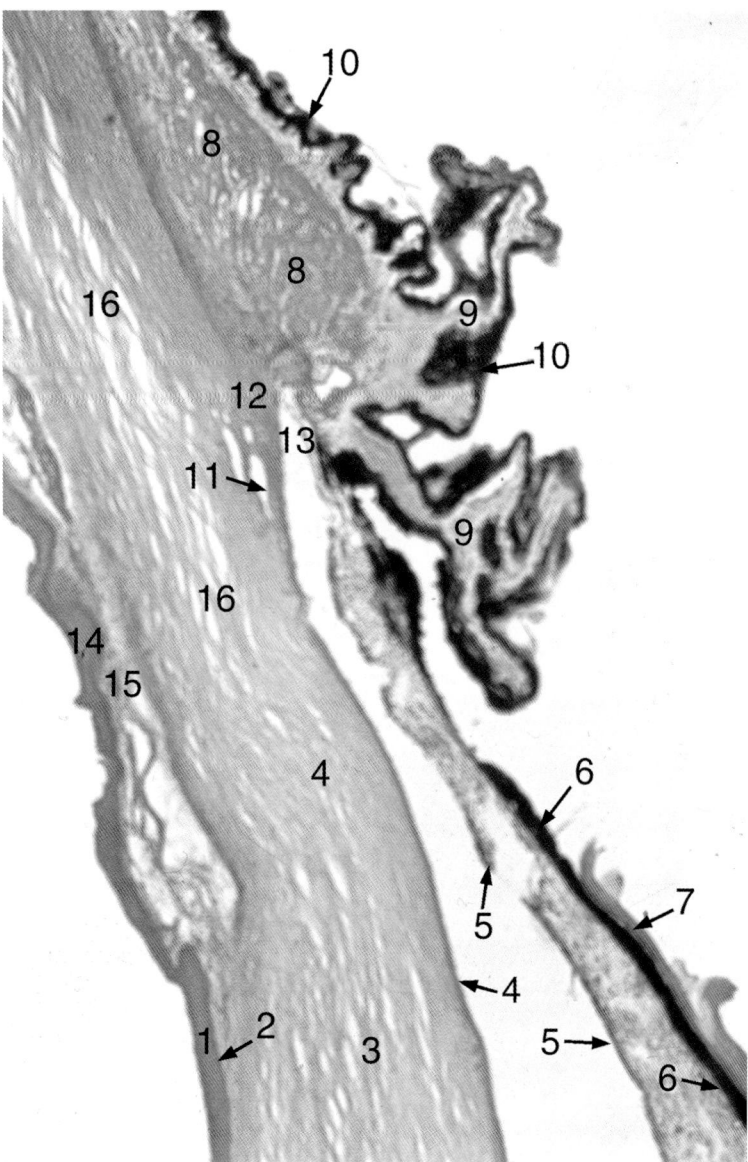

1. Epithelium of cornea
2. Bowman's layer of cornea
3. Substantia propria of cornea
4. Endothelium of cornea bordering the anterior chamber
5. Anterior border of iris
6. Posterior epithelium of iris
7. Fragment of lens attached to the iris
8. Ciliary body, pars plana
9. Ciliary processes (pars plicata)
10. Pigmented epithelium of the blind retina
11. Schlemm's canal
12. Pectinate ligament of iris (uveal trabeculae)
13. Iridocorneal (drainage) angle
14. Conjunctiva
15. Episcleral tissue
16. Sclera

Fig. 2.40

Histology of the iris. A, B – light micrographs, Azan staining, C – semithin section, toluidine blue staining, D – scanning electronmicrograph of the posterior surface. Courtesy of I. Süveges and J. Tóth (C).

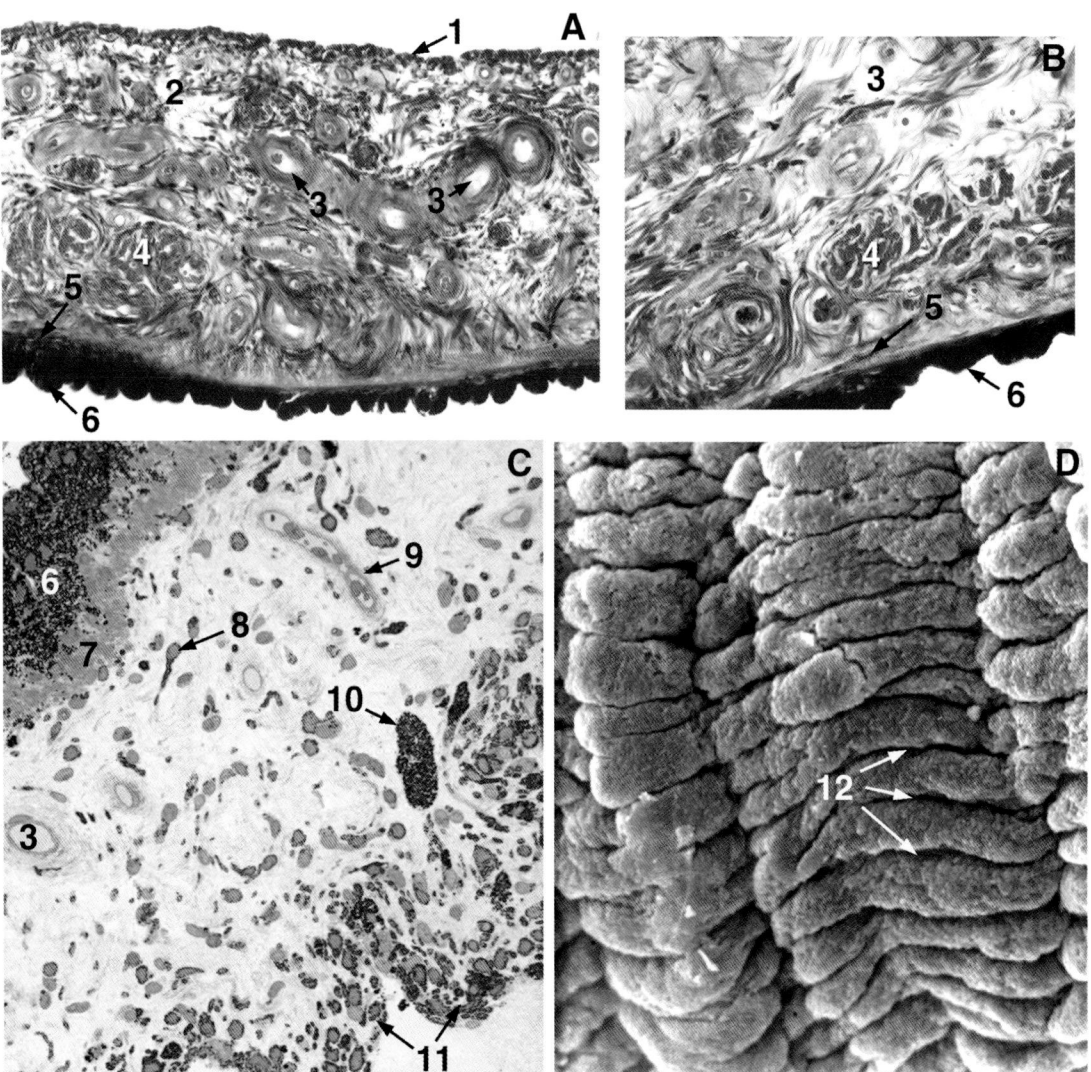

1. Anterior border
2. Stroma
3. Arteriole
4. Sphincter pupillae muscle
5. Fibers of the dilator pupillae muscle
6. Posterior pigmented epithelium
7. Anterior pigmented epithelium
8. Melanocyte in the stroma
9. Capillary in the stroma
10. Clump cell
11. Crypt of the anterior border covered by fibroblasts and melanocytes
12. Contraction furrows of the posterior epithelium

Fig. 2.41

Histology of the pars plicata of the ciliary body. A – light micrograph of ciliary process, Azan staining, B – semithin section of ciliary process, toluidine blue staining, C – transmission electron micrograph, D – scanning electron micrograph (the lens was removed). Courtesy of I. Süveges and J. Tóth (B, C).

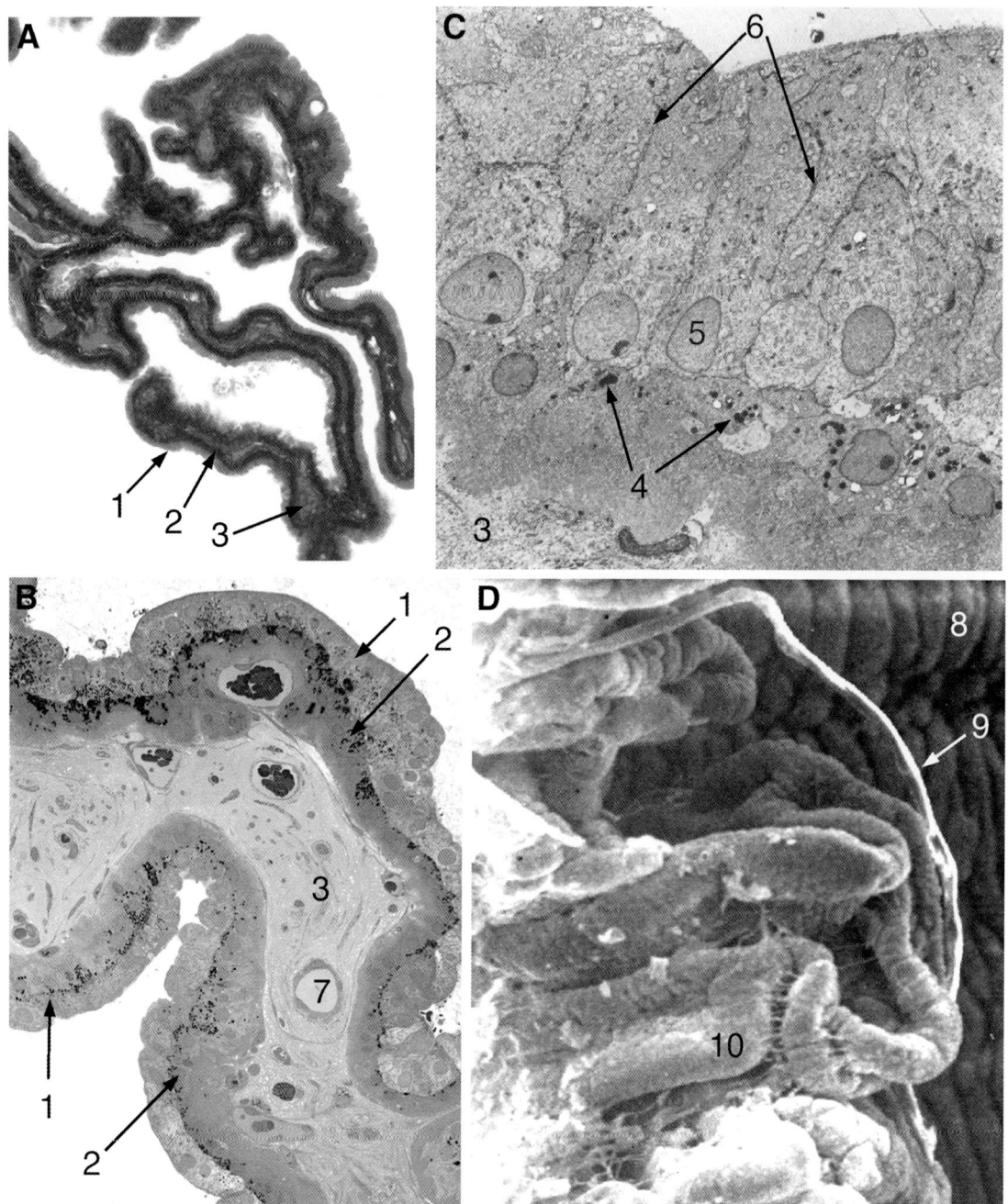

1. Non-pigmented epithelium
2. Pigmented epithelium
3. Stroma
4. Melanin pigment granules
5. Nucleus of non-pigmented epithelial cell
6. Intercellular borders of non-pigmented epithelial cells
7. Capillary
8. Posterior surface of iris
9. Zonule fiber
10. Ciliary process

ATLAS—EYE

Fig. 2.42

Histology of the drainage angle. A – light micrograph, Azan staining, B – scanning electron micrograph, C, D – transmission electron micrographs. Courtesy of I. Süveges and J. Tóth (C, D).

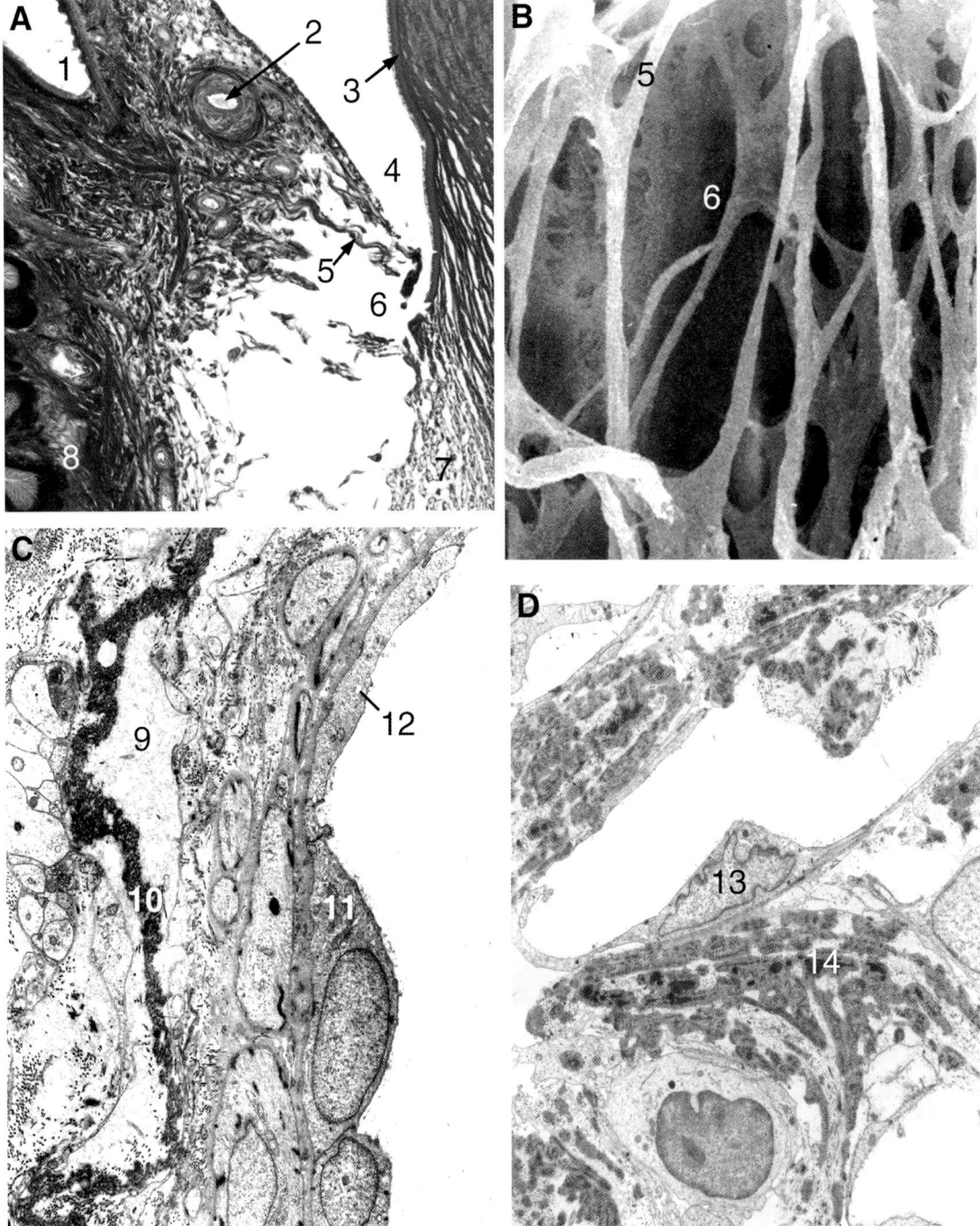

1. Posterior chamber
2. Major arterial circle of iris
3. Cornea
4. Angle recess
5. Uveal trabecula
6. Spaces of Fontana
7. Corneoscleral trabeculae
8. Ciliary body
9. Juxtacanalicular connective tissue
10. Collagen fibers
11. Endothelial lining of Schlemm's canal
12. Vacuoles
13. Trabecular cell bordering the uveal trabecula
14. Collagen fibers in the core of trabecula

Fig. 2.43

Light and electron micrographs (EMs) of the human lens. A – Light micrograph, B – Low-power EM, C – high-power EM. Courtesy of I. Süveges and J. Tóth (B, C).

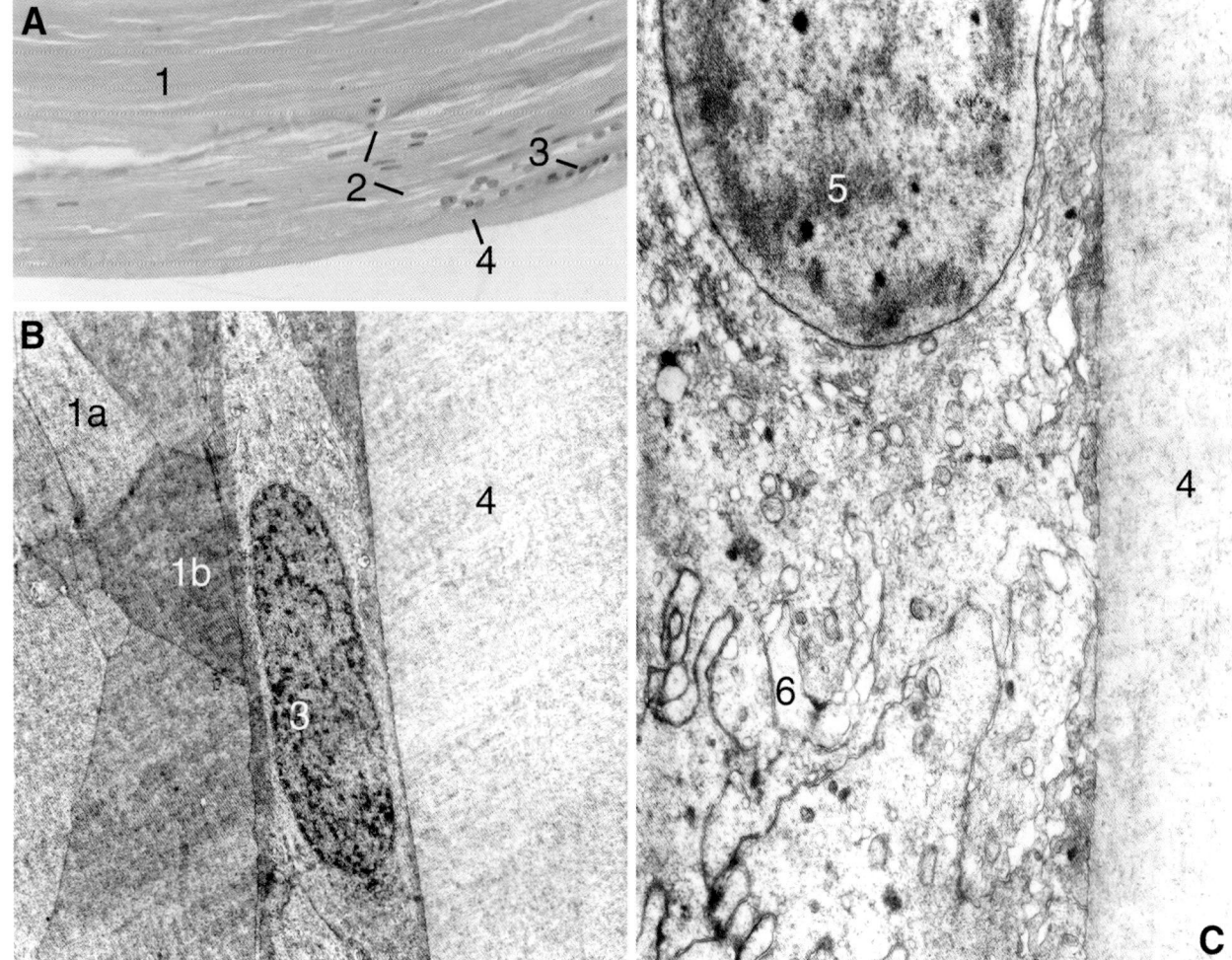

1. Lens fibers
   a. Lens fiber in longitudinal section
   b. Lens fiber in cross section
2. Lens bow
3. Anterior epithelium
4. Lens capsule corresponding to modified basement membrane
5. Nucleus of lens epithelium
6. Endoplasmic reticulum

Fig. 2.44

Histological section demonstrating the retinal layers together with the underlying choroid and sclera. HE staining.

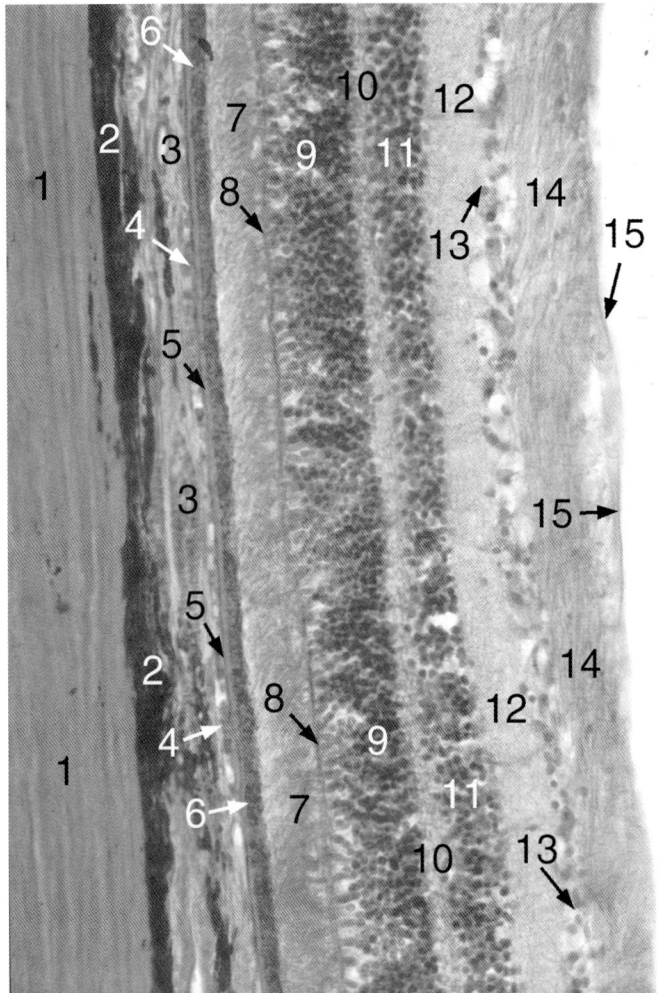

1. Sclera
2. Suprachoroidal layer (lamina fusca sclerae)
3. Vascular layer of choroid
4. Choriocapillaris
5. Bruch's membrane
6. Pigment epithelium of retina (RPE)
7. Rods and cones
8. External limiting membrane (ELM)
9. Outer nuclear layer (ONL)
10. Outer plexiform layer (OPL)
11. Inner nuclear layer (INL)
12. Inner plexiform layer (IPL)
13. Ganglion cell layer (GCL)
14. Optic fiber layer (OFL)
15. Internal limiting membrane (ILM)

Fig. 2.45

Semithin section of the retina, toluidine blue staining. Inset: scanning electron micrograph of the retinal layers taken through a tear in the tissue.

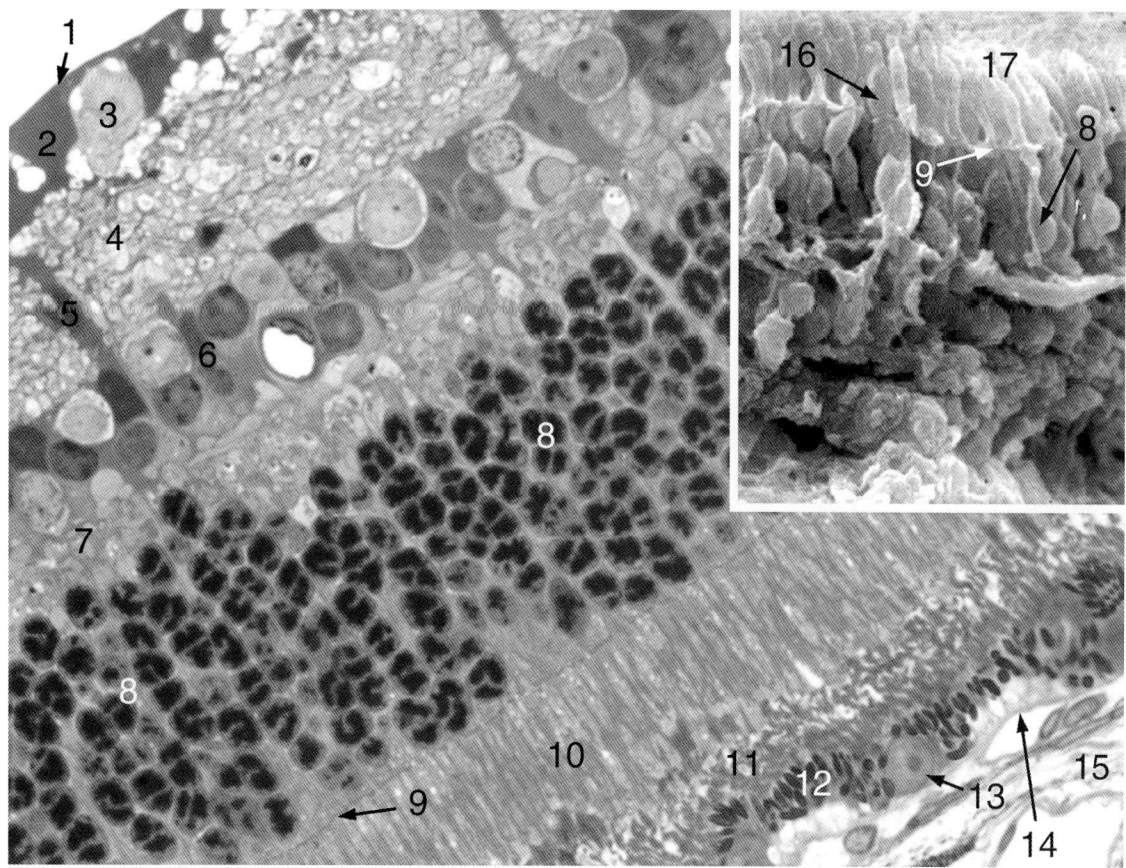

1. Internal limiting membrane (ILM)
2. Optic fiber layer (OFL)
3. Retinal ganglion cell in the ganglion cell layer (GCL)
4. Inner plexiform layer (IPL)
5. Radial process of Müller cell
6. Inner nuclear layer (INL)
7. Outer plexiform layer (OPL)
8. Outer nuclear layer (ONL)
9. External limiting membrane (ELM)
10. Rods and cones (inner segments)
11. Rods and cones (outer segments)
12. Pigment containing processes of RPE
13. Nucleus of RPE
14. Bruch's membrane
15. Choriocapillaris
16. Cone inner segment
17. Rod inner segment

Fig. 2.46

Histology of the retinal pigment epithelium. A – light micrograph of flat-mount unstained specimen, B – electron micrograph (courtesy of P. Röhlich), C, D – scanning electron micrographs following removal of the other retinal layers.

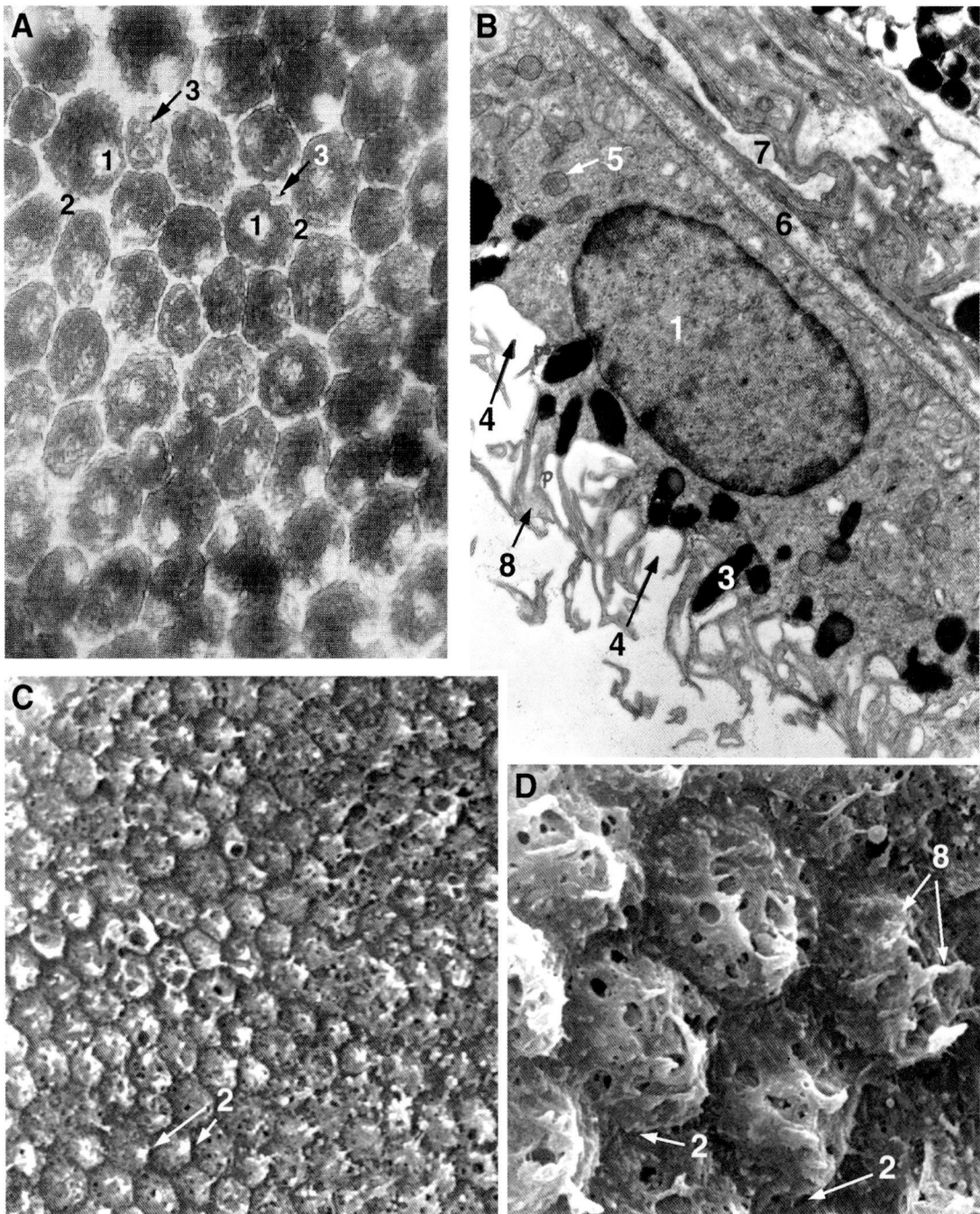

1. Nucleus
2. Intercellular space
3. Melanin granules
4. Spaces for the outer segments of rods and cones
5. Mitochondrion
6. Bruch's membrane
7. Choriocapillaris
8. Surface processes engulfing the outer segments of rods and cones

Fig. 2.47

Electron micrographs of retinal photoreceptors of the macaque monkey. A – Foveal cone, magnification x35,000. The structure resembles that of rods but the discs of the outer segment are attached to the plasma membrane (arrow); B – Peripheral cone, magnification x25,000; C – Longitudinal section from the outer segment of a rod. Dark-adapted animal, magnification x210,000. The discs consist of double membranes and their average thickness is 12.5 nm. Courtesy of P. Röhlich.

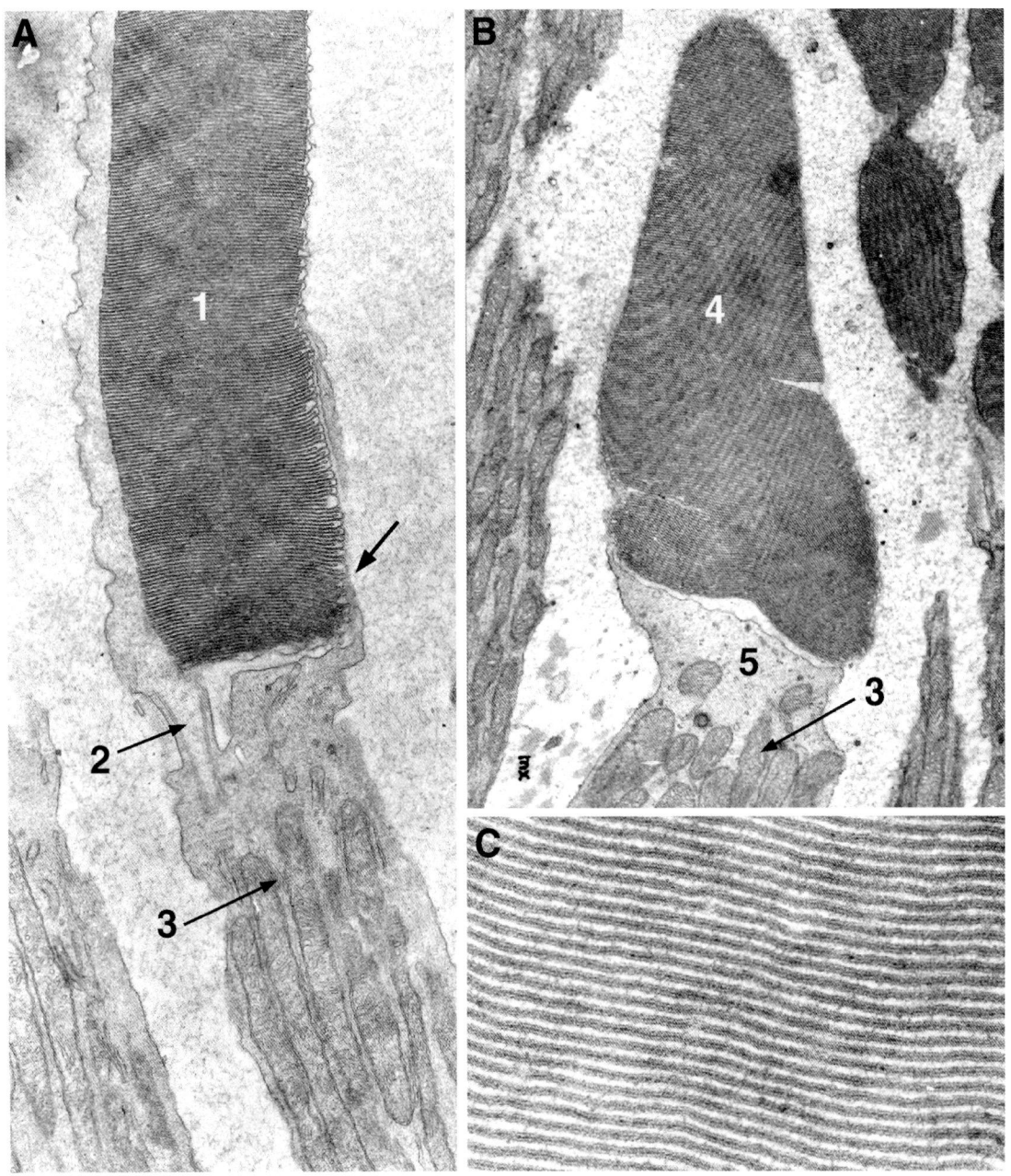

1. Outer segment of foveal cone with discs
2. Cilium in the isthmus (connecting segment)
3. Mitochondria in the ellipsoid of inner segment
4. Discs in the outer segment of peripheral cone
5. Isthmus

Fig. 2.48

Electron micrographs of the retina of the macaque monkey. A – Tangential section at the level of the outer segments; B – Tangential section at the level of inner segments; C – Perpendicular section from the level of the external limiting membrane (the retina is detached from the pigment epithelium); D – Section from the external plexiform layer. Courtesy of P. Röhlich (A, B) and I. Süveges and J. Tóth (C, D).

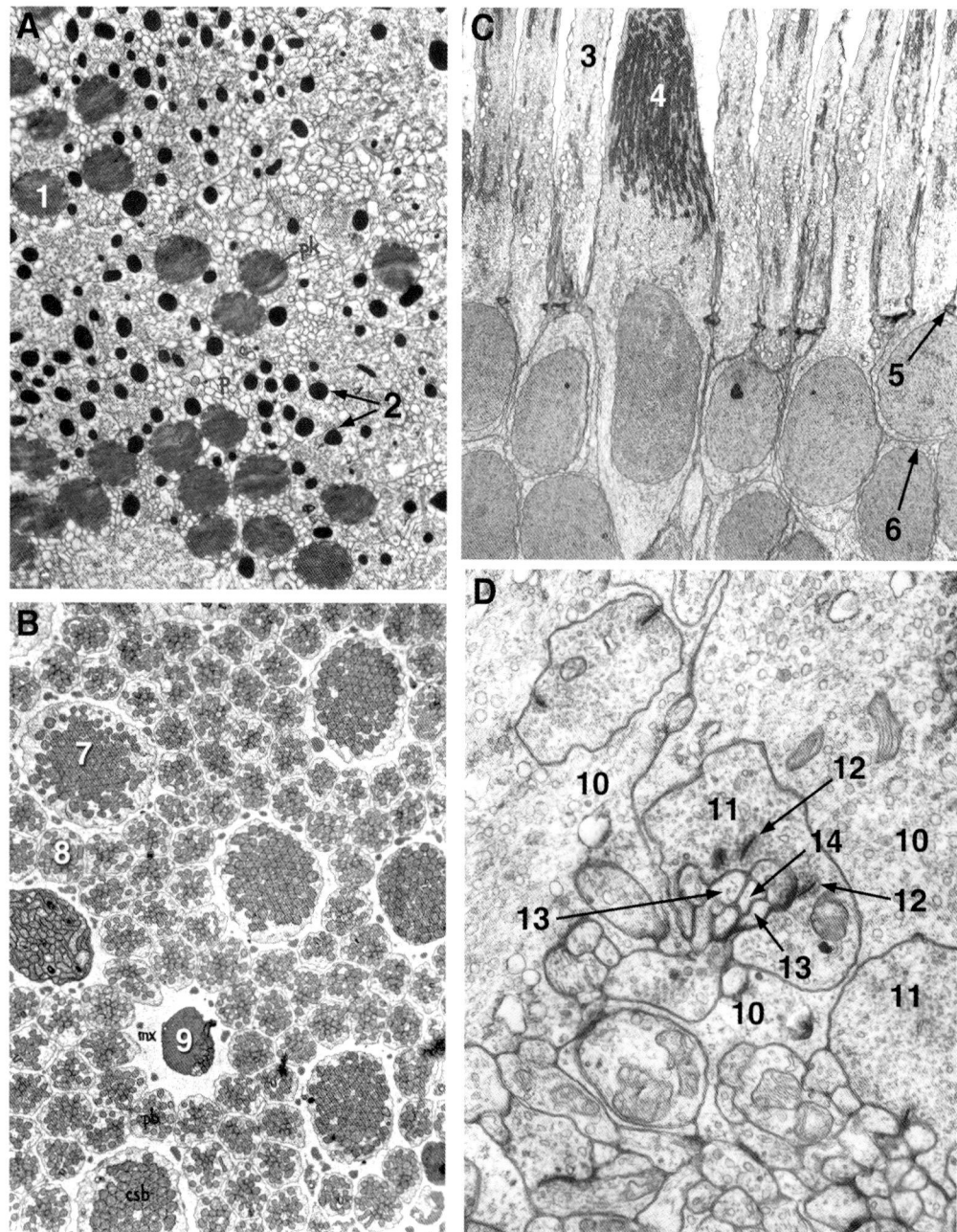

1. Rod outer segment
2. Pigment granule in the process of pigment epithelium
3. Inner segment of rod
4. Innner segment of cone
5. Zonulae adherentes of the external limiting membrane
6. Nuclei and cell bodies in the outer nuclear layer
7. Cone inner segment
8. Rod inner segment
9. Outer segment of a cone (presumably a blue-sensitive one because these are usually shorter than the other photoreceptors).
10. Processes of Müller cells
11. Cone pedicles
12. Ribbon synapses
13. Horizontal cell process forming a synaptic triad in the arciform density
14. Bipolar cell dendrite forming a synaptic triad in the arciform density

Fig. 2.49

Photoreceptor mosaic in the primate retina. A – Consecutive tangential sections derived from the perifoveal region of the monkey retina were reacted with antibodies specific for blue-sensitive cones (OS-2), red/green-sensitive cones (COS-1) and rods (AO), respectively. Identical areas of the sections were photographed to show the staining pattern of each cone. Note the complementarity of the three receptor populations. Each receptor is stained by only one antibody. B – Superposition picture of the immunoreactions with digital color coding. Each immunopositive element was marked by a different color in the digital images. The images of consecutive sections were then merged to show the photoreceptor distribution. Blue and red/green-sensitive cones are rendered blue and red, respectively. Rods are coded green. Owing to the high degree of molecular homology between red- and green-sensitive cones, these two cone types cannot be distinguished. Courtesy of Á. Szél.

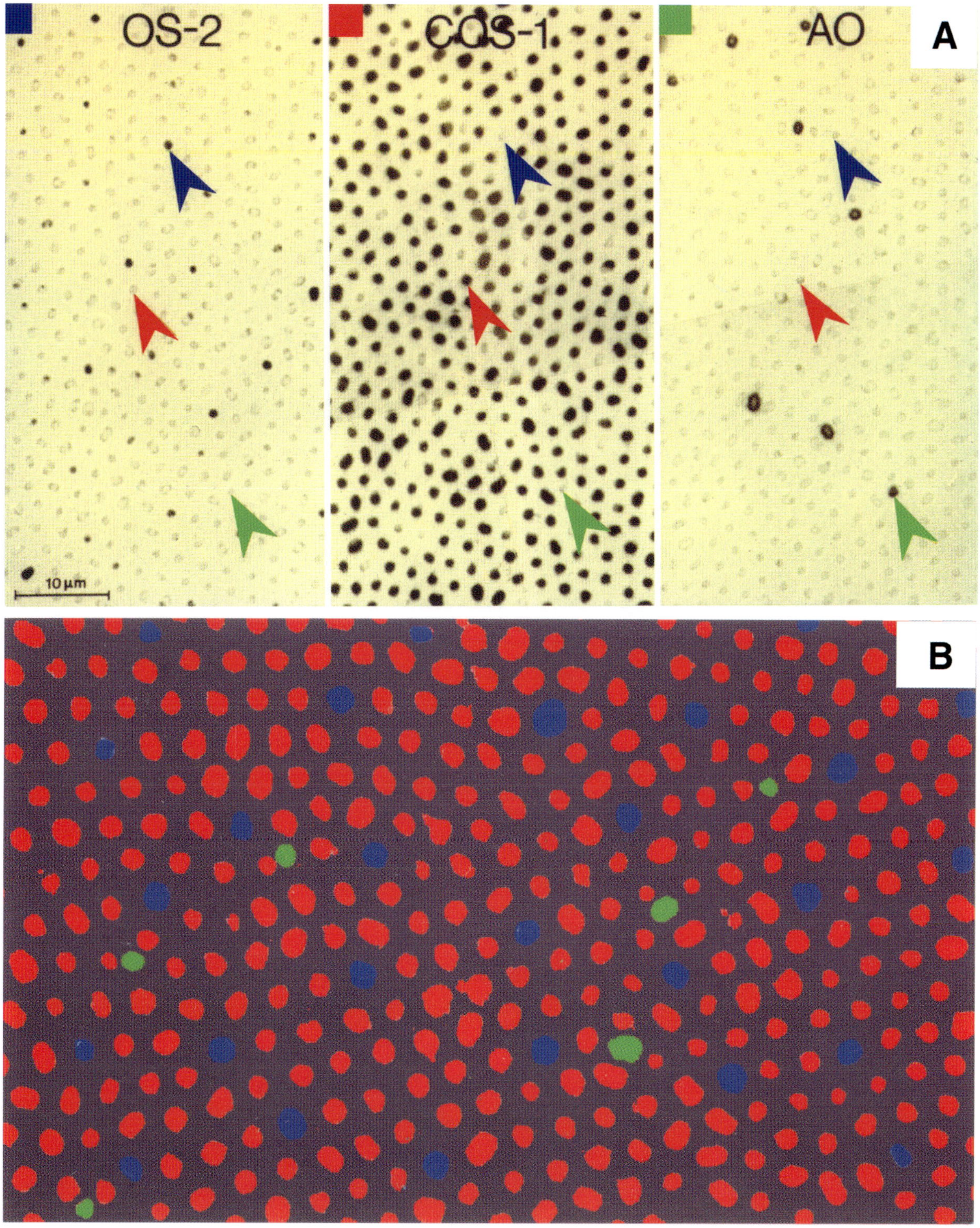

ATLAS—EYE

Fig. 2.50

Histology of the extraocular muscles and the episcleral space. Azan staining.

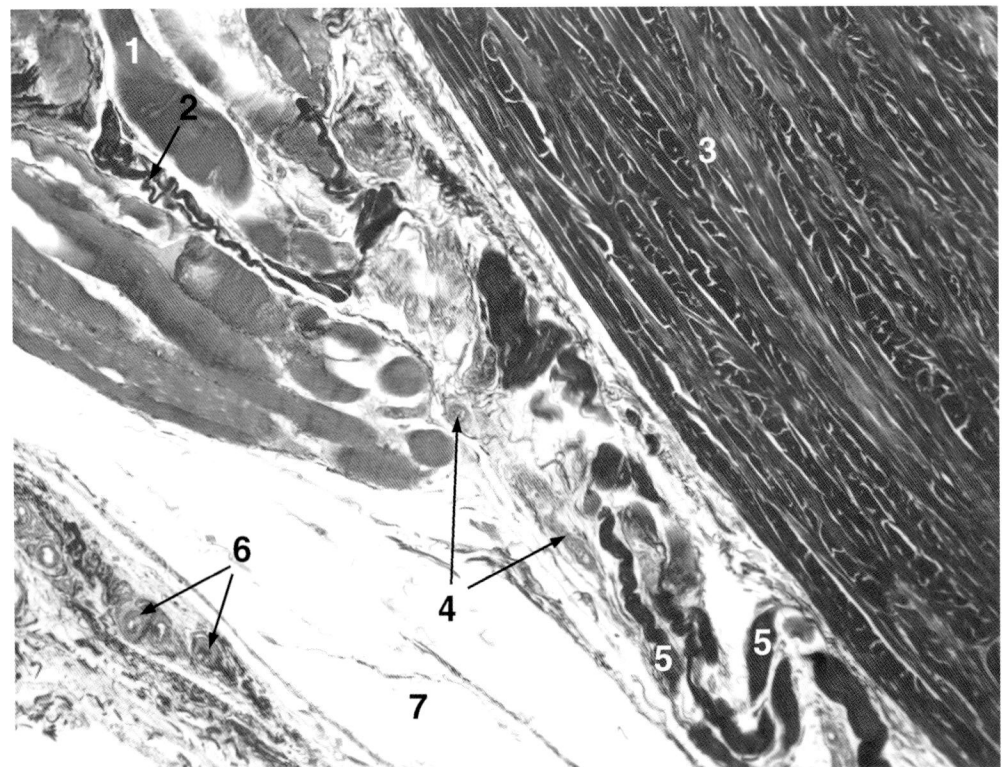

1. Striated muscle fibers of rectus muscle
2. Elastic fibers
3. Sclera (dense connective tissue)
4. Branches of anterior ciliary artery
5. Collagen fibers (tendon of rectus muscle)
6. Episcleral branches of anterior ciliary artery
7. Loose areolar tissue of episcleral space

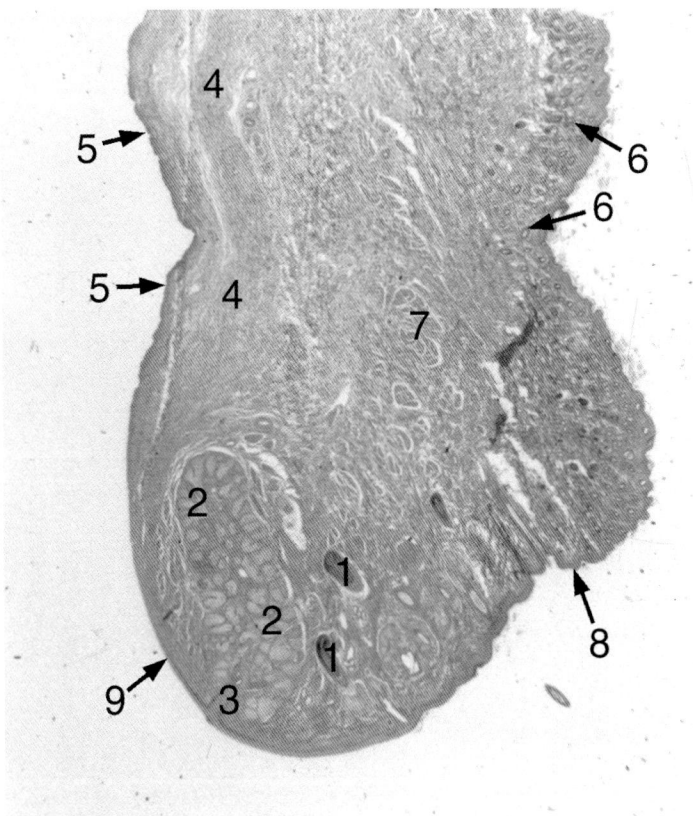

Fig. 2.51

Histology of the eyelid. Low-power light micrograph, HE staining

1. Eyelashes
2. Meibomian gland
3. Excretory duct of Meibomian gland in tangential section
4. Tarsal plate (dense fibro-elastic connective tissue)
5. Palpebral conjuctiva (stratified squamous non-keratinized epithelium)
6. Rudimentary hair follicles
7. Orbicularis oculi muscle, palpebral part
8. Anterior palpebral limbus
9. Posterior palpebral limbus (mucocutaneous border)

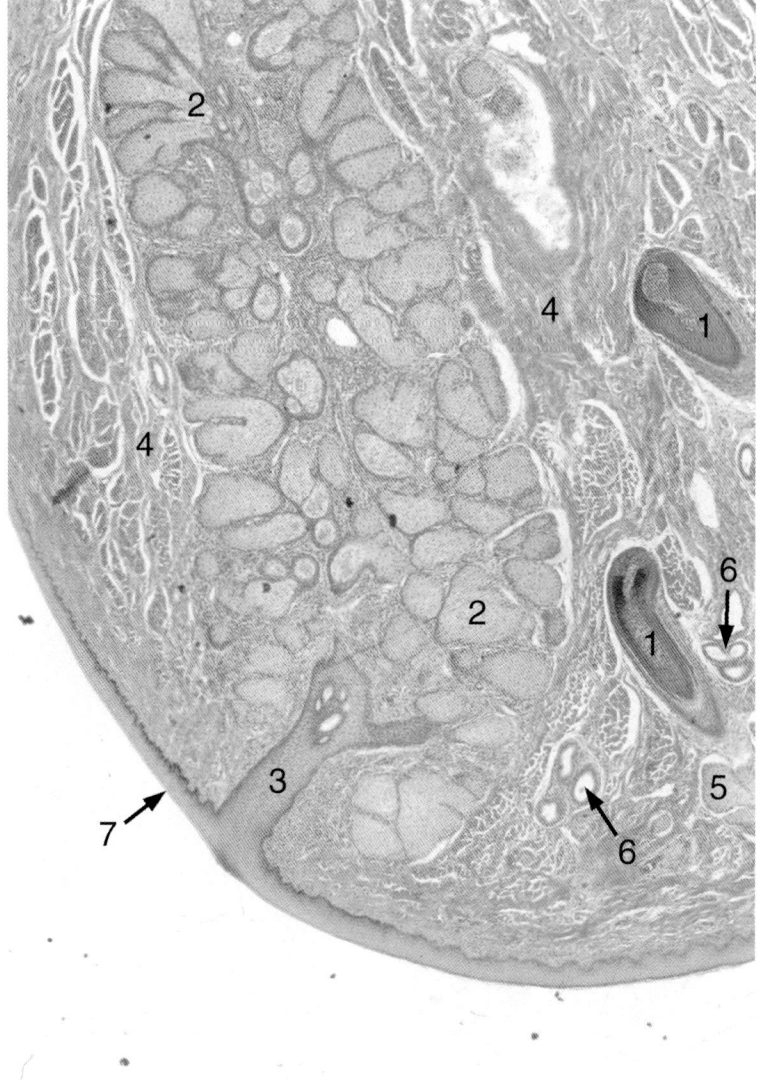

Fig. 2.52

Histology of the eyelid. High-power light micrograph, HE staining

1. Eyelashes
2. Meibomian gland
3. Excretory duct of Meibomian gland in tangential section
4. Tarsal plate (dense fibro-elastic connective tissue)
5. Sebaceous gland (of Zeiss)
6. Glands of Moll
7. Posterior palpebral limbus (mucocutaneous border)

Fig. 2.53

Histology of the lacrimal gland. Light micrographs of different regions (A,B), HE staining. Inset: Azan staining.

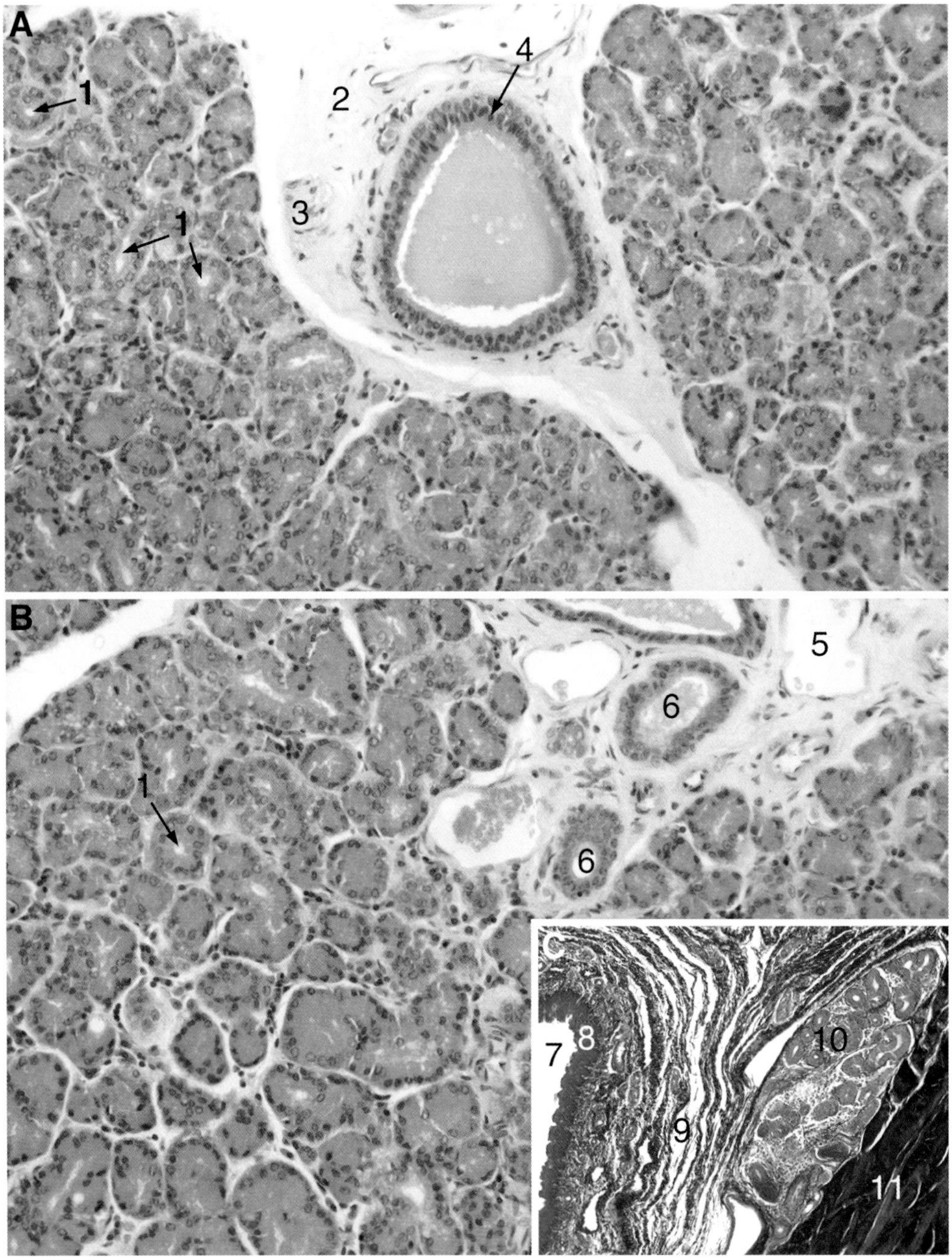

1. Acini with wide lumen
2. Interlobular septum
3. Peripheral nerve
4. Interlobular excretory duct with pseudostratified columnar epithelium
5. Venule
6. Intralobular ducts with cuboidal epithelium
7. Fornix of conjunctiva
8. Epithelium of conjunctiva
9. Episcleral tissue
10. Accessory lacrimal gland (of Krause)
11. Sclera

Fig. 2.54

Histology of the lacrimal gland. A – high magnification light micrograph, HE staining; B, C – electron micrographs (Courtesy of I. Süveges and I. Gábriel)

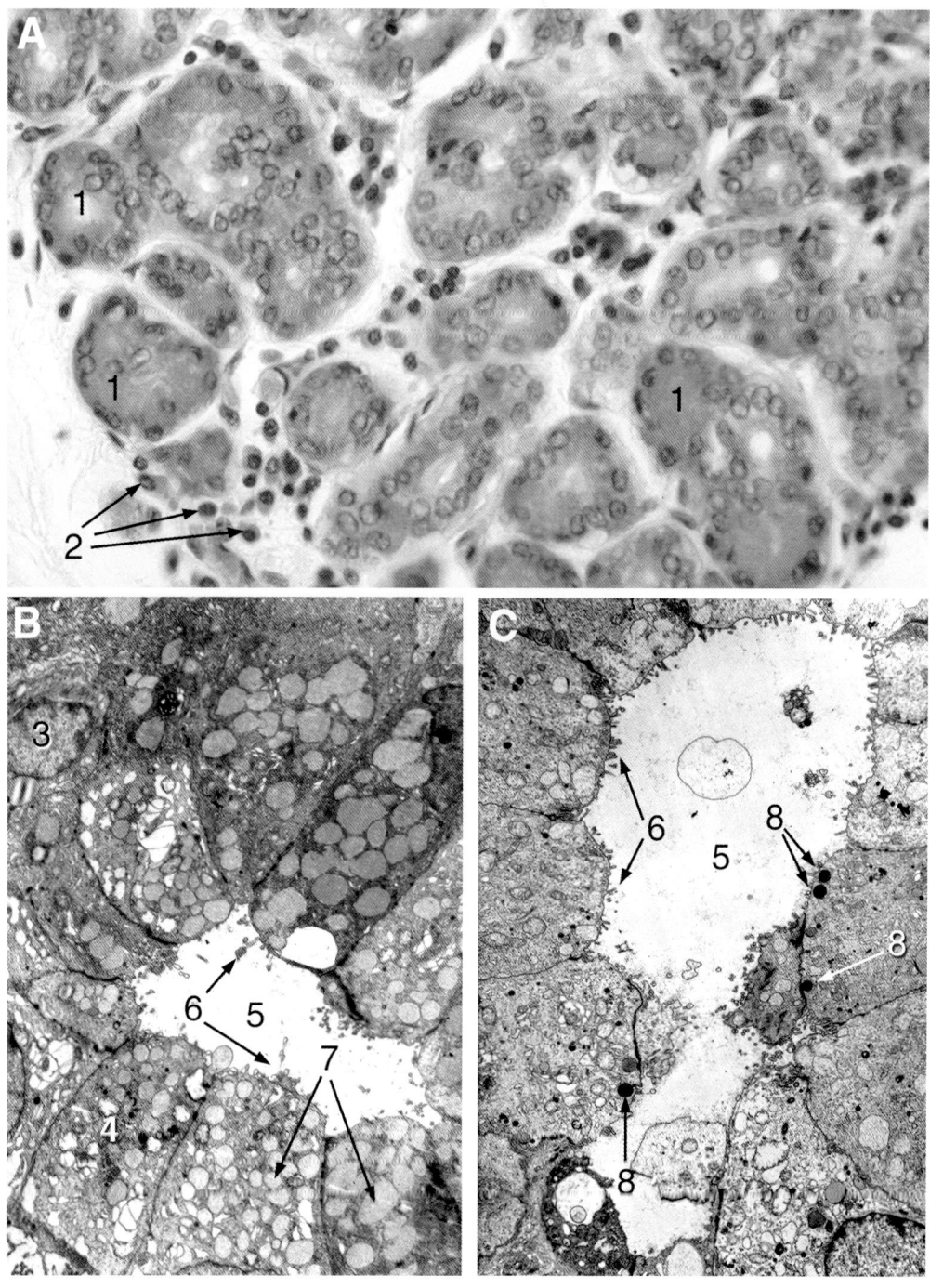

1. Acini
2. Plasma cells in the interstitium
3. Basal cell
4. Acinar secretory cell
5. Lumen
6. Apical microvilli
7. Large and lightly stained secretory granules (presumed mucous secretion)
8. Small and darkly stained secretory granules (serous secretion)

Fig. 2.55

Histology of the lacrimal sac. A – semithin section, toluidine blue staining; B – transmission electron micrograph; C – scanning electron micrograph. Courtesy of I. Süveges and I. Gábriel.

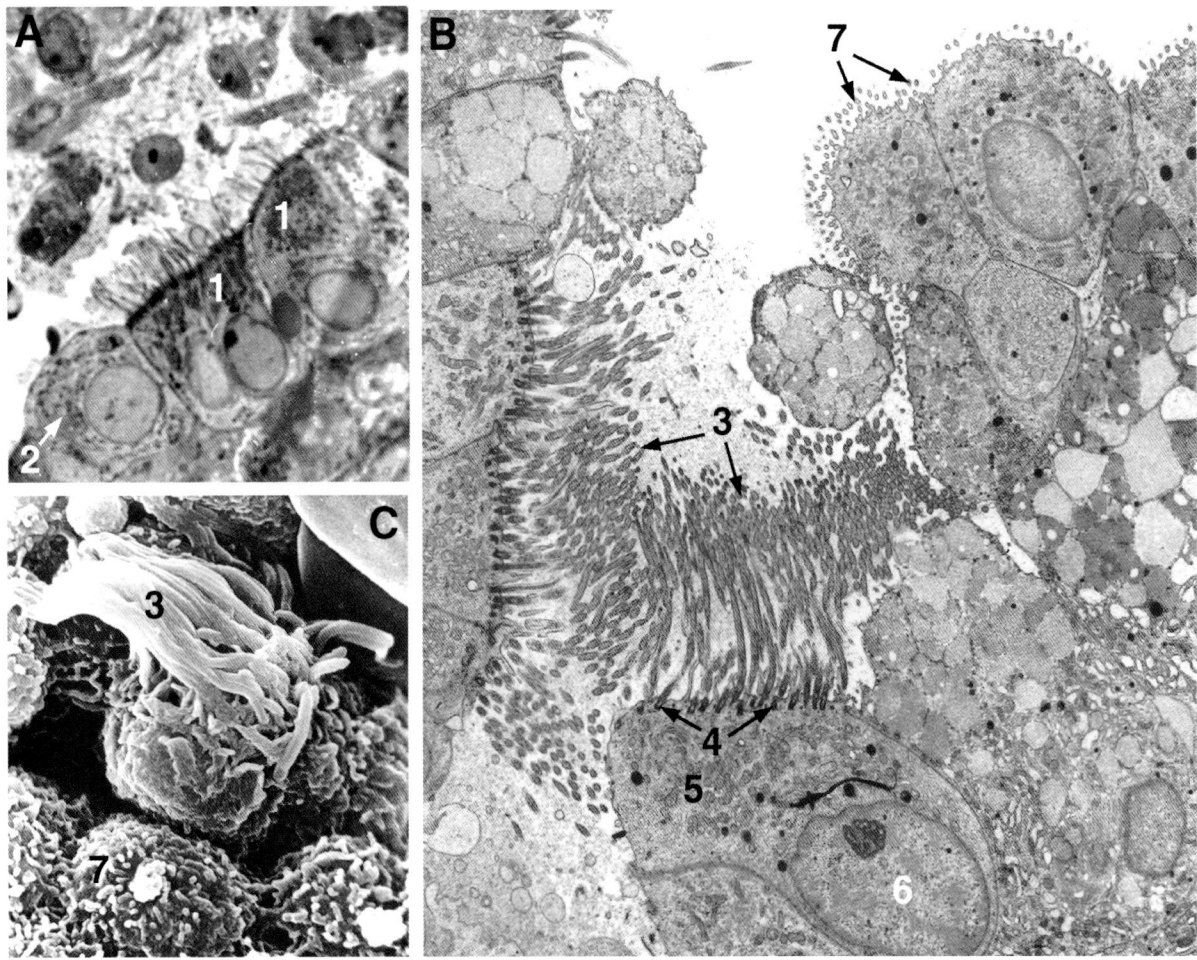

1. Ciliated columnar epithelial cells
2. Non-ciliated epithelial cell
3. Cilia
4. Basal bodies
5. Mitochondria
6. Nucleus
7. Microvilli

THE VISUAL PATHWAY

Optic Nerve and Tract (Figs. 2.56; 2.57; 2.58; 2.59; 2.60)

Given the developmental history of the retina as an extension of brain tissue, the optic nerve has the characteristics of a brain tract rather than a peripheral nerve. It starts as the optic disc visible on the fundus, where the retinal fibers acquire their myelin sheaths. The transition between the optic disc and nerve is quite complex since the optic fibers (together with the central retinal vessels) have to pass through the retinal layers and then the vascular and fibrous ocular tunics. The "retinal part" of the optic nerve head is separated from the intact retina by the astroglial border of Jacoby. The rest of the perforated area (cribriform plate, lamina cribrosa) consists of choroidal and scleral zones. The former is delineated from the choroid by a spur of collagenous tissue, whereas the latter contains fenestrated sheets of collagen, which are continuous with the sclera as well as with the septa of the optic nerve. The next (intraorbital) part of the free optic nerve is invested with the dura mater, arachnoid mater and pia mater, corresponding to the respective meningeal layers of the brain. Accordingly, the optic nerve accommodates a protrusion of the subarachnoid space that extends as far forward as the optic disc and is directly exposed to changes of the intracranial pressure. The bundles (fascicles) of optic fibers are separated from each other by collagenous connective tissue septa, which derive from the pia mater and guide the blood vessels supplying the nerve. The optic fibers are surrounded by thin astroglial sheaths, which also complete the gaps in the septa. As a result, the nervous elements remain isolated from connective tissue and blood vessels throughout the optic nerve (blood–brain barrier). Apart from astrocytes, the optic nerve also contains oligodendrocytes for the formation and maintenance of myelin sheaths.

The slightly sinuous course of the optic nerve largely follows the long axis of the orbit to its apex, after which it passes through the optic canal (intracanalicular part) together with the ophthalmic artery and sympathetic plexus to meet with the contralateral optic nerve in the optic chiasma. This last (intracranial) part of the nerve is enveloped only by the pia. The fibers arising from the nasal half of the retina decussate in the chiasma, whereas those originating from the temporal retina pass without decussation. The optic chiasma lies in a groove of the sphenoid and is related posteriorly to the hypophyseal stalk, ventrally to the diaphragma sellae overlying the pituitary gland, and bilaterally to the internal carotid

Fig. 2.56
Histology of the optic nerve head. HE staining.

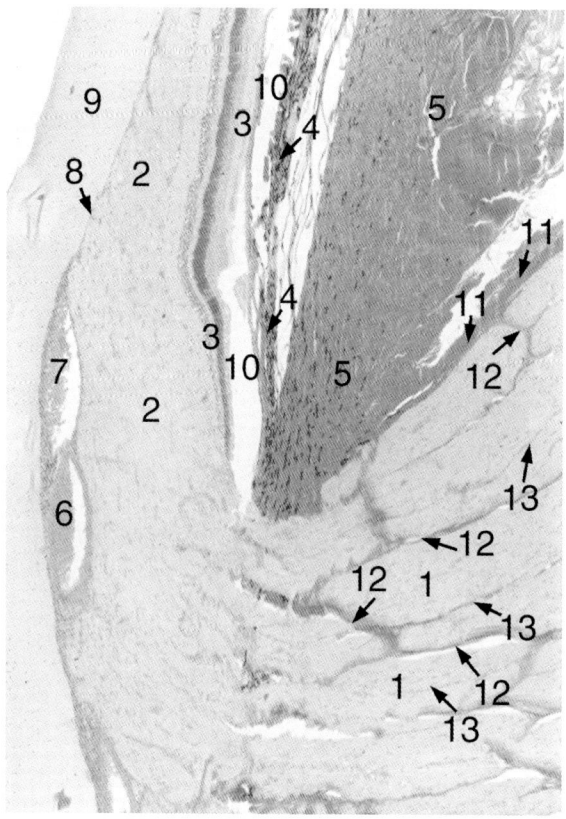

1. Optic fibers
2. Optic fiber layer of retina with unmyelinated fibers
3. Other layers of retina
4. Choroid
5. Sclera
6. Branch of the central retinal artery
7. Branch of the central retinal vein
8. Internal limiting membrane
9. Embedding material (artifact)
10. Gap in tissue (artifact)
11. Dura mater covering the optic nerve
12. Connective tissue septa
13. Glial septa

arteries. Disorders of these topographically related structures may be responsible for injuries of the chiasma region, leading to different visual field deficits. Above the chiasma lie the medial root of the olfactory tract and the anterior perforated substance. Furthermore, the optic chiasma forms the rostroventral border of the third ventricle between the optic and infundibular recesses.

The next part of the main visual pathway is the optic tract passing from the chiasma in a laterally convex arch along the base of the brain to the lateral geniculate nucleus

Fig. 2.57

Histology of the orbital part of the optic nerve. Azan staining. A – section from a segment proximal to the globe, B – section from a segment distal to the globe, inset – high magnification image showing the meningeal envelope.

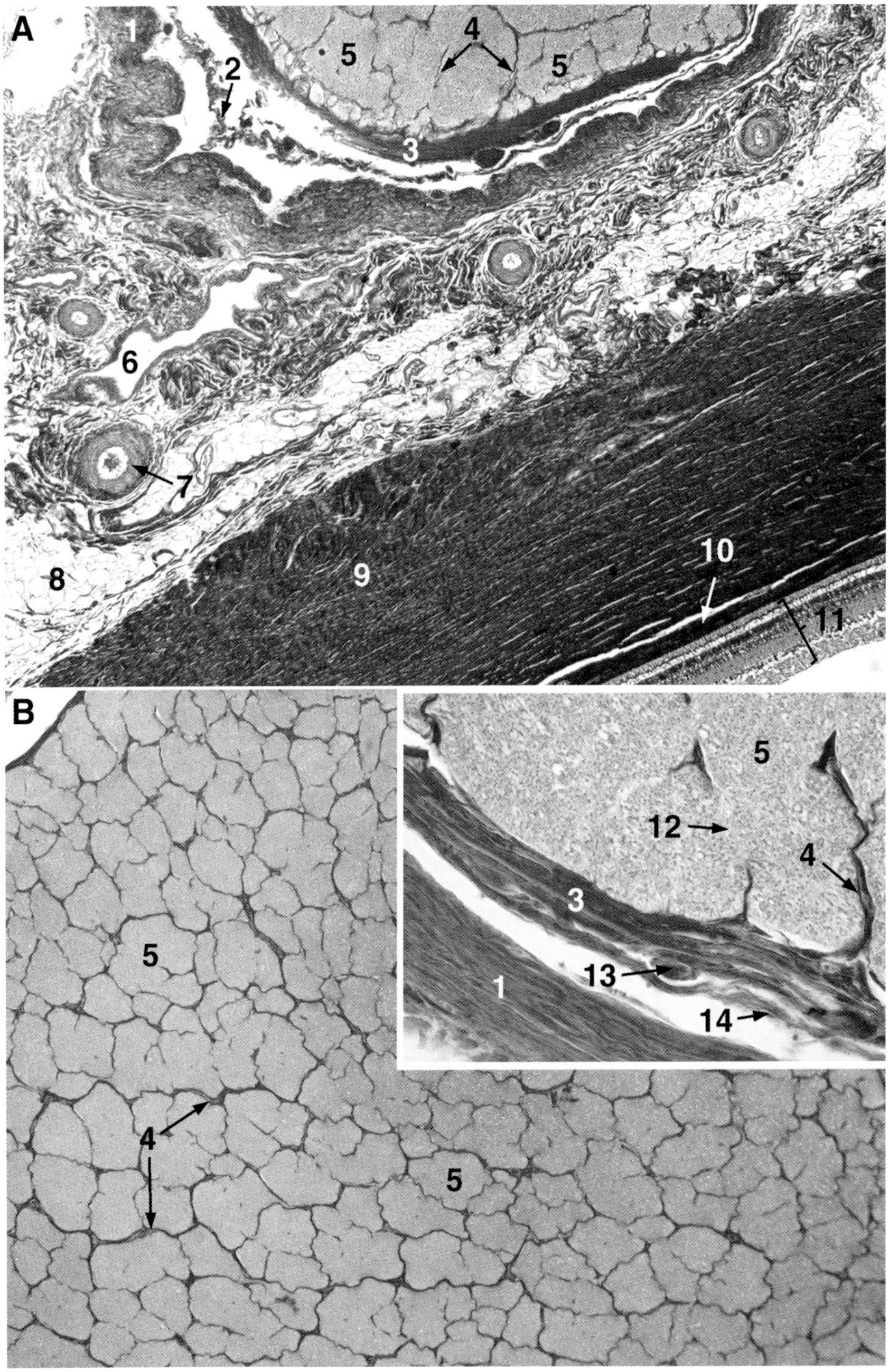

1. Dura mater
2. Arachnoid mater
3. Pia mater
4. Connective tissue septa
5. Optic nerve fibers
6. Vortex vein in retrobulbar space
7. Posterior ciliary artery
8. Retrobulbar (orbital) fat
9. Sclera
10. Choroid
11. Retina
12. Glial septum
13. Arachnoid trabecula in subarachnoidal space
14. Arachnoid (detached)

Fig. 2.58

Anatomical specimens demonstrating the optic pathway. A – Midsagittal section of the head; B – brainstem (ventral aspect); C – brainstem (dorsal aspect) Courtesy of J. Vajda† and L. Patonay.

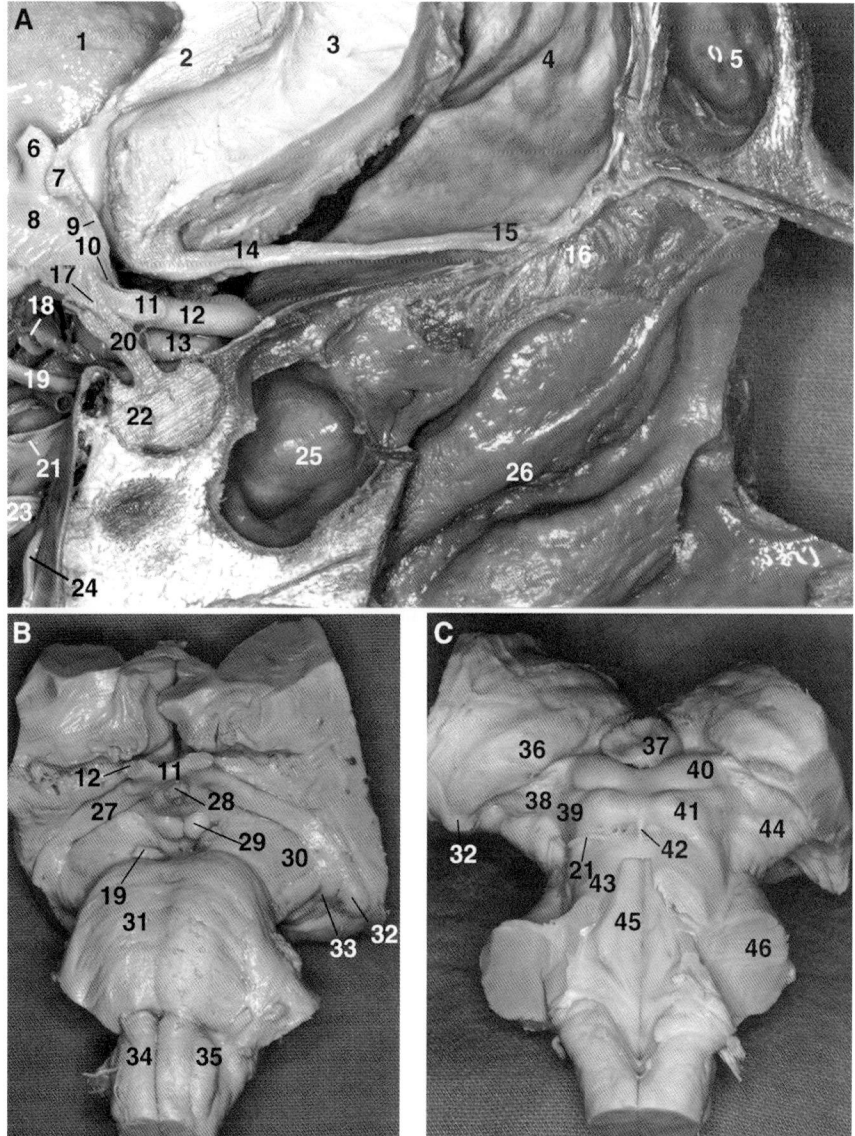

1. Septum pellucidum
2. Corpus callosum (rostrum)
3. Frontal cortex
4. Dura mater
5. Frontal sinus
6. Fornix (column)
7. Anterior commissure
8. Third ventricle
9. Lamina terminalis
10. Optic recess
11. Optic chiasma
12. Optic nerve
13. Ophthalmic artery
14. Olfactory tract (elevated)
15. Olfactory bulb
16. Olfactory region of nasal mucosa
17. Infundibular recess
18. Internal carotid artery
19. Oculomotor nerve
20. Pituitary stalk
21. Trochlear nerve
22. Pituitary gland in sella turcica
23. Trigeminal nerve
24. Abducent nerve
25. Sphenoidal sinus
26. Middle nasal concha
27. Optic tract
28. Tuber cinereum
29. Mamillary body
30. Crus cerebri
31. Pons
32. Lateral geniculate body
33. Superior quadrigeminal brachium
34. Pyramid
35. Olive
36. Pulvinar
37. Pineal gland
38. Medial geniculate body
39. Lemniscal trigone (trigonum lemnisci)
40. Superior colliculus
41. Inferior colliculus
42. Frenulum of superior medullary velum
43. Superior cerebellar peduncle (brachium conjunctivum)
44. Crus cerebri
45. Rhomboid fossa
46. Middle cerebellar peduncle (brachium pontis)

Fig. 2.59

Diagram of optic pathway lesions with corresponding visual field defects. N – nasal hemifield, T – temporal hemifield.

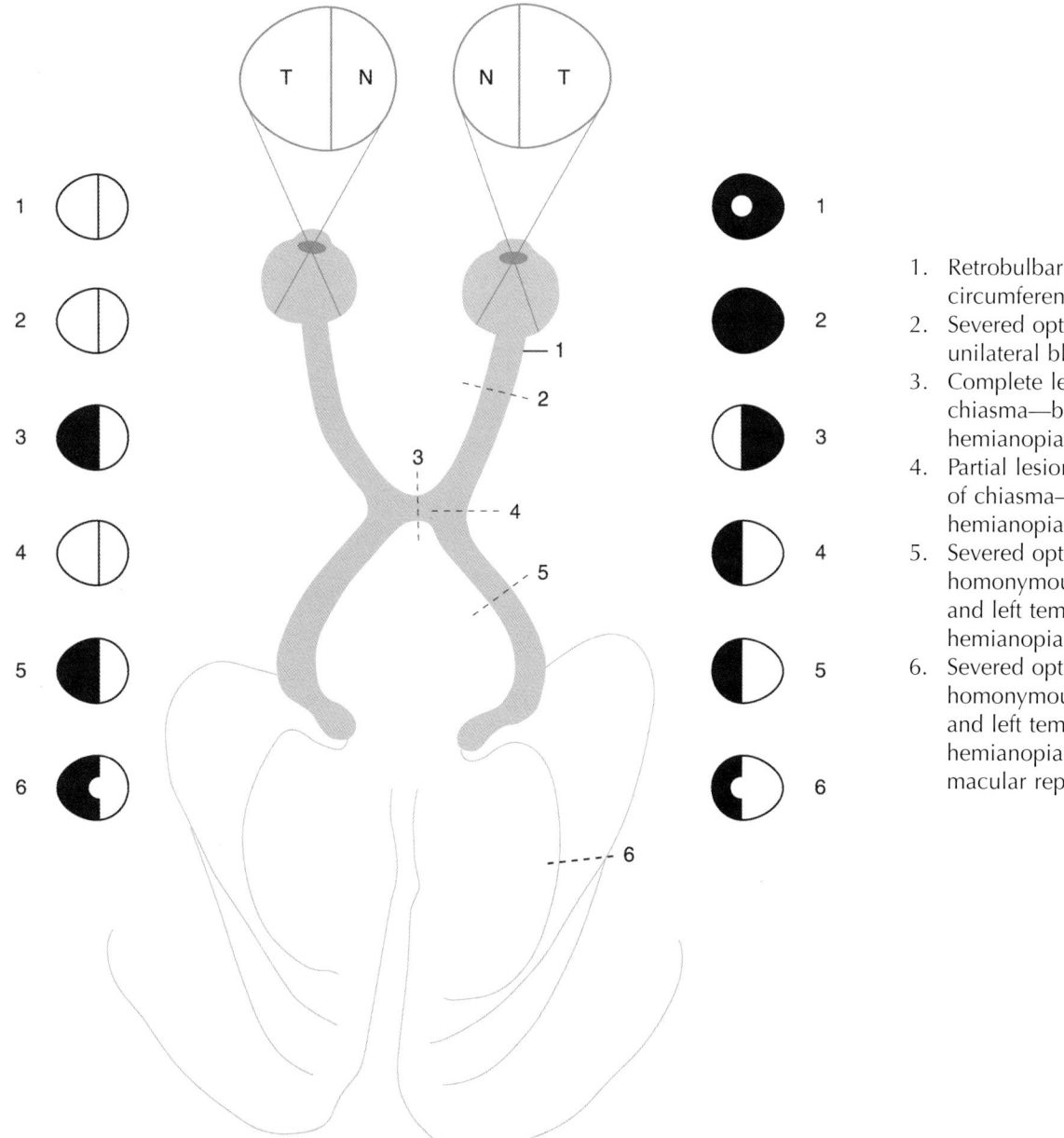

1. Retrobulbar neuritis — circumferential blindness
2. Severed optic nerve—total unilateral blindness
3. Complete lesion of optic chiasma—bitemporal hemianopia
4. Partial lesion of right side of chiasma—right nasal hemianopia
5. Severed optic tract— homonymous (right nasal and left temporal) hemianopia
6. Severed optic radiation— homonymous (right nasal and left temporal) hemianopia with spared macular representation

(LGN). Further destinations of the retinal axons (some disputed or poorly defined in man) comprise the pretectal nucleus (for the pupilloconstrictor pathway), superior colliculus (for additional visual processing and visually driven movements of the eye, head and body), nucleus of the optic tract and other accessory optic nuclei (for the optokinetic reflexes), suprachiasmatic, and ventral infundibular (arcuate) nuclei (for the mediation of light and dark cycles on the circadian rhythm), and the pulvinar of thalamus (retinothalamic system). These accessory retinal connections largely fall beyond the scope of the present work with the exception of those relevant to the pupillary reflexes and the eye movements (see below).

Lateral Geniculate Nucleus (Figs. 2.58; 2.59; 2.60; 2.61; 2.62)

A part of the metathalamus, this nucleus makes a visible protrusion (lateral geniculate body) beneath the posterolateral pulvinar. In coronal sections it has the shape of a (laterally directed) jockey's cap. The LGN is connected to the superior colliculus of the midbrain via the superior quadrigeminal brachium, which passes between the pulvinar and the medial geniculate body. In humans, like in other primates, the LGN consists of six neuronal laminae separated by thin septa of white matter. The laminae are numbered from the ventral concave side (hilum). Four of these laminae (3–6) are small-celled (parvocellular), for the termination of P ganglion cells (cf. the retina), whereas the remaining two layers (1 and 2) have large neurons (magnocellular) and receive the retinal fibres of the M ganglionic neurons. Fibers from the ipsilateral optic nerve terminate in laminae 2, 3, and 5, whereas those from the contralateral eye project to laminae 1, 4, and 6. Each optic fiber branches into five to six terminals synapsing with a corresponding number of relay cells in the same lamina. The macular fibers have a relatively wide representation in the caudal part of the LGN. The optic terminals innervating the LGN follow a strict retinotopy (fibers from the upper retina terminate laterally and those from the lower retina terminate medially).

Phylogenetically, the human LGN corresponds to the dorsal (main) nucleus of the lateral geniculate complex of vertebrates. Its impressive development in primates has been attributed to the highly overlapping visual fields and the importance of binocular vision. The ventral part of the ancient complex is represented by the pregeniculate nucleus medial to the ventral third of the LGN. This nucleus probably has a reduced significance in humans.

The dendrites of the principal relay neurons of the LGN receive input from different excitatory and inhibitory sources. Corresponding to an elaborate processing that occurs within the nucleus, these afferents converge to form synaptic glomeruli. In the ensuing ultrastructural description applied for the cat we follow the terminology by R.W. Guillery. The main glutamatergic excitatory elements are retinal afferents, characterized by large diameter, round vesicles and pale mitochondria (RLP terminals). Corticothalamic terminals from layer VI are small, with round vesicles and dark mitochondria (RDS terminals). Inhibitory terminals, which are GABAergic, can be categorized as extrinsic or intrinsic connections. The former (F1 terminals) arise mainly from the reticular thalamic nuclei, whereas the latter (F2 terminals) belong to the numerous GABAergic interneurons accumulating within the neuronal laminae of LGN. The F2 terminals represent dendritic elements that are often presynaptic to the dendrites of the relay neurons (RD).

Apparently, the presence of an intact corticothalamic input is essential for maintaining the integrity of relay neurons. In the macaque monkey, an experimental unilateral transection of the corticogeniculate pathway leads to

Fig. 2.60
Anatomical drawing depicting the main structures associated with the visual pathway. Ventral aspect.

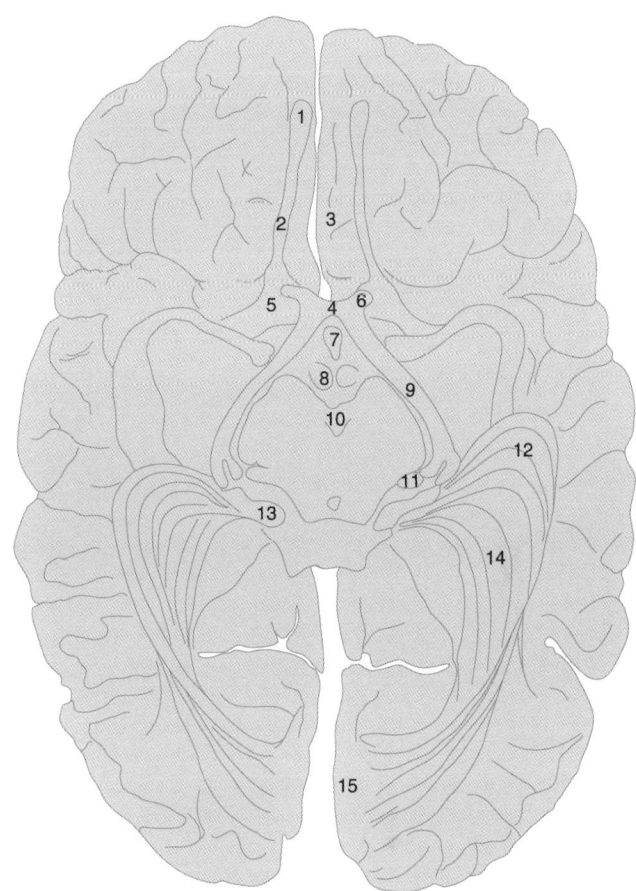

1. Olfactory bulb
2. Olfactory tract
3. Straight gyrus
4. Optic chiasma
5. Olfactory trigone
6. Optic nerve
7. Tuber cinereum
8. Mamillary body
9. Optic tract
10. Interpeduncular fossa
11. Superior quadrigeminal brachium
12. Loop of Meyer
13. Lateral geniculate nucleus
14. Optic radiation
15. Visual cortex

a massive loss of relay neurons in the ipsilateral LGN, whereas the GABAergic interneurons remain intact.

Geniculocalcarine Tract (Optic Radiation of Gratiolet) (Figs. 2.59; 2.60)

The axons of the relay cells of LGN project along a pathway called the optic radiation (of Gratiolet) to the primary visual cortex straddling the calcarine fissure of the occipital cortex. The fibers of the optic radiation are directed first rostrally, toward the temporal pole, as a narrow bundle. Then they arch around the temporal horn of

Fig. 2.61

Experimental histological observations in the primate (Macaque monkey) lateral geniculate nucleus (LGN). Light micrographs of vibratome sections stained with Azure II-methylene blue (A,C) or immunostained for the inhibitory neurotransmitter GABA (B,D,E). A – parvocellular and magnocellular layers of the LGN of the control (right) side; B – GABA immunoreactive interneurons (darkly stained) accumulating in the layers of LGN. The interlaminar borders are indicated by dashed lines; C – LGN of the deafferented side (4 months after transection of the corona radiata in the left hemisphere). The nucleus is reduced in size and the relay cells are absent. D – In the parvocellular layer of the control LGN both GABA immunoreactive interneurons (arrows) and non-immunostained relay neurons (arowheads) are present. E – In the parvocellular layer of the deafferented LGN only the GABA immunolabelled interneurons are present, whereas the relay neurons have degenerated. Courtesy of J. Takács, L. Moiseeva, V. Silakov and J. Hámori.

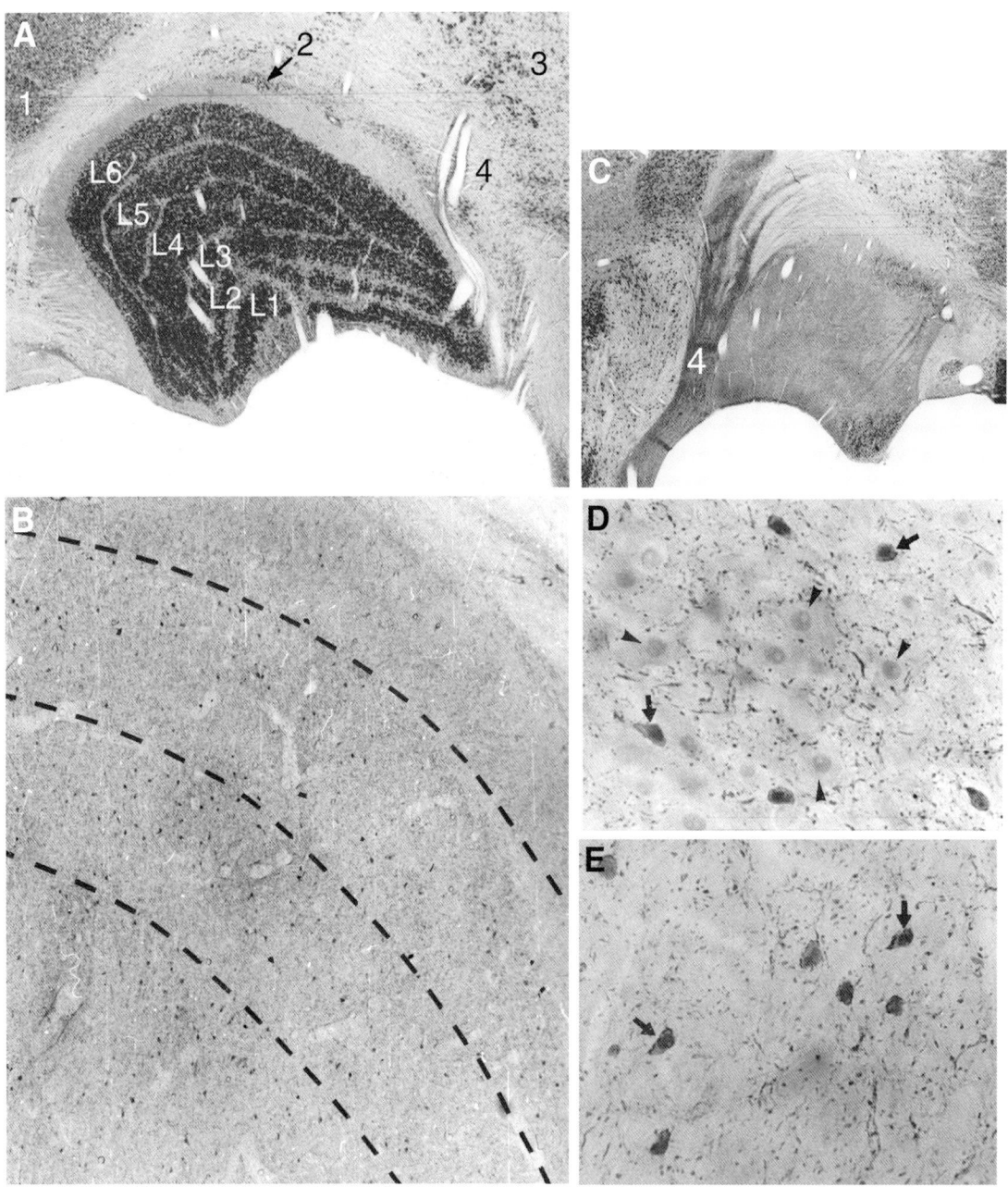

1. Putamen
2. Pregeniculate (GLv) nucleus
3. Subthalamus
4. Cerebral peduncle
L1—L2 Magnocellular layers
L3—L6 Parvocellular layers

Fig. 2.62

Electron micrograph of the LGN of cat following ultrastructural immunostaining against GABA, using the postembedding immunogold method. The presence of GABA is indicated by 15 nm colloidal gold precipitate, appearing as black dots. Abbreviation of terminal types according to R.W. Guillery. Courtesy of J. Takács.

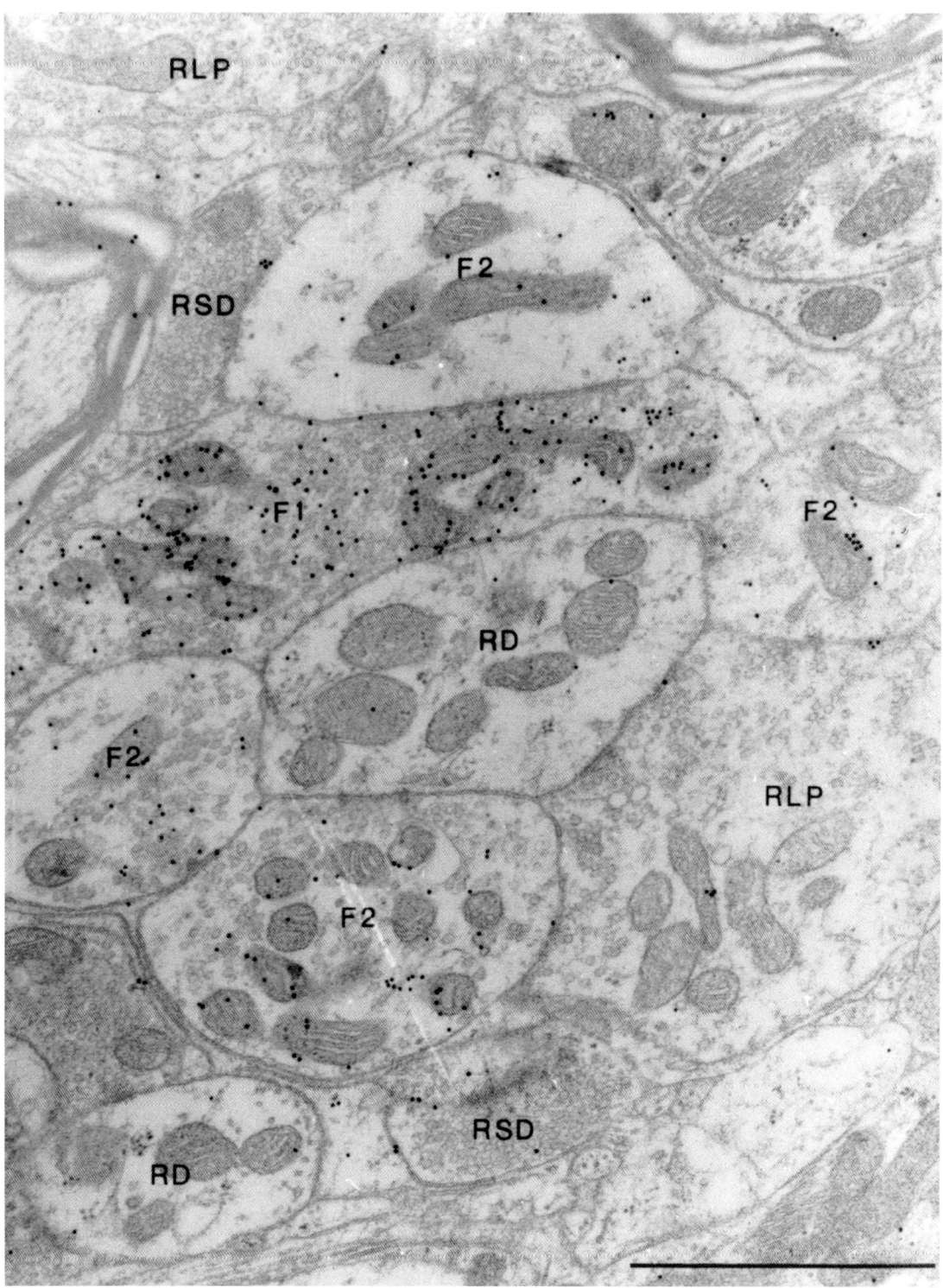

F1 – axon terminal from the reticular thalamic nucleus
F2 – dendritic terminal from local interneuron
RD – relay cell dendrite
RLP – retinal axon terminal
RSD – corticogeniculate axon terminal
(from layer VI of cortex)

the lateral ventricle in a sharp bend (known as the loop of Meyer) and fan out to form a wide and twisting band near the lateroventral border of the ventricle, now following a caudal course. The long and tortuous pathway is rather exposed to injuries (hemorrhage affecting the lateral ventricle or the internal capsule (the optic radiation passes through the posterior sublenticular part of the internal capsule medial to the acoustic radiation), or destructive tumors of the temporal lobe.

Visual Cortex (Figs. 2.63; 2.64; 2.65; 2.66; 2.67; 2.68)

The primary visual cortex (area 17 of Brodman, V1) is situated in the occipital gyri bordering the calcarine fissure and also extending to the cuneus above and the lingual gyrus below (Fig. 2.63). This region is also termed striate cortex after the white line (stria, stripe) of Gennari, which is visible in layer IV of the cortex and corresponds in part to the fibres of the optic radiation and in part to intracortical connections. The stria subdivides the original layer IV into laminae IVA and IVC (the stria itself being lamina IVB).

The upper and lower halves of the retina are represented above and below the calcarine fissure, respectively. Note that the opposite is true for the visual fields. The uniocular visual field (a narrow temporal crescent-shaped field seen by one eye only) is represented in the extreme rostral part of V1, below the calcarine fissure. The representation of the macula is disproportionately wide, occupying the entire posterior part of V1 and extending around the occipital pole onto the lateral surface of the hemisphere. The massive extent of macula-associated fibers inside the geniculocortical tract as well as in the striate cortex may explain the clinical experience that the macular field is often spared on the defective side in hemianopias owing to unilateral lesions of the optic radiation or striate cortex (but not the optic tract!). However, other authors prefer the explanation that the posterior cerebral artery (whose obstruction represents a frequent cause of striate cortical lesions) communicates with the middle cerebral artery via an anastomosis near the occipital pole, that is, the region representing the macula. Apart from this vascular theory, the possibility of a truly bilateral representation of the macular field in the visual cortex (or perhaps already in the LGN) has also been raised but not yet substantiated in humans.

The striate cortex is composed of a mosaic of functional units called ocular dominance columns (Fig. 2.64). Corresponding to the cortical modules described by Szentágothai, ocular dominance columns are clusters of interconnected neurons spanning the entire width of the cortex. Unit recordings from the cells of lamina IVC have demonstrated that they are driven monocularly, alternating between the ipsi- and contralateral eye from column to column. These layer IVC neurons are not orientation-selective but they project within the respective columns to other cells of increasing complexity. "Simple" cells in layer IVB as well as "complex" and "hypercomplex" cells in layers I, III, V, and VI are sensitive to line stimulation (slits, bars, borders of light) of a given orientation (angle) within the visual receptive field, rather than a diffuse patch of light. In addition, complex and hypercomplex cells are also driven binocularly. Within each ocular dominance column there are zones that selectively respond to visual stimuli of given orientation. In the monkey, the shift in orientation selectivity is about 10° by 20 µm, hence the cycle of 180° is completed over a distance of about 360 µm. The higher organizational unit comprising a pair of (left and right) ocular dominance columns, each covering a cortical segment required for a complete orientation cycle of 180° was designated by the Nobel laureates Hubel and Wiesel as a "hypercolumn." Using the mitochondrial enzyme cytochrome oxidase as a marker of neuronal activity, the hypercolumns of orientation-sensitive neurons correspond to the cytochrome oxidase-poor "interblob" regions (Fig. 2.65). The cytochrome oxidase-

Fig. 2.63

Projection of the left (L) and right (R) retinae on the right lateral geniculate nucleus (LGN) and primary visual cortex (V1).

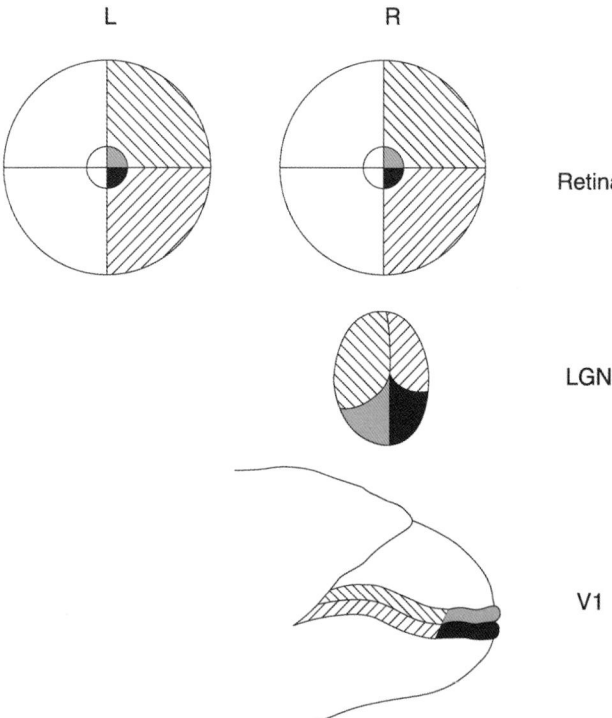

CHAPTER 2 / THE EYE

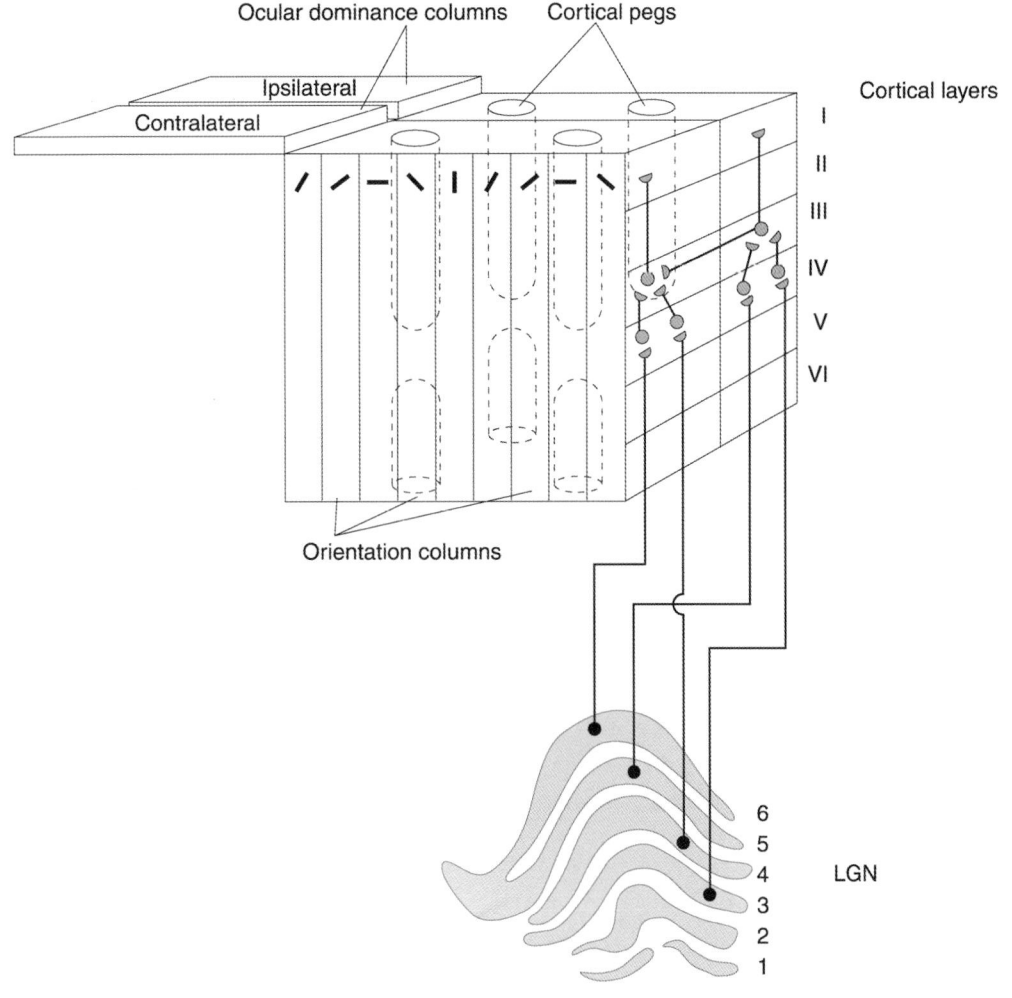

Fig. 2.64

Block diagram of primary visual cortex with the geniculocortical afferents. Redrawn and modified after Kandel ER and Schwatz JH (1985) *Principles of Neural Science, Second Edition*, Elsevier New York

rich "blobs." on the other hand, contain wavelength-selective cells (originally termed in the Hubel–Wiesel model as cortical pegs concerned with color vision). However, it should be noted that these cells are not truly color-sensitive in the sense that they would convey the same perceptual information regardless of the conditions of illumination (e.g., the precise wavelength spectrum of light source). Instead, wavelength selectivity means that such neurons are activated whenever a particular wavelength is represented in sufficient intensity either in their uniform circular receptive field or as a central stimulus surrounded by that of a complementary wavelength (the latter are termed "red on/green off" cells). Color-selective cells coding the genuine perceptual character of color information ("color memory") do occur in the higher order (prestriate) visual cortex, in particular V4.

In non-human primates, the magnocellular layers of LGN project to the sublamina IVCα and thence to IVB, whereas the parvocellular layers project to the sublamina IVCβ and thence to IVA of the striate cortex. Projections to laminae III and VI also occur, to a lesser extent. The cells of layer IVB are binocular and selective for the direction in which the stimulus is moving. This faculty is presumably due to the existence of widespread horizontal connections within the layer. The cells of layer IVA project on the cytochrome oxidase-rich 'blobs' and cytochrome oxidase-poor 'interblobs' of layers II and III, where they converge with the projections arising from layer IVB. The superficial (II-III) layers are also rich in horizontal connections, linking mainly blobs with blobs and interblobs with interblobs. However, there are cells that receive input from both blobs and interblobs. In

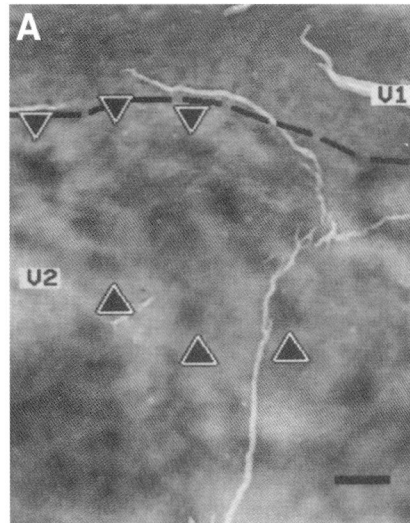 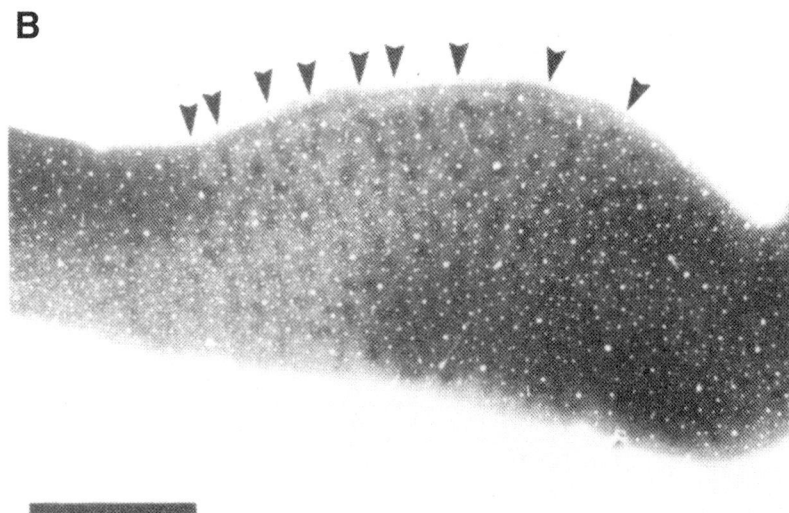

Fig. 2.65

Cytochrome oxidase (CO) activity in the human visual cortex. A – Topography of CO from flattened cortex, mostly within layer 4. Patchy CO-rich stripes are visible in V2, running from upper left to lower right. Three prominent stripes are highlighted by downward-pointing and upward-pointing triangles. Calibration bar: 2 mm. Reproduced from Tootell RBH, Born RT and Ash-Bernal R (1993) in: Balazs G, Ottoson D, Roland PE (Eds) Functional Organization of the Human Visual Cortex, Pergamon, Oxford, p. 69. B – Tangentional section demonstrating CO-rich blobs (puffs) in area 17, forming rows (arrowheads) orthogonal to the 17/18 border. Calibration bar: 5 mm. Reproduced from Wong-Riley MTT (1993) in: Balazs G, Ottoson D, Roland PE (Eds) *Functional Organization of the Human Visual Cortex*, Pergamon, Oxford, p.168.

addition to numerous intracortical connections, the pyramidal neurons of the laminae II and III project to prestriate cortical areas (see below), those of lamina V to the superior colliculus of tectum, and those of lamina VI to the LGN, including its intralaminar zones. In humans, the main recipient of geniculocalcarine fibers seems to be lamina IVC and all available evidence indicates that a separate processing for the M and P systems exists also in the human visual cortex.

Earlier descriptions of the visual association areas have distinguished the parastriate cortex (corresponding to area 18 of Brodmann) and peristriate cortex (area 19 of Brodmann) as belt regions surrounding the primary visual cortex (Fig. 2.66). The former is adjacent to the area 17 of the occipital lobe, but lacking the stripe of Gennari, whereas the latter extends to the posterior parietal and inferior temporal cortices. More recent studies, mainly in the cat and the monkey, have further refined the localisation of higher order visual regions, collectively called prestriate cortex. In the monkey, five major visual regions (V1–V5) can be distinguished, the numbers represent increasing distances from the V1, as well as differences in cyto- and myeloarchitecture. The prestriate cortex spreads as far as the medial temporal cortex, the caudal portion of the superior temporal sulcus, the inferior temporal cortex (reaching the temporal pole) and the intraparietal sulci. The receptive fields of prestriate neurons are generally larger than those of the striate cortex, and the strict retinotopy, characteristic for V1 and more or less for V2, tends to dissipate by V4 and V5.

Processing of visual information in the prestriate cortex is based upon a certain degree of functional specialization. Orientation coding is localized in V2 and V3, color coding is characteristic for V4, whereas V5 is associated with motion and direction sensitivity.

Evidence for a similar functional segregation in the human cortex is scarce and somewhat controversial. The difficulty is that the regional distributions established mainly by single-cell responses in animal experiments may not harmonize with the data obtained from human clinical experience or neuroimaging studies. Despite a tempting analogy, the color-vision deficit of V4-lesioned monkeys, for example, is quite different from the human clinical condition known as cerebral achromatopsia (failure to perceive the world in color following lesion to the caudal part of the fusiform gyrus). Animals with a damage in V4 also show form discrimination deficits that are at least as severe as the color-vision deficits. Conversely, an impairment of form discrimination does not invariably accompany achromatopsia in human subjects. Therefore, a functional equivalence of the V4 of monkey with the caudal fusiform gyrus of man is questionable. Nevertheless, an involvement of the fusiform gyrus in color processing is strongly suggested on the basis of regional cerebral blood flow (rCBF) studies combined with positron emission tomography (PET) imaging.

Both the striate and prestriate cortices of the right and left hemispheres are profusely connected by commissural pathways via the corpus callosum (Fig. 2.67). The significance of callosal connections is to reunite the representations of the two retinal hemifields separated by the optic chiasma. Accordingly, the corpus callosum connects those parts of the respective striate and prestriate cortices in which the midline of the hemifield is represented. One crucial function involving binocular processing is the estimation of depth (stereopsis) by the disparity of visual cues arising from the two retinae. Points in the visual fields further away than the fixed image (i.e., beyond the horopter) generate "divergent disparity" signals, whereas those points nearer than the fixed image (within the horopter) elicit "convergent disparity" signals. Such fine disparity information is detected first by selective neurons of the striate cortex, which in turn project on other cells of the prestriate cortex (primarily V2) that are sensitive to binocular disparities. An activation of the polar striate cortex, representing a rather small central/paracentral visual field, as well as of the neighboring extrastriate cortex in the analysis of binocular disparity tasks was also demonstrated in human rCBF/PET studies.

Higher order processing in the prestriate cortex involves parallel and separate handling of different modalities of the visual image (which is why the visual brain is so fundamentally different from any man-made photographic or image capturing device, no matter how plausible it may seem to draw analogies). Using gross simplification, such modalities can be grouped into those relevant to object recognition (the "what" system) and those subserving spatial localisation (the "where" system) (Fig. 2.68). The "ventral pathway" concerned with object recognition follows the inferior longitudinal fasciculus, connecting striate, prestriate, and inferior temporal areas. The most rostral part of this system, which carries exclusively visual information, is the anterior part of the inferior temporal cortex near the temporal pole (TE). Multiple connections with the limbic system, frontal cortex, and hippocampus may explain why lesions to the ventral pathway lead to deficits in visual discrimination and recognition, and a loss of ability to learn such tasks. The clinical syndrome called prosopagnosia (faulty identification of human faces) may also be associated with the ventral (object recognition) pathway because this condition is

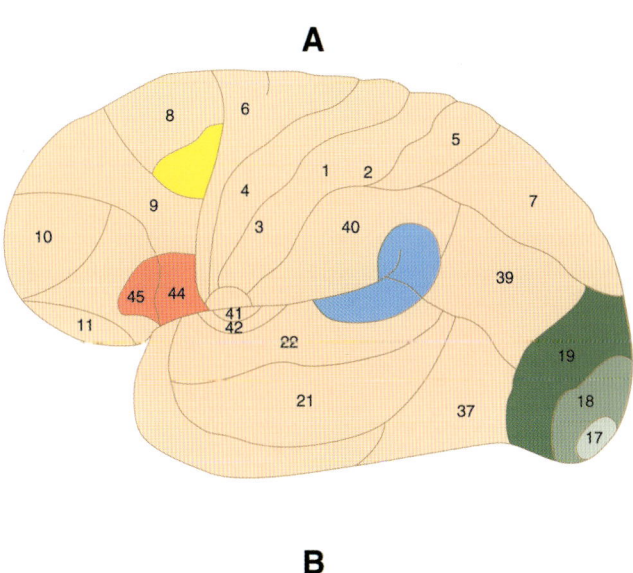

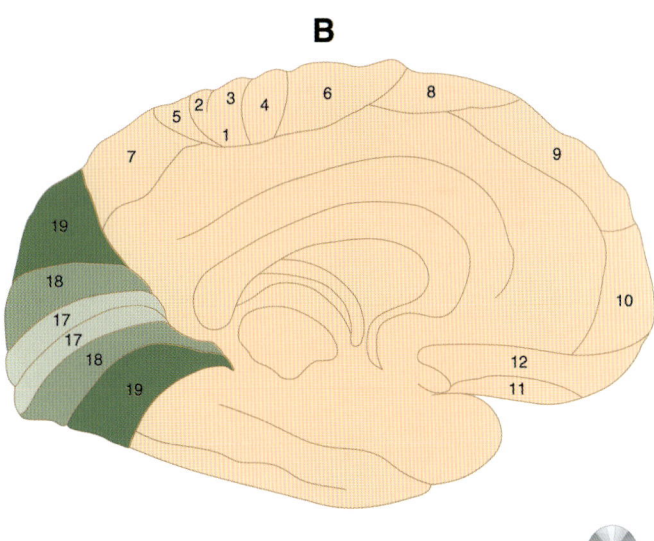

Fig. 2.66

Areas of functional localization on the lateral (A) and medial (B) surface of the cerebral hemisphere. Numbers refer to the areas of Brodmann. Yellow – frontal eye field, red – motor speech area of Broca, blue – Wernicke's sensory speech area, light green – striate cortex, green – parastriate cortex, dark green – peristriate cortex.

observed after bilateral damage to the occipitotemporal and parahippocampal cortices. Notably, cells selectively responsive to complex patterns such as hands or facial expression (highly significant cues in social behavior) have been detected in the floor and banks of the superior temporal sulcus of the monkey. Here again, an anatomical or functional homology with the relevant human cortical areas is not yet evident. In fact, the area just behind the superior temporal sulcus, at the junction between areas 19 and 37 of Brodmann was identified in rCBF/PET studies as the "motion center" (V5) of the human brain. Further PET activation studies in human subjects have indicated

Fig. 2.67

Visual areas of human cortex in relation to callosal afferents (stippled). A – Inferomedial view of a left occipital lobe (hatched in brain inset), orientation as shown in the cube (side = 10 mm). The subdivisions are also represented schematically on a model of flattened cortex. FG: fusiform gyrus, LG: lingual gyrus. B – The heavily myelinated region on the occipital convexity, corresponding to area V5/MT as seen on a lateral view of the left occipital lobe. From Zilles K and Clarke S (1997) in: Rockland et al. (Eds) *Cerebral Cortex*, Vol. 12, Plenum, New York, p. 718. Adapted from Clarke S and Miklossy J (1990).

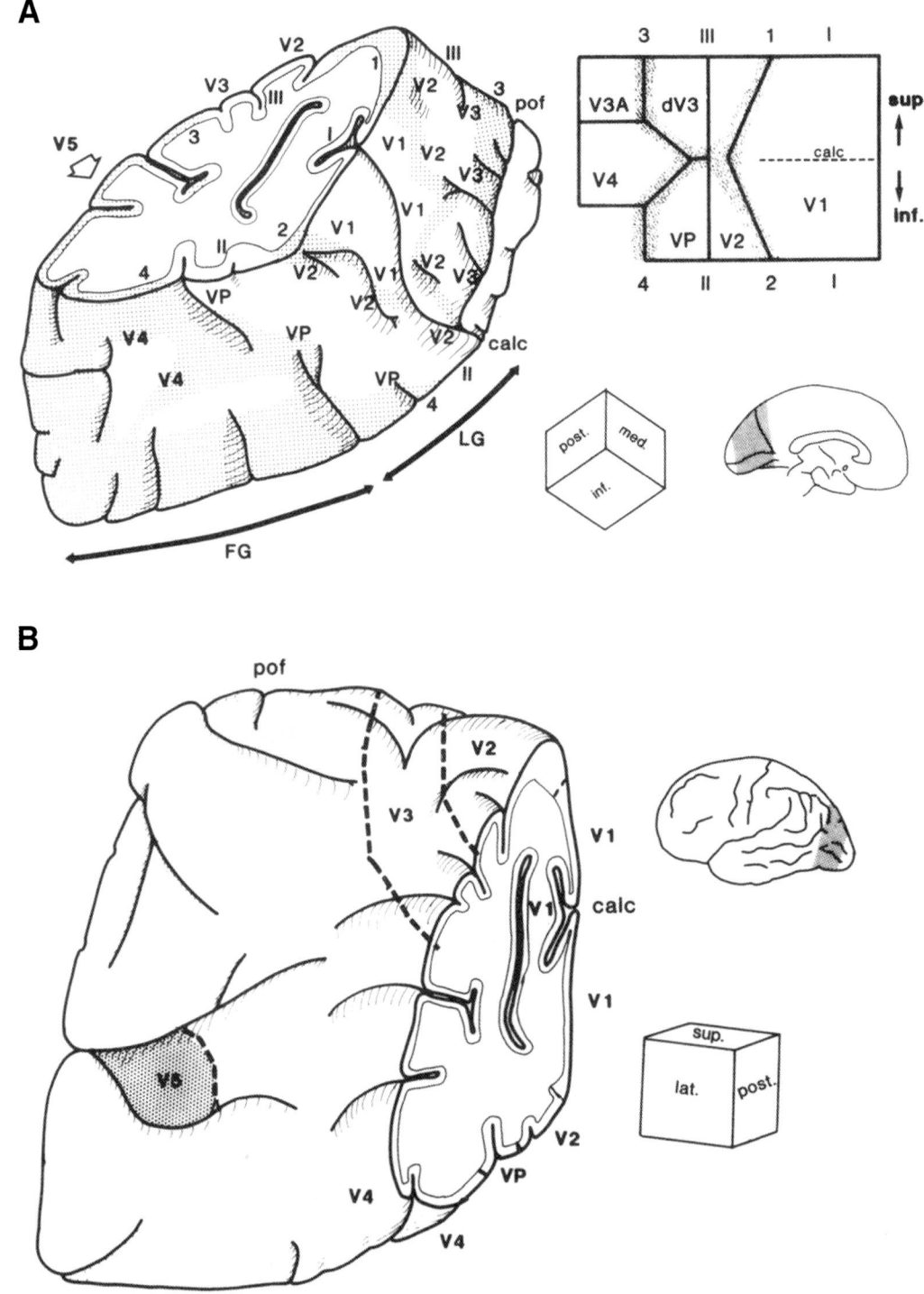

that the regions primarily activated during face recognition and recall were the fusiform and lingual (lateral and medial occipitotemporal) gyri, the parahippocampal gyrus, as well as the anterior temporal lobe. This finding, which fits in rather nicely with the known sites of lesions in prosopagnosic patients, suggests that a specialisation in visual processing areas exists also in humans, though the localization and extent of the specific areas do not necessarily follow the pattern learned from animal experiments.

The dorsal pathway (Fig. 2.68) follows the course of the superior longitudinal fasciculus and involves the striate and prestriate cortical areas as well as the inferior parietal cortex. These regions are probably essential for the construction of spatial maps, also requiring the involvement of other (e.g., tactile) modalities. The parietal area involved (PG) has a multisensory capability of combining visual input from the exterior ("macrospace") with the touch sensation arising from the hand ("microspace"). Furthermore, the perception of visual position (subjective visual axes) is dependent on the anterior parietal region, as evidenced by studies of human neuropathology. Visual length and distance estimation, as well as visuospatial neglect observed in pathological conditions, are closely associated with the parieto-occipital transition zone. It has to be noted that the parietal cortex, including the anterior part of intraparietal sulcus, as well as the posterior insula are also principal targets for vestibular input. Thus, the parietal association cortex presents one example (of, no doubt, many) how the visual modality of perception comes to merge with other sensory modalities relevant to the representation of complex and multimodal entities such as space. Moreover, some of the parietal regions may well serve as long-term "storage sites" for visual images. Human rCBF studies combined with PET imaging showed that a number of parietal cortical areas, including the posterior precuneus, lower intraparietal sulcus, superior parietal lobule, and angular gyrus, are activated during the recall of learned and stored images. According to these PET studies, visual imagery of large-field, stationary, and color patterns does not require the activation of the striate and prestriate cortices (V1–V4). This is at variance with other studies, using event-related potentials (ERP), rCBF, or functional magnetic resonance imaging (fMRI), claiming that mental imagery (constructing a visual representation of a stimulus from memory) also involved visual occipital regions.

It is tempting to relate the ventral ("what") system to the function of the parvocellular LGN and the dorsal ("where") system to the activity of the magnocellular LGN. However, there is no compelling evidence for such a direct link between the two levels of visual processing. Instead, there seems to be an extensive functional overlap between the M and P systems in both visual processing pathways.

Pupillary Light Reflex (Fig. 2.69)

The pupils of both eyes constrict (miosis) involuntarily, simultaneously, and equally on increased illumination of the retina of one eye. This phenomenon is termed consensual pupillary light reflex. The afferent pathway of the reflex arc follows the optic nerve, most probably as reflex collaterals of the ganglionic neurons (despite previous assumptions that they might come from special retinal receptor structures dedicated to the pupillary reflexes). The reflex afferents decussate in the optic chiasma, following the known pattern (see above), and then they follow the optic tract but fail to reach the LGN. Instead they pass to the pretectal nucleus (crossing again in the posterior commissure) along a dorsal course and, followed by a synaptic relay, the secondary axons descend in the medial longitudinal bundle to the accessory oculomotor nucleus of Edinger and Westphal. Thence, the efferent pathway, composed of parasympathetic preganglionic fibers, follows the third cranial (oculomotor) nerve back to the orbital cavity and passes via the short root to the ciliary ganglion for the last relay. Finally, the postganglionic fibers reach the iris sphincter via the short ciliary nerves.

Pupillary constriction is not confined to light-induced miosis. It is also observed when the gaze is switched from a distant to a near object (known as the near reflex). This response is accompanied by accomodation and convergence. The efferent pathway for the near reflex runs parallel to that of the light reflex and it also involves the innervation of the ciliary muscle (cf. nerves of orbit). The afferents, however, follow a separate (ventral) approach course to the pretectal nucleus, which explains selective sparing of the near reflex, even when the light reflex is lost, in certain clinical conditions affecting the dorsal tectum (Parinaud's syndrome).

The sympathetic pathways controlling pupillary dilation (mydriasis) are less well known. Concerning the efferent limb of the reflex arc, the preganglionic fibers emerge from the lateral gray columns of the first and second spinal thoracic segments, pass via the white communicating rami to the sympathetic trunk and ascend to the superior cervical ganglion. Postganglionic fibres from the ganglion join the carotid plexus and enter the orbit as the sympathetic plexus of the ophthalmic artery. Then, through a connection called the sympathetic root, the fibers reach the ciliary ganglion but pass without relay on

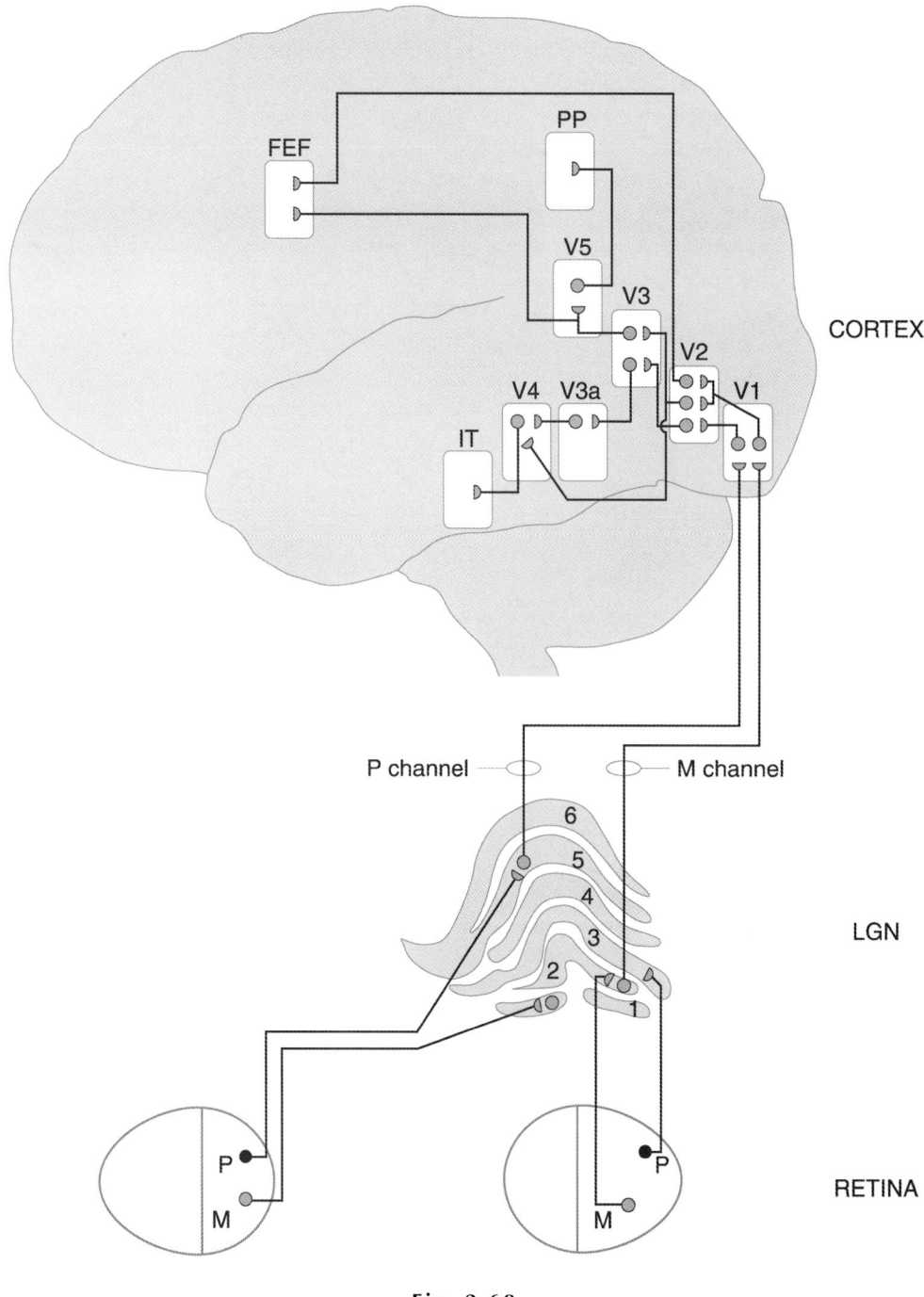

Fig. 2.68

Diagram of ascending visual input channels and main intracortical pathways. FEF – frontal eye field, IT – inferior temporal area, PP – posterior parietal area

to the long ciliary nerves for innervation of the pupillary dilator. The afferent limb of the reflex is poorly understood. Access to the relevant sympathetic cell groups of the lateral gray column (ciliospinal center) presumably involves descending sympathetic pathways from the hypothalamus and/or midbrain tegmentum, following the anterolateral tract of spinal cord. How exactly retinal afferents connect with these pathways, remains to be investigated. Interestingly, mydriasis can also be elicited by a variety of nonvisual stimuli, such as stress or pain. Furthermore, the pupillary dilation response is likely to be due, at least in part, to the inhibition of the nucleus of Edinger and Westphal.

CHAPTER 2 / THE EYE

Neural Control of Eye Movements
(Figs. 2.58; 2.70; 2.71)

This highly complex issue belongs mainly to the realm of the motor systems, which fall outside the scope of this book. However, it is still relevant to the function of the eye and is also closely associated with the vestibular system, discussed in another chapter.

For a continuous search for novel objects of interest in the visual field, for the fixation of such objects, and for keeping track of the visual environment, the gaze as well as the position of the head and body must be optimized and maintained at all times.

At the different levels of hierarchy, the main regions involved in the neural control of eye movements comprise cortical areas, the superior colliculus, midbrain eye field, pontine eye field, and perihypoglossal nuclei (Fig. 2.70).

Three main cortical regions are involved in the higher organization of ocular movements. The frontal eye field in Brodmann 8 is mainly responsible for voluntary oculomotor control, focusing the gaze on objects worthy of attention, while ignoring others. Projections from here pass in the anterior limb of the internal capsule and form a dorsal tract (with thalamic connections and terminals in the deep superior colliculus), and a ventral tract, which descends mainly to the pontine paramedian reticular formation (PPRF). The occipital cortex (mainly areas 18 and 19) is responsible for controlling oculomotor reflexes, including the fixation reflex. This region has ample connections with the pretectum, superior colliculus, and posterior thalamus. The parietal cortex, primarily area 7, is likely to play a role in maintaining the attention to visual targets. Its afferents come from the limbic forebrain (cingulate gyrus), frontal and occipital cortical areas, and it is also reciprocally connected to the superior colliculus and pretectum.

A principal target of the retinotectal pathway, the superior colliculus is a paired elevation in the rostral quadrigeminal plate of midbrain (Fig. 2.58). The nucleus has a laminar structure and a topographic representation of the contralateral visual field. Retinal projections reach the dorsal (sensory) part, which is likely to participate in visual processing at a subcortical level, although this faculty is toned down in primates due to the predominance of the geniculocortical system. The ventral motor portion of the superior colliculus is mainly responsible for orientation movements of the eyes and head in response to novel visual stimuli. An important function is rotational "remapping" of the visual field, whereby the collicular representations have to cross the midline via the tectal commissure. The chief descending projection is the crossed tectospinal tract.

The premotor areas of the brainstem comprise the vertical eye movement field of midbrain (rostral intermediate

Fig. 2.69

Simplified diagram of the pathways concerned with the pupillary light reflex and visuomotor reflexes.

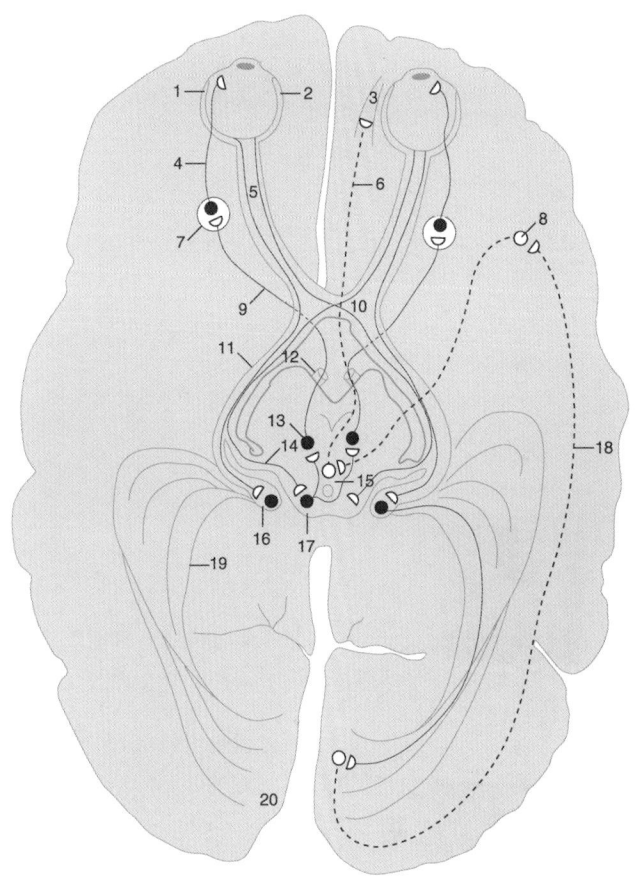

1. Temporal retina
2. Nasal retina
3. Medial rectus muscle
4. Short ciliary nerve (postganglionic fibers for the iris sphincter)
5. Optic nerve
6. Oculomotor branch for the medial rectus muscle
7. Ciliary ganglion
8. Frontal eye field
9. Oculomotor root (preganglionic fibers for the iris sphicter)
10. Optic chiasma
11. Optic tract
12. Oculomotor nerve
13. Accessory oculomotor nucleus (of Edinger and Westphal)
14. Superior quadrigeminal brachium
15. Oculomotor nuclei and premotor area
16. Lateral geniculate nucleus
17. Pretectal nucleus
18. Superior longitudinal fasciculus
19. Optic radiation
20. Visual cortex

Fig. 2.70

Schematic drawing of the brainstem nuclei associated with the rostral medial longitudinal fasciculus and the control of eye movements. Sagittal section.

1. Rostral intermediate area of the medial longitudinal fasciculus (riMLF)
2. Nucleus of Darkschewitsch
3. Superior colliculus
4. Interstitial nucleus of Cajal
5. Oculomotor nucleus
6. Inferior colliculus
7. Mamillary body
8. Third cranial nerve (oculomotor nerve)
9. Trochlear nucleus
10. Pontine paramedian reticular formation (PPRF)
11. Cerebellum
12. Pons
13. Abducent nucleus
14. Vestibular nuclei
15. Perihypoglossal nuclei

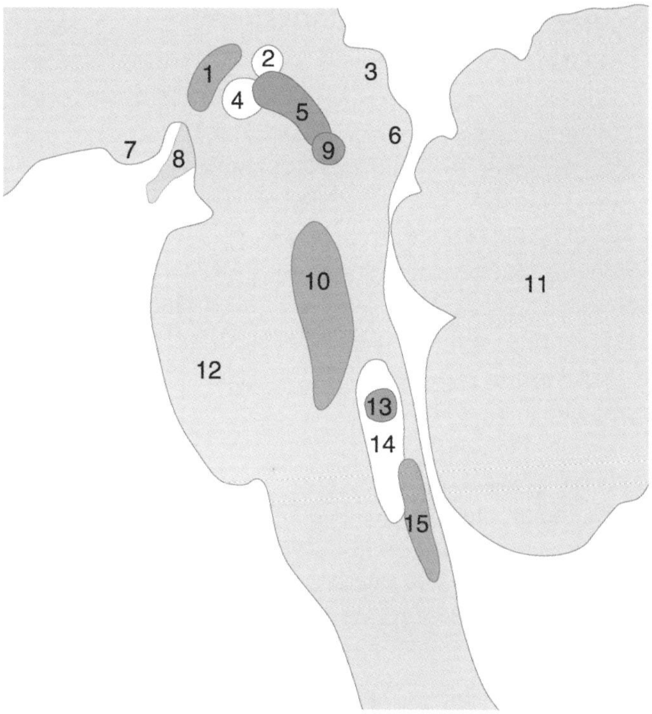

area of the medial longitudinal fasciculus [riMLF]), the horizontal eye movement field of pons (PPRF) and the perihypoglossal nuclei of medulla oblongata. The latter nuclei (prepositus, intercalatus, and sublingual) are responsible for an integration of vertical and horizontal eye movements. All these coordinating centers are linked to the relevant oculomotor nuclei (of cranial nerves III, IV, and VI) as well as to the vestibular nuclei and the vestibulocerebellum. The most important long connection between these centers is the MLF (Fig. 2.71). Descending axons of this fiber stream originate from the superior colliculus, interstitial nucleus of Cajal (INC), reticular formation, and the medial vestibular nucleus, reaching the motoneurons of the upper cervical segments. The ascending connections arise from the vestibular nuclei (mainly superior and lateral) and terminate in the nuclei of the oculomotor, trochlear, and abducent nerves. There are also internuclear connections between these centers within the MLF.

Apart from a voluntary gaze control, two reflexes are associated with oculomotor activity, known as the vestibulo-ocular and optokinetic reflexes. The vestibulo-ocular reflex is instrumental in adjusting the position of the eyes according to the position of the head. Turning of the head is followed by an eye movement in an opposite direction (counter-rolling). The input signal comes from the receptors of the labyrinth via the vestibular nerve. For example, turning the head to the right elicits an increased firing in the right lateral ampullar nerve. The right vestibular nuclei send an excitatory signal to the contralateral (left) abducent nucleus (via the MLF), which projects to the lateral rectus muscle of the left eye. The oculomotor nucleus innervating the medial rectus of the right eye receives fibers back-crossing from internuclear neurons of the abducent nucleus (apparently the nuclear portion innervating the medial rectus does not receive direct vestibular or reticular input). The result is a conjugated horizontal shift of both eyes to the left. This type of eye movement (nystagmus) consists of a slow following response in a direction opposite to the head turn and a rapid "rebound" component (see also the chapter on equilibrium sensation). Vertical eye movements can be explained by a similar, if more complex mechanism involving the oculomotor and trochlear nuclei.

The optokinetic reflex is a response of the eye to the rapid movement (e.g., rotation) of the visual field. The optokinetic nystagmus consists of two components: first the eye follows the moving scene and then it rapidly jumps back with a saccade to fix a fresh target. The stimulus for the optokinetic reflex arises mainly from the peripheral retina and the final effector system is largely shared with that of the vestibulo-ocular reflex, thus the two systems act in a complementary fashion in gaze stabilization, depending on the postural change. Individuals prone to sea-sickness often have an easier ride if they watch the waves because the unpleasant vestibular signals triggered by the motion of the boat are somewhat dampened by visual anticipation and muscle control. Conversely, people in an evenly rotating chair tend to close their eyes because the vestibular stimulus is rapidly attenuated (the cristae ampullares are sensitive to angular acceleration or deceleration but not to an even rotation; see also the relevant chapter), whereas the moving visual scene continues to cause dizziness by the optokinetic mechanism. The optokinetic reflex involves a connection between the retinal afferents and the vestibular nuclei via the accessory optic nuclei. The latter comprise the nucleus of the optic tract (NOT) and further target nuclei in the pretectum and rostral midbrain, which have descending connections to the pontine and perihypoglossal nuclei, the inferior olive, vestibular nuclei, and vestibulocerebellum. Essentially, the accessory optic system is geared to compensate for a shifting of the retinal image (retinal slip) owing to postural changes and locomotion.

Whether voluntary or driven by reflex, the movements of the eye belong to one of the following physiological categories: saccade, smooth pursuit or vergence. Saccades are rapid eye movements seen, e.g., in the second phase of nystagmus or in the rapid eye movement phase of sleep. The first (pulse) component of the saccade brings gaze to the new position. The neural centers for horizontal and vertical saccades are in the PPRF and riMLF, respectively, that is, the premotor nuclei already mentioned. The neurons selective for saccadic motor activity discharge "bursts," whose firing rate depends on the velocity of eye movement. The pulse component is followed by a step component, where the new position is maintained by a tonic input from the neurons of PPRF. Apparently, some neurons of the PPRF and the brainstem reticular formation are "programmed" to encode the eye position. Other units in the reticular formation or certain Purkinje cells in the flocculus of cerebellum encode "gaze velocity" information, firing proportionally to the speed of tracking.

Smooth pursuit or tracking movements are aimed at maintaining a moving image on the fovea. This function is controlled by the parieto-occipital cortex and the frontal eye field, and also supported by the superior colliculus.

Vergence movements are elicited by a disparity of retinal images, when the object falls on noncorresponding retinal points of the two eyes. This type of movement is rather slow by comparison with the other discussed movements and it involves simultaneous innervation of the medial rectus on both sides. The precise origin of the premotor command is yet to be determined. Based on unit recordings, the most likely source is a group of neurons lateral to the oculomotor nuclei. Higher control of vergence involves areas 19 and 22 of the occipital cortex, whereas the sensory input is likely to arrive from disparity-sensitive units of the visual cortex.

Fig. 2.71

Simplified diagram of the vestibulo-ocular reflex pathway (exemplified by horizontal eye movement) and the descending cortical pathways (exemplified by voluntary gaze control). FEF – frontal eye field.

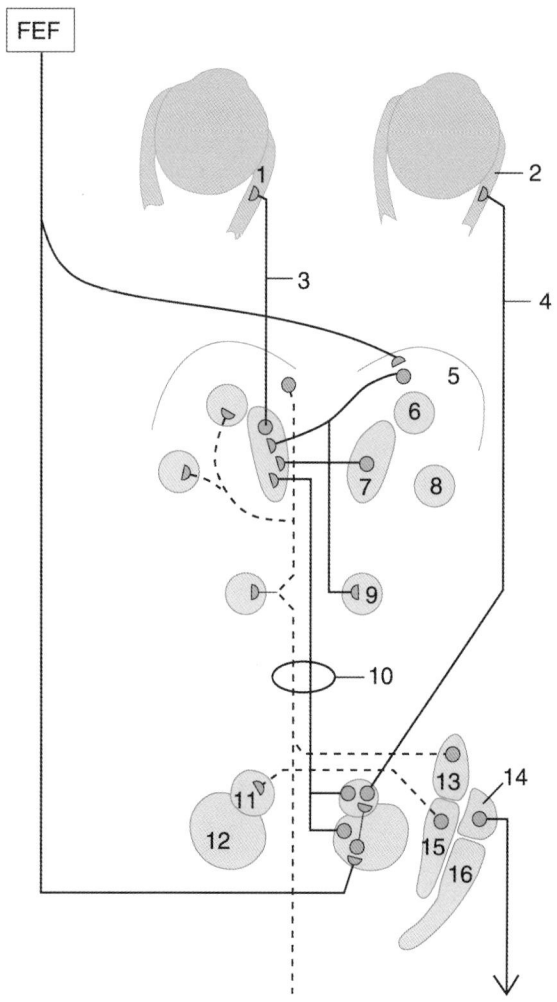

1. Medial rectus of left eye
2. Lateral rectus of right eye
3. Oculomotor branch for medial rectus
4. Abducent nerve for lateral rectus
5. Vertical eye field of midbrain
6. Nucleus of Darkschewitsch
7. Oculomotor nucleus
8. Interstitial nucleus of Cajal
9. Trochlear nucleus
10. Medial longitudinal fasciculus
11. Abducent nucleus
12. Horizontal eye field of pons
13. Superior vestibular nucleus (of Bechterew)
14. Lateral vestibular nucleus (of Deiters)
15. Medial vestibular nucleus (of Schwalbe)
16. Inferior vestibular nucleus (of Roller)

RECOMMENDED READINGS

Textbooks and Handbooks

1. Bannister LH, Berry MM, Collins P, Dyson M, Dussek JE, Ferguson MWJ. Gray's Anatomy, 38th ed., Churchill Livingstone, Edinburgh, 1999.
2. Bron AJ, Tripathi RC, Tripathi BJ. Wolff's Anatomy of the Eye and Orbit (8th ed.), Chapman & Hall Medical, London, 1997.
3. Fini ME (Ed.). Vertebrate Eye Development, Springer, Berlin, Heidelberg, New York, 2000.
4. Gulyás B, Ottoson D, Roland PE (Eds). Functional Organisation of the Human Visual Cortex, Pergamon Press, UK, 1993.
5. Hinrichsen KV (Ed.). Humanembryologie. Lehrbuch und Atlas der vorgeburtlichen Entwicklung des Menschen, Springer, Berlin, 1990.
6. Hogan MJ, Alvarado JA, Weddell JE. Histology of the Human Eye, WB Saunders, Philadelphia, 1971.
7. Kandel ER, Schwartz JH, Jessell TM (Eds.). Principles of Neural Science, 4th ed., McGraw-Hill/Appleton & Lange, 2000.
8. Smith C.U.M. Biology of Sensory Systems, John Wiley & Sons, Chichester, 2000.
9. Snell RS, Lemp MA. Clinical Anatomy of the Eye, 2nd ed., Blackwell Science, US, 1998.

Reviews and Research Reports

1. Clarke S. Association and intrinsic connections of human extrastriate visual cortex. Proc R Soc Lond B 1994;257:87–92.
2. Clarke S, Maeder P, Meuli R, et al. Interhemispheric transfer of visual motion information after a posterior callosal lesion: A neuropsychological and fMRI study. Exp Brain Res 2000;132:127–133.
3. Di Virgilio, G, Clarke S. Direct interhemispheric visual input to human speech areas. Human Brain Mapping 1997;5:347–354.
4. Hubel DH, Wiesel TN. Functional architecture of the macaque monkey visual cortex. (Ferrier Lecture), Proc R Soc Lond (Biol) 1977;198:1–59.
5. Szentágothai J. The 'module-concept' in cerebral cortex architecture. Brain Res 1975;95:475–496.
6. Wong-Riley M, Hevner R, Cutlan R, et al. Cytochrome oxidase in the human visual cortex; distribution in the developing and the adult brain. Visual Neurosci 1993;10:41–58.
7. Zilles K, Clarke S. Architecture, connectivity, and transmitter receptors of human extrastriate visual cortex. Comparison with nonhuman primates, In: Rockland et al., Eds, Cerebral Cortex, vol. 12, Plenum Press, New York, 1997:673–742.

3 The Organ of Olfaction

András Csillag

ANATOMICAL OVERVIEW OF THE ORGAN OF OLFACTION

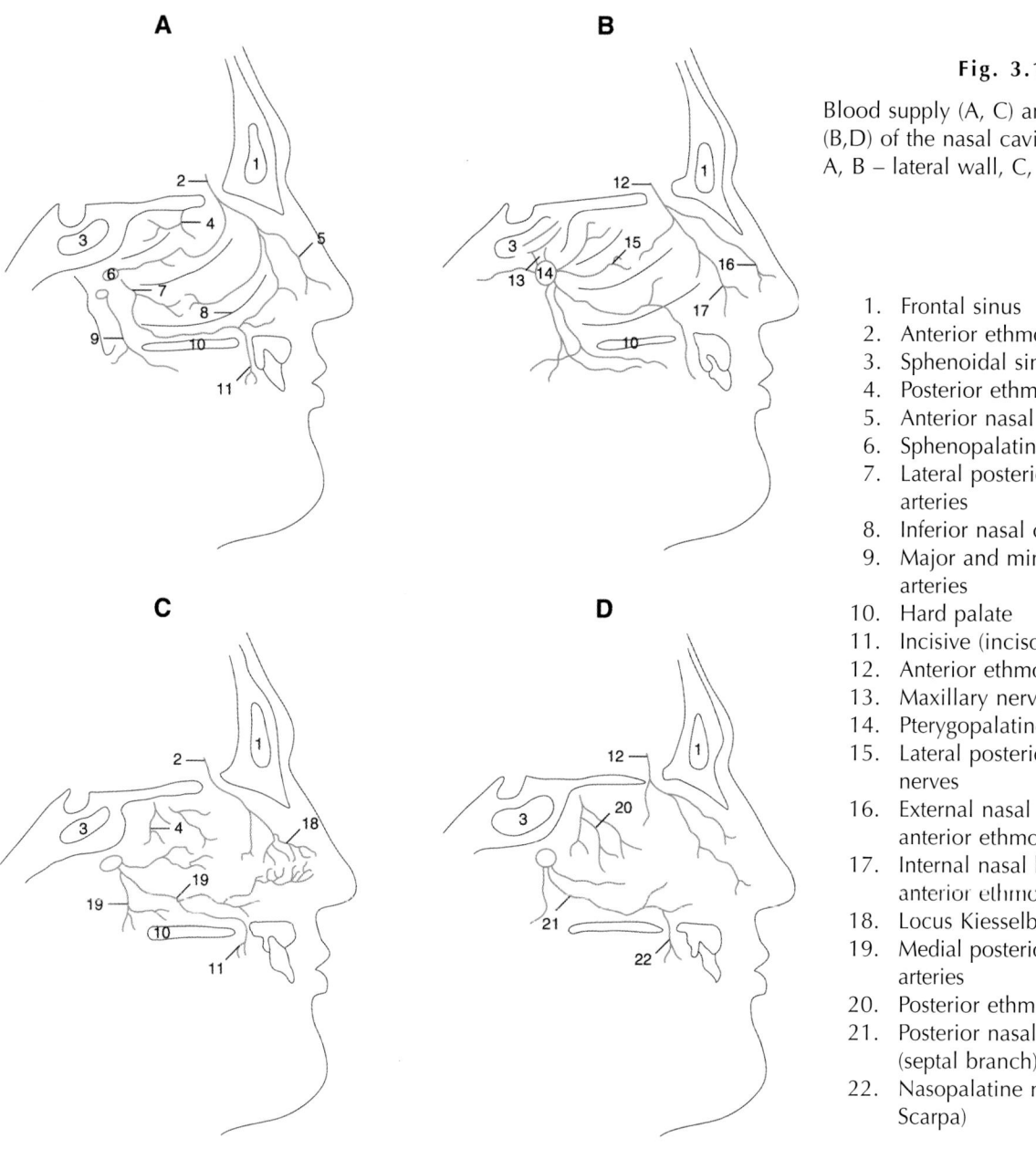

Fig. 3.1
Blood supply (A, C) and innervation (B,D) of the nasal cavity.
A, B – lateral wall, C, D – septum.

1. Frontal sinus
2. Anterior ethmoidal artery
3. Sphenoidal sinus
4. Posterior ethmoidal artery
5. Anterior nasal artery
6. Sphenopalatine artery
7. Lateral posterior nasal arteries
8. Inferior nasal concha
9. Major and minor palatine arteries
10. Hard palate
11. Incisive (incisor) artery
12. Anterior ethmoidal nerve
13. Maxillary nerve
14. Pterygopalatine ganglion
15. Lateral posterior nasal nerves
16. External nasal branch of anterior ethmoidal nerve
17. Internal nasal branch of anterior ethmoidal nerve
18. Locus Kiesselbachi
19. Medial posterior nasal arteries
20. Posterior ethmoidal nerve
21. Posterior nasal nerve (septal branch)
22. Nasopalatine nerve (of Scarpa)

Olfactory sensation belongs to the chemical senses, in particular for the detection of volatile molecules that are traveling freely in the atmosphere. Conversely, the other main chemical sensory function, taste, is based on the detection of compounds that are dissolved in liquid. This difference, however, is not as sharp as it appears because in both cases the molecules are ultimately dissolved in a mucous film overlying the receptor cells. If anything, taste and smell are far more separated by the way their signals reach the brain: olfactory information is the only modality in vertebrate animals that can access the telencephalon directly. All the other sensory inputs, taste included, reach the telencephalic centers via a diencephalic (thalamic) relay. This situation underscores the paramount importance of odor signals in the animal kingdom and, albeit to a lesser degree, also in humans.

The primary sensory apparatus of smell detection is situated in the olfactory mucosa of the nasal cavity. In man, this highly specialized region occupies a relatively small proportion of the nasal epithelium in the vicinity of the roof of the nasal cavity. Yellowish in color (as opposed to the reddish tint of respiratory mucosa), the olfactory region extends to the superior turbinate, spheno-ethmoidal recess, the opposite part of the septum and the arching roof connecting the lateral and medial regions, underlying the cribriform plate. The vacular supply and innervation of the region in relation to the nasal cavity are shown in Fig. 3.1. Overall, the olfactory region represents an area of about 500 mm^2 in a corner of the nasal cavity, less exposed to the flow of air during quiet breathing. Active exploration of odors, however, is accompanied by pulses of deep inhalation called sniffing.

Fig. 3.2

Block scheme of olfactory epithelium. Adapted and modified after Bannister et al., 1999, *Gray's Anatomy*.

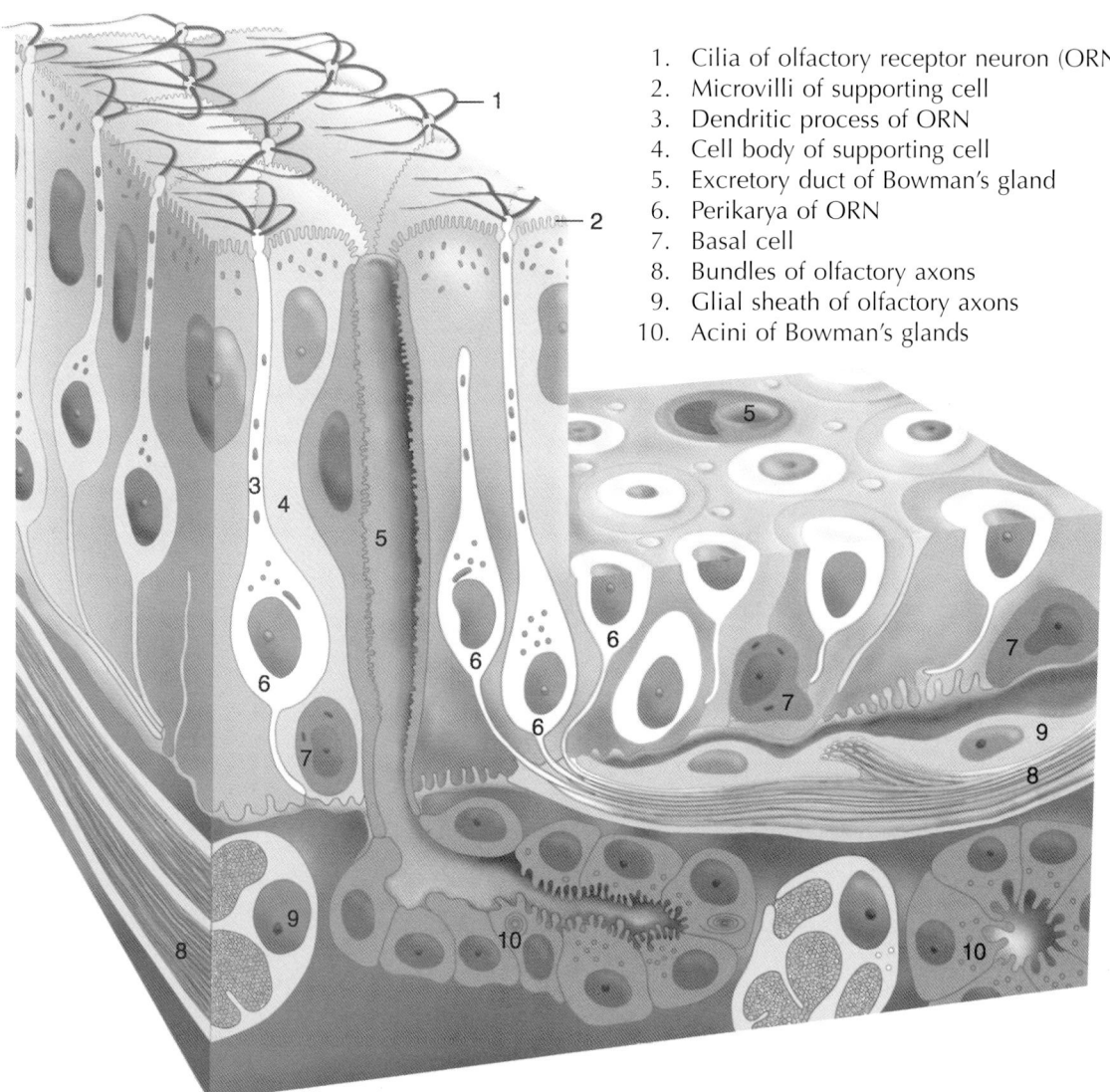

1. Cilia of olfactory receptor neuron (ORN)
2. Microvilli of supporting cell
3. Dendritic process of ORN
4. Cell body of supporting cell
5. Excretory duct of Bowman's gland
6. Perikarya of ORN
7. Basal cell
8. Bundles of olfactory axons
9. Glial sheath of olfactory axons
10. Acini of Bowman's glands

Histologically, the olfactory epithelium corresponds to pseudostratified columnar ciliated epithelium and, in this respect, it is similar to the surrounding respiratory epithelia (Fig. 3.2). However, the olfactory mucosa can be clearly distinguished by its greater thickness (taller epithelial cells) and the presence of the olfactory glands of Bowman (Figs. 3.8; 3.9). The principal elements (i.e., the olfactory receptor neurons [ORN]), are themselves less conspicuous in routine histological specimens because their slender processes are obscured by the supporting (sustentacular) cells. The nuclei of cells are arranged in three main layers, those of the sustentacular cells lying above, those of basal cells below, and those of ORNs occupying an intermediate position. The ORNs are bipolar cells whose bodies and nuclei are loosely packed because of the large number of supporting cells interspersed between them. The peripheral process of each ORN ends in a dendritic knob (also termed rod or olfactory vesicle) that reaches the surface and puts out a number of very long hair-like cilia. The cilia are embedded in a film of mucus, secretion product of the subepithelial glands of Bowman, covering the surface of the epithelium. The central process of the ORN leaves the epithelium in a basal direction (the basement membrane is inconspicuous) and joins other processes to form nerve bundles. These bundles have no myelin sheath but they are encapsulated by a special type of glia known as olfactory ensheathing glia, resembling Schwann cells but clearly belonging in a different category. This accompanies the fibers as they pass through the cribriform plate of the ethmoid (the visible bundles of fibers are termed olfactory filaments, fila olfactoria) as far as the olfactory bulb, where they terminate (Fig. 3.13). The olfactory glial envelope of the axons is replaced by central glia just before the axons synapse within the bulb.

Under the electron microscope the dendritic swellings of olfactory neurons are packed with dark mitochondria but are otherwise relatively poor in subcellular organelles, such as endoplasmic reticulum and Golgi apparatus (Figs. 3.11; 3.12). The cilia have basal bodies and a regular ("9 + 2") system of microtubular doublets, whereas in the distal parts only the central pair is visible. Based on their ultrastructure and functional studies, the cilia of ORNs are presumably nonmotile, at least in mammals. The receptor cells are connected to the sustentacular cells with tight junctions near the surface and zonulae adherentes below. The presence of numerous desmosomal contacts is further supported by the observation that the olfactory epithelium (including Bowman's glands) is intensely immunoreactive to the major desmosomal protein, desmoplakin. The supporting cells have a dense cytoplasm containing numerous organelles (endoplasmic reticulum, microtubules, lysosomes, pigment granules) and their surface is studded with microvilli (i.e., cytoplasmic outgrowths without the structural characteristics of cilia). The basal cells of the olfactory epithelium are capable of proliferation throughout life. They can transform into new ORNs, which occupy their definitive position in the epithelium as substitutes for lost receptor cells. Even under normal conditions ORNs are renewed in every 30 to 60 days. Apparently, the repair capacity of the olfactory mucosa tends to decline in old age, with a gradual attenuation of the sense of smell.

The subepithelial glands of Bowman are tubuloalveolar glands of seromucous nature (Figs. 3.9; 3.11). The secretion product is discharged via thin tubular ducts opening on the epithelial surface. The lamina propria is also rich in blood vessels and lymphatics, however it does not contain the massive venous plexus characteristic for the respiratory mucosa of nasal cavity.

Since the central neurites of ORNs are capable of regenerating even in adult age, and the fibers have to reach the olfactory bulb to form new synapses, the olfactory system is one of the best models of postembryonic axonal pathfinding. The presence of the cellular adhesion molecule (nCAM) in the olfactory nerve fibers of adult humans is indicative of the fibers' ability to continually seek new targets in the CNS (Figs. 3.10; 3.13).

ORNs are capable of responding to odor signals by the generation of action potentials propagating down their central neurites. Similarly to other sensory modalities (e.g., vision), olfactory information is composed of elementary signals, usually representing classes of relatively simple chemical compounds. In most terrestrial animals including man such compounds can be alcohols, aromatic groups, esters, fatty acids, or steroids. However, olfactory perception always involves more complex identification of odorants that collectively form "odor objects" (e.g., rose oil, jasmine fragrance, or the unmistakable stench of the skunk). Thus, the olfactory system faces a similar task to that of the visual system in the case of object or face recognition. Indeed, macrosmatic species (e.g., trained dogs) can recognize a single individual as one odor object, a feat obviously too daunting for microsmatic humans but even they are able to identify many well known odors as objects.

A long-debated issue has been whether ORNs and their central projections are arranged topographically, according to some kind of "odor map." Results from previous studies and current research suggest that ORNs that express a particular receptor gene are organized roughly in zones of the olfactory mucosa. More interestingly still, it now seems increasingly evident that all those ORNs expressing identical or similar receptor genes tend to project upon a single or very few neighboring glomeruli of the olfactory bulb. Thus, one glomerulus—or, at best, a handful of closely related glomeruli—would be responsible for the processing of one particular piece of odor information. Modern activity-mapping studies (such as the incorporation of radioactive 2-deoxyglucose) have clearly indicated that the distribution of "hot spots" of neuronal activity within the bulb reflects the type of odorant molecule. In apparent contradiction with this notion is the fact that all or most vertebrate ORNs are "generalists" rather than "specialists" (the latter is more common in insects), that is, they respond to a class of odorant and not just to a single molecular structure. In addition, odorant detection is concentration dependent and the response of ORNs is prone to generalization, meaning that more units are recruited when the stimulus is sufficiently intense. This can dramatically alter the nature of the perceived odor, too. For example, indol in high dilution has a florid fragrance but it appears putrid when undiluted.

Receptor Genes

The molecular mechanism of odor detection is based on a host of olfactory receptors determined by a surprisingly large family of genes. In the human genome, approx 900 olfactory-relevant genes have been identified. Of these, 350 are functional. The mouse genome contains a similar number of olfactory genes but here all 900 genes are functional. The receptors themselves belong to the extensive group of G protein-coupled receptors with seven transmembrane units (7TM). The binding sites are exposed in "binding pockets," whose molecular constituents representing individual stimulus sources are called epitopes. Epitope variability is based on permutations of gene expression. Binding to the olfactory receptors may be mediated through odor-binding proteins, produced by the olfactory glands, and the receptors have a broad spectrum of sensitivity (molecular receptive range), at least in vertebrates. The whole process of olfactory receptor binding resembles the molecular mechanism and pattern variability that occurs in the binding of antibodies in immune responses.

Receptor Mechanisms

Following odorant binding, the activation of olfactory receptors brings about a cascade of intracellular events within the cilia. Several second messenger mechanisms have been described as instrumental in signal transduction, including cyclic AMP, inositol-triphosphate (IP_3) and cyclic GMP systems, the latter being active in the formation of gaseous transmitters (nitric oxide and carbon monoxide). Cyclic nucleotides or IP_3 open specific (cyclic nucleotide-gated [CNG]) channels of the plasma membrane, enabling the passage of cations including calcium. The entry of calcium leads to depolarization (also assisted by an increased outflow of chloride via the chloride channels), spreading first to the dendritic knob and then to the whole ORN. Summation of depolarization exceeding the threshold value results in the generation of action potentials at the axon hillock, whereby the intensity of the olfactory signal is now encoded in the frequency of spikes propagating down the olfactory nerves. On the other hand, calcium is a powerful inactivator of CNG channels and 7TM receptors, which explains the adaptation observed in lasting exposures to odorants, despite the fact that the olfactory cilia themselves are not prone to desensitization.

DEVELOPMENT OF THE NASAL AND ORAL CAVITIES IN RELATION TO OLFACTORY AND TASTE SENSATION

The following components participate in the development (see also the development of tongue in Chapter 4):

- condensation of prechordal mesenchyme surrounding the site of the oropharyngeal membrane
- occipital myotomes
- prechordal plate (oropharyngeal membrane)
- the placodes surrounding the rostral edge of the neural tube, from front to back: the common olfactory placode and the paired lens placode, trigeminal placode, otic placode, epibranchial placode
- cranial end of neural crest
- head mesoderm derived from neural crest and constituting the facial swellings and branchial arches. The paired facial swellings (maxillary and mandibular) correspond to a smaller dorsal and a larger ventral portion of the first branchial arch.

In addition to the olfactory mucosa, the common olfactory placode forms also the adenohypophysis and the vomeronasal organ. Owing to the growth of the brain and facial swellings, the three primordia separate, the unpaired hypophyseal anlage forming Rathke's pouch deep to the oral cavity.

The proliferation of mesenchyme around the olfactory placode raises the medial and lateral nasal swellings resulting in the invagination of olfactory pits (Fig. 3.3). The sensory epithelial cells of the future olfactory mucosa extend their neurites to form the olfactory nerves entering

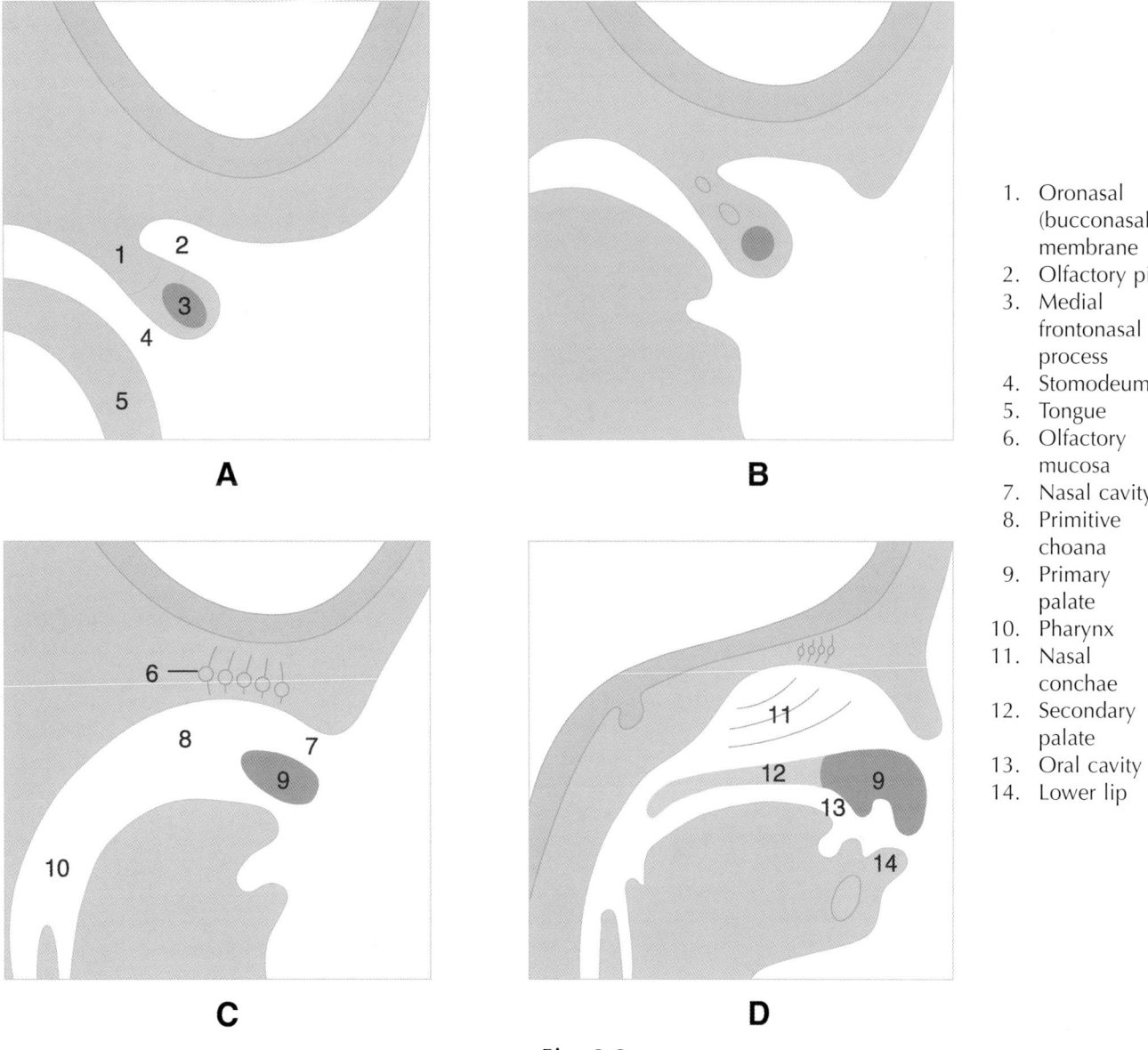

Fig. 3.3

Development of the nasal cavity (sagittal section). A – week 6, B – perforation of the oronasal membrane, C – week 7, D – week 9.

the prosencephalon. The anterior part of the definitive palate (future incisive bone) is formed by a fusion of the medial nasal swellings in the midline.

The primitive oral cavity (stomodeum) is visible once the maxillary and mandibular processes have been formed. The ectodermal lining of the cavity is separated from the endoderm of the pharynx by the oropharyngeal membrane, soon to be perforated.

Communication between the primitive nasal and oral cavities is blocked by the primary palate at first but then the posterior part of the palate (known as the oronasal membrane) becomes perforated to form the primitive choanae.

Following a rapid growth of lingual tissue, the tongue comes to occupy the entire oral cavity, protruding even into the primary choanae (Fig. 3.4). Later, the longitudinal growth of the embryo and the formation of the neck bring about a caudad shift of the heart, thymus, and thyroid, and also a similar displacement of the floor of the oral cavity together with the tongue. The latter movement makes way for the upward (horizontal) deflection of the palatine processes, which were formerly hanging down on both

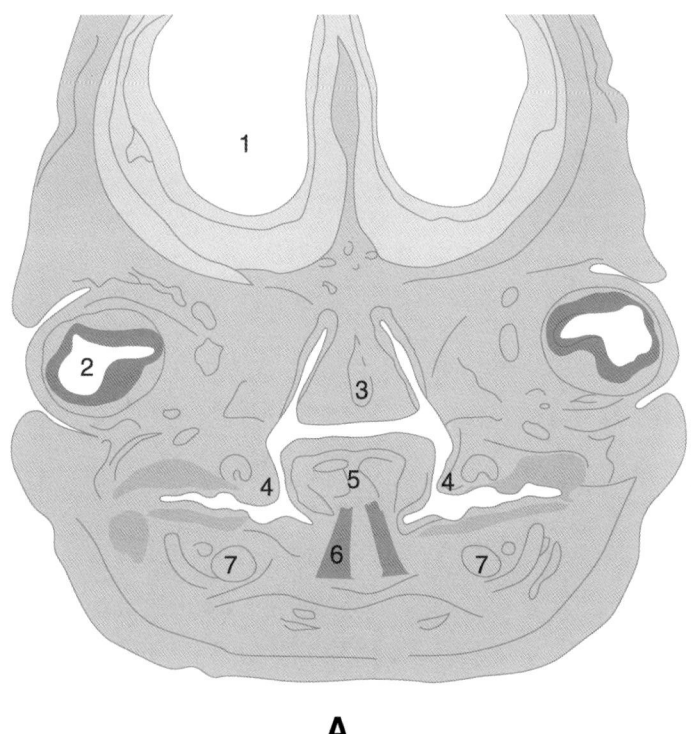

Fig. 3.4

Development of the nasal and oral cavities (after Zenker). Schematic drawing based on a histological specimen.
A – Rostral aspect B – Caudal aspect

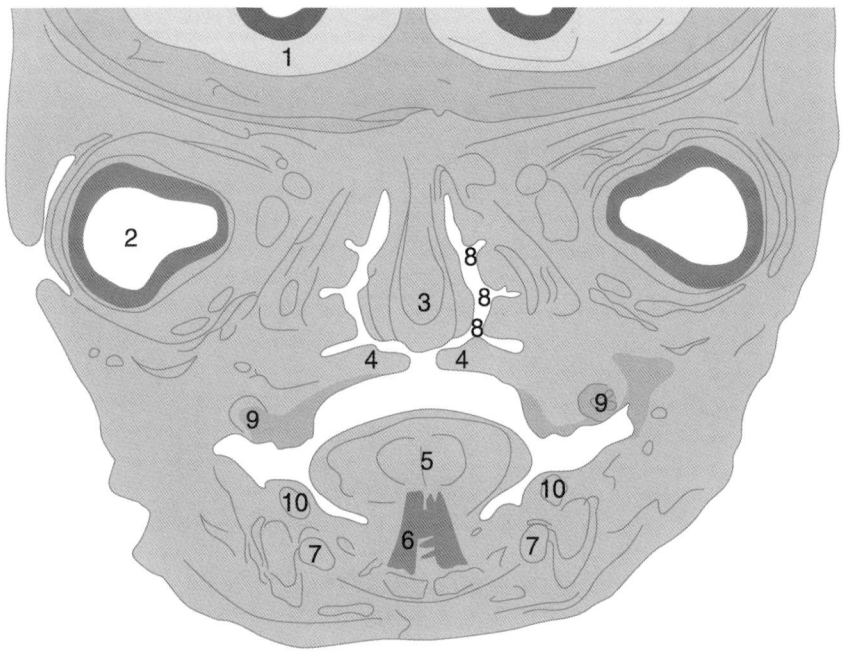

1. Brain vesicle
2. Primordium of eye
3. Nasal septum
4. Palatine process
5. Tongue
6. Hyoglossus muscle
7. Meckel's cartilage
8. Primitive nasal conchae and meatuses
9. Dentogingival lamina of maxilla
10. Dentogingival lamina of mandible

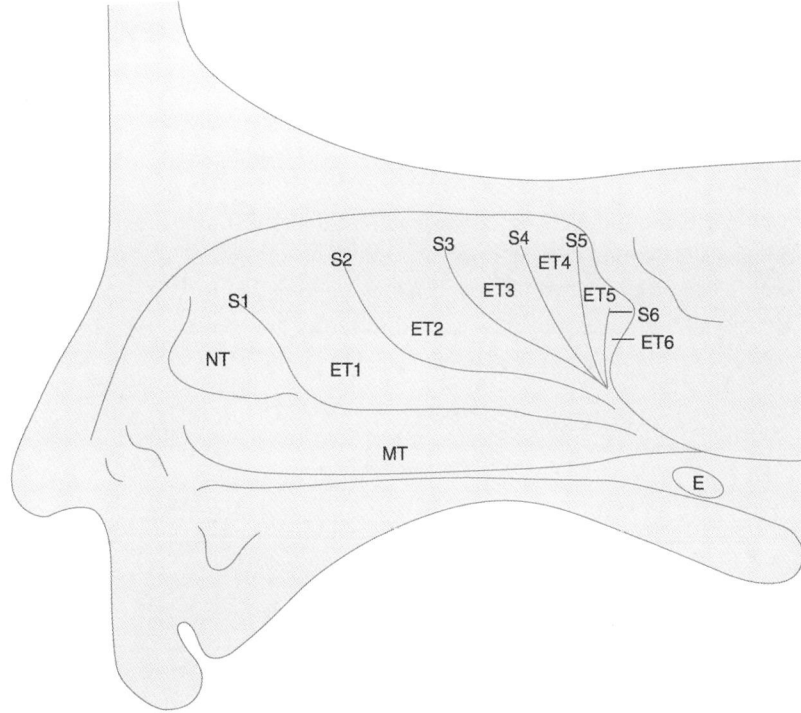

Fig. 3.5

Development of the lateral wall of the nasal cavity including the nasal conchae, Human fetus, week 36–40. After Killian and Peter (1902).

ET – ethmoturbinale,
NT – nasoturbinale,
MT – maxilloturbinale,
E – opening of the Eustachian tube,
S1–S6 principal furrows

sides of the tongue. This elevation of the palate is likely supported by some kind of "erectile" mechanism as a result of an increase of water content of the tissue.

The following fusion lines appear between the above described swellings.

- The mylohyoid raphe and the median lingual groove from fusion of the mandibular processes on the oral floor.
- The buccal raphe (sometimes signified by a prominent scar-like crest, known as torus villosus, on the inner side of the cheek), derived from fusion of mandibular and maxillary processes.
- The fusion line between the maxillary and frontal processes, stretching along the nasolacrimal duct and on both sides of philtrum.
- The fusion line between the two palatine processes and the primitive nasal septum (a remnant persisting between the nasal pits) marks the formation of the secondary palate. The latter will be further enlarged by incorporation of a part of the maxillary processes. The primary palate persists in the region of the incisive bone. The fusion between the nasal septum and palatine processes begins anteriorly during the ninth week and is completed posteriorly by week 12 of embryonic age.

Roles in the fusion mechanism are attributed to migration of tissue from the neural crest, cellular death near the surfaces and alteration of junctional complexes between surface cells. Cleft lip, cleft palate, and other deficiencies may develop in case of incomplete or aborted fusion.

The secondary palate ends in the choanae with a free edge, thus allowing both the nasal and oral cavities to communicate with the pharynx. The upper part of the original stomodeum becomes the postero-inferior part of the definitive nasal cavity, whereas the antero-superior part corresponds to the primitive nasal cavity. Hence, the nasal cavity is supplied by both the anterior ethmoidal artery and nerve (from the neurovascular system of the primordial frontal swelling) and the maxillary artery and nerve (from the maxillary process). The boundary between the two supply regions, stretching approximately from the sphenopalatine foramen to the incisive foramen, corresponds to the position of the oronasal membrane.

Derivatives of the head mesoderm, the precursors of the bony skeleton of the nasal and oral cavities are the cartilaginous nasal capsule and the first two (Meckel's and Reichert's) branchial cartilages. The ethmoid bone, inferior nasal concha, and the hyoid bone develop by chondrogenic ossification, whereas the nasal cartilages persist in their original form. The maxilla, palatine bone, medial plate of pterygoid process and nasal bone develop by external apposition to the nasal cartilage, yet following intramembranous ossification. The vomer ossifies as two separate plates on both sides of the primitive septum (Fig. 3.7). Once the latter disappears, the two bony plates fuse. The anterior part of the mandible (rostral to the mandibular foramen) develops by intramembranous ossification from connective tissue enveloping the Meckel's cartilage. The proximal part of the mandible represents a secondary outgrowth. The only part of the mandible to develop by chondrogenic ossification directly from Meckel's cartilage is the chin (mental protuberance). The body of sphenoid is derived from the condensation of prechordal mesoderm followed by the basisphenoid (hypophysial) cartilage.

Of the paranasal sinuses, only the ethmoid air cells and the maxillary sinus appear prior to birth. The remaining sinuses develop postnatally by gradual pneumatization. Occasionally, the cavities may extend as far as the orbital process of palatine bone (known as "palatine sinus"), dangerously close to the optic nerve. At times the pneumatization of the clinoid processes or the entire orbital roof plate can also be observed, mostly without clinical symptoms.

Initially, five nasal conchae (turbinates) start to develop, the lowest in the region of maxilla (maxilloturbinale), whereas the other four belong to the ethmoid (ethmoturbinale, ethmoidoturbinale). The definive turbinates derive from the lowest three primordia, although occasionally a fourth one above persists as the supreme nasal concha (Figs. 3.5; 3.6).

Fig. 3.6

Formation of the lateral wall of the nasal cavity. After Killian and Peter (1902).

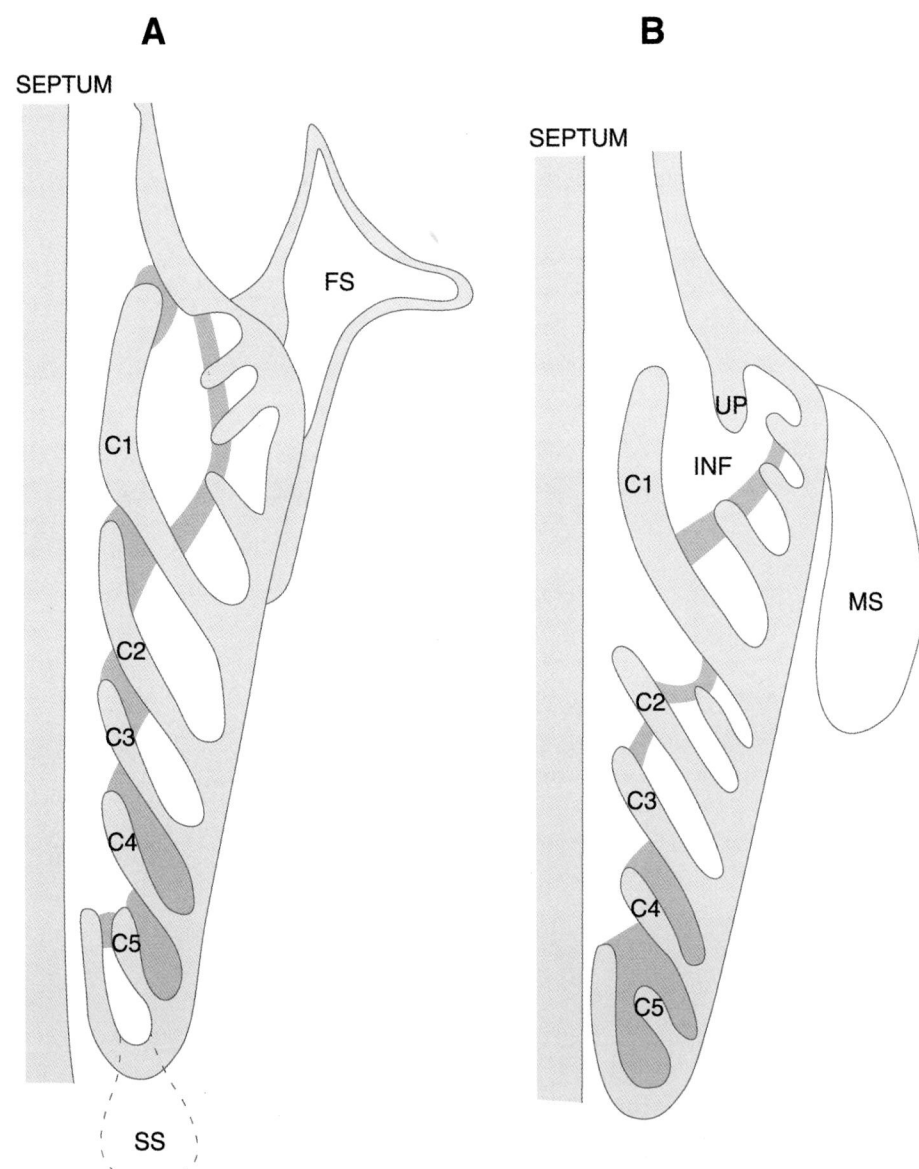

A – superior level,
B – inferior level,
C1-C5 – principal conchae,
FS – frontal sinus,
SS – sphenoidal sinus,
Up – uncinate process
INF – Infundibulum
MS – Maxillary sinus

ATLAS PLATES

Fig. 3.7

Histological specimens demonstrating the development of the nasal cavity. Coronal sections from rat embryo, Azan staining. A – rostral level, B – caudal level, C – enlarged field from A.

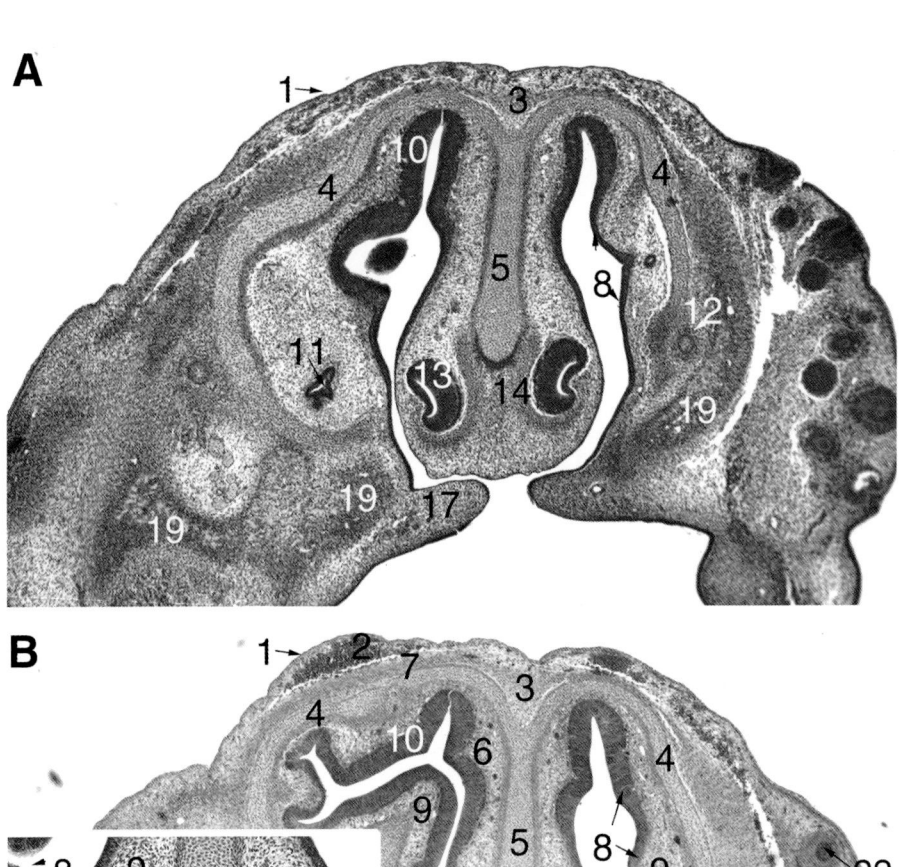

1. Epidermis
2. Hemorrhage
3. Mesenchymal condensation (nasal bone)
4. Cartilage of nasal capsule (future ethmoid bone)
5. Cartilage of nasal septum
6. Perichondrium
7. Opening of nasal bone with passing nerve branch
8. Nasal conchae
9. Mesenchyme of nasal mucosa
10. Olfactory epithelium (note the prominent thickness as compared to the other epithelia)
11. Primordium of paranasal sinus
12. Lacrimal canal
13. Primordium of vomeronasal organ
14. Mesenchymal condensation of future vomer (vomeronasal cartilage)
15. Primordium of buccinator muscle
16. Primordium of muscle of mastication
17. Palatine process after elevation but prior to fusion
18. Low epithelium earmarked for fusion
19. Mesenchymal condensation of maxilla
20. Dentogingival lamina
21. Vestibulum
22. Developing hair follicles (of mystacial vibrissae)

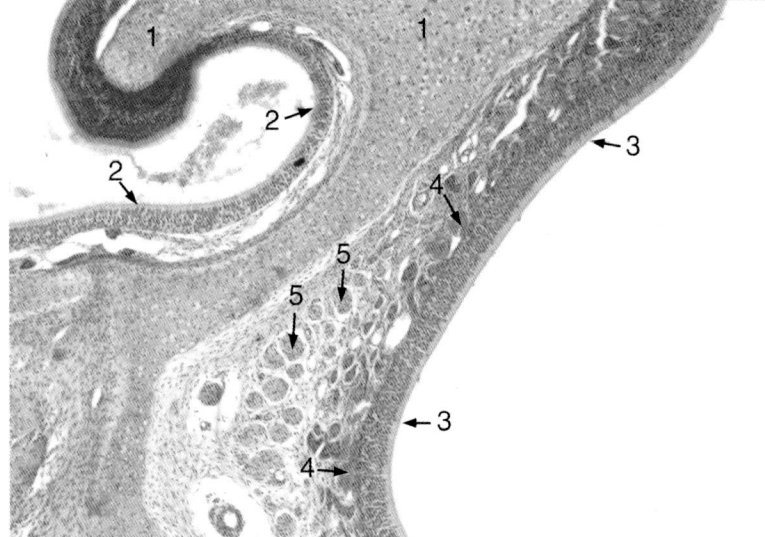

Fig. 3.8

Nasal concha, HE staining.

1. Nasal concha, still cartilagineous
2. Respiratory epithelium (pseudostratified columnar, with goblet cells)
3. Maxillary sinus, arrow point at the epithelial lining
4. Venous plexus
5. Nerves in the lamina propria

Fig. 3.9

Light micrographs comparing the histological characteristics of the respiratory (A) and olfactory (B) mucosae. Paraffin embedded sections from human nasal cavity, HE staining.

1. Mucus
2. Nuclear-free surface layer of columnar epihelial cells
3. Goblet cell
4. Basal cell layer
5. Basement membrane (prominent only in respiratory epithelium)
6. Veins of subepithelial venous plexus
7. Collagen fibers in lamina propria
8. Nuclei of supporting cells
9. Cell bodies and nuclei of olfactory receptor neurons (ORN)
10. Dendritic process of ORN
11. Bowman's glands
12. Cell-rich lamina propria

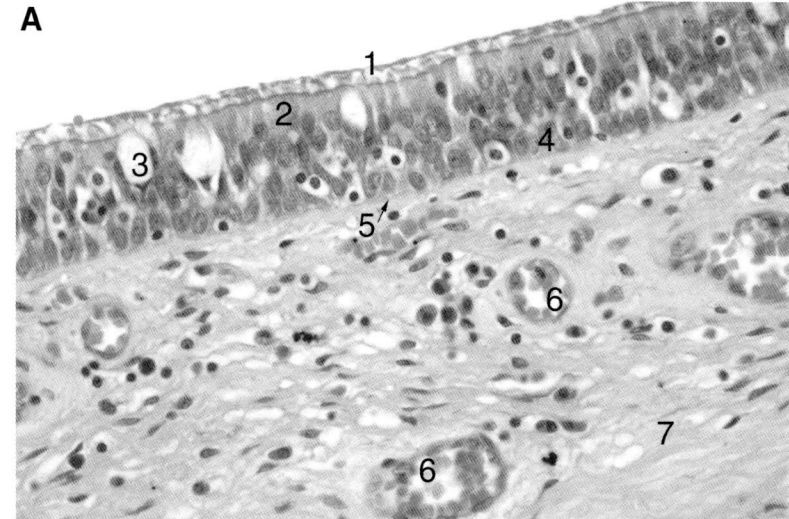

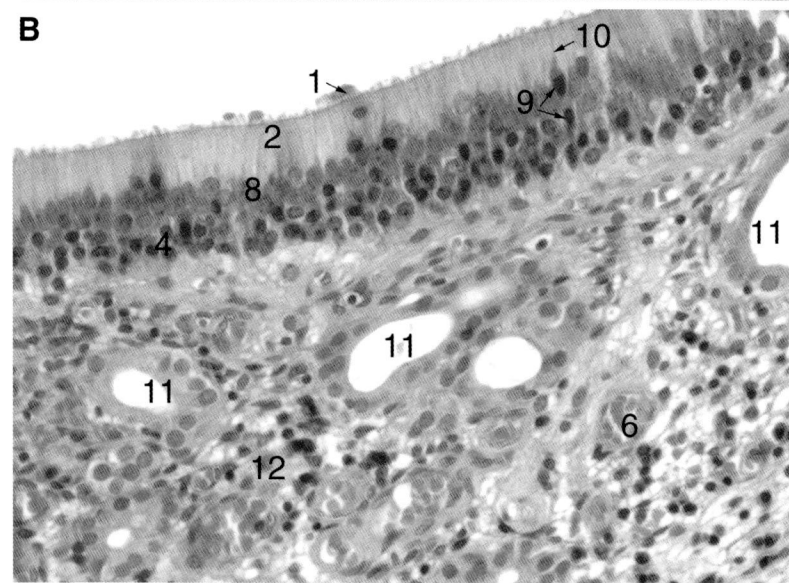

Fig. 3.10

Medium- (A) and high- (B) power micrographs of the olfactory mucosa following immunoreaction against neural cell adhesion molecule (nCAM), using the immunoperoxidase method and haemalaun counterstaining. Paraffin-embedded sections from decalcinated human tissue. The olfactory nerve fibers are highlighted by the immunostaining.

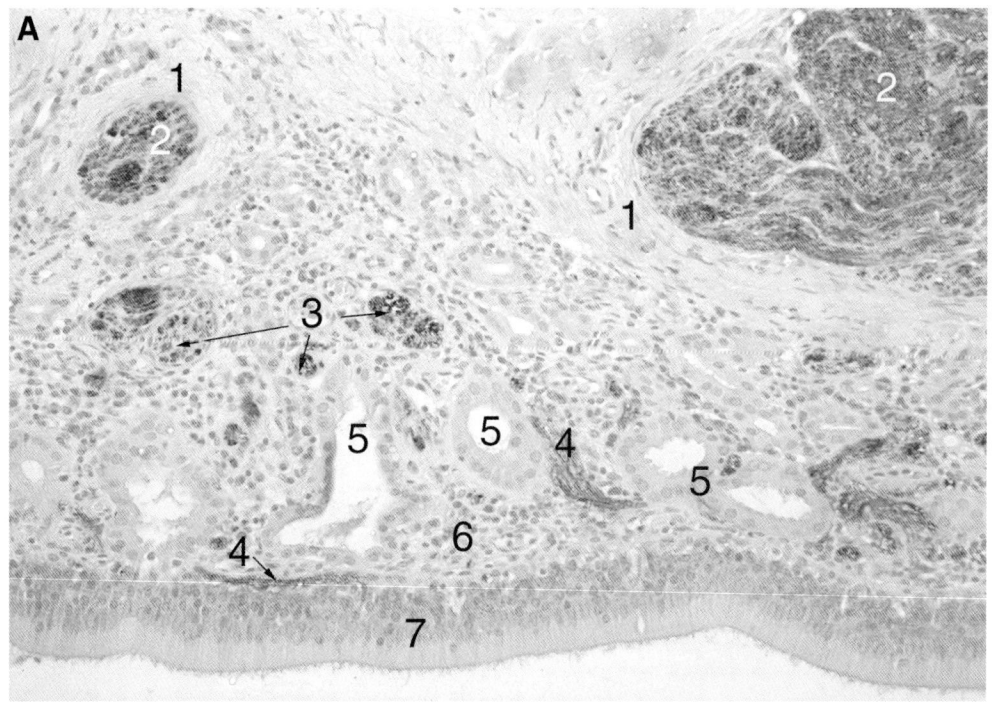

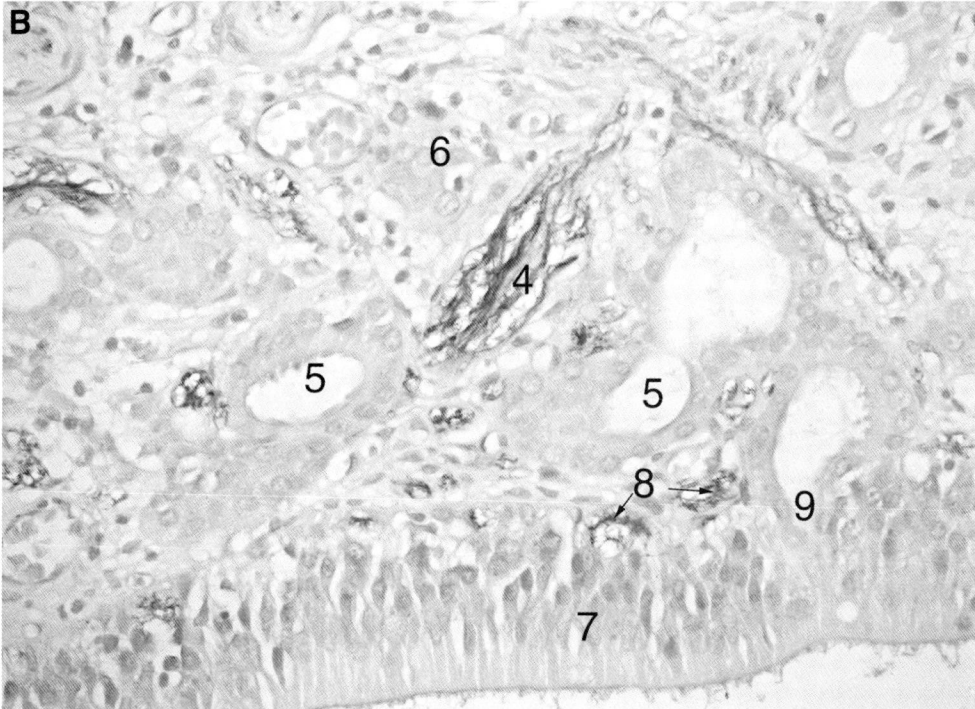

1. Perineurium
2. Large olfactory nerve bundles surrounded by fibrous perineurial coating associated with the periosteum/ endosteum of the cribriform plate
3. Small olfactory nerve fibers without perineurial coating, in cross section
4. Small olfactory nerve fibers without perineurial coating, in longitudinal section
5. Bowman's glands
6. Lamina propria
7. Olfactory epithelium
8. Small nerve fibers emerging from the epithelium (central processes of ORNs)
9. Excretory duct of Bowman's gland

ATLAS—NOSE

Fig. 3.11
Electron micrographs of the olfactory mucosa of the cat.
A – Surface layer of epithelium, B – Deep layer of epithelium, C, D – Subepithelial layer

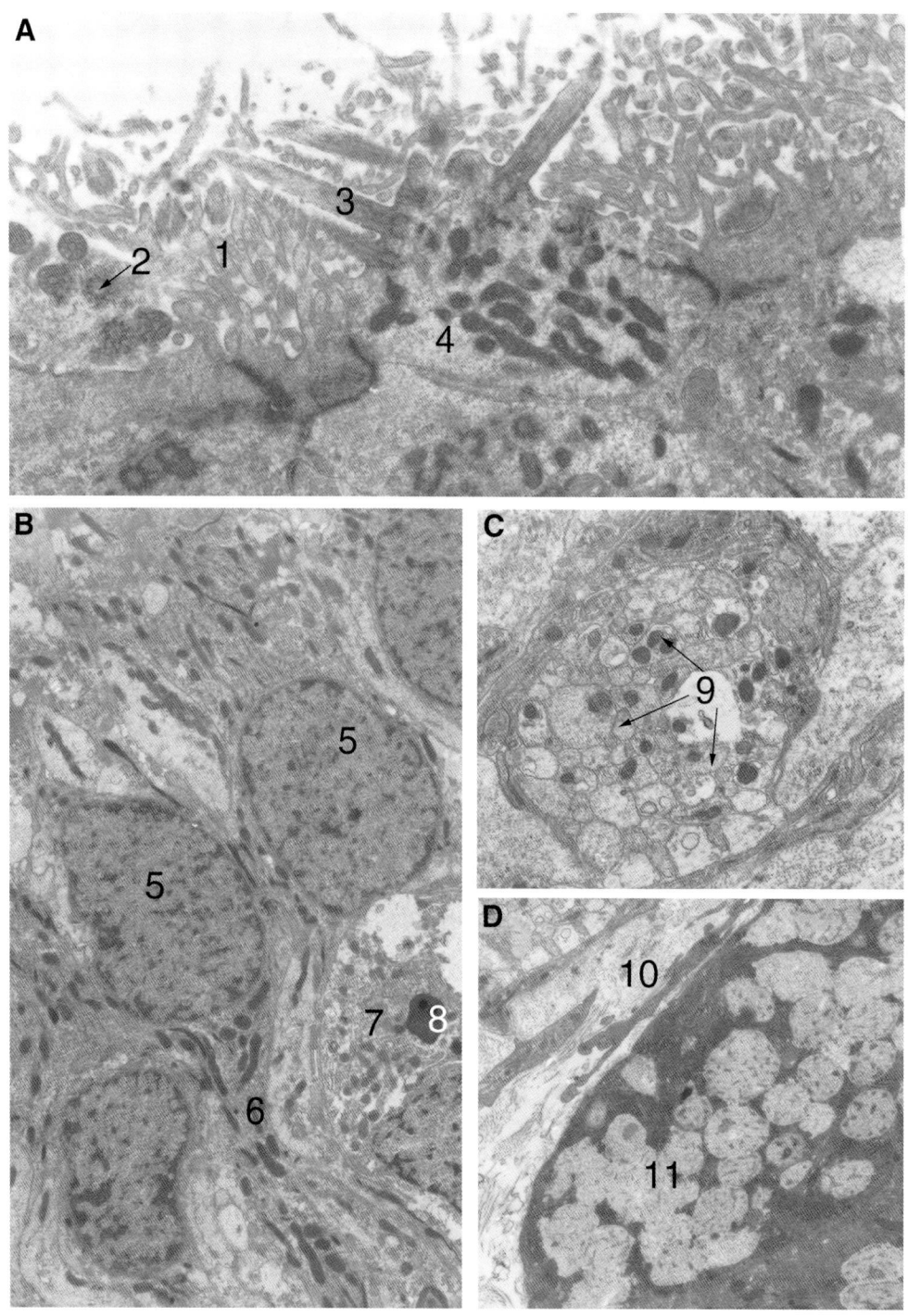

1. Microvilli of supporting (sustentacular) cell
2. Cilia of olfactory vesicle (dendritic knob) in cross section
3. Cilia of olfactory vesicle in longitudinal section
4. Olfactory vesicle
5. Nuclei and perikarya of olfactory receptor neurons (ORN)
6. Central process of ORN
7. Cytoplasm of basal cell
8. Lipofuscin granule
9. Olfactory nerve axons
10. Collagen fibers of lamina propria
11. Mucous droplets in an acinar cell of Bowman's gland

Fig. 3.12

Electron micrographs of the olfactory mucosa of the cat.
A – medium-power view, B – high-power view.

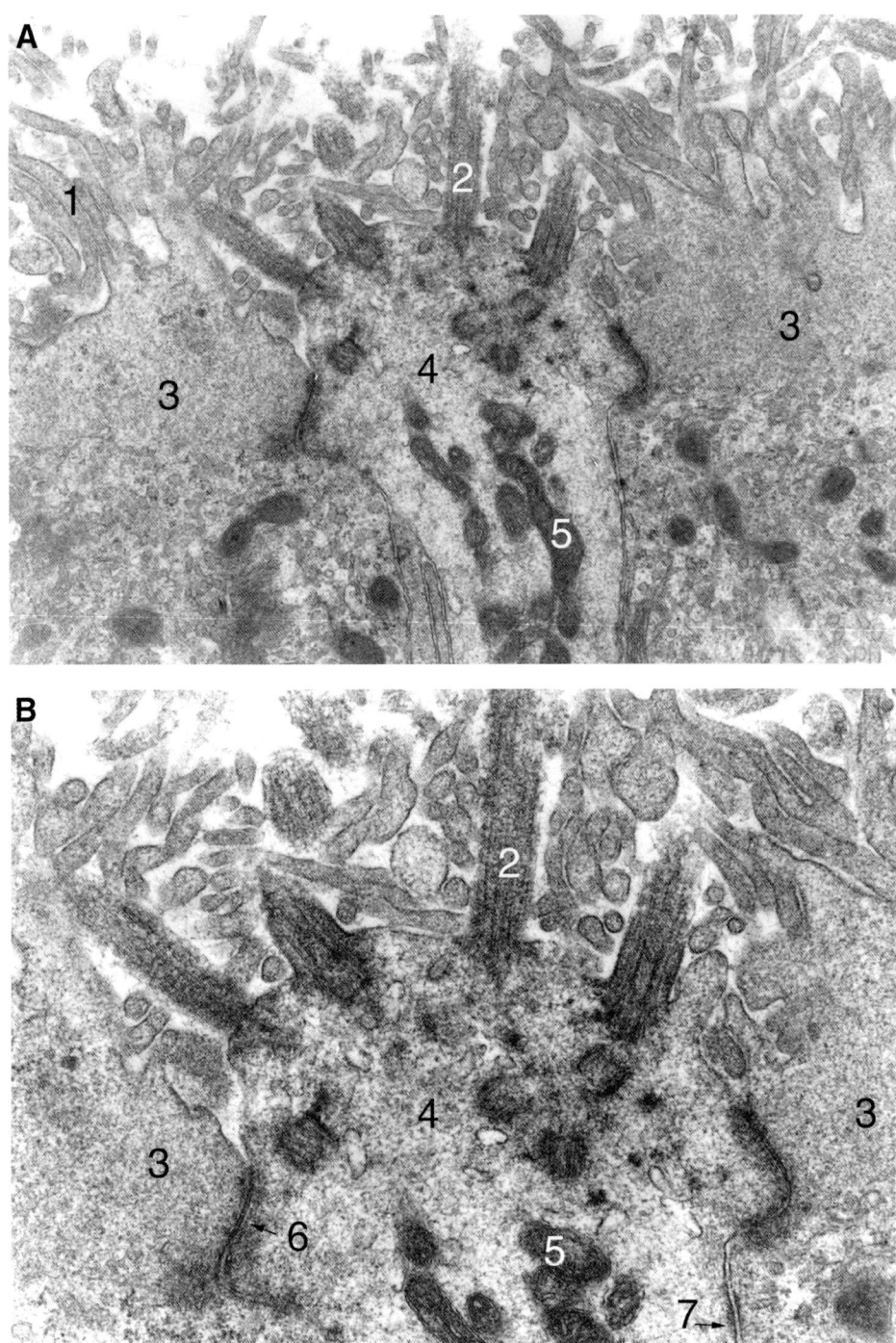

1. Microvilli of supporting (sustentacular) cell
2. Cilia of olfactory vesicle
3. Cytoplasm of supporting cell
4. Cytoplasm of olfactory vesicle
5. Mitochondria
6. Zonula occludens
7. Macula adherens

Fig. 3.13

Low-power light micrographs demonstrating the olfactory mucosa in relation to the cribriform plate and the olfactory bulb. Paraffin-embedded sections from decalcinated human tissue. A, B – HE staining, C – immunoreaction against neural cell adhesion molecule (nCAM) using the immunoperoxidase method, haemalaun counterstaining.

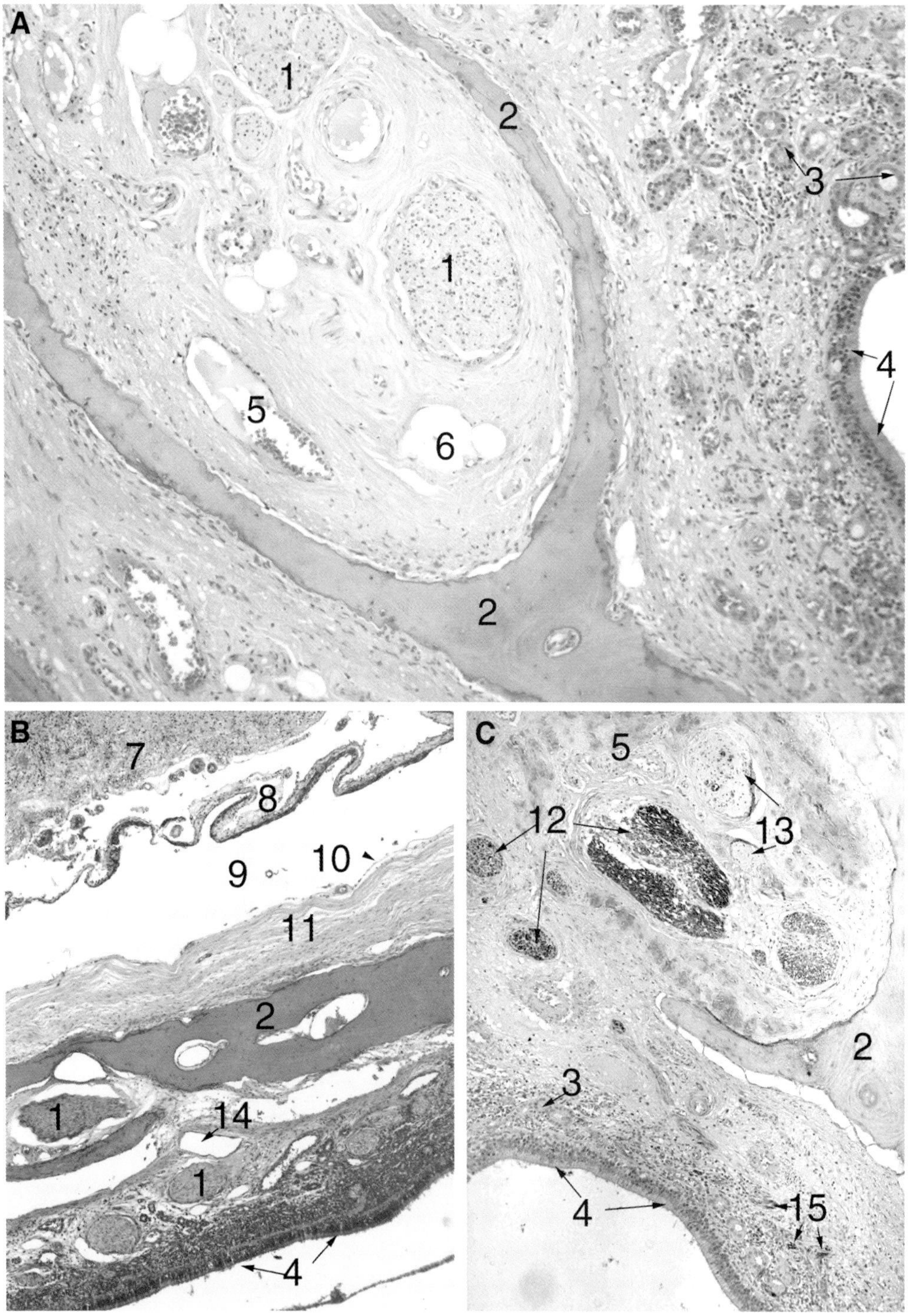

1. Olfactory nerves (fila olfactoria) surrounded by the endosteum of cribriform plate (lamina cribrosa)
2. Bone tissue of cribriform plate
3. Bowman's glands in the lamina propria
4. Olfactory epithelium
5. Blood vessel (vein) passing through the cribriform plate
6. Adipocytes
7. Olfactory bulb
8. Pia mater
9. Subarachnoid space
10. Arachnoid (remnants only)
11. Dura mater
12. Nerve bundles immunoreactive to nCAM, corresponding to olfactory nerves
13. Nerve bundles non-immunoreactive to nCAM, presumably corresponding to trigeminal (anterior ethmoidal) branches
14. Lymph vessel
15. Subepithelial nCAM immunoreactive fibers (central processes of ORNs)

178   ATLAS—NOSE

Olfactory Pathways

OLFACTORY BULB (FIGS. 3.13; 3.14)

The first stage of central olfactory processing resides in the olfactory bulb. A derivative of the phylogenetically ancient paleocortex, this structure is characterized by a strict layering of its neural components, not dissimilar from that of the retina. In fact, the intricate synaptic network and elaborate processing of primary sensory information occurring in the bulb has often been compared to similar features of the retina. Starting from the outer (pial) surface the layers are as follows.

I. Olfactory nerve layer
II. Glomerular layer
III. External plexiform layer
IV. Mitral cell layer
V. Granule cell layer
VI. Fibers of the olfactory tract

The presence of an internal plexiform layer as distinct from the granule cell layer has been noted in some descriptions but this classification is not shared by the majority of authors. The olfactory nerves forming the outermost layer terminate in large cellular clusters of globular shape called glomeruli. Each olfactory glomerulus is a synaptic complex, where the olfactory nerve terminals synapse with the dendrites of mitral and tufted cells, with an average convergence ratio of 100:1. Information transfer within the glomeruli can be modified by input from the periglomerular neurons whose dendrites synapse with other dendritic elements within the parent glomerulus, whereas the axons form synaptic contacts with mitral and tufted dendrites belonging to neighboring glomeruli. Periglomerular cells receive synapses from the ORNs as well as from centrifugal efferent fibers. Mitral cells (termed so because of their peculiar shape resembling the bishop's traditional headdress, the mitre) constitute a separate layer of the bulb as the chief relay neurons. Tufted cells are smaller than mitral neurons and their somata usually lie external to the mitral layer, however, their branching pattern and synaptic connectivity are largely similar to those of mitral cells. Hence, the two relay neurons are often collectively referred to as M/T cells. The layer between the

Fig. 3.14

Schematic illustration of the neural layers and connections of the olfactory bulb (OB). OE – olfactory epithelium; GRL – granule cell layer; MCL – mitral cell layer; EPL – external plexiform layer; GL – glomerular layer; CP – cribriform plate with olfactory axons; OT – olfactory tract.

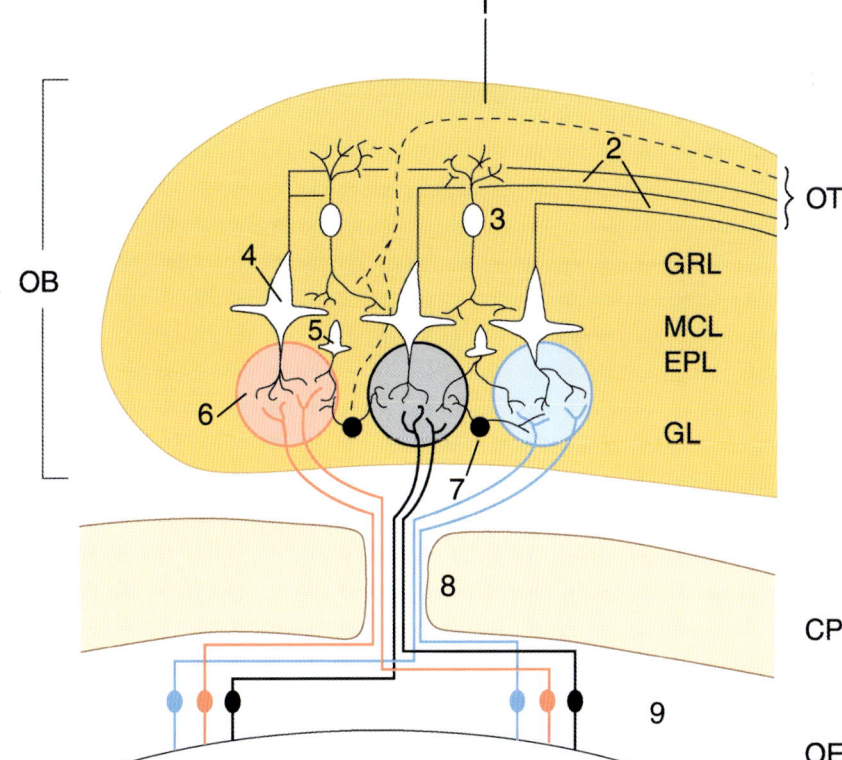

1. Centrifugal (efferent) fibers
2. Centripetal axons of mitral and tufted (M/T) cells
3. Granule cell
4. Mitral cell
5. Tufted cell
6. Glomerulus
7. Periglomerular cell
8. Cribriform plate
9. Olfactory receptor neurons

glomeruli and the mitral cells is known as the external plexiform layer (EPL). The EPL is an intricate meshwork of synapses connecting the dendrites of mitral neurons and those of granule cells. A relatively rare type of neuronal contact, dendrodendritic synapses are common occurrence in the bulb and they presumably underlie the generalization of olfactory signals. Although, as we have seen above, one glomerulus likely receives input only from one—or, in any case, a very limited number of—ORN type, the powerful "switchboard" of densely interconnected EPL may be able to compute a kind of "weighted average" at any moment of olfactory sensation. Thus, the most salient information is transmitted to the higher brain centers, whereas others remain suppressed. Granule cells, whose dendrites participate in the synaptic complex of EPL are smaller than M/T neurons and do not possess genuine axons. They receive input from collaterals of M/T cells as well as from centrifugal efferents and their main role is a demarcation of stimuli by contrast enhancement. At the level of the glomeruli, a similar contrast-enhancing effect is attributed to the periglomerular cells, based on collateral inhibition. As a result, the final output leaving the bulb via the M/T axons gathering in the olfactory tract has undergone filtration and focusing.

OLFACTORY TRACT AND ANTERIOR OLFACTORY NUCLEUS (FIGS. 2.58; 3.15)

The olfactory tract is a flat band of white matter underlying the olfactory groove (sulcus olfactorius) between the straight gyrus and the orbital gyri of the frontal lobe. Its caudal end forms a triangular thickening called the olfactory trigone, continuing in the lateral and medial olfactory striae on the respective sides. The aterior olfactory nucleus (AON) is a small cluster of neurons embedded in the tract behind the bulb. The AON receives collaterals from the M/T axons of the tract and it is also one of the targets for noradrenergic (from locus coeruleus), serotonergic (from dorsal raphe), and cholinergic (from the nucleus of diagonal band) fibers, descend-

Fig. 3.15

Block diagram of the olfactory pathway.

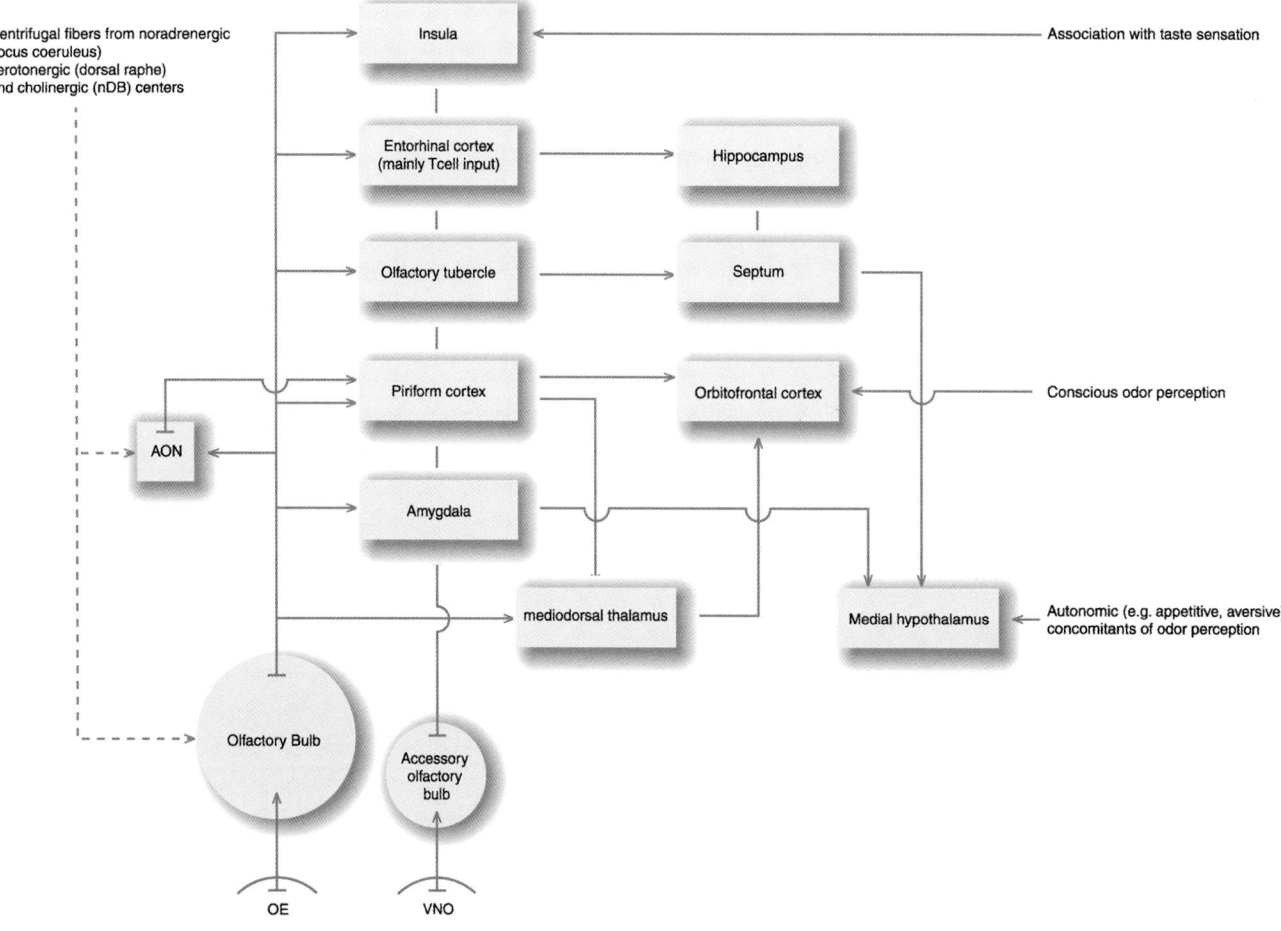

ing to the olfactory bulb as centrifugal modulatory pathways. The bilateral AON are connected via the commissural pathway formed by the medial olfactory striae.

PRIMARY OLFACTORY PROJECTION SITES (FIG. 3.15)

Apart from the AON, the M/T axons of the olfactory tract synapse within the following main target areas: olfactory tubercle, corticomedial amygdala, piriform cortex, entorhinal cortex, and insula. The olfactory tubercle is a small elevation behind the olfactory trigone by the termination of the intermediate olfactory stria. This small nucleus, receiving mainly the axons of tufted cells, lies within the anterior perforated substance, and its role is probably less significant in humans. The piriform and entorhinal cortices as well as the cortical amygdala belong to the paleocortical regions, as the main receptive fields of olfactory input. However, it has been shown that a neocortical field, the insular cortex, also receives fibers of olfactory origin directly via the lateral olfactory stria and indirectly through intracortical connections from the above mentioned regions (which are themselves profusely interconnected). The importance of insular representation may be that it forms an anatomical link with taste sensation. Recent studies in primates have indicated that convergent odor–taste perception (flavor processing) is a function of the caudal orbitofrontal cortex and adjacent agranular insular cortex. Event-related functional magnetic resonance imaging studies in human subjects confirmed the activation of the amygdala as well as the caudal orbitofrontal, insular, and anterior cingulate cortices on the administration of both taste and odor stimuli. In addition, more refined analysis has indicated that combined olfactory and taste stimuli brought about activation mainly in the anterolateral orbitofrontal cortex, whereas the anteromedial part of the orbitofrontal cortex is involved in the "pleasantness" of such sensations. It is well known from common experience that taste and smell walk hand-in-hand when one enjoys the flavor of quality wine or the zest of good food, for example. Often, it is almost impossible to separate the two modalities. We all know how a head cold (leading to nasal congestion and the loss of olfactory perception) can also destroy taste sensation, rendering almost everything we eat bland and tasteless.

Another important target of M/T axons, at least in some mammals, is the mediodorsal thalamic nucleus, albeit this, too, has more indirect connections via the piriform cortex. As mentioned before, olfactory sensation is special because, unlike all the other sensory modalities, it can reach the cortex directly, bypassing the specific sensory nuclei of the diencephalon. However, this does not imply that thalamic nuclei have no part in the relay of olfactory stimuli at all. The mediodorsal thalamic nucleus is reciprocally connected to the orbitofrontal cortex, which is likely to represent the highest (conscious) level of odor perception (see above).

Further second-order connections include the septum (via the olfactory tubercle and medial olfactory stria), hippocampus (via the entorhinal cortex and septum), and the medial hypothalamus (via the septum and amygdala). These connections are likely involved in odor memory and in the appetitive or aversive autonomic responses (appetite, hunger, satiety, disgust) associated with the perception of attractive or repulsive odors.

It is interesting to note that although women outperform men in most olfactory identification tasks (ironically, they tend to use expensive perfumes which men cannot even smell, let alone appreciate), functional studies using positron emission tomography (PET) revealed no gender-specific difference in the pattern of cerebral activation (bilaterally in the amygdala, piriform, and insular cortices). Therefore, the reported differences are presumably manifested at a cognitive, rather than perceptive level of olfactory processing. Further PET activation studies have also pointed to an involvement of other, hitherto unreported, brain regions in parallel and hierarchical olfactory processing. Such regions comprise the cingulate cortex in single odor identification, the cerebellum in the assessment of odor intensity and the caudate nucleus and subiculum in odor quality. Finally, odor recognition memory also engaged the temporal and parietal cortices. Odor signals are processed both ipsi- and contralaterally, with an apparent preponderance of the right hemisphere regardless of the stimulated nostril.

THE VOMERONASAL ORGAN (FIGS. 3.7; 3.16; 3.17)

This highly specialized receptor organ, also known as the organ of Jacobson, has likely developed independently in land vertebrates. Despite unquestionable similarities to the main (turbinal) olfactory epithelium, the vomeronasal organ (VNO) has peculiar characteristics in the molecular composition of receptors and also in its histology and central processing. The paired VNOs are blind mucosal pockets in the lower part of the nasal septum whose epithelia are exposed to air passing through the common nasal meatuses. In mammals (e.g., rodents), the VNO appears as a tubular organ, whose medial wall is composed of thick stratified neuroepithelium, containing neurosensory, supporting and basal cells, whereas the lateral wall is formed by a thinner non-neural pseudostratified columnar epithelium (Fig. 3.16). Histologically, the components of the epithelium of VNO are largely similar

Fig. 3.16

Light microscopic images of the developing vomeronasal organ in different mammalian species. A – mouse, B – cat, C,D – rat. A, C, D – Semithin sections from resin-embedded tissue, toluidine blue staining, B – immunostaining against calcitonin gene-related peptide (CGRP). Courtesy of L. Dénes and J. Takács.

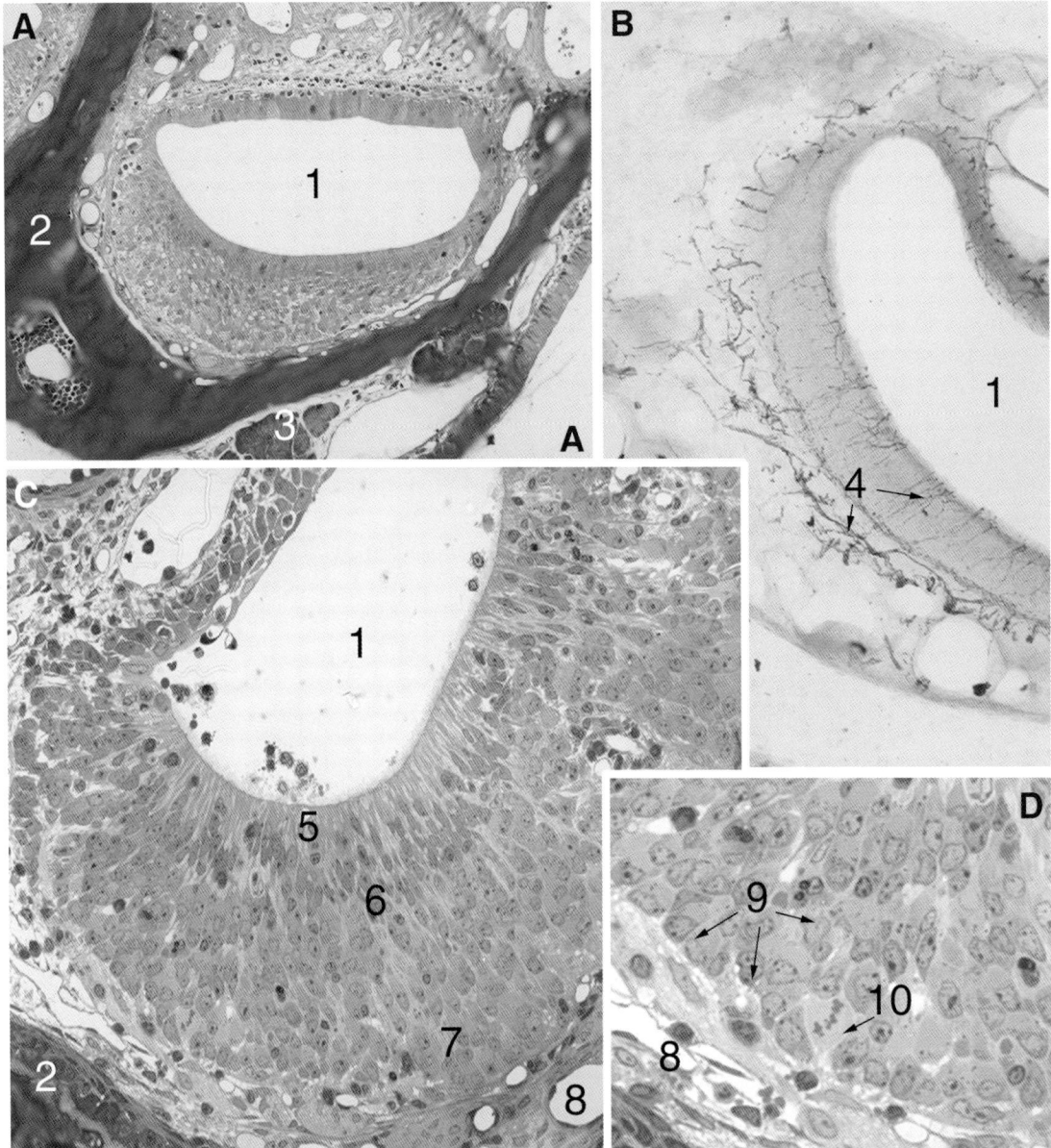

1. Vomeronasal tubule
2. Vomer
3. Nasal glands
4. CGRP immunoreactive fibers representing bipolar sensory neurons and their central processes within the vomeronasal nerve
5. Supporting cell layer of vomeronasal epithelium
6. Bipolar sensory neuron layer of vomeronasal epithelium
7. Basal cell layer of vomeronasal epithelium
8. Venous plexus of lamina propria
9. Basal cells in resting state
10. Mitotic basal cell

to those of the turbinal olfactory epithelium. The only major difference is that the dendritic bipolar cells (sensory neurons) possess microvilli rather than true cilia, and the dendritic knobs contain numerous cytoplasmic vesicles and endoplasmic reticulum, as compared to the scarcity of such cytoplasmic organelles in the ORNs (Fig. 3.17).

The central processes of receptor neurons terminate in the accessory olfactory bulb (AOB), whose glomeruli are altogether less well delineated than those of the main olfactory bulb. From there, the axons of relay neurons reach the corticomedial amygdala and the anterior-ventral hypothalamic nuclei (i.e., the central representation of VNO input is more restricted than that of the main olfactory system; Fig. 3.15). Hence, the connectivity of VNO is indicative of an affective, rather than cognitive, role of this system.

 With regard to the molecular pattern of receptors, two main types of genes encoding putative receptor proteins have been described in the VNO. One is a 7TM G protein-coupled receptor gene, distinct from those determining similar receptors in the main olfactory epithelium. The other gene determines a receptor protein that has an unusually long extracellular N-terminal and is related to the metabotropic glutamate receptor. These proteins are likely to represent functional pheromone receptors. Furthermore, M10 genes were found to be expressed only in the VNO and not in other olfactory epithelia. Such genes are known to determine polypeptides involved in antigen presentation and it is assumed that a similar molecule may also be instrumental in the binding of pheromones to the specific receptor (V2R). Single VNO receptor neurons can only express one subtype of M10 gene, which also determines their zone of termination within the AOB. The response properties of VNO receptor neurons are also clearly different from the properties of ORN. They are sensitive to extremely low concentrations of the bound molecule and their specificity is substantially greater, making them genuine specialists of chemical detection.

Although the VNO has been described in many mammalian species, its presence and functional significance in adult humans has long been a matter of controversy. Recent studies on human fetuses have demonstrated that the VNO appears around midpregnancy as a pair of tubular structures (with an open end anteriorly and a closed end posteriorly) on both sides of the septum. The organ reaches its maximum extent (4–8 mm in length) in the sixth fetal month. At this stage, the lumen is lined by neuroepithelial cells, which are similar to the ORNs of the main olfactory epithelium. The axons of these cells converge to form the vomeronasal nerve. Later, the organ may regress, leaving small cystic remnants. Regression is supported by the lack of neuron-specific enolase staining after the seventh month. The VNO is accompanied by vomeronasal cartilages, essentially long strips of cartilage between the lower edge of the ethmoidal perpendicular lamina and the vomer. As the vomeronasal cartilages fail to regress during embryonic life, they may determine further development of the anterior-ventral part of the human septum.

Earlier on, the VNO in man was believed to be present only in fetal age but its presence is now firmly established in adults. An opening of the VNO could be located in about 40% of individuals, and histological remnants of VNO structures were evident in about 70% of the cases investigated. Notably, however, failure to find the VNO could have been due to insufficient tissue preservation or processing. More compelling still, in an extensive study involving more than 400 subjects, vomeronasal pits were observed in all individuals except those with pathological conditions affecting the septum. Electron microscopic investigation revealed the presence of microvillar (receptor) cells and unmyelinated intraepithelial axons in the epithelium lining the vomeronasal tubules. What boils down from this information is that, although its degree of development may vary among individuals, the VNO can no longer be considered a vestigial organ in most humans, even though the accessory olfactory bulb regresses after the fetal period, and alternative neural connections from the VNO to the brain have not been convincingly demonstrated.

Concerning the role of VNO, this organ has long been implicated in the detection of pheromones, chemical substances that convey sexually relevant signals, such as attracting members of the opposite sex, timing of estrus, or inclination to mating. Often, steroids by chemical nature, pheromones are present in the urine or sweat, and they are also secreted by the scent glands of the skin (e.g., in the axillary or anal regions). Although pheromones are best known to attract insects, mammals also have an elaborate system for pheromone detection. Surprisingly, rodents such as laboratory rats can only recognize conspecifics of the opposite gender by vomeronasal signals (i.e., no visual or acoustic input alone can serve as a proper "badge of masculinity"). This is known from experiments on mutant animals lacking specific vomeronasal receptors. Such an exclusive significance of vomeronasal signals is certainly not observed in primates. However, electrophysiological studies on human subjects have demonstrated gender-specific functional responses to putative pheromones in the VNO region. Contrary to common belief, however, there is no unambiguous exper-

ATLAS OF THE SENSORY ORGANS

Fig. 3.17

Electron microscopic images of the vomeronasal organ of the rat. A – superficial layer of vomeronasal epithelium, B – vomeronasal nerve bundle, C – high magnification of vomeronasal nerve fibers. Courtesy of L. Dénes and J. Takács.

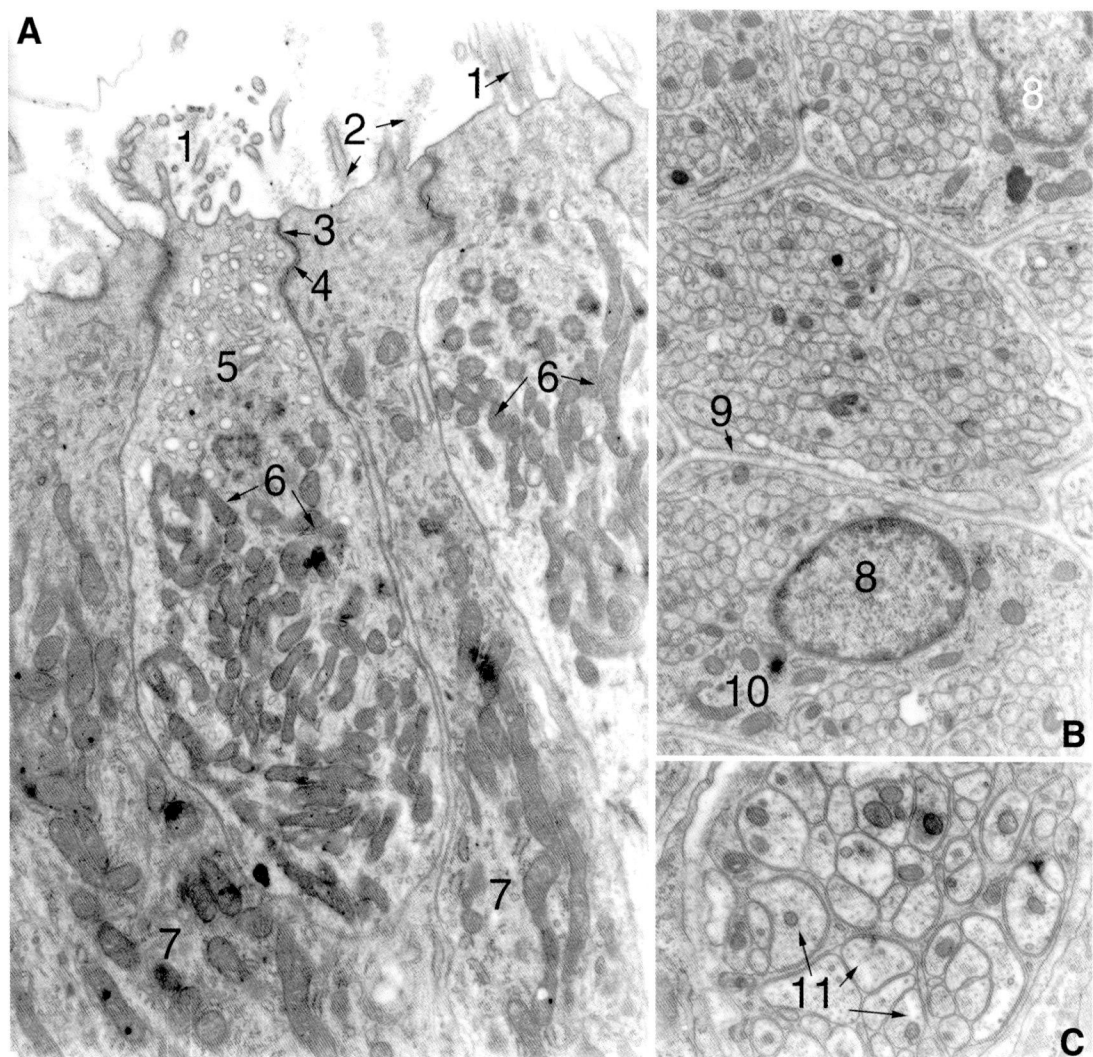

1. Microvilli of the dendritic knob of receptor neurons
2. Microvilli of supporting epithelial cell
3. Zonula occludens
4. Macula adherens
5. Cytoplasm of receptor neuron
6. Mitochondria
7. Cytoplasm of supporting cell
8. Nuclei of olfactory glia resembling Schwann cells
9. Olfactory glial sheath
10. Cytoplasm of olfactory glia
11. Thin non-myelinated nerve fibers (central processes of sensory receptor neurons)

imental evidence, at least in mammals, to support that the VNO acts only as pheromone detector, or that the VNO is *the* sole pheromone detector. In fact, alternative pathways via the main olfactory bulb have been demonstrated in ferrets or pigs. In a recent comparative PET activation study on male and female subjects it was found that women smelling an odorless androgen-like compound (extracted from male axillary skin) activate the hypothalamus (centering on the preoptic and ventromedial nuclei) with no activation in the "olfactory" brain regions (amygdala, piriform, insular, and orbitofrontal cortices). Conversely, men activate another part of hypothalamus (dorsomedial and paraventricular nuclei) when smelling an odorless estrogen-like substance. Thus, the patterns of activation showed reciprocal features in the two genders. Although the reported experiments strongly suggest the existence of specific pathways for the detection of pheromone-like substances in humans, they are not conclusive as to whether such pathways are indeed VNO-related.

RECOMMENDED READINGS

Textbooks and Handbooks

1. Bannister LH, Berry MM, Collins P, Dyson M, Dussek JE, Ferguson MWJ. Gray's Anatomy, 38th ed., Churchill Livingstone, Edinburgh, 1999.
2. Drews U. Color Atlas of Embryology, Thieme, Stuttgart, New York, 1995.
3. Hinrichsen KV (Ed.). Humanembryologie, Lehrbuch und Atlas der vorgeburtlichen Entwicklung des Menschen. Springer, Berlin, 1990.
4. Smith C.U.M. Biology of Sensory Systems, John Wiley & Sons, Chichester, 2000.
5. Squire LR, Bloom FE, McConnell SK, Roberts JL, Spitzer NC, Zigmond MJ. Fundamental Neuroscience, Academic Press, Imprint of Elsevier Science, USA, 2003.

Reviews and Research Reports

1. Araujo IET de, Rolls ET, Kringelbach ML, McGlone F, Phillips N. Taste-olfactory convergence, and the representation of the pleasantness of flavour, in the human brain. Eur J Neurosci 2003;18: 2059.
2. Bengtsson S, Berglund H, Gulyas B, Cohen E, Savic I. Brain activation during odor perception in males and females. NeuroReport 2001;12:2027-2033.
3. Iwema CL, Schwob JE. Odorant receptor expression as a function of neuronal maturity in the adult rodent olfactory system. J Comp Neurol 2003;459:209-222.
4. Johnson A, Josephson R, Hawke M. Clinical and histological evidence for the presence of vomeronasal (Jacobson's) organ in adult humans. J Otolaryngol 1985; 14:71-79.
5. Monti-Bloch L, Jennings-White C, Dolberg DS, Berliner DL. The human vomeronasal system. Psychoneuroendocrinology 1994;19:673-686.
6. Ophir D. Intermediate filament expression in human fetal olfactory epithelium. Arch Otolaryng Head Neck Surg 1987;113:155-159.
7. Savic I, Berglund H, Gulyas B, Roland P. Smelling of odorous sex hormone-like compounds causes sex-differentiated hypothalamic activations in humans. Neuron IEP Published Online, Aug 3, 2001.
8. Savic I, Gulyas B. PET shows that odors are processed both ipsilaterally and contralaterally to the stimulated nostril, NeuroReport 2000;11:2861-2866.
9. Savic I, Gulyas B, Larsson M, Roland P. Olfactory functions are mediated by parallel and hierachical processing, Neuron 2000;26:735-745.
10. Shepherd GM. Discrimination of molecular signals by the olfactory receptor neuron, Neuron 1994;13:771-790.
11. Stensaas LJ, Lavker RM, Monti-Bloch L, Grosser BI, Berliner DL. Ultrastructure of the human vomeronasal organ, J Steroid Biochem Mol Biol 1991;39:553-560.

4 The Organ of Taste

Andrea D. Székely and András Csillag

ANATOMICAL OVERVIEW OF THE ORGAN OF TASTE

Taste sensation is a form of chemical sense, specialized for the detection of compounds (tastants) dissolved in the saliva. Broadly speaking, taste is just one specific type of visceral sensation that is particularly relevant to food ingestion. Taste signals trigger a host of behavioral and autonomic responses, appetitive or aversive, most of which are visceral reflexes (salivation, gastrointestinal activation, swallowing, gagging, vomiting), or patterns of involuntary locomotor activity (orofacial movements, disgust responses). However, unlike many other visceral signals, taste is also accompanied by conscious perception, recognition, hedonic quality, and memory formation (i.e., predominantly cortical functions). Although these features make taste sensation a truly elaborate faculty, deserving a place among the five principal human senses, the intimate link with visceral sensation is reflected in the remarkably diffuse character of both taste perception and the processing of gustatory (taste-relevant) input.

Taste receptor cells (TRC) are distributed in the epithelia of the tongue, soft palate, incisive papilla of hard palate, epiglottis, aryepiglottic folds, pharynx, and the upper third of the esophagus. They are secondary sensory cells (remember, olfactory receptor elements are neurons!), contacted by—but not continuous with—afferent nerve fibers. Like other sensory receptor cells, TRCs used to be regarded as the derivatives of neurogenic ectoderm, in particular the epibranchial placode. However, current research on genetic markers has confirmed the possibility that TRCs may stem from the local epithelium (which may be of ectodermal or endodermal origin, depending on location), whereas their sensory nerves and ganglia indeed originate from the neural crest/placodal ectoderm. The sensory cells are grouped in specific end organs called taste buds (gemma gustatoria) (Fig. 4.1). In the tongue, taste buds are situated in papillae, whereas in other regions (palate, larynx, pharynx, esophagus) they are freely distributed in the epithelium of mucosa.

The mucosa of the tongue possesses four types of papillae: filiform (on the dorsal surface) (Fig. 4.4), fungiform (antero-laterally) (Fig. 4.4), foliate (postero-laterally) (Fig. 4.5), and circumvallate (along the sulcus terminalis). With the exception of the filiform papillae, all the other types have taste buds. The largest number and most characteristic type of taste buds are found within the circumvallate papillae (Fig. 4.6). These represent a circular area of lingual mucosa that is demarcated from the surrounding parts by a deep moat. The latter can capture a considerable amount of liquid and, as a result, the tastants may linger in it eliciting a lasting signal. The detection of tastants is assisted by the secretion of von Ebner's glands (Fig. 4.6), which open into the trenches of the cicumvallate papillae. Unlike other glands of the tongue, von Ebner's glands are serous and they may supply specific proteins for the protection of gustatory epithelium and perhaps also for the binding of sapid molecules (tastants). Such proteins of the lipocalin family, which are able to bind lipophilic molecules by enclosure within their structure, thereby minimizing contact with solvent, have been described in the secretion of von Ebner's glands.

Approximately 250 taste buds (Fig. 4.1; 4.6) are situated in each circumvallate papilla, 8 to 10 of which line up in front of the sulcus terminalis of the human tongue. Taste buds are poorly delineated from the surrounding epithelium in which they are embedded. Under the light microscope, they appear slightly more translucent and their cells are arranged perpendicularly to the surface, rather like orange slices, as opposed to the horizontal arrangement found in the tongue mucosa. Each taste bud opens to the surface with a pore (porus gustatorius) surrounded by microvilli of neighboring cells. The cellular composition of taste buds is not uniformly described, most probably because the cells are remarkably short-lived (the turnover time is hardly more than 10 days) and any histological observation is merely a snapshot of

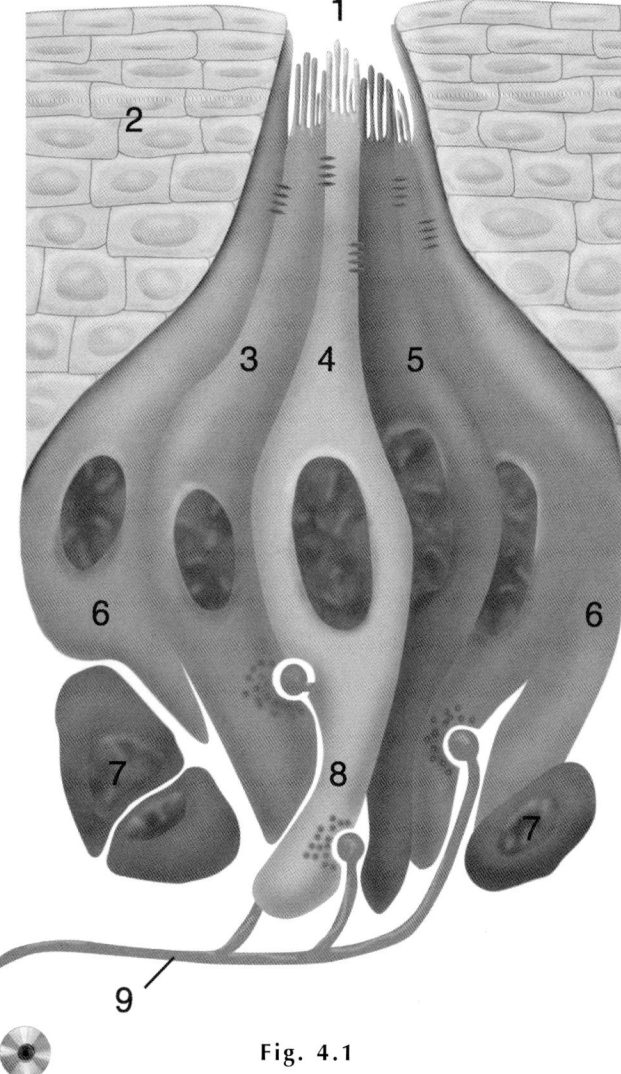

Fig. 4.1

Schematic illustration of taste bud.

1. Gustatory pore
2. Epithelium of tongue
3. Intermediate taste cell
4. Light taste cell
5. Dark taste cell
6. Sustentacular cell
7. Basal cell
8. Synapse of afferent taste fiber
9. Afferent taste fiber

dynamic changes "frozen in time." When observed under the light microscope, three types of cells—dark, light, and basal—can be distinguished (Fig. 4.6). Under the electron microscope, most descriptions mention four cell types. Type I (dark) cells, constituting 55–75% of all cells, extend from the base to the apex of bud. They are characterized by an irregular nucleus, long and branched apical microvilli, and the presence of dense cytoplasmic granules and rough endoplasmic reticulum (RER). Type I cells have membranous extensions that ensheathe unmyelinated nerve fibers and neighboring cells. In some (but not all) species, such cells were observed to receive synapses.

Type II (light) cells (20%) also extend from base to apex but they have short microvilli, large oval nucleus, and smooth endoplasmic reticulum. Type III (intermediate) cells (5–15%) are largely similar to type II cells except that they possess numerous dense core vesicles near their base. Synapses with afferent nerve terminals have been found mainly on type III cells. Type IV cells corrrespond to basal cells because they do not reach the taste pore. These are also rich in RER and dense core vesicles. The role of basal (type IV) cells is as yet unclear: they may represent precursor cells or sensory elements not directly relevant to taste (e.g., mechanoreceptors). Furthermore, it is difficult to reconcile the above categories of cells with those previous (otherwise "logical") assumptions that the taste buds should contain both sensory and supporting (sustentacular) cells. If the presence of visible synapses were taken as the only landmark for TRCs, then the sensory cells would comprise type I and type III, whereas type II cells could be categorized as sustentacular cells. However, because of the propensity of type I cells to ensheathe other cells or nerve fibers, some consider these, too, to be supporting elements. The apices of all three types of cell are joined by tight junctions. However, it is increasingly evident that several taste substances (e.g., KCl) can stimulate not only the internal (receptor) membrane of the sensory cells but also the outer (basolateral) surface by permeating the tight junctions (zonula occludens).

It has long been established that taste sensation is a combination of elementary tastes, four of which (bitter, sweet, sour, and salty) are generally accepted. More recently a fifth taste termed "umami" was added to the list, although its elementary character is still somewhat disputed. The Japanese word originally means "delicious taste" and in fact it represents the savory taste of monosodium-L-glutamate, which is present in meat extracts (e.g., chicken broth) or fish dishes. Monosodium glutamate is also used as a food additive mainly to enhance the "meaty" flavor of the meal. Surprisingly, the taste of water is also a likely candidate and some even distinguish "metallic" as a potential elementary taste. Taste cells contain specific receptors that are present in the membrane close to the gustatory pore. Upon binding to the tastant molecules, the receptors trigger a cascade of intracellular events leading to depolarization and the formation of action potentials that is transmitted to the afferent nerve terminals via synapses. The receptors for salty and sour tastes are ionic channels specialized for the binding of Na^+ or H^+ ions. The detection of sweet or bitter tastes requires far more elaborate receptor and intracellular messenger mechanisms.

In particular, the sodium channel active in the detection of salt is not related to the voltage-gated channel instrumental in the action potential. It can be blocked by

Fig. 4.2.

Development of the tongue. I-IV – branchial arches.

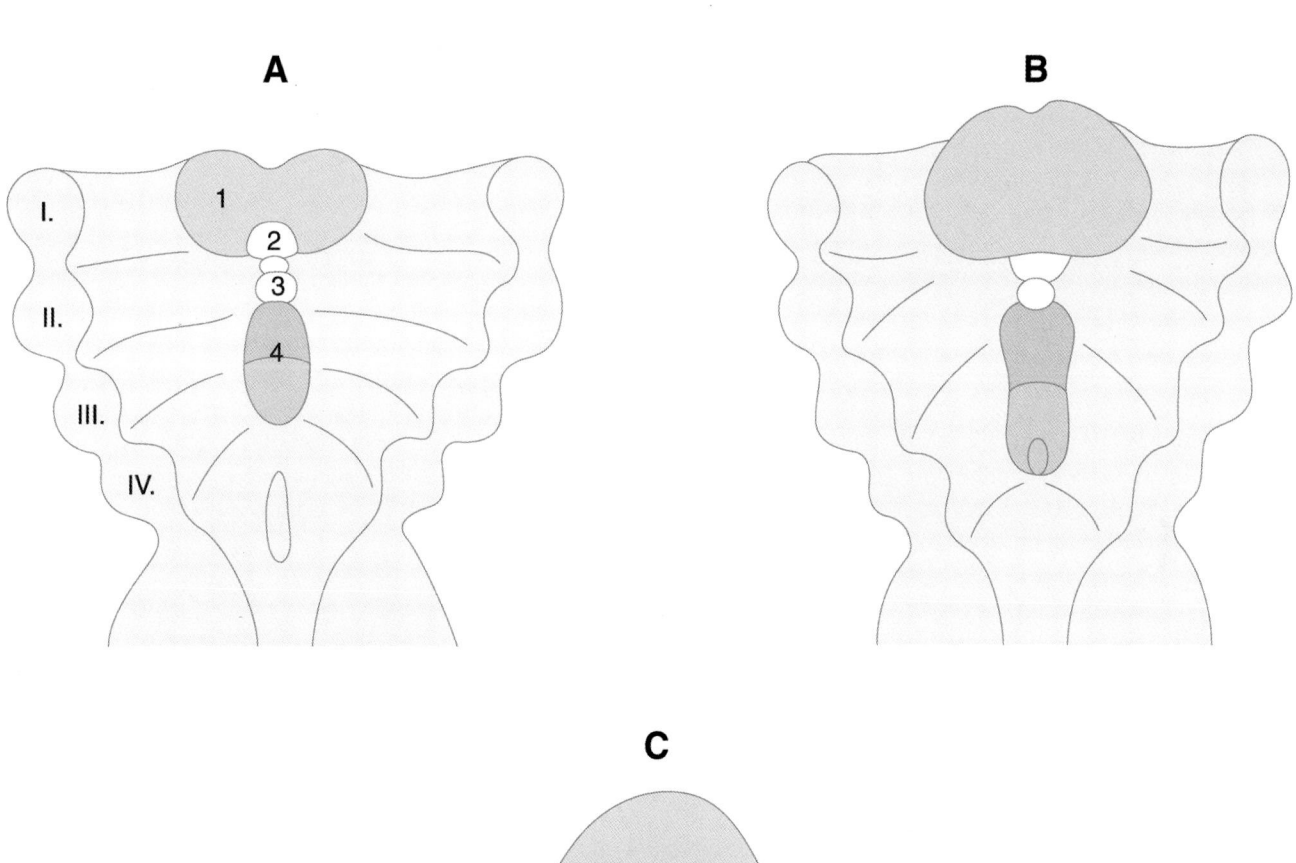

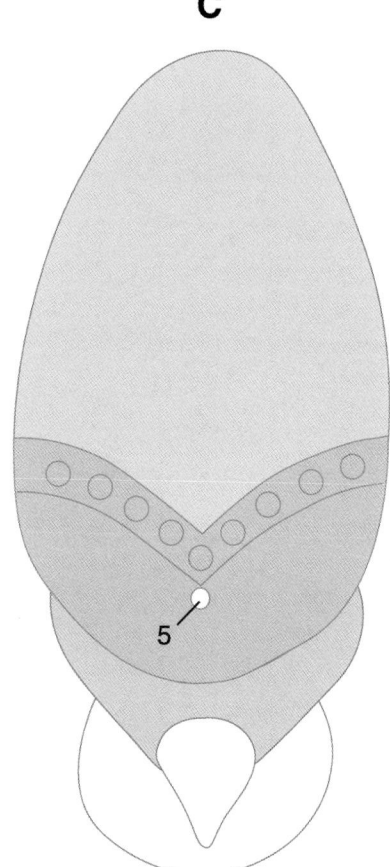

1. Lateral tubercle
2. Tuberculum impar
3. Copula
4. Hypobranchial eminence
5. Ductus thyroglossus

amiloride and is also sensitive to aldosterone and vasopressin (i.e., hormones influencing salt and water equilibrium). The influx of sodium opens Ca^{2+} channels and leads to depolarization of TRC and the discharge of synaptic vesicles, activating also the afferent boutons synapsing with the cell. Other salts such as KCl may also interact with the (apically situated) amiloride sensitive Na^+ channels or other sodium sensitive channels of the basolateral membrane. Amiloride sensitive Na^+ channels are also activated by H^+ ions, which serves as the basis for the detection of sourness. Thus, this taste, too, can activate the membrane of TRC directly. Sweetness is detected by two different transduction pathways. The natural tastant, sucrose, is bound to a 7TM receptor (TR1) through a long extracellular N-terminal. TR1 is related to the metabotropic glutamate receptor. Binding to TR1 induces the release of cAMP, which, in turn, closes K^+ channels (via protein kinase A) and opens Na^+ and Ca^{2+} channels. The ensuing membrane depolarization leads to the influx of Ca^{2+} and the release of transmitter from synaptic vesicles. Other sweet tastant molecules, such as artificial sweeteners, activate the TRC via phospholipase C (PLC) and IP_3 leading to the release of Ca^{2+} from intracellular stores and, ultimately, membrane depolarization as before. Bitter taste is detected by another family of receptors known as TR2 receptors, whose sequence is partially homologous with that of TR1 receptors. Binding of bitter tastants to TR2 receptors brings about the activation of G proteins called gustducins (the name resonates with transducins, a similar family of molecules occurring in olfactory cilia and the outersegments of photoreceptors). Gustducins activate cAMP-phosphodiesterase and, by so doing, a transformation of cAMP into 5'AMP. Elimination of cAMP removes the suppression of Na^+ and Ca^{2+} channels and, as a result, the membrane of TRC depolarizes. The depolarization of TRCs and subsequent synaptic transmission triggers a similar depolarization in the afferent nerve fibers. The frequency of spikes propagating along the sensory nerves reflects the intensity of taste signals.

The Development of Tongue

The primordia of the tongue (Fig. 4.2) appear on the floor of the stomodeum and foregut (see also in Chapter 3). The tuberculum impar (between the branchial arches 1 and 2) arises first, followed by the paired tubercula lateralia (from branchial arch 1). Caudal to these, the copula is formed by the fusion of the ventral (hypobranchial) mesodermal anlages of branchial arches 2 to 4. However, the definitive tongue derives only from the lateral tubercles and from the components of arch 3 (the latter virtually overgrowing the components of arch 2), whereas the components of arch 4 are separated as the furcula to form the epiglottis. Accordingly, the somatosensory innervation of the lingual mucosa comprises the mandibular nerve (arch 1) and glossopharyngeal nerve (arch 3), whereas the motor innervation of the tongue muscles is supplied by the hypoglossal nerve (occipital myotomes) (Fig. 4.3).

Notably, the distribution of taste fibers of the tongue overlaps only partially with that of the somatosensory nerves. In particular, the taste fibers for the root of the tongue (pars follicularis) derive from the glossopharyngeal nerve but those for the anterior part of the tongue (pars papillaris) are borrowed from the chorda tympani, a branch of the facial nerve (the nerve of branchial arch 2). This way, the hidden second branchial arch has a special representation in the tongue mucosa after all. Interestingly, the taste buds of the vallate papillae are innervated by the glossopharyngeal nerve despite them lying anterior to the terminal sulcus (i.e., in the terrain of trigeminal nerve innervation). This reflects the extent of the invasion of the components of branchial arch 3, overgrowing and displacing the components from neighboring arches.

Fig. 4.3

Sensory innervation of the tongue.

1. Lingual nerve
2. Chorda tympani nerve
3. Glossopharyngeal nerve
4. Vagus nerve
5. Vallate papillae innervated by the glossopharyngeal (taste) and lingual (somatosensory) nerves

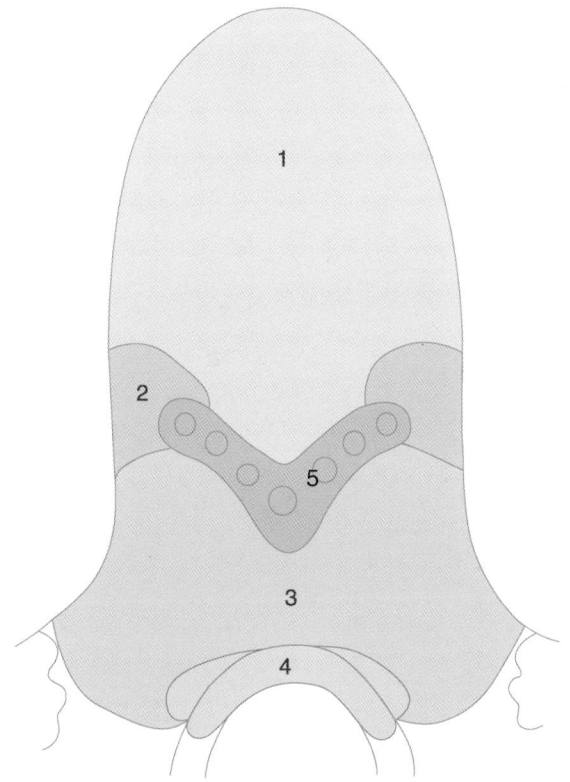

ATLAS PLATES

Fig. 4.4

Filiform and fungiform papillae of the tongue, HE. The plane of section does not follow the axis of papillae, therefore, the connective tissue core is seen either in transverse or oblique sections. The system of epithelial ridges and connective tissue papillae can be clearly observed. Both types of papillae display weak keratinization on the surface, more prominently in the filiform papilla. This epithelium belongs to the "non-keratinized" type, some textbooks (mainly of oral morphology), however, classify it as keratinized epithelium.

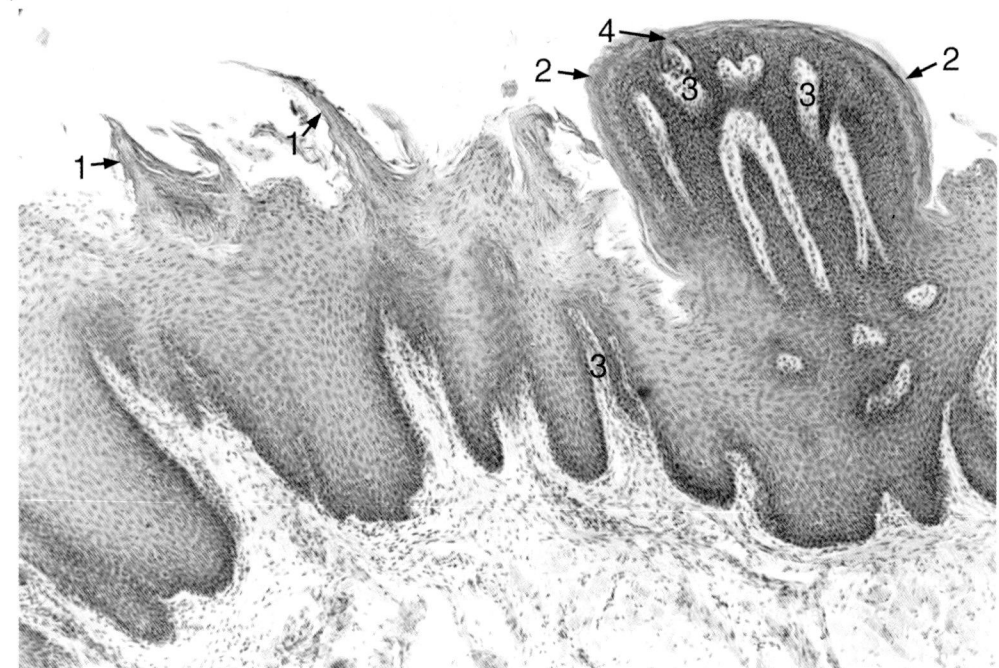

1. Filiform papilla
2. Fungiform papilla
3. Connective tissue (dermal) papilla
4. Taste bud, arrow points to the gustatory pore

Fig. 4.5

Foliate papillae of the tongue HE.

1. Taste buds in the side of the papilla
2. Lymph vessel, usually distended
3. Small salivary gland, of serous type
4. Duct of a small salivary gland, opening in the space between the papillae
5. Striated skeletal muscle fibers

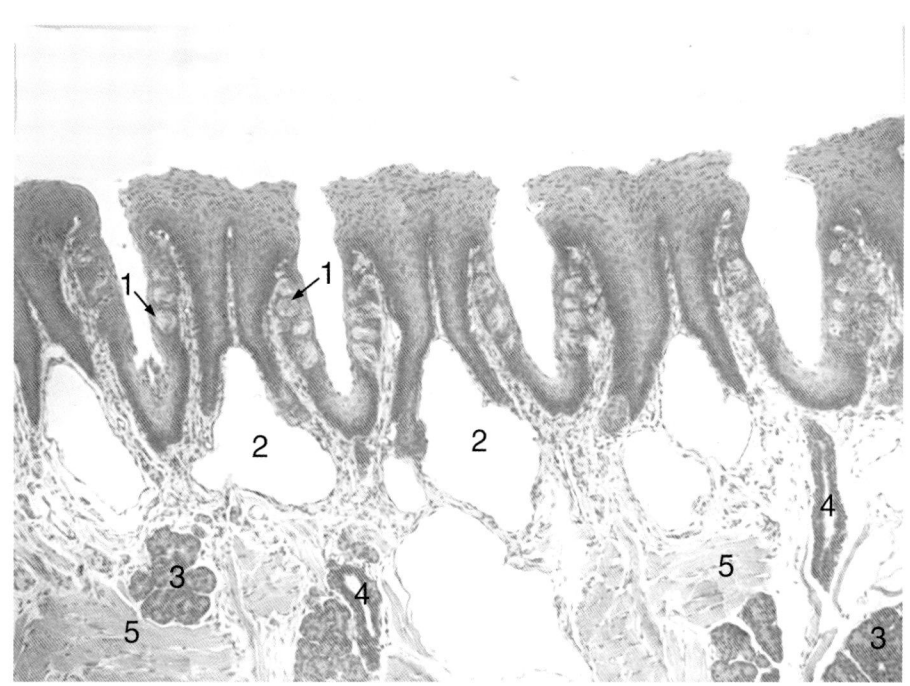

Fig. 4.6

Vallate papilla and taste buds of the tongue. A – Low-power image of vallate papilla, HE, B – Low-power image of fungiform papilla, HE; C – High-power image of the epithelium mucosae of tongue, HE.

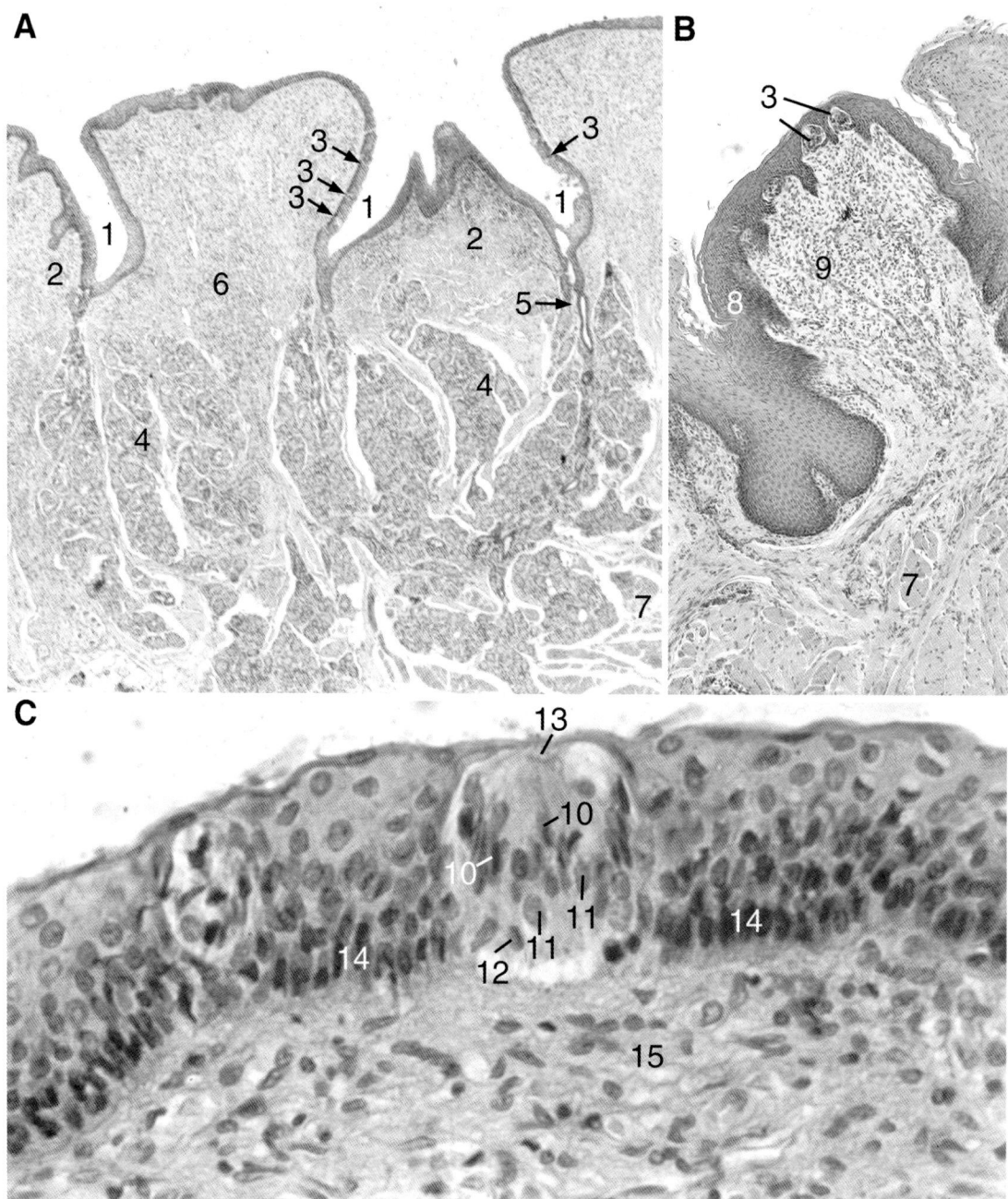

1. Groove around the vallate papilla
2. Mucosal elevations between the grooves
3. Taste buds
4. von Ebner's serous glands
5. Duct of serous gland, opening in the groove around the papilla
6. Nerve fibers in the papilla
7. Striated skeletal muscle fibers in cross section
8. Stratified squamous epithelium
9. Connective tissue core of papilla containing an abundance of nerve fibers
10. Dark and thin cell with an elongated nucleus, in the upper part of taste bud
11. Light and conical cell with a round nucleus, in the lower part of taste bud
12. Reserve (basal) cell with a round nucleus in the basal part of cell
13. Gustatory pore
14. Stratum basale of epithelium
15. Lamina propria

Gustatory Pathways

Corresponding to the fact that the mucosal epithelia harboring TRCs have derived from different branchial arches, gustatory signals from the periphery travel along three different neural pathways to the CNS.

ROUTE 1 (CT-GSP ROUTE)

The anterior two-thirds of the tongue, containing the taste buds of fungiform papillae, is innervated by the chorda tympani (CT) nerve, a branch of facial nerve (branchial arch 2) that joins the lingual branch of the mandibular nerve (third division of trigeminal nerve, branchial arch 1) supplying somatosensory fibers to the same area. The taste buds of the soft palate and the nasoincisor ducts are supplied by the greater superficial petrosal (GSP) nerve, another branch of the facial nerve, whose fibers pass through the pterygopalatine ganglion without synapsing to enter the lesser palatine nerve (for the soft palate) or the nasopalatine nerve (for the nasoincisor ducts). The perikarya of taste fibers lie in the geniculate ganglion of the facial nerve. Central neurites of its pseudo-bipolar neurons enter the pons to terminate in the rostral aspect of the nucleus of the solitary tract (NST).

ROUTE 2 (GL ROUTE)

The posterior third of tongue, including the taste buds of vallate and foliate papillae, receives its innervation from the glossopharyngeal nerve (GL, branchial arch 3). Afferent gustatory fibers, whose cells are located within the petrosal ganglion of the IXth cranial nerve, enter the medulla oblongata to terminate in the intermediate portion of NST.

ROUTE 3 (SL ROUTE)

The taste buds scattered on the laryngeal surface of epiglottis, aryepiglottic folds, and in the upper esophagus are innervated by the internal branch of the superior laryngeal nerve (SL) and other branches of the vagus (branchial arches 4–5). The cell bodies belonging to afferent taste fibers are found in the nodose ganglion of the vagus nerve, and the central neurites enter the medulla oblongata to terminate in the caudal portion of the NST.

Electrophysiological (mainly single-fiber) recordings from the afferent nerves revealed differential responses to standard sets of tastants in laboratory rodents. The CT nerve proved most sensitive to salty (NaCl) stimuli, followed by sweet (sucrose) and sour (HCl) tastes, however, its sensitivity to bitter taste was very low. Apparently, the GSP nerve is even more specific for the detection of sweet taste, which resonates rather well with the common notion that the palate would be particularly involved when one enjoys sweets, or, for that matter, any type of tasty food (hence the expression "palatable").

Incidentally, people of different cultures must have long wondered where exactly taste is felt in the mouth. This uncertainty (not found with any other principal sense) is reflected in the various expressions used in the different languages for the experience of tasting. For example, although good food is "pleasing to the palate" of an English speaker (cf. *avoir le palais fin* in French), those proud (or unfortunate) chocolate fans (whose group the editor of this book is also a modest member of) are referred to as having "sweet tooth" in the same language. Similarly, teeth or gums are often mentioned as the supposed site of tasting in other languages, for example, Hungarian (incorrectly, we may add). The palate is the site of tasting in modern Hebrew, whereas in Hindi and several southern Indian languages, the whole mouth is implicated. Ironically, some of the correct sites (epiglottis or, generally speaking, the throat) are neglected, the most surprising being the tongue itself, which is not considered an organ of taste in most European and Asian languages. However, it is implicated in tasting in Japanese, albeit in a more sophisticated way. In Japanese cuisine, texture (the mechanical features of food at biting and chewing) is highly respected as a significant component of foods. This means that the Japanese speaker traditionally separates the sense of taste (belonging to tongue) and the sense of texture (relevant to the palate, gums and teeth) in addition to flavour (which also has an olfactory component), and this separation is made rather strictly. The importance of texture in the enjoyment of a meal is expressed also in Russian, where the word for "tasty" (*vkusno*) is a derivative of the verb meaning "to bite." In Slavic thinking, food is considered tasty when it feels good to bite into (this can be compared with a similar implication of gums and teeth in other languages). How exactly this feeling is generated is discussed in the section on the sensory innervation of periodontium.

Recordings from the GL have shown that this pathway is particularly sensitive to bitter (quinine) and sour (HCl) tastes and far less to other taste qualities. Thus, the GL route seems to be primarily relevant to aversive (nociceptive) gustatory responses, warning to poisonous or unripe foods, whereas the GSP and CT systems convey chiefly signals of appetitive significance.

The SL route was found to be sensitive only to electrolyte (NaCl, HCl) and water stimuli, that is, neither of the tastes requiring complex receptor mechanisms (sweet or bitter) elicited responses of this system. It is assumed that the vagus system does not participate in taste discrimination, it is only relevant to the maintenance of normal pH and ionic equilibrium of the larynx, hypopharynx, and pharyngo-esophageal transition (a likely component of the sensation of thirst, protection of airways). With this in mind, it is just as well that the epiglottis region (the throat, in general) is not mentioned as a taste sensitive area in common expressions.

Central processing of taste information follows two separate pathways: thalamocortical and limbic. The former is an ascending projection from the NST to the parvicellular ventroposteromedial thalamic nucleus (VPMpc) followed by a thalamocortical projection to the gustatory necortex (agranular insular cortex in rodents). In many species, such as laboratory rodents the NST also engages the parabrachial nucleus (PbN) of the pons, which sends an ascending pathway to the VPMpc. The second ascending pathway is directed to limbic forebrain areas relevant to feeding, such as the central nucleus of amygdala, bed nucleus of stria terminalis, and lateral hypothalamus.

Descending connections from the insular cortex and other forebrain regions reach the NST and PbN, which are also profusely connected to the cranial nerve nuclei involved in orofacial and pharyngeal motor patterns (trigeminal, facial, glossopharyngeal, vagus, accessory, hypoglossal) as well as some related premotor neuronal groups in the reticular formation of brainstem. Together, these pathways control a host of reflexes and (mainly involuntary) motor activities that are known to be associated with feeding: chewing, secretion of saliva, tears or mucus, swallowing, gagging, vomiting, gaping, tongue protrusion, facial distortion (expressing surprise or disgust), more evident in their pure form in the animal kingdom (e.g., pecking in birds). In man, such responses are largely "humanized," subject to social experience and culture (e.g., facial expressions), but there seems to be a common repertoire in all humans for the expression of culinary enjoyment (licking of the lips, chewing, swallowing). As for the opposite, it is possible to keep a straight face in a decent company when one is exposed to unusual or repulsive food but no good manners can stop the secretion of excess saliva or even tears when triggered by a hot curry.

It should be emphasized that the descending pathways from forebrain areas are not essential for the motor patterns related to taste. Such activities can be elicited even in decerebrate animals, meaning that they are maintained by local connections between the NST/PbN and the motor nuclei of cranial nerves via the reticular formation of the brainstem. The existing motor activity patterns can then be modified by the descending pathways.

Complex taste information is expressed as a triad of intensity, quality, and pleasantness (hedonic component). As with other sensory modalities, a more intense stimulus can increase the spike frequency of afferent nerve fibers and recruit more sensory units. Quality coding has been a long-disputed issue. Basically, most neurons of NST or PbN are broadly tuned, that is, they respond to a variety of tastes (if to a different degree, see earlier discussion) rather than to one sharply defined taste. This fact, together with the notion that gustatory information is not topographically represented in the brain, points to the likelihood of population coding. This means that a concerted action of many taste units is required for the coding of a given piece of gustatory information. In this mechanism, any given sensory neuron plays only a limited role.

There are, however, arguments also for the alternative hypothesis, known as labeled line coding. Here, groups of functionally different neurons would selectively respond to each principal taste. Indeed, sensory neurons with characteristic response patterns, peaking at a "preferred" taste, have been identified in the NST and PbN of the hamster. Such differential responses have already been detected at the level of single afferent nerve fibers of the CT or GL in Pfaffmann's pioneering study (1955). However, current results indicate that, although each neuron type is required for the repesentation of a particular taste, they cannot serve in isolation as a labeled line. The only (albeit incomplete) separation in quality coding is evident between the appetitive and aversive pathways. The former involves sweet-preferring and (mostly) bitter-insensitive neurons of the NST, conveying the message of sweetness to the forebrain, leading to ingestive behaviors. The latter pathway arises from bitter-preferring and (mostly) sweet-insensitive neurons of the NST, which transmit the information of bitterness to the forebrain, engaging the mechanisms of food rejection. The proposed scheme is also relevant to the question of the hedonic value of taste information.

MAKING SENSE OF THE TEXTURE OF FOOD: PERIODONTAL SENSATION (FIGS. 4.7; 4.8)

The periodontal membrane is composed of a connective tissue network surrounding the dental roots within the alveolar bone (Fig. 4.7). At the cervical rim of the tooth it is continuous with the lamina propria of the gingival mucosa. The periodontal membrane subserves multiple functions including the firm yet elastic anchoring of the dental roots within the alveolar sockets (gomphosis). This connection lends the teeth a certain, although limited, extent of mobility to protect the hard tissues against pressure arising from mastication. The periodontal membrane also participates in the restructuring of the surrounding tissues under physiological or artificial circumstances (e.g., teething and orthodontic treatments, respectively). Numerous blood vessels are present in the periodontium,

Fig. 4.7

Schematic illustration of the periodontal ligaments. A – longitudinal section, B – cross-section.

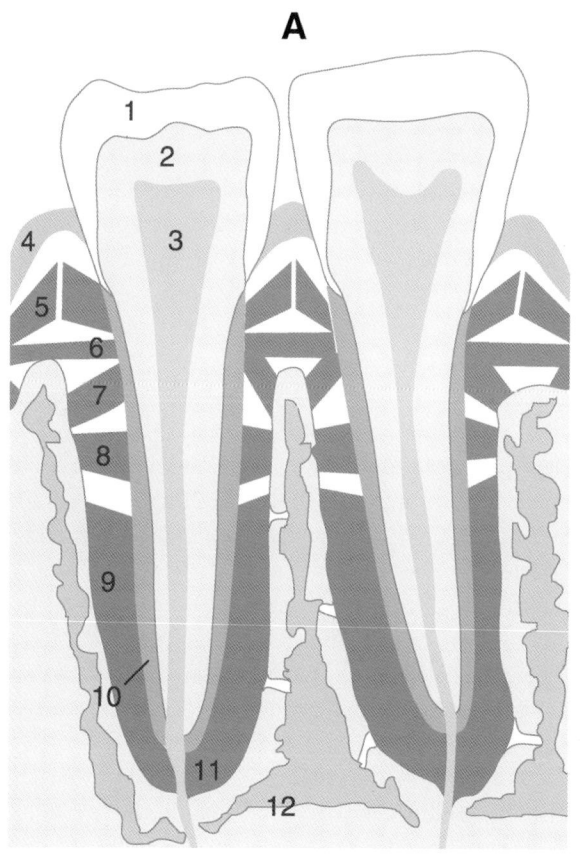
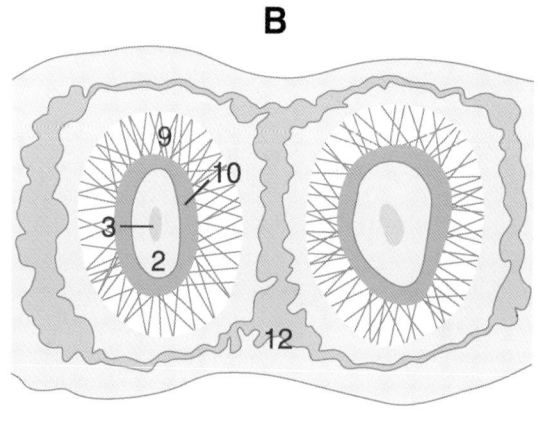

1. Enamel
2. Dentin
3. Pulp chamber
4. Interdental papilla
5. Dentogingival fibers
6. Trans-septal fibers
7. Superior part
8. Horizontal part
9. Oblique part
10. Cementum
11. Periapical part
12. Alveolar bone

whereas the rich net of sensory nerve endings is responsible for mechanosensation and nociception.

Morphology of the Periodontal Membrane

The space between the external surface of dental roots and the wall of the alveolar sockets resembles a fine champaigne glass whose wall varies between 0.18 and 0.25 mm in thickness. The total volume of the space ranges between 35 and 150 mm^3 depending on form and location (i.e., incisor, premolar or molar). The profuse web of blood vessels represents a mere 1 to 2%, whereas the 70% of the periodontal mass is composed of connective tissue cells, collagen, and oxytalan fibers. The rest is built up by nerves, interstitial, and migrating cellular elements. The periodontal space is reduced with age because the lower pressure or reduced impact of mastication result in the degeneration of periodontal ligaments.

Histology of the Periodontal Membrane

The periodontal tissue consists of blood vessels, nerves, oriented bundles of fibers, and connective tissue cells embedded in an amorphous matrix. The perivascular zones are rather poor in fibrous components, here the cells are surrounded by a strongly viscous gel, containing mucopolysaccharides (glycosaminoglycane), glycoproteins, and lipids.

Cellular Elements

Fibroblasts are most frequently encountered along the fibers. They have a large and euchromatic nucleus with a dark nucleolus, and the cytoplasm is filled with vacuoles containing collagen fibrils or proteolytic enzymes. Several processes protrude from the cytoplasm to join in desmosomes with other neighboring fibroblasts, thus providing the three-dimensional framework for ligament regeneration. *Cementoblasts* are only seen when cellular cementum is formed, they are readily identified by their cuboidal or oval shapes and basophilic cytoplasm. *Osteoblasts* are always present in the peripheral parts (i.e., close to the alveolar wall). These cuboidal cells seem to form clusters exclusively in those regions where bone restructuring is needed. *Osteoclasts* are always seen together with the osteoblasts. These multinuclear huge cells participate in the remodeling of the osseous wall, and their enzymatic activity is associated with the Howship's lacunae. *Dentoclasts* represent a subclass of osteoclasts, resorbing cementum and dentin. The epithelial bodies of *Malassez* are developmental remains of the Hertwig sheath, later they may give rise to cysts. Certain migrating immune cells, *lymphocytes* and *monocytes*, invade the periodontal space via the wall of the blood vessels. Their number is greatly elevated in case of inflammation.

Ligaments

The alveolodental fibrous system is divided into a primary, or principal, and a secondary portion. The primary ligaments stretch between the cementum and the alveolar wall and always exhibit certain orientation, whereas the secondary fibers form a looser network around blood vessels and nerves. The ligaments are mainly composed of collagen (types I and III), with a minor contingent of oxytalan. The oxytalan fibers arise from the cervical cementum to follow the course of the collagen fibers. They never seem to reach the alveolar surface but rather terminate in the vicinity of blood vessels. It is thought that their role might involve the regulation of local circulation by recording the position of teeth.

The primary fibers of Sharpey form five groups according their course(Fig. 4.7A):

1. Pars superior: the fibers arise from the supraalveolar cementum to reach the rim of the alveolar socket.
2. Pars horizontalis: the shortest fibers stretch between the superior one-third of the cementum and the opposing alveolar wall.
3. Pars obliqua: these fibers arise from the largest part of the cementum and run upward in appox 45° to reach the alveolar bone.
4. Pars periapicalis: a radial set of fibers arising from around the apical foramen. The direction of the fibers may be horizontal, oblique or longitudinal.
5. Pars interradicularis: these fibers radiate between the bifurcation or trifurcation of multiple roots and the interradicular septum of the alveolar socket.

Such arrangement of ligaments, best seen in cross-section (Fig. 4.7B), protects the teeth from rotation because the fiber bundles do not simply exhibit a fan-shaped radiation, but also cross each other. The structure is maintained by additional thinner fibers binding the aforementioned bundles.

Development of the Periodontal Membrane

Periodontal structures derive from the dental sac (ectomesenchyma) where the developing cementum meets the primordium of the alveolar bone. The formation of the periodontal membrane is slightly different during the development of deciduous and permanent teeth. The periodontal membrane of deciduous teeth begins to develop together with root development before the tooth erupts. Here, the alveolar socket is already formed before the ligaments start to take up specific orientation and the tooth only needs to penetrate soft tissues during eruption. Permanent teeth rest deep inside the alveolar bone before teething starts. The alveolodental ligaments will start their development only when the overlying hard tissue is removed and the crown reaches the oral cavity. The periodontal ligaments will find their final orientation only when the eruption is complete.

Blood and Nerve Supply

The periodontium of the upper jaw receives its blood supply from the branches of the maxillary artery: superior posterior alveolar, infraorbital, greater palatine, and incisive arteries. The blood vessels of the mandibular periodontium arise from the lingual, inferior alveolar, buccal sublingual, and mental arteries.

The rich network of lymph vessels drains toward the submental, submandibular, and the deep cervical lymph nodes.

Somatosensory innervation of periodontal structures derives from the maxillary and mandibular branches of the tigeminal nerve. The periodontium of the premaxilla receives its innervation from the superior anterior dental plexus of the infraorbital nerve, whereas the more posterior structures are supplied by the medial and posterior superior dental plexuses. The deeper regions of the lower jaw are solely innervated by the inferior alveolar nerve. The nerves follow the course of the arteries and, once inside the periodontal membrane, they terminate in mechanoreceptor end organs, mainly unencapsulated and branched Ruffini-like corpuscles for mechanosensation of teeth, or free endings subserving nociception (Fig. 4.8).

The mechanoceptive nerve fibers of the periodontal ligament follow either of two pathways. Those arising from the apical periodontium are low-threshold fibers belonging to perikarya of the mesencephalic trigeminal nucleus, whereas those innervating the middle of the roots are high-threshold fibers that have their cell bodies in the Gasserian ganglion. It has been suggested that the former group would be responsible for a positive feedback regulation of jaw closing muscles (resulting in harder bite), whereas the latter trigeminal afferents would arrest chewing via inhibitory premotor interneurons. This way, the different stretch receptors of the periodontal ligament can

Fig. 4.8

The innervation of the periodontal ligament. Adapted from Byers MR (J Comp Neurol, 231, 1985, 500-518).

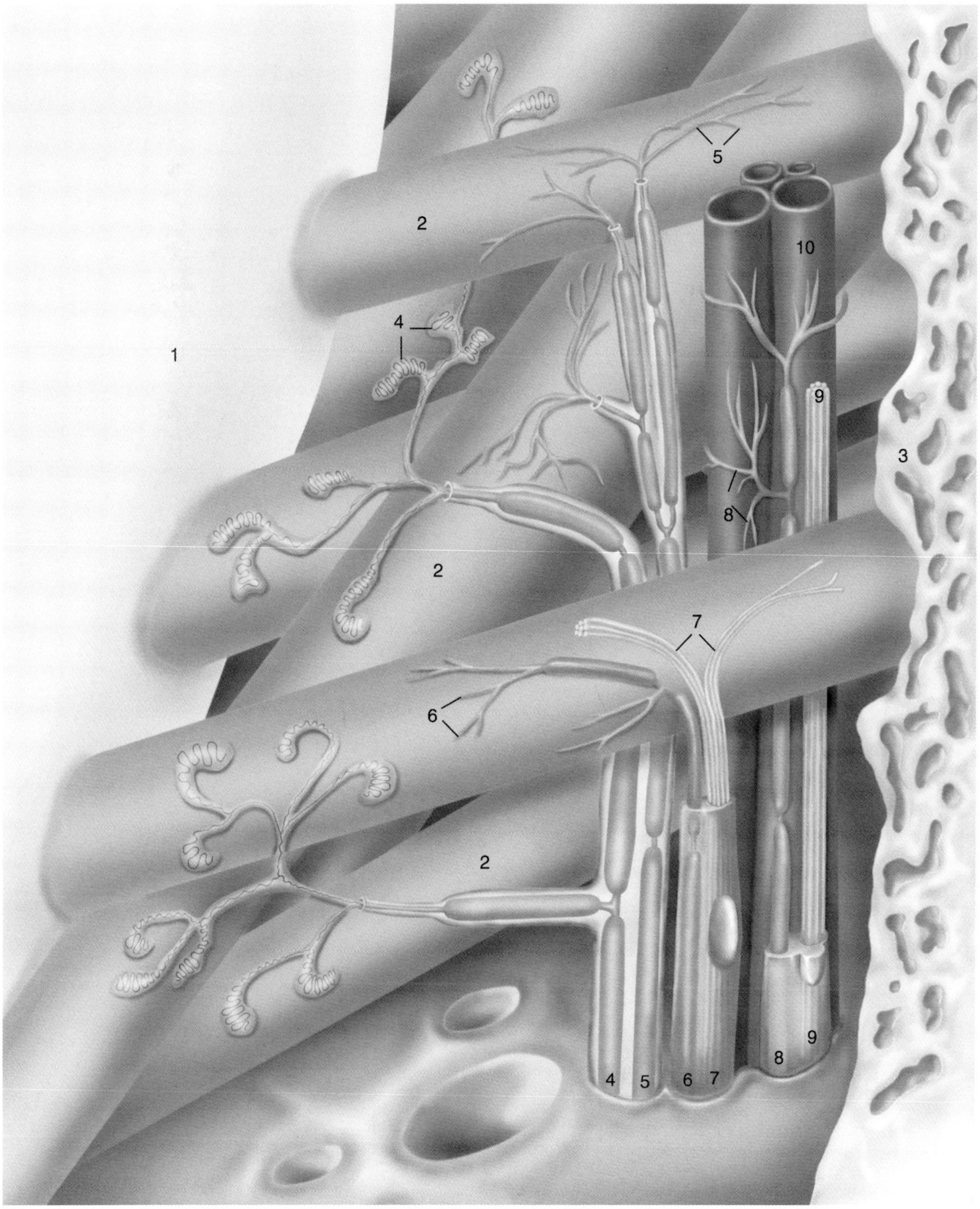

1. Cementum
2. Ligament fibers
3. Alveolar bone
4. Complex Ruffini-like endings, ensheathed preterminal axon
5. Simple Ruffini-like endings, ensheathed preterminal axon, preterminal axons can form paired branches
6. Simple Ruffini-like endings, branching from small free, myelinated axon
7. Bundles of free unmyelinated axons
8. Capillary-associated simple Ruffini-like endings, branching from small myelinated axons
9. Capillary-associated bundles of free unmyelinated axons
10. Capillary

effectively control the force of bite when the teeth come together or in contact with hard food. For conscious evaluation of food texture, a secondary trigeminothalamic pathway (dorsal trigeminal lemniscus) ascends to the thalamic ventroposteromedial nucleus, and from there the third neuron projects to the somatosensory cortex.

Apparently, the mechanosensory nerve fibers of periodontium are not the only elements instrumental in the detection of the textural features of ingested food. Apart from the somatosensory afferents of the tongue mucosa (filiform papillae), a rich innervation of the hard palate has also been described. Sensory nerve fibers were found predominantly in the papilla incisiva and the rostralmost rugae palatinae (i.e., areas that come in close contact with the tongue during chewing of food). In the Rhesus monkey, five types of nerve endings have been reported. Free nerve endings were ubiquitously found in the epithelium and lamina propria. Merkel cells and terminals occurred in the basal epithelial layer, Meissner corpuscles in the connective tissue papillae, whereas small lamellated corpuscles were observed beneath the epithelial pegs and Ruffini corpuscles in the deep layer of lamina propria. Such variety of receptor end organs indicates that the hard palate has an elaborate role in monitoring the mechanical properties of food.

This function is further assisted by proprioceptive afferents from the muscle spindle receptors (see also Chapter 5) of the muscles of mastication. The latter fibers reach the pontine trigeminal nuclei via the mesencephalic trigeminal nucleus (e.g., bypassing the Gasserian ganglion), similarly to the apical periodontal afferents. The ascending trigeminothalamic and thalamocortical pathways have already been mentioned.

RECOMMENDED READINGS

Textbooks and Handbooks

1. Bannister LH, Berry MM, Collins P, Dyson M, Dussek JE, Ferguson MWJ. Gray's Anatomy, 38th edition, Churchill Livingstone, Edinburgh, 1999.
2. Drews U. Color Atlas of Embryology, Thieme, Stuttgart, New York, 1995.
3. Hinrichsen KV (Ed.). Humanembryologie. Lehrbuch und Atlas der vorgeburtlichen Entwicklung des Menschen, Springer, Berlin, 1990.
4. Smith C.U.M. Biology of Sensory Systems, John Wiley & Sons, Chichester, 2000.
5. Squire LR, Bloom FE, McConnell SK, Roberts JL, Spitzer NC, Zigmond MJ. Fundamental Neuroscience, Academic Press, Imprint of Elsevier Science, USA, 2003.
6. Norgren R. The gustatory system, In: G. Paxinos, ed. The Human Nervous System, Academic Press, 1990.

Reviews and Research Reports

1. Byers MR. Sensory innervation of periodontal ligament of rat molars consists of unencapsulated Ruffini-like mechanoreceptors and free nerve endings. J Comp Neurol 1985;231:500-518.
2. Byers MR, Dong WK. Comparison of trigeminal receptor location and structure in the periodontal ligament of differenent types of teeth from the rat, cat and monkey. J Comp Neurol 1989;279:117-127.
3. Gilbertson TA, Damak S, Margolskee RF. The molecular physiology of taste transduction. Current Opinion in Neurobiology 2000;10/4:519-527.
4. Halata Z, Baumann KI. Sensory nerve endings in the hard palate and papilla incisiva of the rhesus monkey. Anat Embryol 1999;199:427-437.
5. Lindemann B. Taste reception. Physiological Reviews 1996;76:718-766.
6. Smith DV, St. John SJ. Neural coding of gustatory information. Curr. Op. Neurobiol. 1999;9/4:427-435.
7. Smith DV, Margolskee RF. Making sense of taste. Scientific American 2000;284: 32-39.
8. Stone LM, Finger TE, Tam PPL, Tan S-S. Taste receptor cells arise from local epithelium, not neurogenic ectoderm. Proc Natl Acad Sci USA 1995;92:1916-1920.
9. Stone LM, Tan S-S, Tam PPL, Finger TE. Analysis of cell lineage relationships in taste buds. J Neurosci 2002;22:4522-4529.
10. Yoshida A, Fukami H, Nagase Y, et al. Quantitative analysis of synaptic contacts made between functionally identified oralis neurons and trigeminal motoneurons in cats. J Neurosci 2001;21: 6298-6307.

5 The Skin and Other Diffuse Sensory Systems

Mihály Kálmán and András Csillag

ANATOMICAL OVERVIEW OF THE SKIN AND OTHER DIFFUSE SENSORY SYSTEMS

This chapter summarizes those sensory structures that have not been covered in Chapters 1 through 4. First, we provide a general survey of the receptors discussed, then we describe the skin as a specific organ of sense, followed by the neural pathways of tactile sensation as exemplified by the mystacial vibrissae.

Receptors

The classification of receptors is based on the following:

- source of stimulus
- sensation and its biological effect
- morphological structure
- level of perception (conscious or subconscious)
- characteristics of their innervation.

According to their principal structure, the diffuse endings belong to one of the following three categories:

1. free nerve endings (the major component)
2. encapsulated endings (possessing a lamellar or non-lamellar connective tissue capsule)
3. epidermal (more generally: cell-associated) endings

Interestingly, the other classifications seem to be independent of the morphological structure.

According to the source of stimuli, these receptors are represented in all three main groups: exteroceptors, visceroceptors, and proprioceptors (the latter two collectively forming the group of interoceptors).

The exteroceptors (receiving stimulus from the environment) are represented here by the skin and periorificial mucosal receptors. For anatomical reasons, some of the oral mucosal receptors, as well as the corresponding proprioceptors relevant to the textural attributes of food, have been described in the chapter on taste sensation.

Cutaneous (skin) receptors are responsible for a multitude of "skin" sensations. In addition to the classical "touch," "warm," "cold," and "pain" qualities, we distinguish hair-bending, pressure, vibration, stretching, itching, "discomfort," stroking, tickling, titillation, and wetness. However, the morphological variety of receptors seems to be poor when compared to the impressive register of sensations. The following explanations may be inferred.

1. the combination of receptor activities (actually, deep pressure, stretching and vibration cannot be registered without the participation of the proprioceptive system, see below);
2. morphologically similar receptors—mainly free nerve endings—can be used for different stimuli and sensations;
3. sensation-processing neural systems may modify the elementary sensations, mainly concerning their pleasant or unpleasant modalities. The involvement of the limbic system is to be mentioned here.

Proprioceptors comprise the receptors of the locomotor system. The term refers to the fact that these receptors are activated by the active or passive movements of the same muscles that respond to their stimuli that is, they are "own" (*proprius* in Latin) receptors of the muscles. The vestibular organ may also be classified as a member of this group, owing to its functional relations. Histologically, proprioceptors comprise neuromuscular spindles, neurotendineus organs of Golgi, Pacinian corpuscles, and bare nerve endings of joints.

Visceroceptors are distributed in the wall of hollow organs (including blood vessels), the capsule of parenchymal organs, and the suspension of viscera. They react to tension and/or muscular contraction rather than touch. Irritant receptors respond to noxious chemical, mechanical, or thermal stimuli. Histologically, they correspond to encapsulated (Pacinian, Krause) or free endings. It has to be noted, however, that the direct effect of the stimulus can be similar in the aforementioned systems. Pain can arise in all of them. Heat can affect the visceral systems, too, and the mechanical stimuli can always be narrowed down to compression or stretching. This may explain why one can find morphologically similar structures in all three classes of receptors. Functional data also suggest similar mechanisms of action, e.g., integrin $\alpha_2\beta_1$ may be a linking agent between mechanical stress in the extracellular matrix and the modulation of neuronal response. Visceral chemoreceptors are integrated in the humoral, rather than the nervous system. Important exceptions are the oxygen saturation receptors of the aortic and carotid bodies.

According to the electrophysiological characteristics, the receptive field of receptors may be small or large (depending on their superficial or deep position), the threshold may be low or high (the latter is characteristic for pain), and the adaptation may be slow or rapid (registering situation or alteration).

The myelinated low threshold cutaneous afferents belong to four main classes: simple rapidly adapting (RA), Pacinian corpuscle-associated, rapidly adapting (PC), slowly adapting, type I (SAI) and SAII.

Free (Bare) Nerve Endings (Figs. 5.22; 5.23)

These endings occur in all types of connective tissue: dermis, meninges, capsules of joints and glands, ligaments, fasciae, perimysium, tendons, Haversian and Volkmann canals of bones, periosteum, perichondrium, all layers of hollow viscera, vascular adventitia, and the parietal serosa. They also innervate the epithelium of the skin (Fig. 5.22D,E), mucosa (Fig. 5.23B), and especially frequently, the cornea. Bare endings terminate either in the corium (or submucosa) or penetrate the epithelium. They may form glomeruli. For their demonstration, metallic (mostly silver) impregnation has been the method of choice until the last decade, when immunohistochemical methods (detection of neurofilament proteins, recently protein 9.5, etc.) appeared in our arsenal. Free endings always belong to the thinnest nerve fibers, either myelinated (e.g., Aδ) or unmyelinated (C). The ending itself, however, which usually branches and forms a plexus, is always devoid of Schwann cells. Based on electrophysiological evidence, several sensory modalities are associated with free nerve endings: mechanoreceptors (of rapid adaptation), warm or cold thermoreceptors and nociceptive receptors, either uni- or polymodal (i.e., sensitive to one or several noxious stimuli). In the cornea, dentine, and periosteum, all such receptors are nociceptive, and therefore every stimulus is registered as pain in these locations.

Cold receptors are slender myelinated fibers that penetrate the basement membrane of epidermis, lose their Schwann cell investment and branch among the cells of the basal layer. They respond best to 25–30°C.

Warm receptors appear to be unmyelinated fibers responding maximally at 40–42°C. The integrity of non-noxious thermal sensation systems is essential for the normal perception and recognition of thermal pain.

Pain receptors are similar to thermal ones, with round clear and large dense-core vesicles. Nociceptors can be specific (Aδ), or polymodal (C), containing calcitonin gene-related peptide (CGRP) and substance P (SP). Both types also contain glutamate. These receptors are affected by algogens (potassium ions, serotonin, histamine, bradykinin), whose effect can be potentiated by sensitizing agents (eikozanoid-prostaglandins, leukotrienes, SP, CGRP). These substances, evoking axonal reflexes, vasodilation and inflammation, are released from damaged cells. Nociceptors are slowly adapting and may be hyperirritable.

Cell-Associated Nerve Endings

Hair Follicle-Associated Terminals (Figs. 5.1; 5.16; 5.36; 5.37; 5.38)

A characteristic type of hair follicle-associated terminal is the lanceolate or palisade ending (Fig. 5.37), whose parent fibers approach the follicle from several directions below the junction of the sebaceous gland, and run in the outer follicular layer in a circumferential, longitudinal, or palisade arrangement. Lanceolate endings are RA mechanoreceptors, registering the movement and deformation of the follicle. Their receptive field is large because one axon supplies several hundred follicles.

A special receptor system (essentially a sensory organ) surrounds the long vibrissal hairs (whiskers) of some animal species. Similar receptor function is attributed to the skin appendages around the nose and upper lip of walruses and moles. Owing to its importance in the general understanding of somatosensory functions, the vibrissal receptor organ is discussed later.

Merkel Cells and Endings (Tactile Menisci, Disc Endings) (Figs. 5.2; 5.17; 5.30; 5.37)

These receptors are found in the basal layer of epidermis, just above the basal lamina of glabrous skin, but they also occur in hairy skin, mainly around the apical end of hair follicles. The intraoral (e.g., palatal) occurrence of Merkel cells has also been reported. Their nerve fibers expand into a disc (Merkel ending) apposed to the base of a Merkel cell. The latter are "clear" cells with a lobulated nucleus, and their cytoplasm is thrown into protrusions (resembling a floating sea mine) penetrating the surrounding keratinocytes, to which they are connected by desmosomes. Merkel cells contain large dense-core vesicles, suggesting a synaptic mechanism, although transmitter release is yet to be proved. The Merkel cells belonging to one branching nerve fiber may form groups at a dome-like epidermal formation (touch dome, touch corpuscle) of hairless skin in the bottom of epidermal ridges, or at a specialized hair follicle (tylotrich, tactile hair discs).

The SAI nerve fibers belonging to Merkel discs are relatively thick and myelinated, of the Aβ type, and they branch and lose their myelin sheaths about 10 μm from the epidermal border. They then perforate the basal lamina and form Merkel discs: axonal endings containing clear vesicles. Merkel cells are secondary receptors with several attributes of neurons (so-called paraneurons), containing neuron-specific enolase, nerve growth factor-receptor, VIP, and met-enkephalin. They express keratins (8, 9, 19) that are characteristic of embryonic rather than mature epithelial cells. Apart from its receptor function, the Merkel cell is a putative spatial organizer in the epidermis, has neuroendocrine functions, and displays features of amine precursor uptake and decarboxylation (APUD) cells. The origin of Merkel cells is a topic of controversy. Several sources suggest an epidermal origin because these cell are situated in the epidermis, rather than in the dermis, preceding the ingrowth of nerve endings, and proliferate in vitro without neural effects. Other results on transgenic mice (neural crest cells marked with β-galactosidase) suggest a neural crest origin. Villin is an excellent marker of Merkel cells. A stretch-activated Ca^{2+}-influx channel is known to operate in Merkel cells, and recent results suggest that these cells are responsible, at least in part, for the SAII type receptor activity, instead of Ruffini corpuscles.

Encapsulated Endings

This group exhibits a considerable variety, but the common feature is that the nerve ending is encapsulated by nonirritable cells. Besides the different cutaneous receptors, this group may also comprise the neuromuscular and neurotendinous organs. The capsule operates as a filter of mechanical impulses, and it is responsible for the delay of adaptation. The deeper the receptor, the larger its receptive field (see below).

Encapsulated receptors fall into four main categories: deep and RA (PC); deep and SA (Ruffini endings); super-

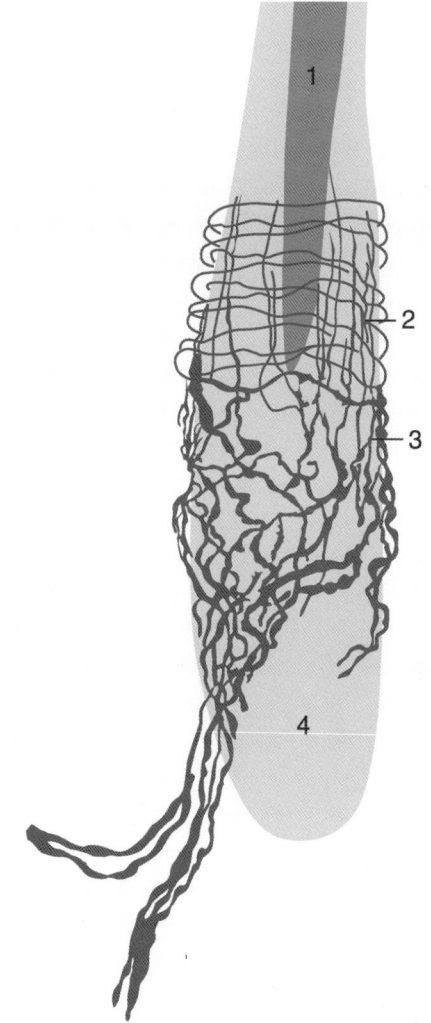

Fig. 5.1

Innervation of hair follicle in the facial skin of cat. Redrawn after an original silver impregnated specimen of Ábrahám.
1. Hair shaft
2. Circularly arranged nerve fibers
3. Longitudinally arranged nerve fibers
4. Root sheath

ficial and RA (Meissner's corpuscle); superficial and SA (Merkel cell).

MEISSNER'S TACTILE CORPUSCLES (FIGS. 5.3; 5.18)

With dimensions of 80–150 μm by 30–50 μm, these encapsulated lamellar structures are situated in the dermal papillae. They occur mainly in thick and hairless skin, all parts of the hand (fingerpad: 24/mm²) and foot, the palmar side of the forearm, lips, oral mucosa, palpebral conjunctiva, nipple, and prepuce. Their connective tissue

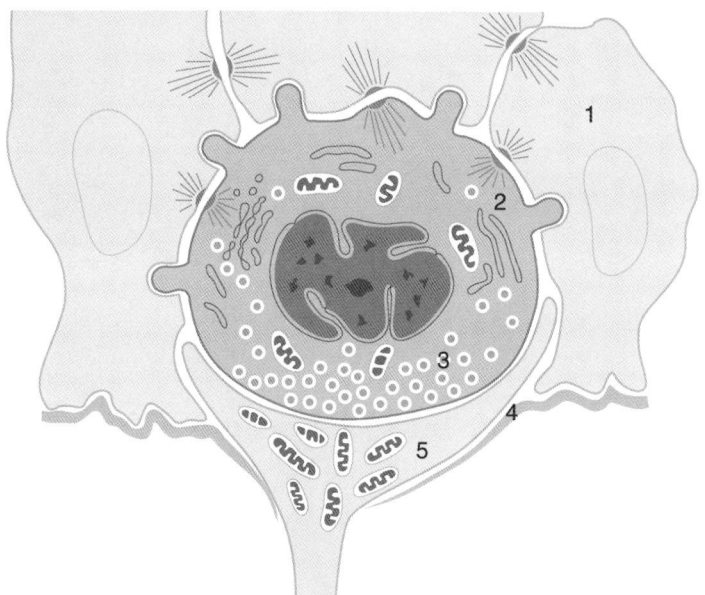

Fig. 5.2

Schematic drawing of the Merkel cell. Redrawn and modified after Röhlich.

1. Keratinocyte
2. Merkel cell
3. Dense core vesicles
4. Lamina basalis
5. Nerve terminal

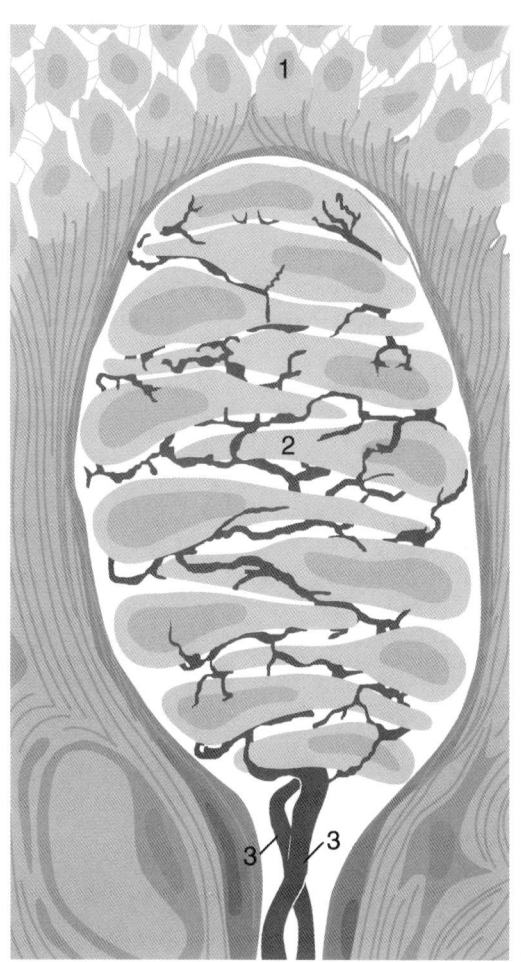

Fig. 5.3

Schematic drawing of the Meissner's tactile corpuscle. Redrawn and modified after Röhlich.

1. Epithelial cells
2. Nerve fibers
3. Myelin sheaths

capsule is continuous with the perineurium of the incoming nerve fiber. Up to seven nerve fibers may enter the capsule and course upward in the inner core, between the flat cells of Schwann cell origin, which are stacked like dinner plates. The Meissner's corpuscles receive at least three types of innervation. Aβ terminals, peptidergic (CGRP, SP) C terminals and non-peptidergic C-terminals. They may have nociceptive capabilities as well.

PACINIAN CORPUSCLES (LARGE LAMELLATED CORPUSCLE OF VATER-PACINI, LAMELLAR BODY) (FIGS. 5.4; 5.19)

These receptors occur in the hypodermis of hairless thick skin, but also in the external genital organs, arm, neck, nipple (all are considered erogenic regions!), periosteum, interosseous membranes, near the attachment of ligaments, tendons and joint capsules, the peritoneum at the attachment of organs, pancreas, and urinary bladder. PCs are oval, spherical, or even curved bodies, with diameters of 2–4 by 0.5–1 mm, just allowing macroscopical dissection. Three zones can be distinguished: a capsule, an intermediate growth zone, and a central core with the axon terminal. The capsule is formed by lamellae (up to 30–80) of overlapping flat perineurial cells, and the interlamellar spaces contain proteoglycan and collagen fibers attached mainly to the external surface of the cells. The intermediate zone containing mitotic cells is inconspicuous in mature corpuscles. The core consists of about 60 concentric lamellae (Fig. 5.4), formed by interdigitating processes of elongated cells (probably modified Schwann cells but could be modified fibroblasts). The supplying nerve is of Aβ type, which loses its myelin

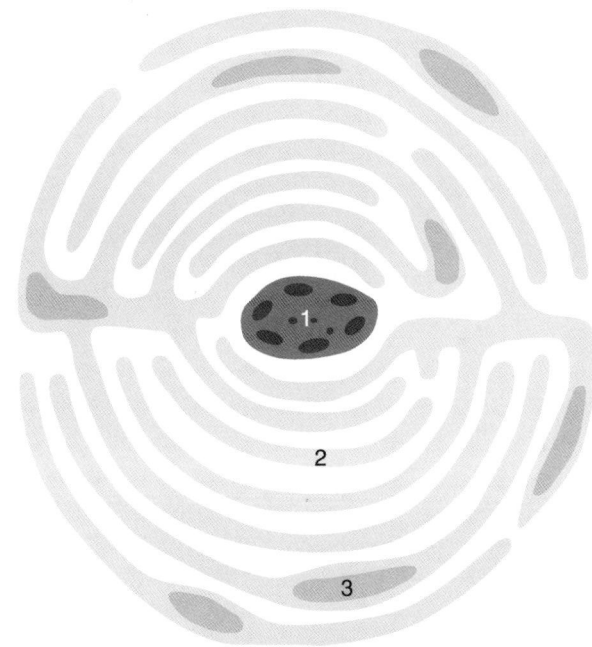

Fig. 5.4

Schematic drawing of the inner capsule of the Pacinian corpuscle. Redrawn and modified after Röhlich.

1. Nerve fiber
2. Concentric cellular laminae
3. Nucleus of cell forming the concentric lamina

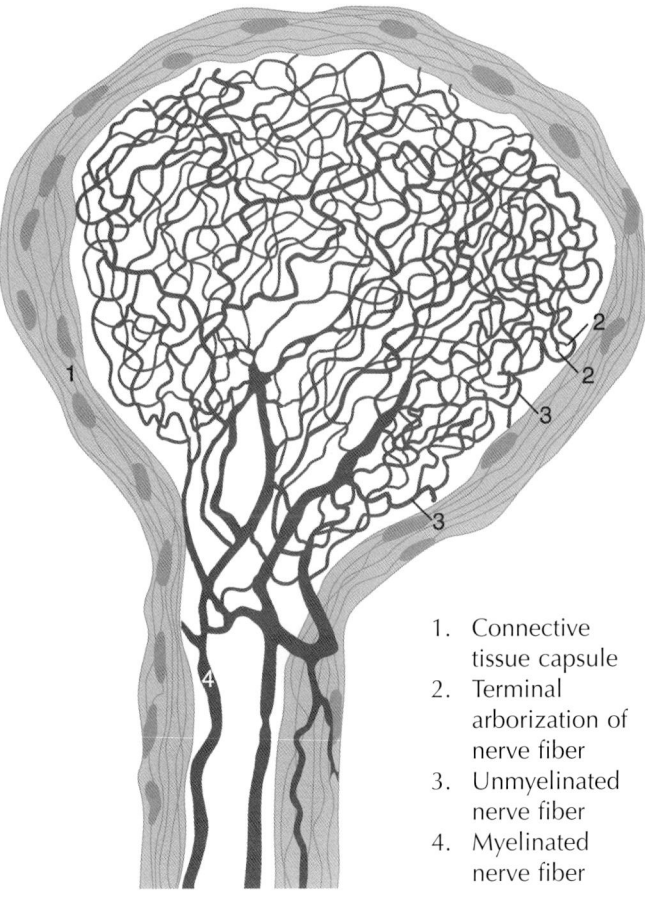

1. Connective tissue capsule
2. Terminal arborization of nerve fiber
3. Unmyelinated nerve fiber
4. Myelinated nerve fiber

Fig. 5.5

Schematic drawing of a genital corpuscle. An essentially similar organization is found in the end-bulbs of Krause and various other bulbous "innominate corpuscles." After Patzelt.

sheath on entering the core. Co-entrance of unmyelinated fibers has also been reported. Pacinian corpuscles develop in the third fetal month, and they may form clusters. Very rapidly adapting mechanoreceptors, they respond only to sudden changes of pressure, especially to vibration. The pressure in the intralamellar fluid provides turgidity, whereas the lamellae act like a "high-pass" frequency filter, dampening slow distortions by fluid movements. There are 200–400 Pacinian endings in the hand, mainly (40–60%) on the fingers, but also (20–50%) in the metacarpophalangeal area, and (10–20%) in the thenar and hypothenar eminences. Both in Meissner's and Pacinian corpuscles, the glial (Schwann cell) origin of inner cells is supported by their immunoreactivity to S-100 and vimentin, but not to cytokeratin.

OTHER ENCAPSULATED ENDINGS (FIGS. 5.5; 5.20; 5.21)

Various other encapsulated endings have been described in humans and in other mammalian species, but the physiological data on their function are scarce.

Paciniform endings (smaller and less organized than proper PCs) occur in the connective tissue sheaths of vibrissae and in joint capsules.

With a size of 0.1 mm, the end-bulbs of Krause (Fig. 5.20) are composed of coiled and branching nerve endings surrounded by a simple connective tissue capsule. They are found mainly in the glans, prepuce, clitoris, conjunctiva, lip, tip of tongue (note the preponderance of erogenic zones), but they also occur in the peritoneum and other parts of the skin (e.g., the footpad of nonprimates). Reportedly one of the most frequent endings, their formerly suggested role as cold-receptor has not been sustained, although there is still uncertainty about their function. The bulbous corpuscles, genital corpuscles (Fig. 5.21), and "innominate" corpuscles (Fig. 5.23), appearing in the different descriptions, seem to be local variants of the end-bulb of Krause.

GOLGI-MAZZONI ENDINGS (FIGS. 5.6; 5.22A)

These receptors are related to either the PC (with less organized lamellae) or to the end-bulb of Krause (with multilayered capsule) as their local variants. Their existence as an independent entity was doubtful until the advent of electron microscopic investigation. Golgi-Mazzoni endings are invested with an outer connective tissue capsule and an inner cellular/lamellar capsule. The nerve

Fig. 5.6

Schematic drawing of Golgi-Mazzoni ending. Redrawn and modified after Röhlich.

1. Lamellar connective tissue capsule
2. Reticular telodendrion (fine network of axonal arborization)
3. Nerve fiber

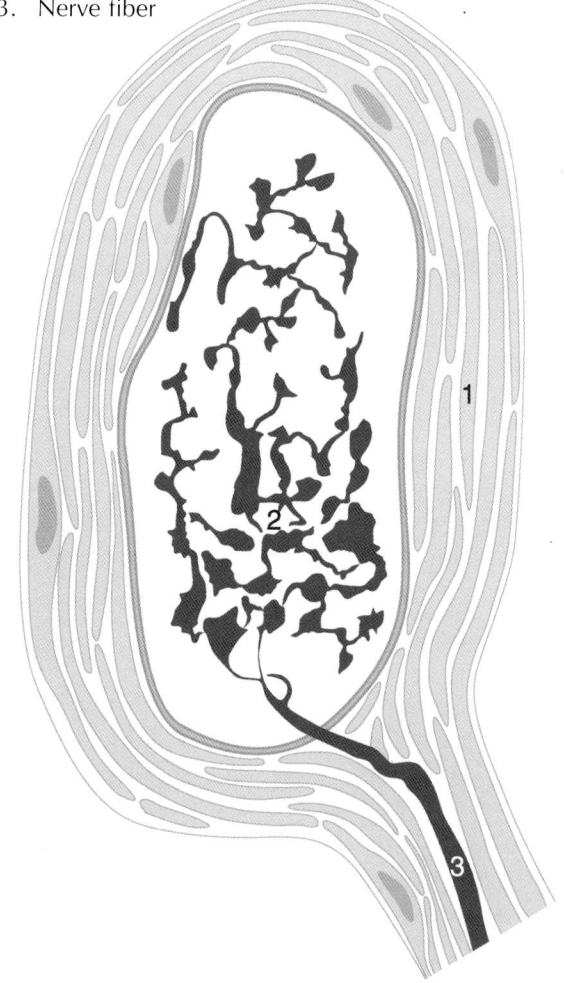

ending is profusely branching and coiled, with a glomerular appearance. These receptors have been observed in several parts of the skin (subcutis) and mucosa, especially in genital organs, conjunctiva and nail bed, but also in the peritoneum, tendons and joint capsule attachments, with a likelihood of proprioceptive function.

RUFFINI ENDINGS—SAII CUTANEOUS MECHANORECEPTORS (FIGS. 5.7; 5.22B).

Ruffini endings occur in the dermis of hairy and glabrous skins, at knuckle joints, nail bed, glans penis, tendon and joint attachments, and also in the gingiva and the periodontal ligament (see Fig. 4.8 in Chapter 4) near the apical part of teeth. These receptors are 0.5–2 mm long, approx 150 μm across, and consist of highly branched, nonmyelinated endings of Aβ-myelinated fibers, which invade bundles of collagen fibers and form a spindle-shaped structure enclosed in a fibrocellular sheath of perineurium. The core consists of collagen fibers and nerve terminals, and the fluid-containing space between core and capsule is occupied by perineurial cells. A single myelinated axon (7–12 μm thick) enters either at the pole or near the equator. Having lost its myelin sheath, the nerve fiber continues as a club-like expansion with short branches. Ruffini endings can be affected by tension transmitted from the environment. Their structure resembles those of the neurotendinous organ of Golgi (Fig. 5.8) and like structures in joint capsules, even in electrophysiological characteristics, being responsive to sustained tension. Ruffini-like spray endings without a capsule have been also described. Another variant, pilo-Ruffini endings, occur in the hair (mainly whisker) follicles (Fig. 5.37) and also in or near joints. Ruffini endings used to be taken for warm receptors, but this view is no longer considered tenable. According to current theory, Ruffini corpuscles are stretch receptors with SAII primary afferents. On the other hand, electrophysiological recordings suggest that SAII fibers represent approx 15% of the myelinated mechanosensitive axons. Only a few SAII afferents are likely to terminate in Ruffini corpuscles, despite numerous and thorough histological investigations. In a comprehensive immunohistochemical study covering all distal phalanges of three individuals of different age, only a single typical Ruffini corpuscle and a few Ruffini-like spray endings were found. All other apparent Ruffini-like structures proved to be innervated blood vessels (indeed, Ruffini had originally described this receptor adjacent to a vessel). Therefore, there are opinions that the specific type of sensory ending that accounts for the SAII-like electrophysiological properties is unlikely to be the long-accepted Ruffini corpuscle. Instead, Merkel endings of

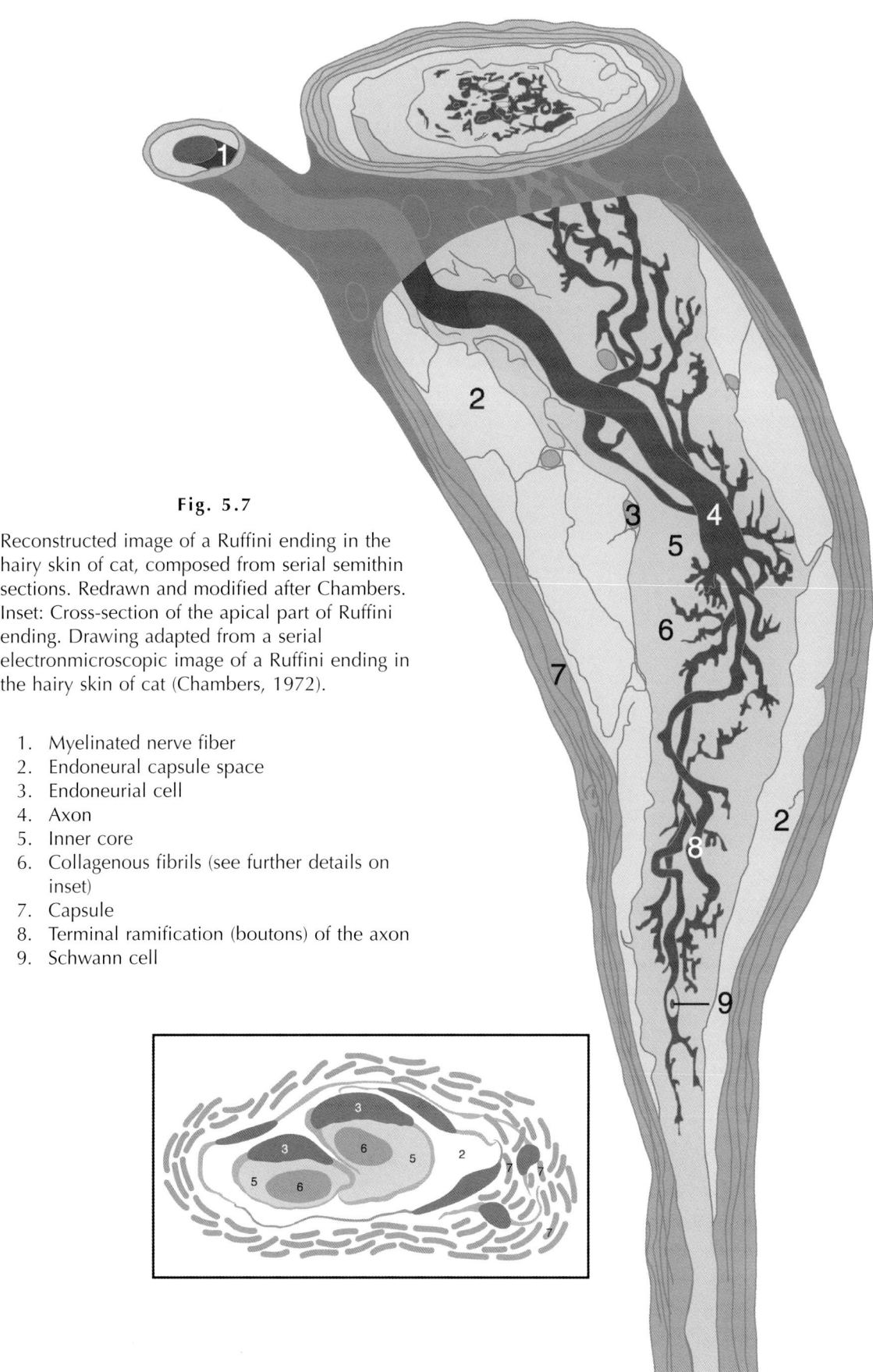

Fig. 5.7

Reconstructed image of a Ruffini ending in the hairy skin of cat, composed from serial semithin sections. Redrawn and modified after Chambers. Inset: Cross-section of the apical part of Ruffini ending. Drawing adapted from a serial electronmicroscopic image of a Ruffini ending in the hairy skin of cat (Chambers, 1972).

1. Myelinated nerve fiber
2. Endoneural capsule space
3. Endoneurial cell
4. Axon
5. Inner core
6. Collagenous fibrils (see further details on inset)
7. Capsule
8. Terminal ramification (boutons) of the axon
9. Schwann cell

different position and innervation may be associated with the physiologically distinct SAI and SAII afferents.

NEUROTENDINOUS ENDINGS (GOLGI TENDON ORGANS) (FIG. 5.8)

These receptors group (about 50) at the myotendinous junction, on both ends of the muscle belly, or at joint capsules. Each ending consists of a bundle of tendon fibers (intrafusal fascicle) in a delicate lamellar cellular capsule. Being 0.5–1 mm in length and 100 μm in diameter, neurotendinous endings are supplied by Aβ myelinated fibers of about 10 μm diameter. Slowly adapting, these receptors respond to an active muscular contraction or passive stretching, and sustained tension in particular.

Receptor Structures of Joints

Type I: Ruffini endings situated in the superficial layers of fibrous capsule (mainly in the joints of high postural significance: hip, knee) continually registering the position and movements of the joint.

Type II: PCs, usually smaller than those of the skin, forming clusters in the deep layers of the fibrous capsule. This is a low-threshold and RA receptor, registering rapid articular movements and stress.

Type III: Identical to the Golgi tendon organ, occurs in ligaments but not in capsules. This receptor is characterized by high threshold and slow adaptation, and it serves to prevent extreme stress (overstreching) via proprioceptive reflexes of the adjacent muscles.

Type IV: Free endings, ramifying in the capsules, correspond to SA high-threshold receptors of articular pain.

NEUROMUSCULAR SPINDLES (FIGS. 5.9; 5.22C)

This end organ contains a few intrafusal muscle fibers with their individual inner investment (possibly a derivative of endomysium) and a common external connective tissue capsule (perimysium). Based on their ultrastructure, histochemistry, and physiology, the intrafusal fibers have two main types:

1. nuclear bag fibers (with a cluster of nuclei in a bulge) may be static or dynamic, with type IA sensory nerve endings, and they respond to both the degree and rate of stretch.
2. nuclear chain fibers (with a straight row of nuclei in the center) belong to various categories (short or long, primary and secondary). They are sensitive only to the degree of muscle stretch, i.e., they are static, with nerve endings of type IA and II.

Sensory innervation of muscle spindles is twofold:

1. Primary, annulospiral nerve endings (of Aα fibers) coil around the nucleated parts of the fibers; they exhibit rapid adaptation;
2. Secondary flower-spray endings (of B fibers), mainly to the nuclear chain receptors, slowly adapting.

Motor innervation of the muscle spindle is supplied by two different Aγ fibers and one Aβ fiber, the former are special fusimotor efferents, whereas the latter is a collateral of the extrafusal slow twitch muscle. These fibers adjust the sensitivity of the receptors via the contraction of intrafusal fibers, whereby they act like a servo mechanism engaging the proprioceptive reflex for fine tuning of muscle movements (cf. the innervation and activity of extraocular muscles, in the chapter on vision).

CAROTID BODY (GLOMUS CAROTICUM) (FIGS. 5.10; 5.24; 5.25)

Brownish structures of 2–5 mm diameter, carotid bodies lie bilaterally behind the carotid bifurcation. They consist of sinusoid blood vessels, afferent and efferent nerve

Fig. 5.8

Schematic drawing of the neurotendinous ending (Golgi tendon organ). Redrawn and modified after Patzelt and Ábrahám.

1. Muscle fiber
2. Collagen fibers of tendon
3. Nerve fiber
4. Terminal arborization of nerve

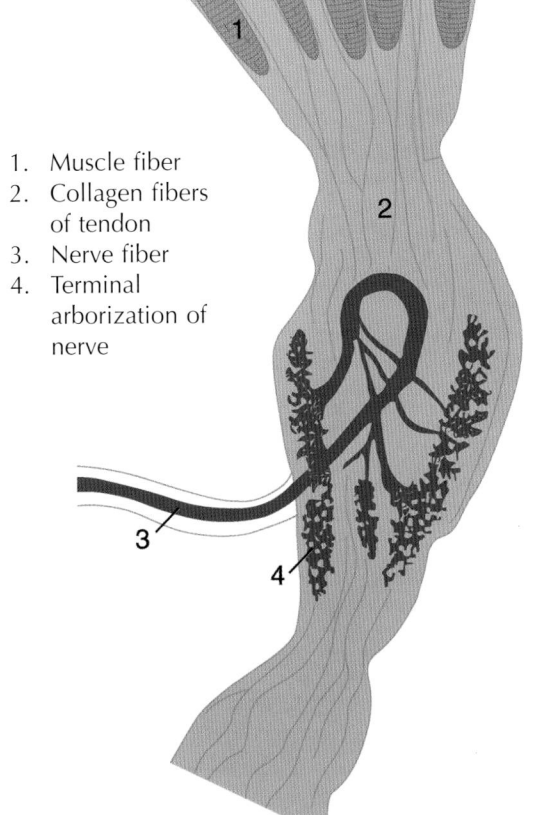

Fig. 5.9

Schematic drawing of the muscle spindle. Redrawn and modified after Bannister et al., Gray's Anatomy, Churchill Livingstone, Edinburgh, 1999

1. α Efferent
2. End plates
3. Dynamic γ efferent
4. Static γ efferent
5. Secondary (flower spray) ending
6. Dynamic nuclear bag fiber
7. Nuclear chain fiber
8. Static nuclear bag fiber
9. Primary annulospiral endings of dynamic nuclear bag fiber
10–11. Extrafusal fibers
12. Primary annulospiral endings of static nuclear bag fiber
13. Primary annulospiral endings of nuclear chain fiber
14. Capsule
15. Secondary (flower spray) ending
16. Primary afferent
17–18. Secondary afferents
19–20. Trail endings
21. Dynamic γ efferent
22–23. Static γ efferents

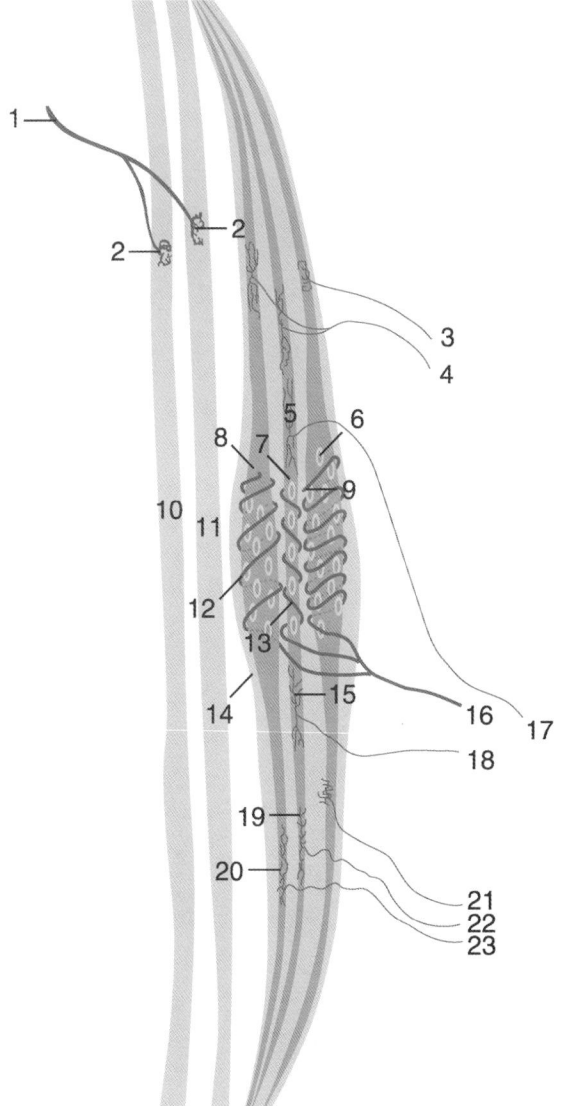

endings, sparse ganglionic cells, and two types of epitheloid cells: glomeral (type I) and sustentacular (type II, or sheath cell). The latter separate the glomeral cells from the sinusoids. Afferent nerves passing between the sustentacular cells form synapses (i.e., postsynaptic terminals) with the (presynaptic) glomeral cells. Efferent parasympathetic (from the glossopharyngeal and vagus nerves) and sympathetic fibers (from the sympathetic trunk) innervate the corresponding ganglionic cells. The glomeral cells exhibit certain signs of endocrine activity, and produce dopamine and other neurotransmitters. They may be regarded as paraneurons, forming dendrodendritic-like synapses among each other, whereas the sustentacular cells may represent glia-equivalents. The exact mechanism of chemosensation (O_2 saturation of the plasma), however, is not yet clear. This receptor organ is demonstrated here as an example of multiple visceroceptors, most of which fall outside the scope of this volume.

THE SKIN AND ITS APPENDAGES (FIGS. 5.11; 5.12; 5.13; 5.14; 5.15; 5.26; 5.27; 5.28; 5.29; 5.30; 5.31; 5.32; 5.33; 5.34; 5.35; 5.36)

The skin (known also as the common integument) forms about 8–16% of the total body mass (the heaviest organ of all), and its surface (varied according to body height and weight) amounts to 2.2 m^2 on average. Its thickness is between 1.5 and 4 mm, depending mainly on the body region. The skin consists of two main layers, the epidermis (epithelium) and dermis (or corium, connective tissue), collectively termed cutis. The term cutis, however,

Fig. 5.10

Schematic architecture of the carotid body (glomus caroticum). Redrawn and modified after Bannister et al., Gray's Anatomy, Churchill Livingstone, Edinburgh, 1999.

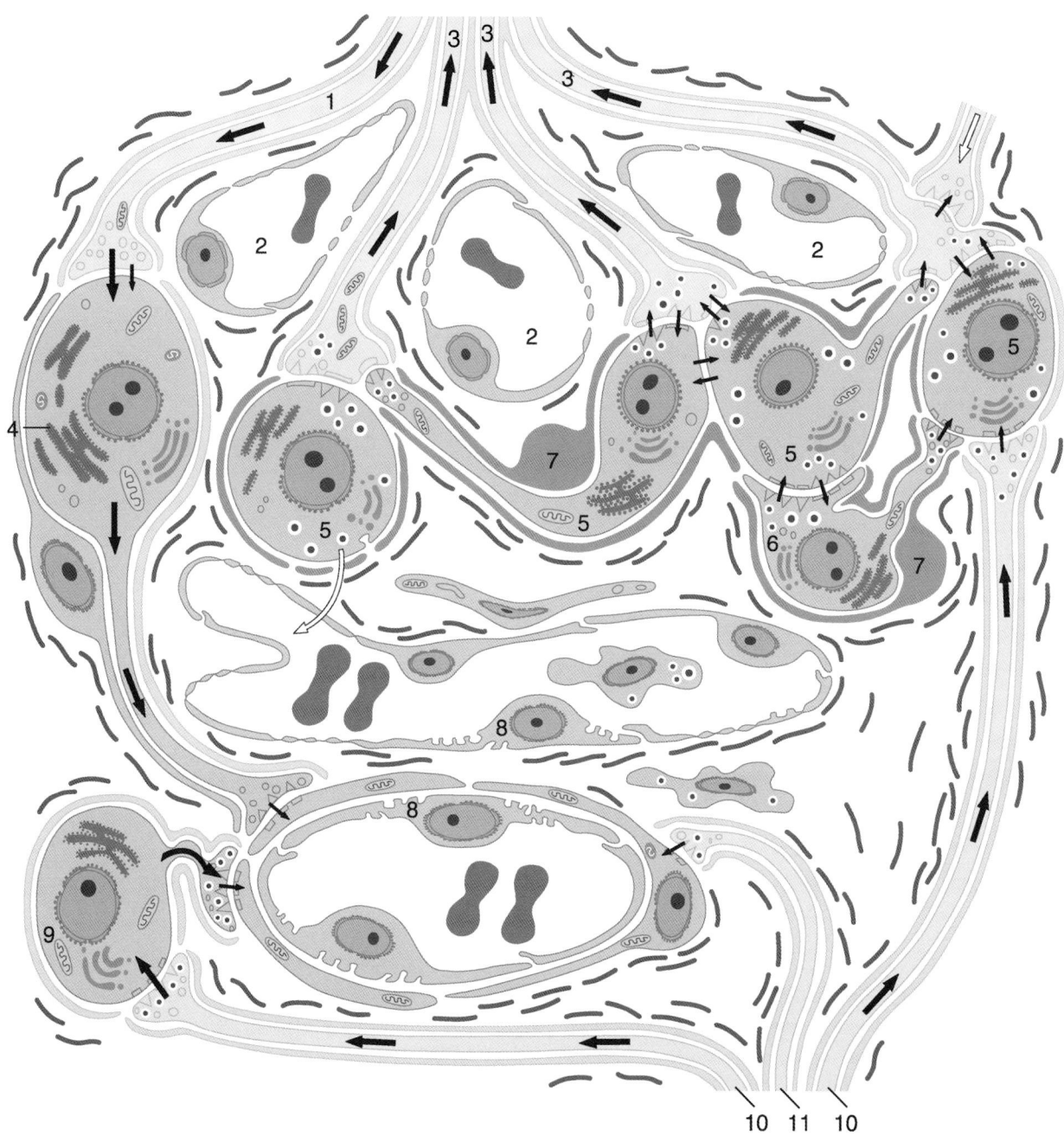

1. Preganglionic parasympathetic fiber
2. Sinusoid
3. Afferent glossopharyngeal axon
4. Parasympathethic ganglion cell
5. Type I glomus cell (A)
6. Type I glomus cell (B)
7. Type II sheath (sustentacular) cell
8. Endothelium
9. Sympathetic ganglion cell
10. Preganglionic sympathetic fiber
11. Postganglionic sympathetic fiber

is sometimes used (incorrectly) as a synonym for the dermis (corium).

The epidermis (at least in amniotes) always consists of keratinizing-stratified squamous epithelium. The epidermis extends several types of appendages: inward the glands (the main types: sudoriferous, sebaceous, odoriferous), outward the hairs, and their relatives and derivatives: nails, various animals' feathers, claws, hooves, different kinds of horns, scales, or even special types of armour (e.g., turtles, ankylosaurs, hedgehogs, armadillos, pangolins). Even the teeth correspond to modified scales.

The skin covering continues in the external auditory meatus, the vestibule of the nose, and the external aspect of the tympanic membrane. Near the orifices, a specialized skin of mucocutaneous junction occurs (including the free margin of eyelids and the puncta lacrimalia). These transitory parts are distinguished by special structures or a specialization of the common structures (free sebaceous glands, other special glands, thin keratinization, special hairs, high connective tissue papillae, pigmentation).

The skin is not a simple cover but an interface with the environment, with self-repairing, self-renewing, self-cleaning (by desquamation), and self-defending mechanisms. It provides communication in both directions.

The functions of common integument can be listed as follows:

- protection against mechanical, chemical, osmotic, thermal, and photic damage, water loss, and microbial invasion. In the latter, the immune surveillance function of Langerhans cells, and the importance of the proper bacterial flora overgrowing alien microbes, are to be noted.
- selective permeability to a variety of chemical substances;
- biosynthetic processes (e.g., of vitamin D from its precursor under the effect of ultraviolet light);
- target of a variety of hormones (regulating pigmentation, secretion, turgidity, etc);
- heat regulation, via circulation, sudoriferous (sweat) glands, adipose tissue;
- reservoir of blood, fat, and water (turgor);
- excretion of metabolites (sweat);
- sociosexual communication, by vascular, muscular (i.e., mimic), and odoriferous signals;
- carries signs of identity (finger-, palm-, sole-, lip-prints, facial expression);
- expresses symptoms of internal diseases, reflects metabolic and vascular states, and general health;
- lactation;
- provides surfaces for gripping and walking (manipulation and locomotion);
- sensory organ and receptive zone of reflexes.

The skin is not uniform throughout the body. Every body part has its characteristic subtype, but overall, two major classes can be distinguished (Fig. 5.11):

- thick, hairless (glabrous) skin, containing pilosebaceous units;
- thin, hairy (hirsute) skin, containing friction ridges.

All the other types are regional varieties of these (Figs. 5.27; 5.28). The underlying dermis determines the main type of the skin. Primarily hairless (glabrous) skin is found on the palm and sole, and on the corresponding sides of the fingers and toes. All the other regions represent variants of the hairy (hirsute) skin.

The hypodermis (or subcutis) consists mainly of adipose tissue. However, the latter is absent in special areas (penis, scrotum, nose, auditory tube, auricle, eyelid). Several descriptions do not consider the subcutis as part of the skin. Other authors emphasize that there are two layers to be distinguished: a "superficial fascia" and panniculus adiposus.

The dermis comprises two layers of connective tissue: the looser stratum papillare, forming the connective tissue papillae (called also dermal papillae in the skin), and beneath it the stratum reticulare, consisting of a three-dimensional network of collagen and elastic fibers.

The protrusions of the stratified squamous epithelium into the underlying connective tissue form a continuous network of epithelial ridges, which surround the separate connective tissue papillae. Therefore, separate epithelial tissue islets are rarely seen in sections, and the term "epithelial peg" is misleading. The border of epithelium with the connective tissue contains a regular basement membrane, consisting of reticular fibers and lamina basalis, to which epithelial cells are attached with hemidesmosomes, and the elastic and collagen fibers are anchored with special (type VII) collagen. This basement membrane, however, is hardly visible without proper staining (e.g., Azan).

The glabrous skin (Fig. 5.11; 5.28; 5.29) forms long dermal ridges (rather than papillae) interlocking with the epidermal ridges (Fig. 5.21A). The dermal ridges determine the individually characteristic pattern of the friction ridges ("fingerprint," toruli tactiles, dermatoglyphes) of the fingertips, palm, and sole. The definitive dermal ridges develop via the fission of primary ridges by the intrusion of sweat gland-carrying epithelial "rete pegs." Therefore, sweat glands open at regular intervals along the summit of friction ridges (Figs. 5.11; 5.29; 5.31B).

The dermal pattern is stable (genetically determined) and is absent only in less than 1% of individuals. The friction ridges increase the gripping and tactile abilities (notably, the touch domes of Merkel cells are apposed to such "rete pegs"). These faculties are further supported by the fact that this type of skin is tightly adherent to the underlying bone or aponeurosis through connective tissue bundles. Conversely, hairy skin (e.g., scalp skin) develops without such epithelial pegs and is loosely associated with the underlying tissue (Fig. 5.26).

The different layers of the epidermis represent different stages in the life course of its epithelial cells (keratinocytes), earmarked for keratinization.

The stratum germinativum (or stratum basale), where the new cells are produced, and the stratum spinosum together form the Malpighian layer (Fig. 5.29). The term stratum spinosum (also called stratum polygonale) refers to the shape of the cells after fixation: they shrink and move apart, except for those points where they are held together by desmosomes ("prickle cells"). Here, the inter-

Fig. 5.11

Three-dimensional illustration of the main structural elements of common integument.
A – Hairy (thin) skin, B – Hairless (thick) skin.

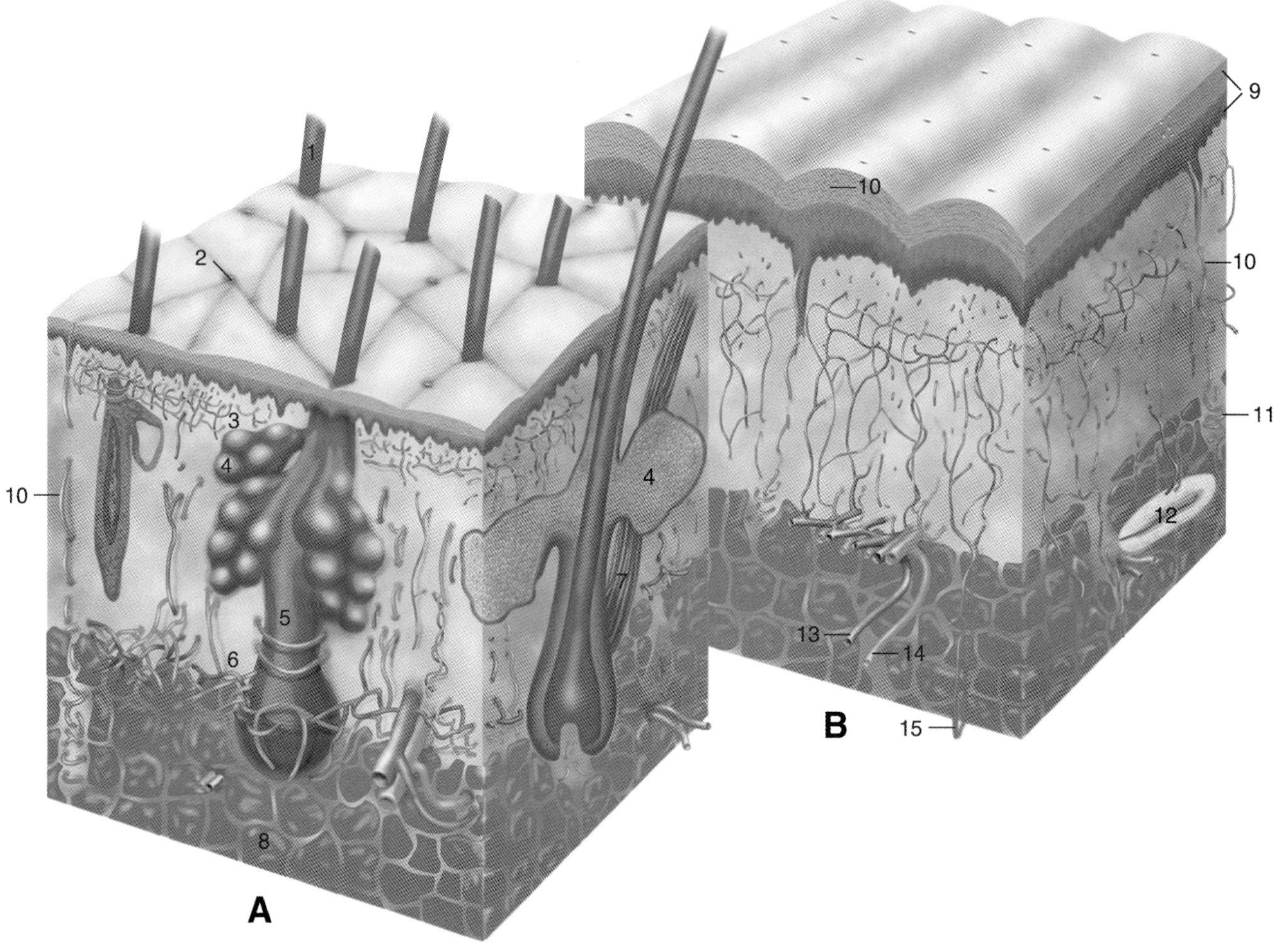

1. Hair shaft
2. Opening of sweat duct
3. Subpapillar neurovascular plexus
4. Sebaceous gland
5. Hair follicle
6. Deep cutaneous neurovascular plexus
7. Piloarrector muscle
8. Adipose tissue of subcutis
9. Epidermis
10. Sweat duct
11. Sweat gland
12. Pacinian corpuscle
13. Artery
14. Vein
15. Nerve

connected parts are elongated and form spine-like processes. The desmosomes are attached to tonofibrils (keratin filaments).

The characteristic staining of the stratum granulosum and stratum lucidum is the result of the different stages of keratinization (i.e., the accumulation of keratohyaline and eleidin, respectively). Where the latter two stages are skipped (e.g., in the hair and the nail), the hard form of keratin is produced, otherwise soft keratin is formed. On the other hand, these layers may appear to be absent where the skin is thin and keratinization is slow.

In the **stratum granulosum** (Fig. 5.29) the nuclei are picnotic, the organelles degenerating, and the keratin filament bundles are compact and aggregated. The characteristic protein of this stage is keratohyalin (filaggrin), a histidine-rich, intensely basophilic phosphoprotein. During further keratinization filaggrin, an intermediary filament protein, forms a matrix in which the keratin filaments are embedded. A further important feature is the presence of lamellar granules, which produce lipids and glycosaminoglycans. The **stratum lucidum** (Fig. 5.29) is conspicuous only in thick glabrous skin. It consists of transitional cells to be keratinized. The nuclei are poorly discernible, although they do still exist. Debris from nuclei and organelles, and densely packed keratin filaments, are embedded in an electron-dense matrix. The lamellar granules release their content (eleidin) into the intercellular space as a cement that forms an impregnable seal or permeability barrier. Owing to the matrix-filled intracellular and sealed intercellular spaces, the layer appears homogenous and highly refringent optically. Originally a reptilian "invention," it is this barrier, rather than the cornified layer itself, that makes our skin truly water proof.

In the **stratum corneum** (Fig. 5.29) the keratinized (cornified) cells are filled with keratin, filaggrin, and lipids, and they gradually flake off by a process called defoliation. Apart from the elimination of "cemented," nonviable cells, this layer also serves as a thick barrier of defense. The term keratinocyte (not to be confused with the keratocyte of cornea!) refers to the cells capable of keratinization. Those cells that have undergone the process of keratinization are called corneocytes.

The epithelial cells of common origin form an epidermal proliferative unit (EPU), supposedly regulated by a Langerhans cell via cytokines (Fig. 5.12).

Apart from keratinocytes, the stratum germinativum also contains melanocytes, Merkel cells and Langerhans cells, but none of these can be identified accurately in H&E-stained specimens. At best they appear as "clear" cells in routine sections (melanocytes only in an inactive stage).

Derivatives of the neural crest, melanocytes form a self-sustaining and locally dividing cellular system (Figs. 5.13; 5.28; 5.33). They possess no keratin or desmosomes. When depleted, they repopulate the epidermis. The regional differences in cutaneous pigmentation are due to the local activity of melanocytes, rather than their frequency, which ranges from 800 to 2300/mm^2. One melanocyte and about 30 keratinocytes together form an epidermal melanine unit (EMU). Melanocytes also occur in mucosal (mainly oral) epithelia. They can be detected by using silver impregnation, enzyme histochemistry

Fig. 5.12

Cartoon of a proposed model of epidermal proliferative unit (EPU). Redrawn and modified after Potten.

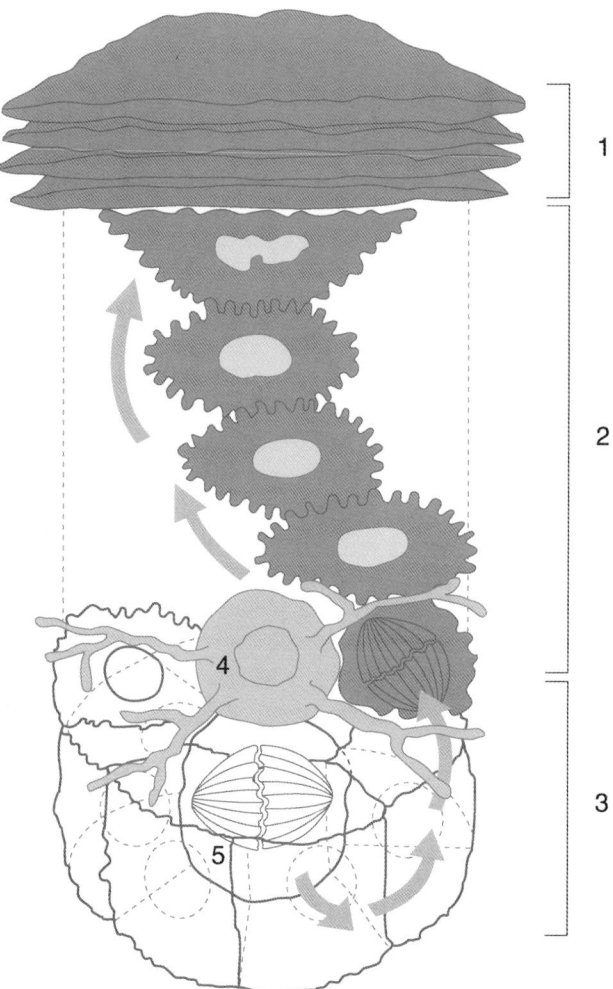

1. Stratum corneum with defoliating corneocytes
2. Migrating column of keratinocytes
3. Epidermal proliferative unit (basal part)
4. Dendritic (Langerhans) cell
5. Basal stem cell

Fig. 5.13

Schematic drawing of the melanocyte and the production and dissemination of melanin pigment. Redrawn and modified after Ross.

1. Melanin pigment granules in keratinocytes
2. Cytocrine secretion
3. Mature melanosomes
4. Cell in stratum basale
5. Lamina basalis
6. Premelanosome
7. Golgi complex
8. Rough endoplasmic reticulum

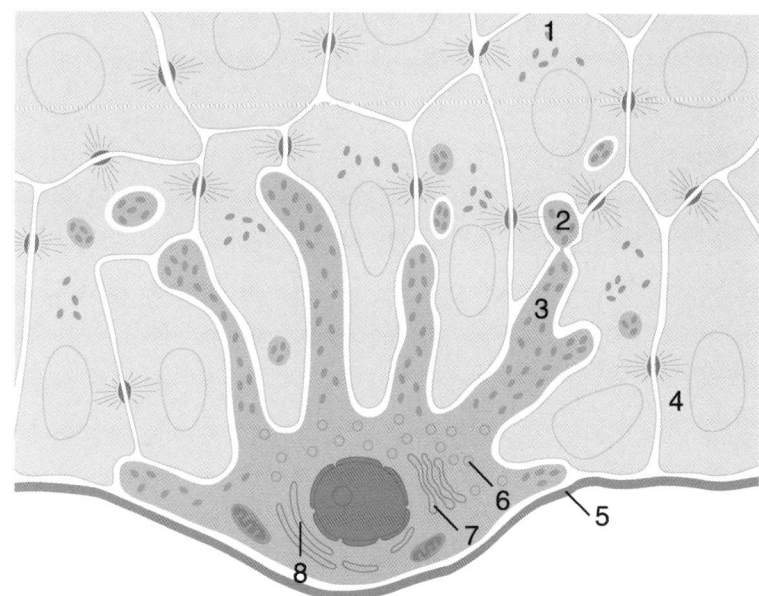

(DOPA reaction) or immunoreactivity against tyrosine hydroxylase. Melanocytes have long dendritic processes by which they distribute the melanin granules (so-called melanosomes) into the cells of the superficial layer of skin. Such direct cell-to-cell passage of molecules by exo- and endocytosis is known as cytocriny. The pigment melanin can occur as eumelanin (the common form) or phaeomelanin (in red hair), and is eliminated by melanophages and melanophore cells.

Langerhans cells (dendritic cells) are responsible for the immunological surveillance of the skin (Figs. 5.12; 5.30). These immunocompetent, antigen-presenting cells, essentially specialized macrophages of bone marrow origin, are regularly distributed in the stratum basale and spinosum, also in the stratified squamous (or transitional) epithelia of viscera, including the conjunctiva but not the cornea. Langerhans cells can be visualized by impregnation with gold chloride, osmium-zinc iodide, and also by adenosine triphosphatease reaction, lectin labeling, or specific antibodies (e.g., CD1A, S-100) (Fig. 5.30). Their density is surprisingly high, 500–1000/mm^2 (i.e., these cells constitute about 2–3% of all epidermal cells). They are continually replenished from the bone marrow. Extraepidermal occurrence of dendritic cells has been revealed in the dermal stroma and in lymphatic organs. Langerhans cells belong to the system of skin-associated lymphatic tissue (SALT) (together with cutaneous lymphocytes, lymph vessels, and nodes). For the presentation of phagocytosed antigens (signified by the Birbeck's granules seen in mature cells), they migrate to lymph nodes. Keratinocytes mutually interact with Langerhans cells by means of cytokine production, see also EPU).

The overlapping units SALT, EMU, EPU are designated as epidermal symbionts.

Innervation

There are two plexuses: a deep reticular plexus for the sweat glands, hair follicles, and larger arterioles, and a papillary plexus at the border of the papillary zone; from here fibers pass to receptors and free endings of the dermis and epidermis (Fig. 5.11). The first nerve fibers of the skin appear in the eighth week of pregnancy, whereas mature Meissner's and PCs are evident in the fourth month.

Circulation (Figs. 5.11; 5.31; 5.36)

Blood supply of the skin is highly variable according to demand. The difference between the maximum and minimum perfusion rates can be as great as 20-fold. There is a deep reticular plexus for the adipose tissue, sweat glands, and hair follicles, and a superficial papillary plexus with loops to the papillae. The veins are arranged in a similar manner, except that there is a flat intermediate plexus in the reticular layer in addition to the superficial and deep venous plexuses. The arterial and venous streams provide a heat-preserving counter-current exchange. In the deeper layer, arteriovenous anastomoses (glomera) are found, with thick muscular coats (Fig. 5.31B inset). These occur mainly at the acral parts, serving as short-circuits to divert blood away from the superficial plexus, reduce heat loss, or support central redistribution of circulation. The dermal microvascular unit comprises: capillary cells, mast cells, pericytes, T lymphocytes, and perivascular dendritic cells (veil cells).

CHAPTER 5 / THE SKIN

The area of skin belonging to a single small artery is called the angiosome. The arrangement of lymphatics is similar to that of blood vessels.

Hairs (Pilosebaceous Units) (Figs. 5.14; 5.15; 5.32; 5.33; 5.34; 5.35; 5.36)

For the sensory function of the skin, the most important appendage is the hair. However, hairs have a number of other functions and exhibit some modifications accordingly. Such hair-associated functions are heat-preservation, protection against irradiation (especially of the head), protection against insects (orifices), or sweat (eyebrow, armpit), mechanical defense (scalp), secondary sexual sign (pubic and facial hair), emotion, and communication. The shape and distribution of hair is characteristic for race, gender, and body region. The density of hairs ranges between 60 and 600/cm^2.

Hairs occur in developmental-functional pilosebaceous units, which comprise a hair with its follicle, sebaceous glands, and an arrector pili muscle, which may raise or lower the hair for heat conservation, sensory, and communicative function, as well as for promoting the release of secretion from glands. In some regions, odoriferous glands (apocrine sweat glands) may also be present, which develop from a common anlage with the hair follicle. The development and function of sweat glands, however, is completely independent from the pilosebaceous unit, as is demonstrated by their occurrence in glabrous skin.

Around the hair shaft, the surface epithelium declines into the dermis and forms the hair follicle. The cavity of the follicle is called epidermal canal, and it continues as the dermal pilar canal or infundibulum below. The infundibulum narrows into an isthmus, and from this point downward, no gap is found between the hair (here: hair root) and the wall of the follicle (here: root sheath).

The lower end of the root bulges (hair bulb) to form a cup above and around a dermal (hair) papilla, which induces and maintains the hair germ and its growth (like in the case of toruli tactiles, the corium is the decisive factor). Around the dermal papilla, the epithelium of the hair root and the root sheath fuse in the proliferating hair matrix. Above the bulge, the piloarrector muscle is attached to the hair follicle. The hair follicle is surrounded by a vitreous (glassy) membrane (i.e., a thick basement membrane) and a dermal sheath of collagenous fibers. The structure of the hair follicle resembles that of the developing tooth, because both are derivatives of the scales of ancient vertebrates.

The life cycle of the hair consists of anagen (actively growing), catagen (involution, growth ceases, follicle shrinks), and telogen (resting, inferior segment is missing) phases, after which the hair is shed, and then the anagen phase may start again. A hair bulb is found only in the anagen phase. During the lifetime of humans, different types of hair (lanugo, vellus hairs, terminal hairs) develop (Fig. 5.26). The hair is the latest system to reach maturity (about 40 yr).

In the wall of hair follicle, the layers of surface epithelium continue, but in a modified form. The infundibulum of hair follicle is still lined by a slightly modified surface (interfollicular) epithelium, with no epithelial ridges or proliferating cells. Below that level, however, where the sebaceous glands open into the hair follicle (above the isthmus), the interfollicular epidermis is replaced by the layers of the root sheath. The outer root sheath is continuous with the Malpighian layer (stratum basale *plus* stratum spinosum), whereas the inner root sheath appears in the place of (but is not continuous with) the other layers of keratinized-stratified squamous epithelium. Instead, the internal root sheath undergoes fragmentation in the upper part, corresponding to the degenerating cells of the sebaceous glands. The inner root sheath comprises the Huxley's and Henle's layers with a characteristic granular appearance, resembling (but not corresponding to) the stratum granulosum and lucidum. The granulation of the Huxley's layer (resembling the stratum lucidum) consists of trichohyalin instead of keratohyalin. The innermost layer of the inner root sheath is called the cuticle.

The hair shaft has an outer epithelial layer of hard keratin (cortex), similar to that found in the nail, and usually (but not always) a core with soft keratin (medulla), like that of the surface epithelium. The outermost layer of the cortex is the hair cuticle. These cuticles should not be confused with the histological term cuticle, meaning a zone of microvilli on the surface of epithelial cells. The cuticles on the outer surface of the hair and the inner surface of the root sheath are formed by highly keratinized epithelial cells, which overlap each other like shingles. The hair shaft is held in place inside the hair follicle by means of these interdigitating cuticle layers.

In contrast to the surface epithelium, cell proliferation only occurs in the matrix, where the epithelium of the root sheath is continuous with the epithelium of the hair. Otherwise, the hair would continually thicken and the root sheath would become narrower. Proliferation in the matrix maintains a cellular drift, which is tangential to the surface of the root sheath (and of the hair, too) but perpendicular to the surface of the skin. While ascending, the cells pass through a zone of keratinization.

The **sebaceous gland** (Fig. 5.32) develops from a common anlage with the hair follicle, and therefore they are usually found together, except for a few places (such as

Fig. 5.14
Diagram of hair follicle in longitudinal section.

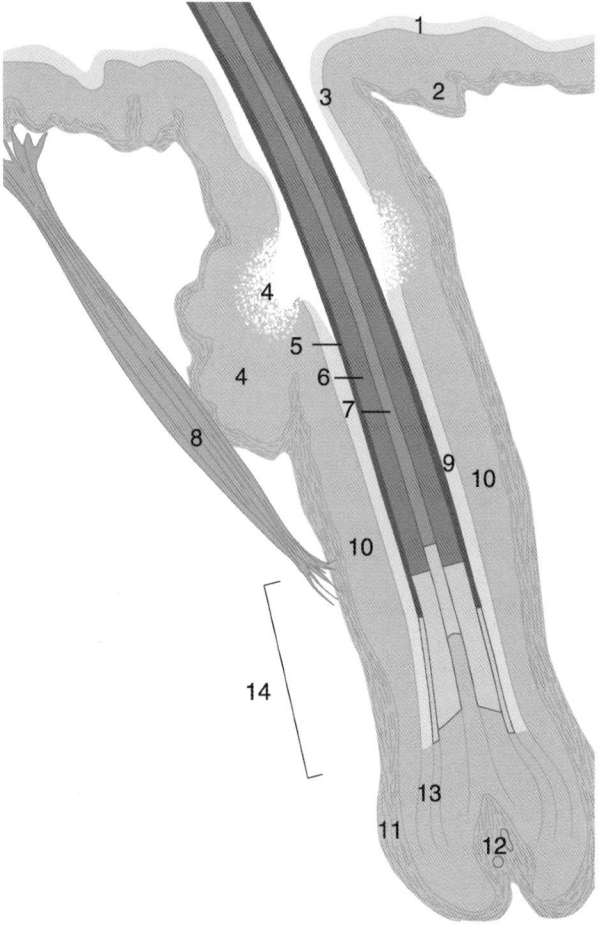

Fig. 5.15
Simplified block diagram of the keratogenous zone of hair follicle.

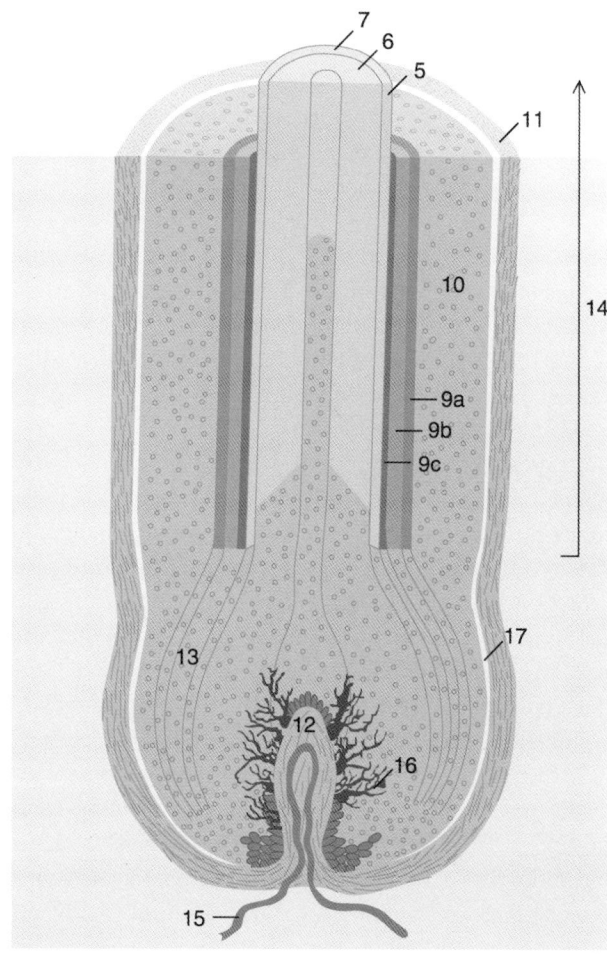

1. Surface layers of epidermis
2. Malpighian layer of epidermis
3. Interfollicular/follicular epidermal transition
4. Sebaceous collar
5. Cuticula pili (apposed follicular cuticle not shown)
6. Cortex of hair
7. Medulla of hair
8. Piloarrector muscle
9. Inner root sheath (Henle + Huxley + cuticle)
   a. Henle's layer
   b. Huxley's layer
   c. Cuticle of root sheath
10. Outer root sheath
11. Connective tissue capsule (separated by glassy memembrane)
12. Dermal papilla of hair
13. Matrix (contour lines highlight the separation of the presumptive layers of hair root and root sheath)
14. Zone of keratinization
15. Vessel
16. Melanocytes
17. Glassy membrane

mucocutaneous junctions, the best example being the labia minora), where sebaceous glands occur without hair follicles (orphan glands). The sebum is a hydrophobic, insect-repellent, odoriferous, and (supposedly) antibacter-ial substance. It is formed by the gradual transformation of live and proliferating epithelial cells into a homogeneous secretory product. This process is known as holocrine secretion. The sebaceous gland opens into the hair follicle, serving as the duct of the gland, and the glandular epithelium is continuous with the epithelium of the follicle.

NEURAL CORRELATES OF TACTILE SENSATION

Having demonstrated the structural elements, we move on to summarize the mechanisms underlying skin sensation, in particular, the integration of different end organs. Pain and thermal perception (nociception in general), although heavily dependent on cutaneous sensation, fall largely outside the scope of our current review. With regard to the generation of pain by extrinsic or intrinsic (e.g., inflammatory) stimuli, primary pain processing in the spinal cord and the spinal trigeminal nucleus, the ascending pain pathways, their effects on limbic and autonomic centers, and conscious pain perception, the reader is referred to specialist textbooks, as well as experimental and clinical reports.

As a specific tactile organ, the skin is capable of detecting two main types of stimulus: discriminative touch and vibration. The former is mainly the function of Merkel cells, Ruffini endings, Meissner's tactile corpuscles and lanceolate end organs, whereas the latter is attributed principally to PCs.

The density of Merkel cells per innervating sensory neuron ranges from 2 to 5 (vibrissae), through 4 to 16 (glabrous skin of fingertips), to 40 to 50 (hairy skin of forearm). Merkel cells show directional selectivity (i.e., they are activated by deflection of a given direction and inhibited by deflection of an opposite direction). Such a selective response is based on the pattern of fine surface protrusions of Merkel cells connected to the surrounding keratinocytes. The sensory neurons belonging to Merkel cells display slowly adapting tonic responses to epithelial dislocation. This means that the firing frequency of the afferent neurons remains stable as long as the stimulus persists, and therefore these neurons are driven by static deformation of skin. Although the impulse frequency is proportional to the amplitude of dislocation, the duration of the volley of action potentials is a function of the duration of stimulus (i.e., the Merkel cell represents a signal-converting analogous transducer).

Ruffini endings are characterized by SA, analogous, and tonic responses, and directional selectivity. However, unlike Merkel cells, these receptors are only sensitive to large amplitude traction or dislocation. One sensory neuron belongs to one end-organ only.

Meissner's corpuscles are composed of the terminal branches of several sensory neurons. In the glabrous skin, one sensory nerve fiber supplies 4 to 16 receptors, but each receptor receives afferents from 2 to 7 neurons. Meissner's corpuscles are rapidly adapting, dynamic receptors that are particularly sensitive to phasic changes such as low frequency (5–50 Hz) vibration, know as flutter. They correspond to a digital transducer: the response is a rapidly declining sharp impulse on both the onset and cut-off of stimulus, regardless of the duration, amplitude or direction vector of stimulus.

Lanceolate end-organs are situated in the wall of hair and vibrissal follicles (substituting for Meissner's corpuscles here). One sensory nerve fiber is connected to one to three lanceolate endings. This rapidly adapting type of touch receptor detects the relative deflection of hairs with no directional selectivity. In this respect, it is similar to Meissner's corpuscles, except that lanceolate endings can also detect vibrations of both high (1000 Hz) and low (200 Hz) frequency ranges.

PCs occupy a deep position in the subcutis. An RA receptor type, PCs are sensitive to the temporal frequency of cutaneous dislocations (i.e., vibration of 100–400 Hz), probably by detecting the the standing waves generated within the multilamellar shell.

Primary Sensory Afferents

The peripheral neurites of sensory neurons terminate as free (bare) endings or specific end organs in the skin or other tissues (Fig. 5.16). The pseudo-bipolar cell bodies are situated in the dorsal root ganglia of the spinal cord (for most parts of the body), or in sensory ganglia of cranial nerves (for the head). The central neurites enter the cord (via dorsal roots) or the trigeminal nuclei of the brainstem. The nerve fibers transmitting impulses from the end organs of Merkel, Ruffini, and Meissner, as well as lanceolate endings and PCs, are all of the Aβ type (with a diameter of 6–12 µm and a conduction velocity of 35–75 m/s). Primary sensory neurons are characterized by their receptive field, that is, that area of skin that is innervated by a single neuron (in other words, the area from which firing of the afferent nerve fiber can be evoked). The size of the receptive field depends on the number of receptors innervated by a single fiber and on the density of receptors. For Merkel discs, the diameter of receptive field ranges from 4 mm (on the fingertips) through 10 mm (palm skin) to 40 mm (forearm). In the latter region, one sensory neuron innervates 40–50 Merkel cells, which are accumulated in a touch dome in the center of the receptive field. By default, all primary afferent neurons of tactile sensation (Aβ) are RA, phasic, and not direction-sensitive. The other specific properties are acquired during the development of specific end-organs. Also, the percep-

Fig. 5.16

Schematic survey of the major types of mammalian cutaneous sensory endings and their afferent fibers. Redrawn and modified after Bannister et al., Gray's Anatomy, Churchill Livingstone, Edinburgh, 1999.

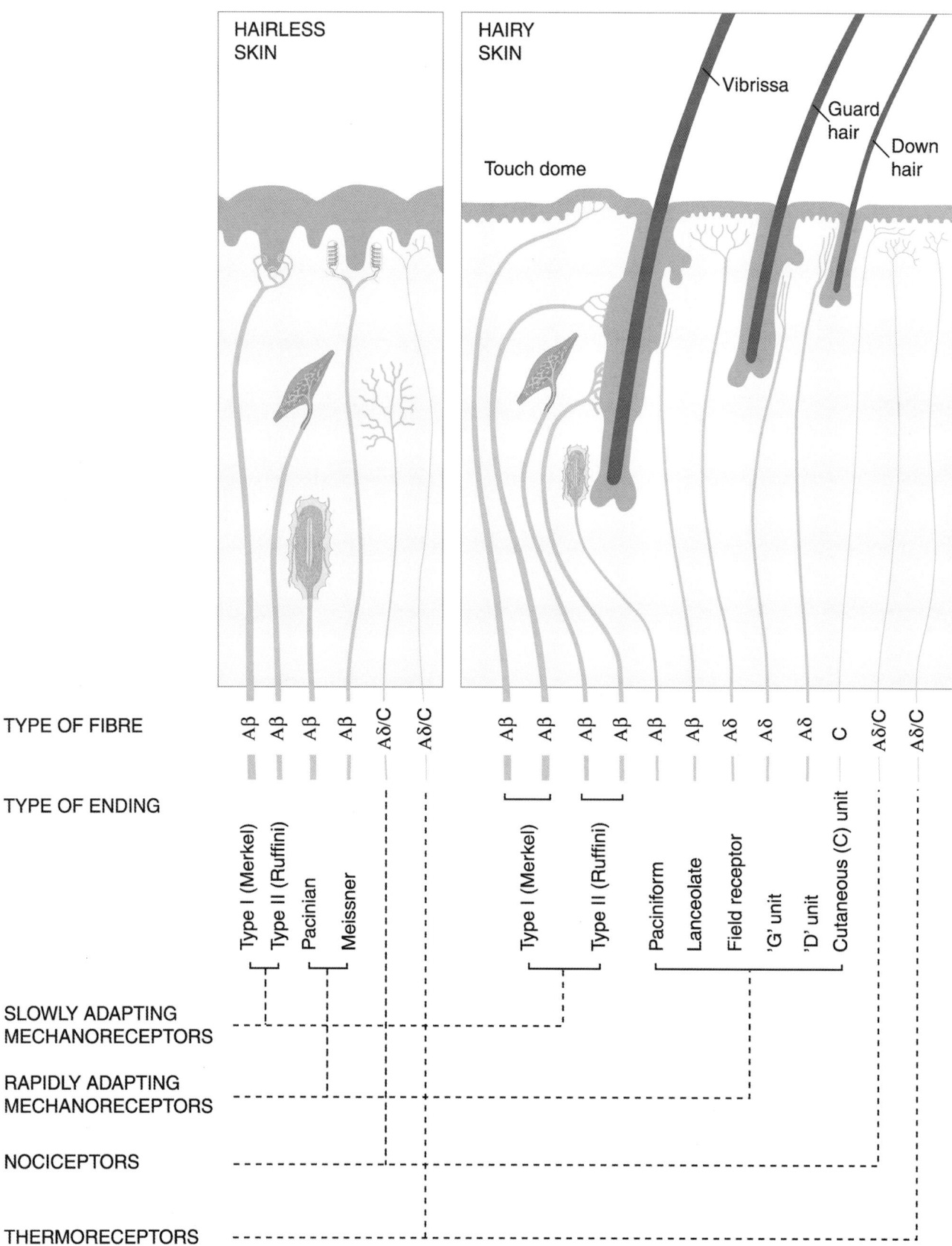

tional significance of fiber responses develops as an adaptive learning process. A volley of action potentials coming from Meissner or Merkel receptors has a very different meaning to the brain (e.g., roughness) than a similar action potential response arriving from PCs (vibration). This organizational principle, known as labeled line coding, also explains the phenomenon of paradoxical cold sensation: if a cold spot of skin is touched with a hot pin (45°C) the subject will report the feeling of cold.

Hairy skin is characterized by a low density of touch receptors (all five described types occur), large receptive field (d = 40–50 mm) and a resolution (two-point limen, the minimal separation between two points that permits both to be perceived) of 40–50 mm. Conversely, the glabrous skin of fingertips has a much higher density of receptors (one Merkel cell and two Meissner's corpuscles/mm^2), yielding a receptive field of 3.7–4.0 mm diameter. Moreover, as evidenced by psychophysiological experiments, the actual resolution (tactile acuity) is even greater: the two-point limen can be as low as 1 mm! This is mainly due to the overlapping receptive fields of Meissner's corpuscles (innervated jointly by two to seven sensory neurons), whereby the resolution is limited only by the absolute density of receptors (roughly one in every millimeter).

Discriminative touch is determined primarily by the concerted action of Meissner's corpuscles and Merkel cells. Owing to their low receptor density and large receptive field, PCs and Ruffini endings play only a negligible role in this process. Whereas the Meissner's corpuscles "scan" the surface of objects by detecting phase-encoded relative differences (edges), leading to a dynamic mapping, Merkel cells provide a static map based on the different parameters (height of elevation, radius of curvature, etc.) of surface patterns, also gauging the direction, pressure, and duration of deformations. Because of their greater overall acuity, Meissner's corpuscles would be better suited for a precise tactile mapping of objects than are Merkel cells; however, they lack the information pertaining to static parameters. The Braille script with its pattern of raised dots is an ideal object for Meissner's corpuscles, although the Merkel system is also required for the evaluation of the height of dots and the pressure on fingertips. Indeed, in physiological experiments the SAI fiber response (associated with the Merkel cells) was found to preserve the pattern of Braille characters more faithfully than did the RA (Meissner-associated) fiber response. However, in a real situation of Braille reading the fingertip is not stationary but is rapidly sliding over the surface. The reading speed of a well-trained (nonblind) person is 13 characters per second when using Braille. Notably, this value drops to 4.5 characters per second when the Braille characters are replaced by raised (embossed) Latin characters. This is because the total surface area of elevations is now increased, and the greater surface/edge ratio renders the script "less dynamic" (i.e., less accessible to the Meissner system). Therefore, the subject must rely chiefly on the Merkel system of lower acuity, at the expense of a diminished reading velocity and accuracy.

The activation threshold of Meissner's corpuscles is lower than that of Merkel cells, partly because the former receptors are closer to the surface. They are most responsive for a relative edge height of 2–4 µm and, most importantly, for moving edges. Thus, Meissner's corpuscles, while sliding over the Braille text, "pick up" the pattern as a low-frequency flutter (approx 5–50 Hz), not dissimilar from the pick-up of an old-fashioned gramophone. Conversely, Merkel cells only detect edges of 8 µm and above, and for an optimum response the surface marking must be a corner (pivot point of deformation) rather than an edge. Both Meissner's and Merkel's tactile receptors show stronger response to curved than to plane surfaces, and to dislocations that are orthogonal to the epithelial ridges. For a precise localization of stimulus on the finger, the Pacinian system is also involved.

The perception of roughness depends on the spatial distribution, height, and diameter of surface markings (irregularities, elevations). Roughness is positively correlated with pressure and negatively correlated with the temperature of surface. Notably, with cold fingers, a rough surface is felt smoother because the afferent nerves are unable to transmit action potentials of sufficiently high frequency (owing to a prolongation of refractory period, known as Wedensky inhibition). Roughness is perceived at its peak when the surface elements are about 4 mm apart (in this case, all receptors are activated and the receptive fields of neighboring sensory neurons are just touching each other). With distances above this value, the surface markings of the object are felt as forms, rather than roughness, because some of the neurons supplying the area remain silent. With distances below 4 mm, the relative difference in the signal intensities of neighboring neurons (adequate stimulus for roughness perception) is diminished owing to the overlapping receptive fields of the neurons supplying Meissner's corpuscles. In general, both Meissner's and Merkel's endings participate in the detection of roughness.

Dislocation of objects is detected as skin deformation with respect to a contacting surface. The direction of dislocation is established by Merkel discs and Ruffini endings, and the receptors are either activated (on-selective) or inhibited (off-selective) by traction. Cortical processing is based on the activity of various specific neuronal types. Opponent direction cells show spontaneous activity that is enhanced by dislocation in one direction and inhibited by dislocation in an opposite direction. Unidirectional cells have no spontaneous activity and they respond only to movements in the on direction, however, this response can be offset by movements in the off direction. Multidirectional cells are activated by dislocations of any direction, although they may have a preferred direction for maximum response. Orientation selective neurons integrate information from the receptive fields of several neurons (which may represent a tactile area as large as a whole finger). The perception of motion appears in somatosensory cortical areas Brodmann (Br) 1 and 2, where the neurons encode spatial information partially processed by areas 3a and 3b. Such complex information, comprising large receptive fields of more than one finger, is relevant to the three dimensional form of objects (stereognosis), and, ultimately, to the optimal grip for object manipulation. Although the latter function requires collaboration with the somatomotor cortex, motion perception is part of a hypercomplex analysis, also including visual and acoustic modalities. This is represented in the posterior parietal associative areas (Br 5, 7), whose neurons perceive motion *as such* in an abstract and multimodal fashion.

In summary, elementary postural information (from muscle spindles, neurotendinous receptors) is processed in the cortical area 3a, whereas discriminative touch (from the end organs of Merkel, Meissner, Ruffini, and Pacini) is represented in area 3b. Converging submodalities and complex information from larger receptive fields are summarized in area 1 (texture, vibration, periodic and aperiodic stimuli) and area 2 (orientation, motion, 3D form and size of objects, judged also by the position of joints). Finally, the association cortex (e.g., areas 5, 7) is the site for converging modalities and abstractions.

VIBRATION

Low-frequency vibration (5–50 Hz, flutter) is detected by the Meissner's corpuscles for encoding *spatial* information relevant to surface patterns (see above). Medium/high-frequency vibration (100–400 Hz) is adequate stimulus for the PCs for encoding *temporal* information, which may also express spatial (surface) characteristics. Very high-frequency vibration (1000 Hz) is picked up by lanceolate endings, especially those attached to hair follicles. This modality is perceived as true oscillation, rather than surface deformation. In addition to light touch, high-frequency vibration detectors (primarily those of the mystacial vibrissae) are also responsive to sound (vibroacoustic sensation).

The firing frequency of the vibration-selective cortical neurons is proportional to the vibratogenic character of the stimulus, rather than to its periodicity. Thus, even aperiodic stimuli can be encoded on the basis of average frequencies. In the first stage of processing (area 3b) the activity of neurons strongly correlates with the periodicity of primary sensory neurons. However, this linkage is gradually diminished in area 1 and, finally, dissipated in the vibration selective neurons of secondary cortical areas.

ATLAS PLATES

Fig. 5.17

Histological specimens demonstrating Merkel cells. A – Skin of finger, HE; B–C – Silver impregnated specimens from palm skin. Photographed from the histological collection of A. Ábrahám, courtesy of Prof. K. Gulya (B-C).

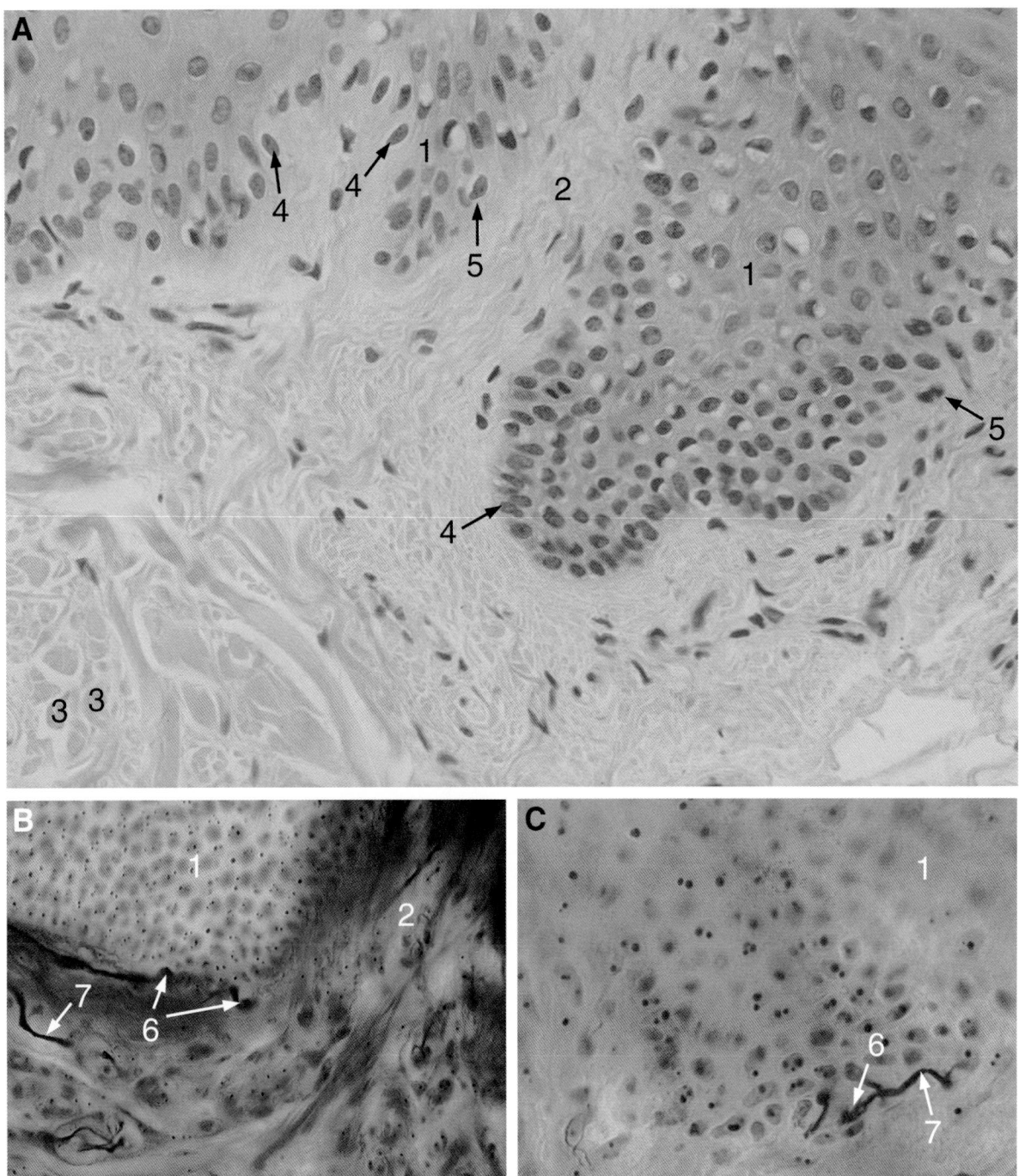

1. Epithelial ridge
2. Connective tissue ridge
3. Collagen fibers in the stratum reticulare of the dermis
4. Stratum basale with columnar cells perpendicular to the surface
5. Merkel cells with irregular and horizontally arranged nuclei, and light cytoplasm
6. Merkel endings (discs) with bulbous swellings beneath the epithelial cells
7. Nerve fibers

Fig. 5.18

Histological specimens of Meissner's tactile corpuscles. A – Finger-pad, HE. B–D – Silver impregnated specimens of palm skin. The non-impregnated structures appear as vague contours. The dark granules in the nuclei of epithelial cells represent nonspecific precipitates. Photographed from the histological collection of A. Ábrahám, courtesy of Prof. K. Gulya (B–D).

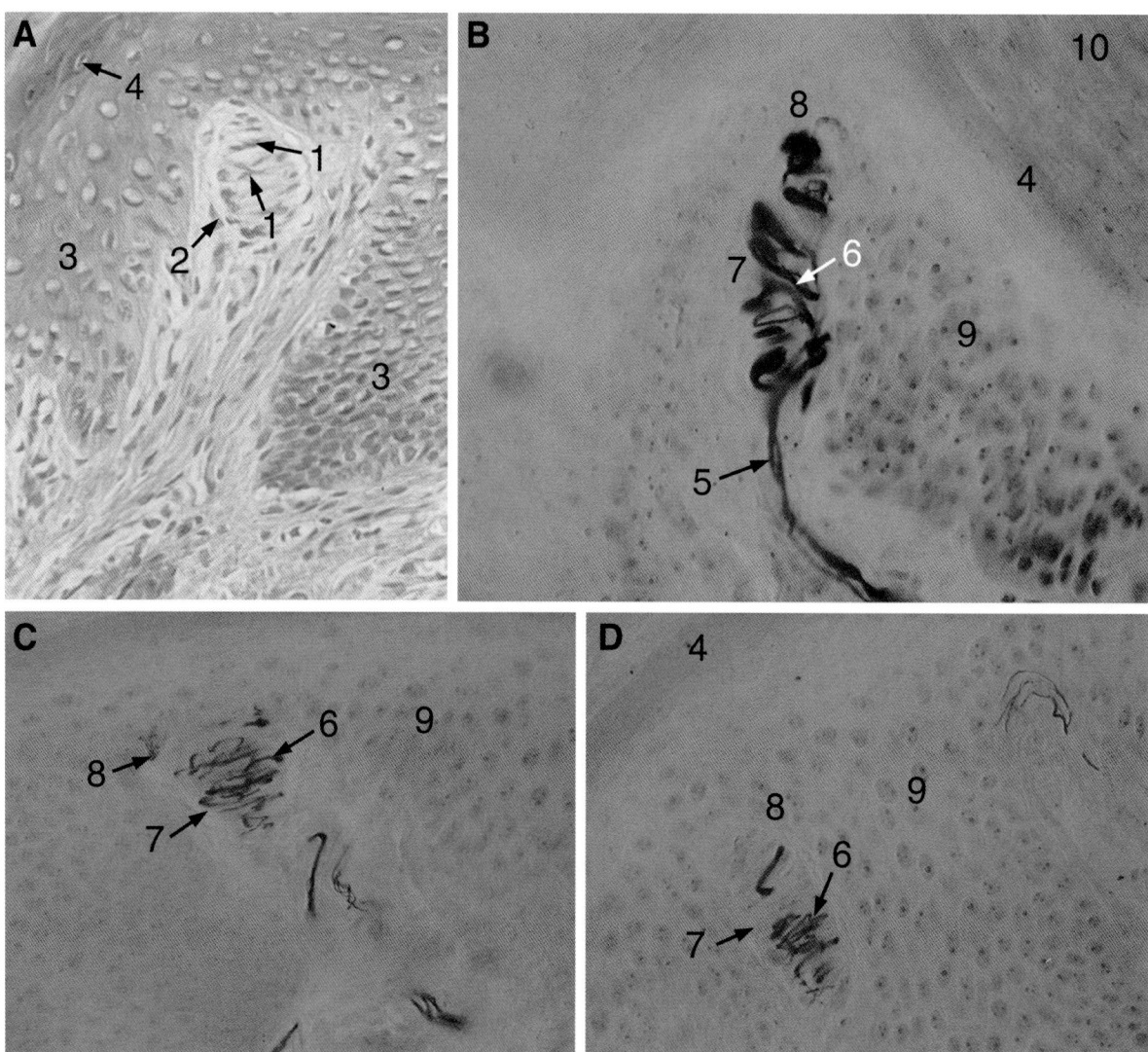

1. Inner flat cells
2. Connective tissue capsule, arrow points to a fibroblast
3. Epithelial ridge
4. Stratum granulosum and stratum lucidum
5. Nerve fibers entering the corpuscle
6. Internal nerve fibers coiling around non-impregnated inner cells
7. Capsule of corpuscle, not impregnated, invisible
8. Stratum basale on the bottom of an epithelial ridge
9. Stratum polygonale
10. Stratum corneum

Fig. 5.19

Specimens demonstrating Pacinian corpuscles. A – Longitudinal section, palm skin, HE; B – Longitudinal section, palm skin, silver impregnation; C – Peritoneal spread specimen; D – Pancreas, HE; E – Microdissection specimen from a human index finger. The Pacinian corpuscles are highlighted by black plastic underlays (dissection by Dr. M. Tóth).

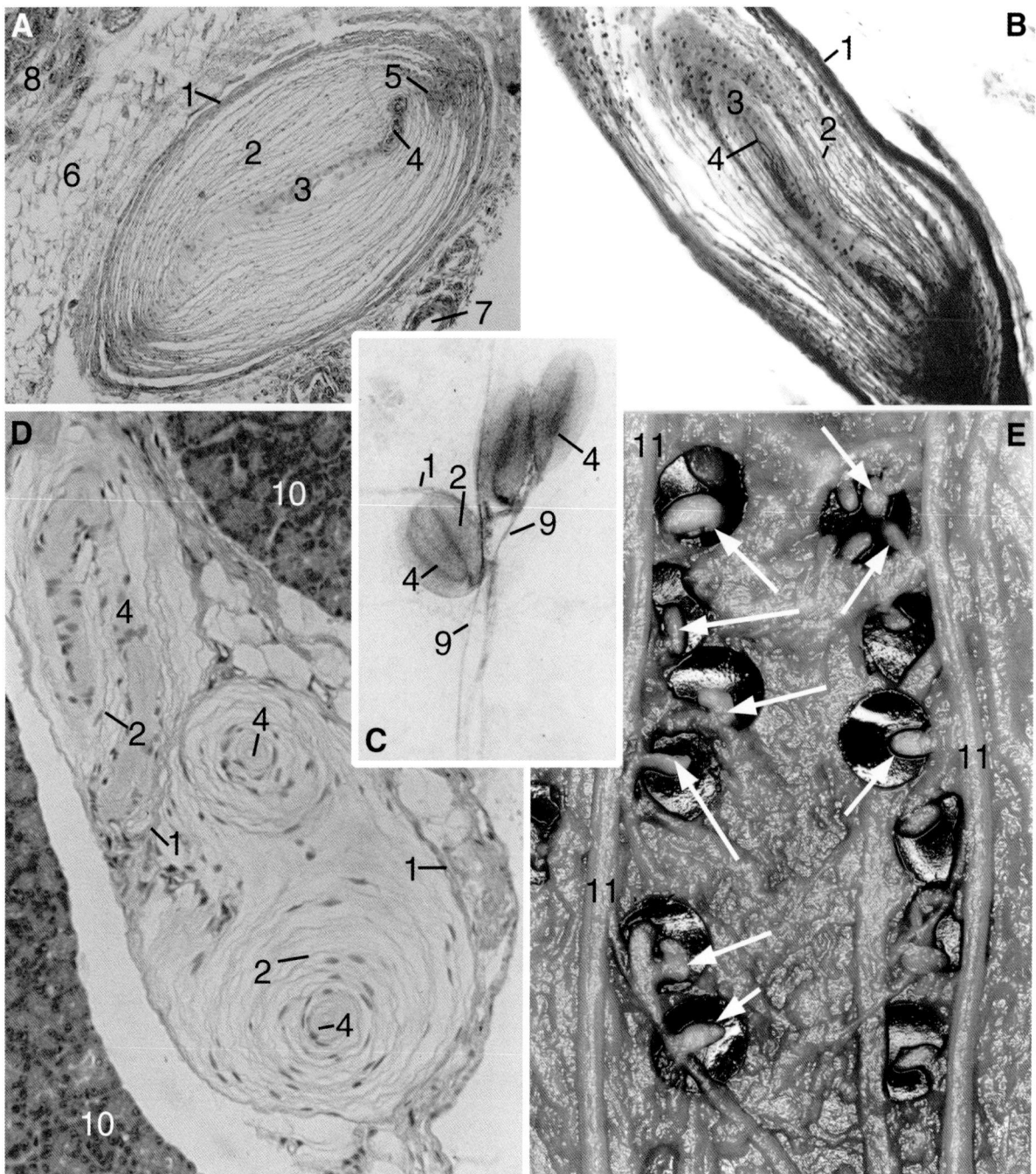

1. Outer fibrous capsule
2. Inner capsule of concentric lamellae formed by flat cells
3. Inner core formed by modified Schwann cells
4. Nerve fiber in the inner core
5. Site of penetration of nerve fiber
6. Adipocytes of hypodermis
7. Arteriole
8. Sweat glands
9. Nerve fiber innervating the Pacinian corpuscle
10. Acini of pancreas
11. Proper palmar digital nerve branches with bulbous endings, representing Pacinian corpuscles (arrows)

Fig. 5.20

Histological specimens of the end bulbs of Krause and related structures. A – Krause-like corpuscle from the skin of finger, HE; B – Krause-like corpuscle from the skin of finger, silver impregnation; C, D – End-bulbs of Krause, silver impregnation (enlarged view in D). Photographed from the histological collection of A. Ábrahám, courtesy of Prof. K. Gulya (B–D).

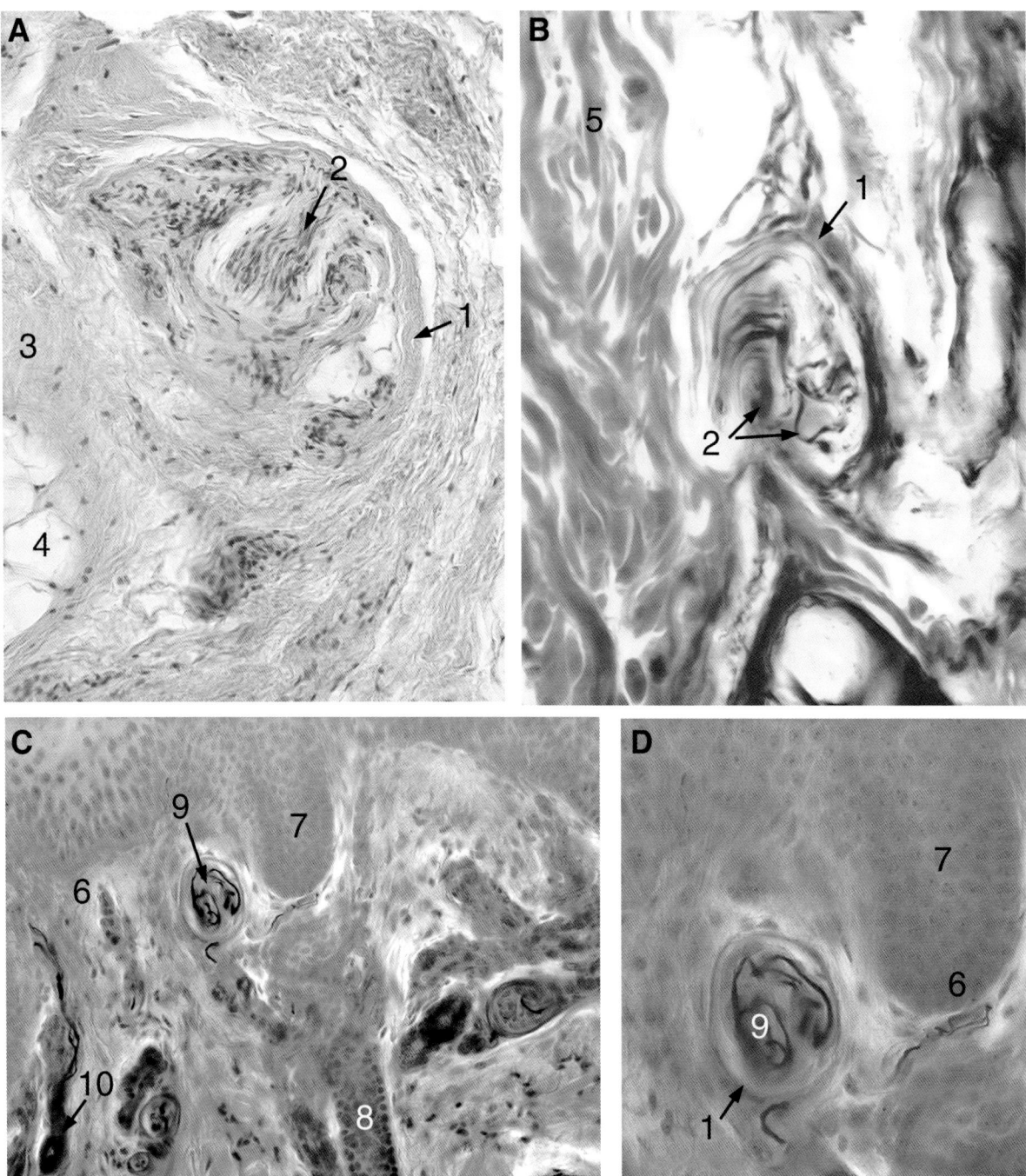

1. Connective tissue capsule
2. Nerve fiber entering and arborizing
3. Dermis
4. Adipocytes of hypodermis
5. Collagen fibers
6. Deep layer of stratified squamous epithelium
7. Epithelial ridge
8. Duct of sweat gland
9. End-bulbs of Krause in the corium (dermis)
10. Nerve fibers in the corium (dermis)

Fig. 5.21

Histological specimens of genital bodies. Silver impregnation with counterstaining (A, B, E) or without counterstaining (C, D, F, G). A, B – Prepuce. The densely arborizing nerve fibers in A seem to coalesce in the thick section. The tangential section (A) reveals the system of epithelial ridges, which appear to be "pegs" in perpendicular sections. The structure in B is similar to the end-bulb of Krause, except that it is larger and the nerve fiber is more profusely arborizing; C, D – Glans penis. Without counterstaining the nerve fibers of the end organs are less masked by the inner cells; E – Glans penis. The internal structure of this body resembles that of the Meissner's corpuscle rather than the end-bulb of Krause. F, G – Clitoris. Photographed from the histological collection of A. Ábrahám, courtesy of Prof. K. Gulya.

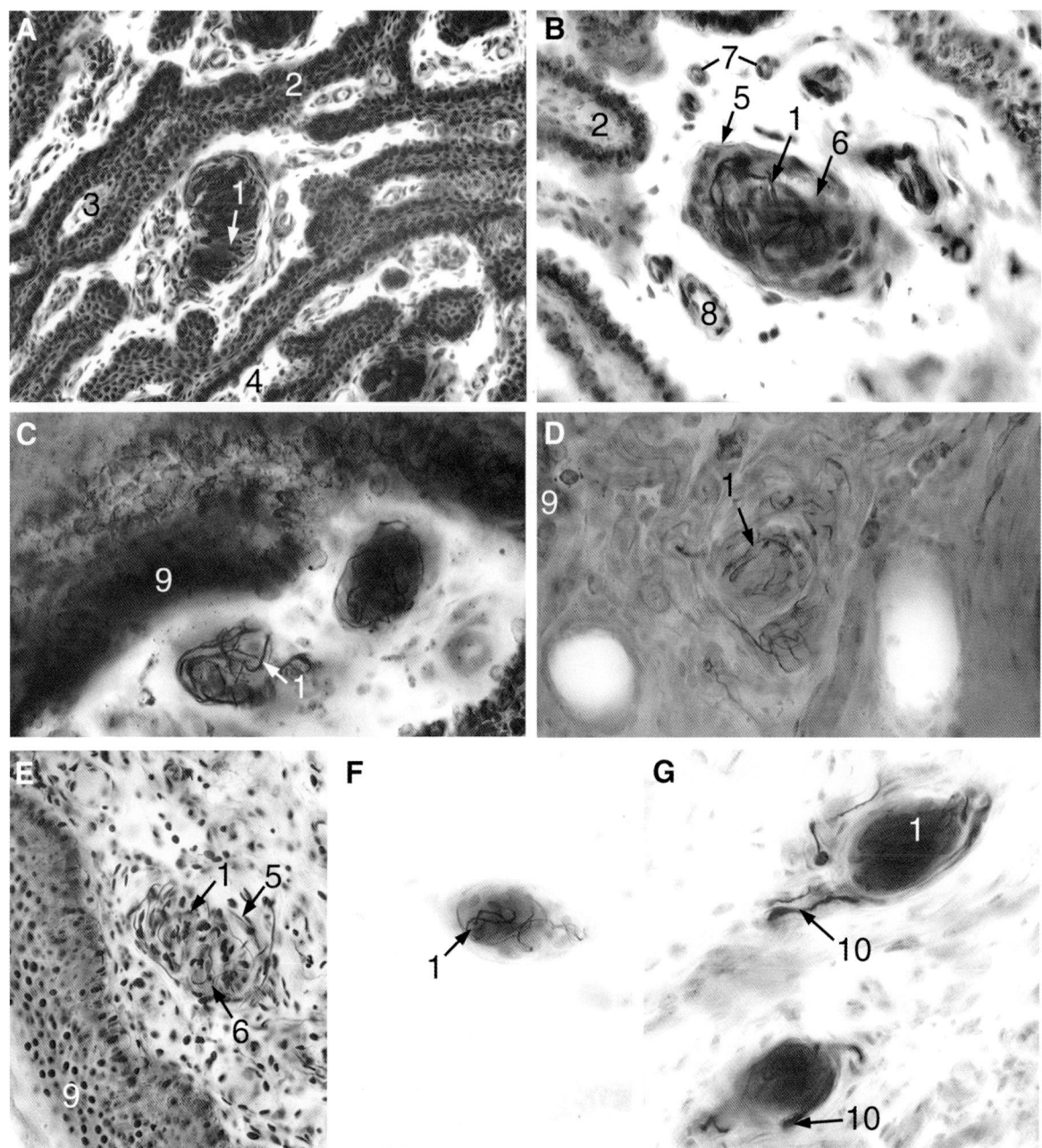

1. Arborizing nerve fiber
2. Epithelial ridge
3. Connective tissue papilla surrounded by epithelial ridges
4. Connective tissue of dermis
5. Connective tissue capsule
6. Contours of inner cells
7. Arteriole
8. Venule
9. Surface epithelium
10. Nerve fiber entering the genital body

Fig. 5.22

Histological specimens demonstrating different types of receptor endings. A – Golgi-Mazzoni ending in palm skin, HE; B – Ruffini ending in palm skin, silver impregnated specimen. The contours of the end-organ are marked by arrowheads; C – muscle spindle, silver impregnation; D – free nerve ending in palm skin, silver impregnation; E – myelinated axon in the dermis, silver impregnation. Photographed from the histological collection of A. Ábrahám, courtesy of Prof. K. Gulya (B–E).

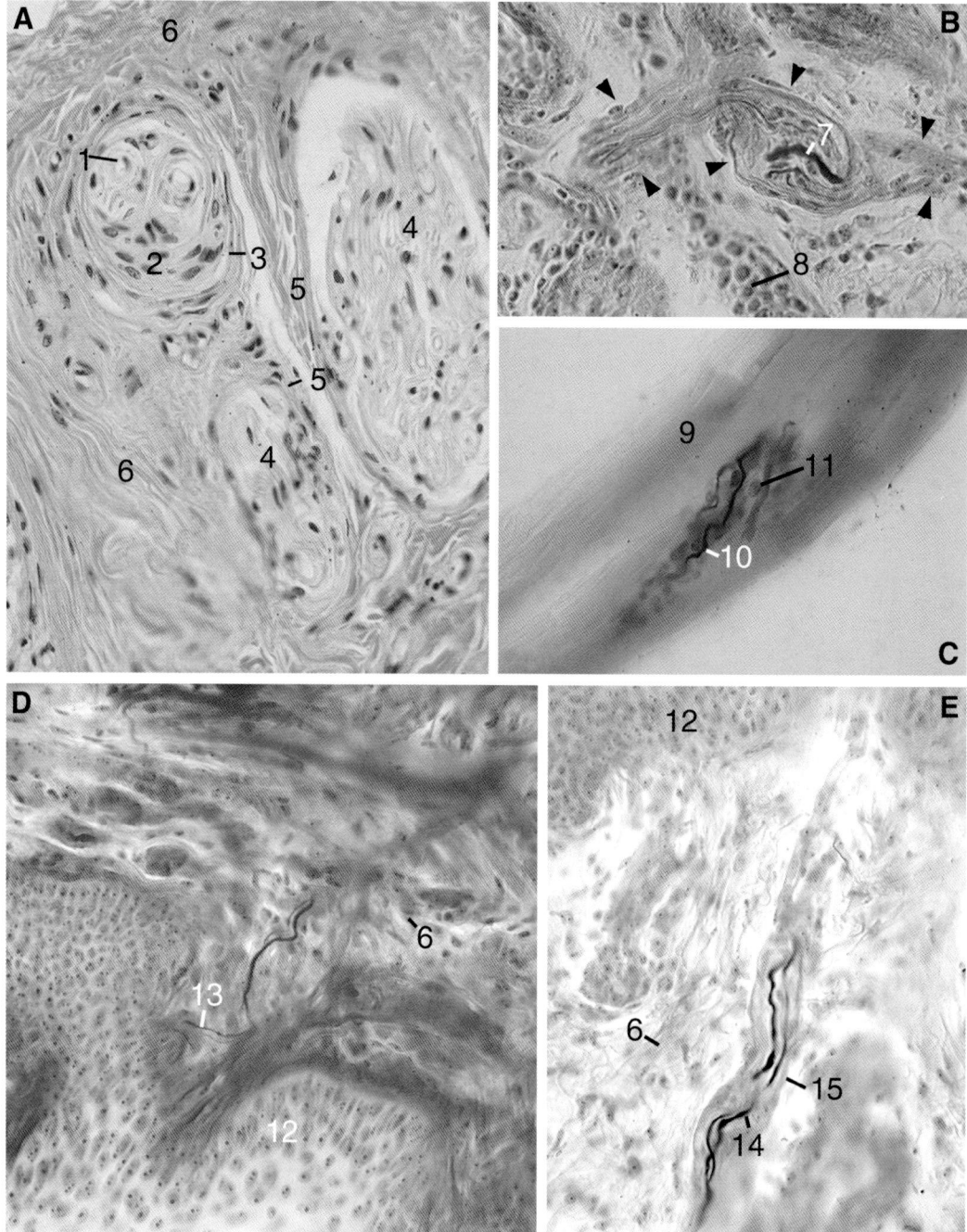

1. Inner nerve fibers
2. Circular capsule of cellular lamellae
3. Connective tissue capsule
4. Small nerve
5. Perineurium
6. Collagenous fibers of the reticular layer of dermis
7. Nerve entering at the centrally located swelling of the end-organ
8. Sweat glands (tangentially sectioned)
9. Contours of striated muscle fibers
10. Nerve fiber
11. Nuclear bag receptor
12. Deep layers of epithelium
13. Bare ending of nerve fiber beneath the epithelium
14. Myelinated nerve fiber traversing the dermis
15. Myelin sheath

Fig. 5.23

Histological specimens demonstrating the innervation of the tongue. A – Fungiform papilla, HE; B – Detail of lingual papilla, silver impregnation; C, D – tactile corpuscles of the tongue, silver impregnation. Photographed from the histological collection of A. Ábrahám, courtesy of Prof. K. Gulya (B, D).

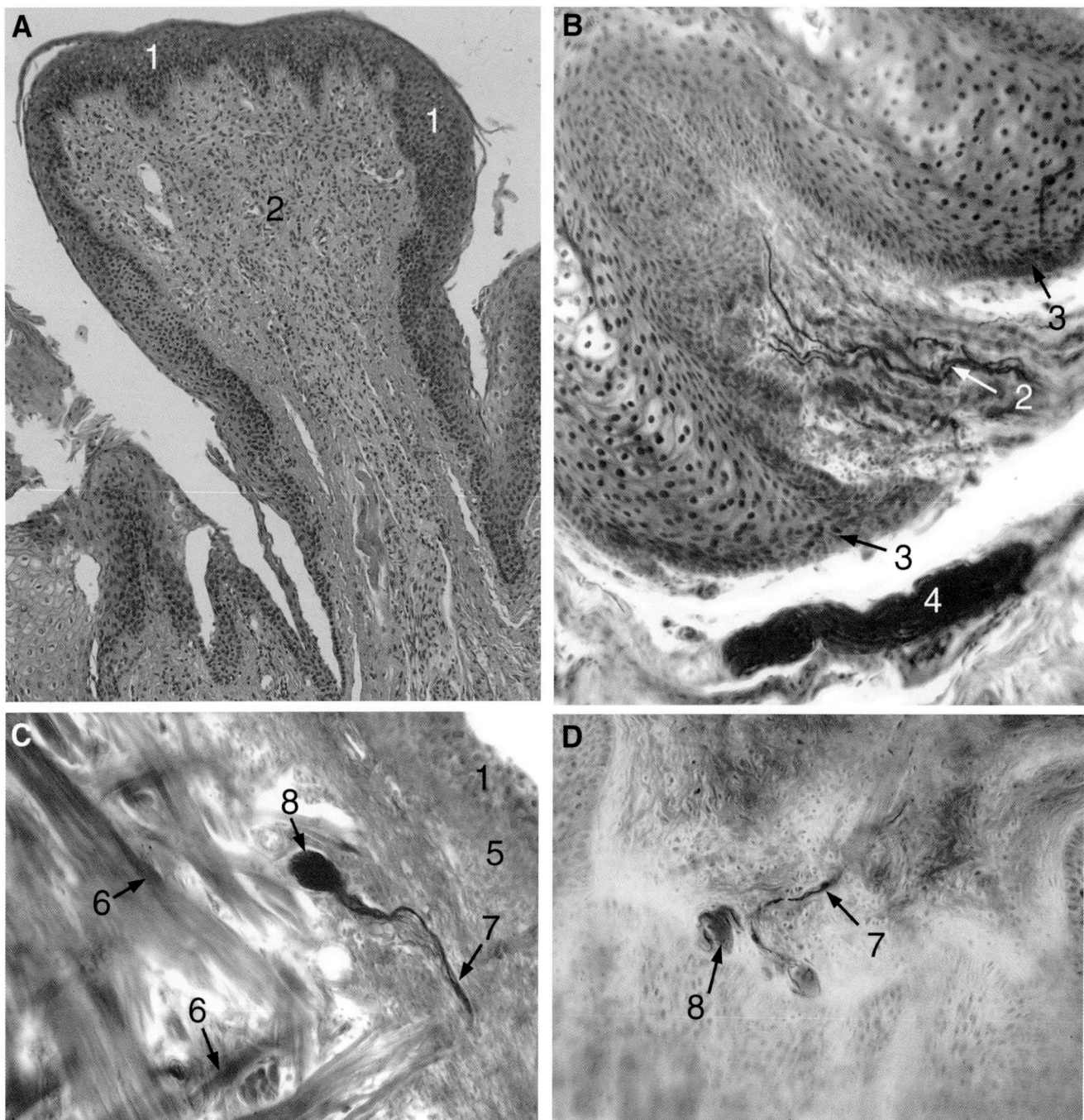

1. Stratified squamous epithelium
2. Connective tissue core of papilla, containing an abundance of nerve fibers
3. Deep layers of epithelium
4. Bundle of nerve fibers
5. Lingual aponeurosis (dense connective tissue layer)
6. Striated muscle of varied orientation
7. Nerve fiber
8. Tactile corpuscle, with a thick tangle of nerve endings, appearing almost homogenous

ATLAS—SKIN

Fig. 5.24

Carotid body (glomus caroticum) of human, Azan. This staining highlights the internal connective tissue septa and the lobular arrangement of epithelial cells. Since the specimen was fixed by immersion fixation, the vessels are full of red blood cells.

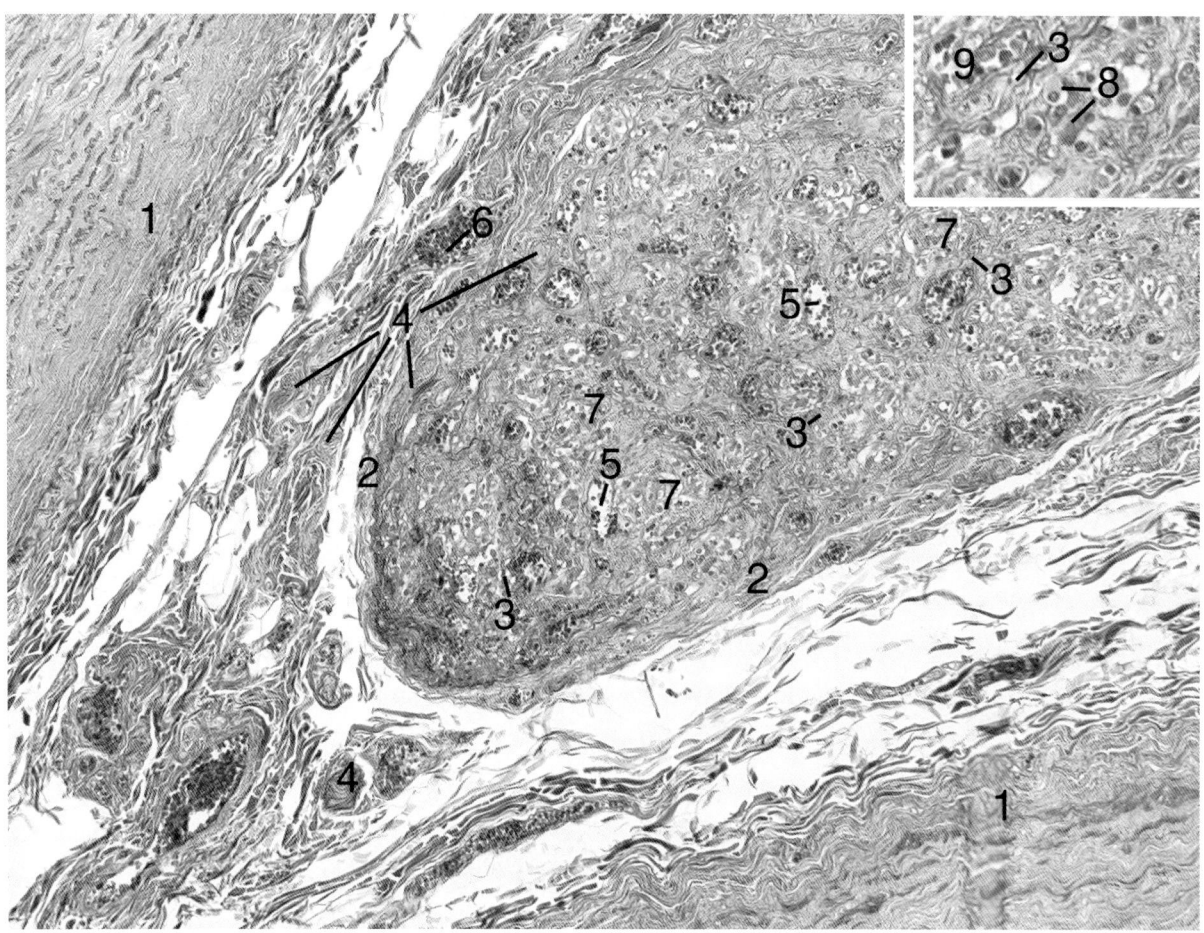

1. Carotid arteries
2. Connective tissue capsule
3. Connective tissue septum
4. Nerves
5. Inner venule
6. Efferent venule
7. Epitheloid cell groups
8. Inset: staining differences between the epitheloid cells, revealed by the azan method
9. Inset: Capillary

Fig. 5.25

Carotid body (glomus caroticum) of rat, HE. A – Low-power image; B – High-power image. Note the distended blood vessels owing to perfusion fixation of the tissue. Some important details, e.g., sympathetic and parasympathetic ganglion cells, the subtypes A and B of glomeral cells, cannot be distinguished.

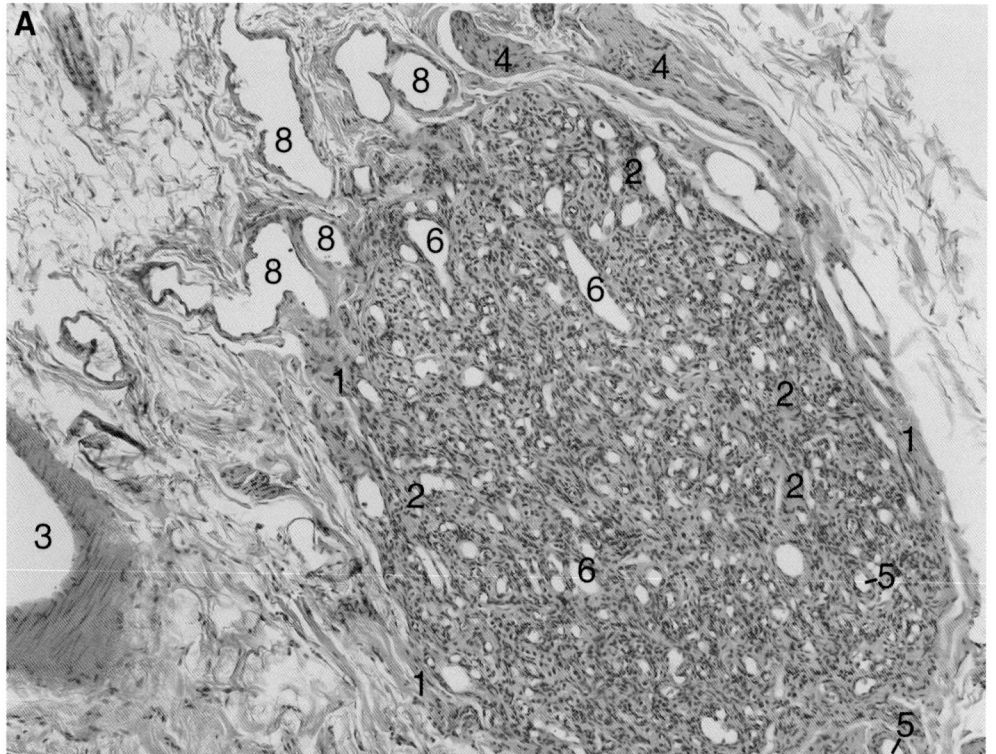

1. Connective tissue capsule
2. Connective tissue septum
3. Adjacent (carotid) artery
4. Nerves
5. Arteriole
6. Inner venule
7. Capillary
8. Efferent venule
9. Tangential section of venule: endothelial cells
10. Epitheloid cell (glomeral cell, type I), frequently grouped
11. Sustentacular cell (type II) attached to the epitheloid cell

Fig. 5.26

Developing scalp skin HE. A – Low-power image; B – High-power image.

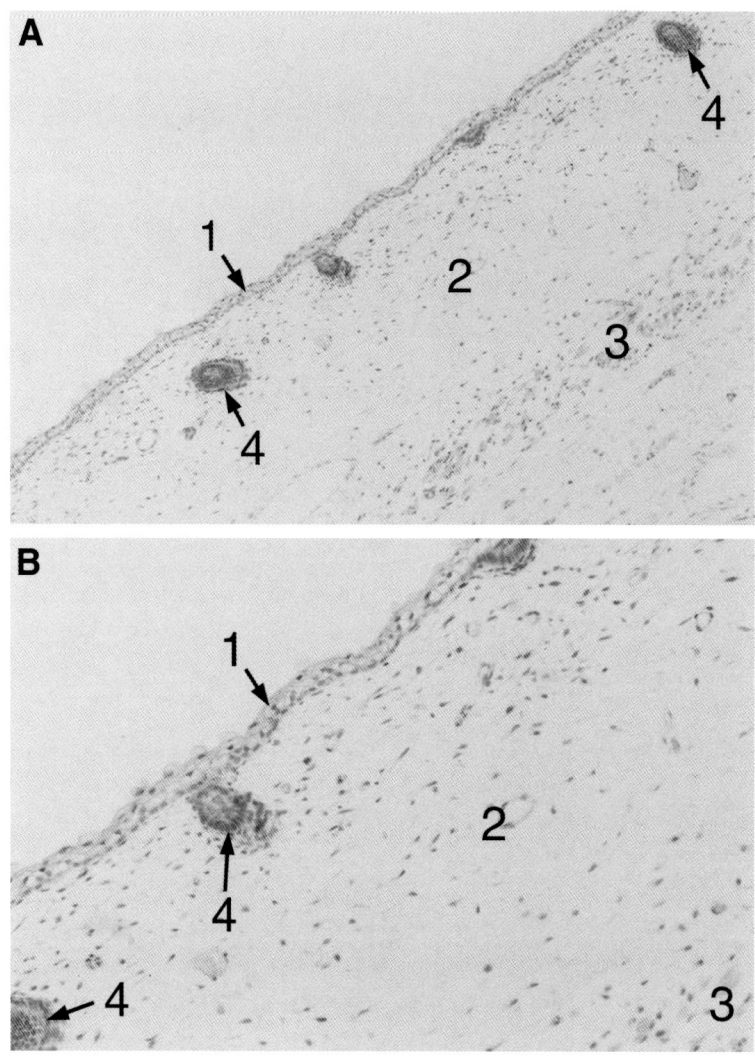

1. Epidermis, note that it consists of only two to three layers with no epithelial pegs
2. Corium and subcutis, the connective tissue layer is rather loose
3. Myoblasts corresponding to the future epicranial muscles
4. Developing hair follicle

Fig. 5.27

Specific types of human skin. I. Axillary skin, HE. The skin of the armpit contains representative examples of glands for all types of secretion: merocrine (eccrine) sweat glands, apocrine sweat (odoriferous, scent) glands and holocrine sebaceous glands.

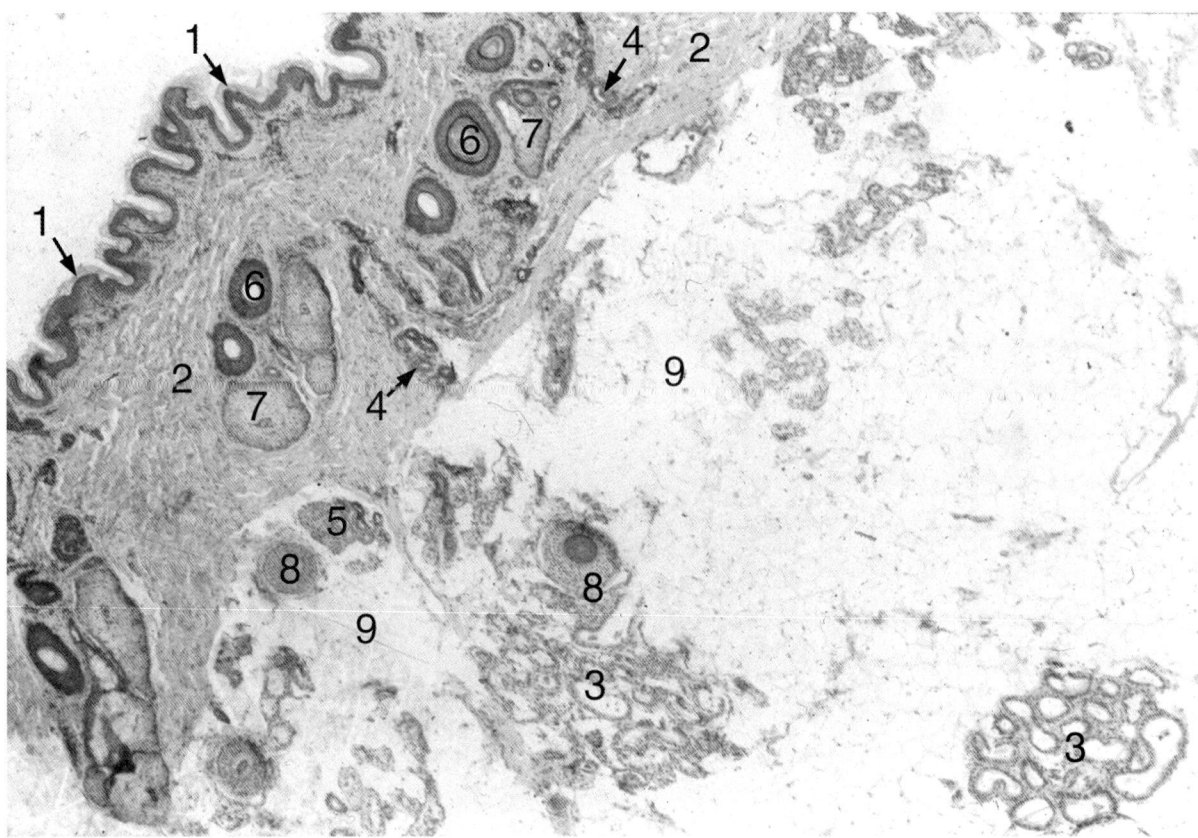

1. Epidermis, stratified squamous keratinizing epithelium
2. Collagen fibers in the stratum reticulare of dermis
3. Odoriferous gland (apocrine "sweat" gland) in the hypodermis (subcutis)
4. Sweat gland, the excretory part has a narrow lumen and is lined by a double layer of densely stained cuboidal cells
5. Sweat gland, the secretory part has a wider lumen and is lined by pale cells
6. Hair follicle (superficial part) in the dermis
7. Sebaceous gland
8. Hair follicle (deep part) in the hypodermis (subcutis)
9. Hypodermal (subcutaneous) adipose tissue

Fig. 5.28

Specific types of human skin. II. Pigmented skin, HE, The melanocytes are represented by long dendritic processes containing and distributing melanin in the form of melanosomes.

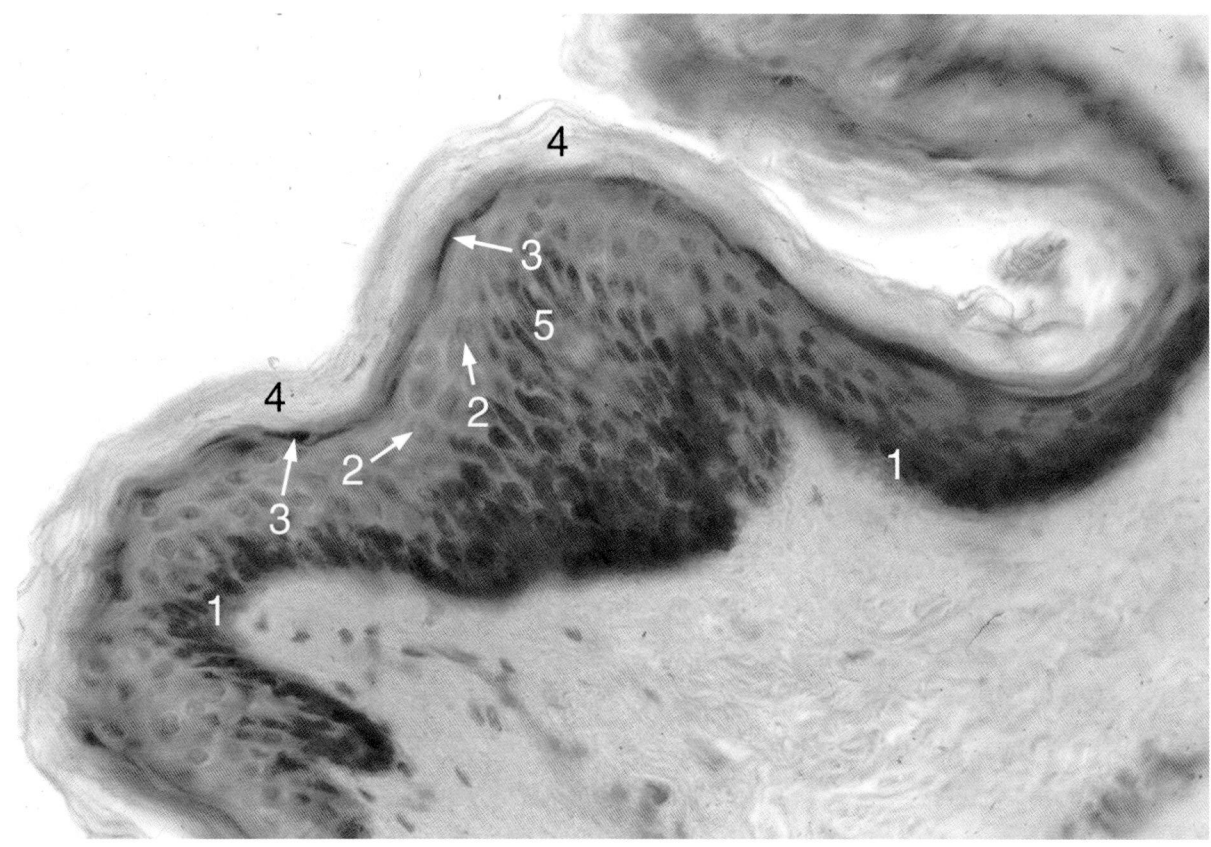

1. Stratum basale (germinativum), with melanocytes
2. Stratum spinosum, the epithelial cells also contain melanin granules
3. Stratum granulosum and lucidum
4. Stratum corneum
5. Stratum basale (germinativum) around a dermal papilla, in tangential section

Fig. 5.29

Stratified squamous keratinized epithelium of glabrous skin, HE.

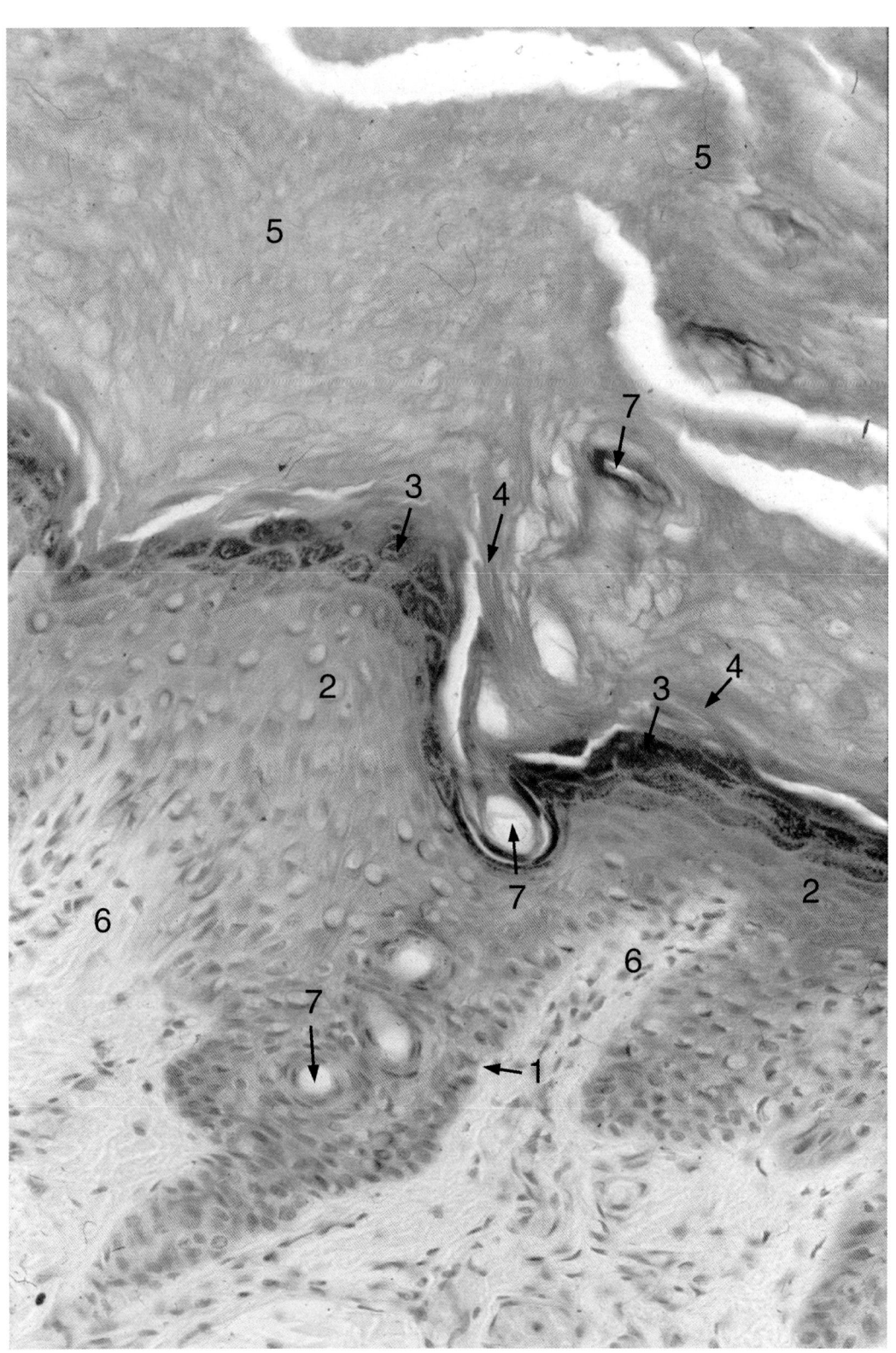

1. Stratum basale (germinativum)
2. Stratum spinosum
3. Stratum granulosum
4. Stratum lucidum
5. Stratum corneum
6. Connective tissue papilla
7. Duct of sweat gland (straight segment of the excretory part)

Fig. 5.30

Langerhans cells in hirsute skin. Immunolabelling against S-100 (A) and CD1a (B), with HE counterstaining. The former also stains Merkel cells, whereas the latter is more specific for Langerhans cells. Courtesy of Dr. A. Iványi.

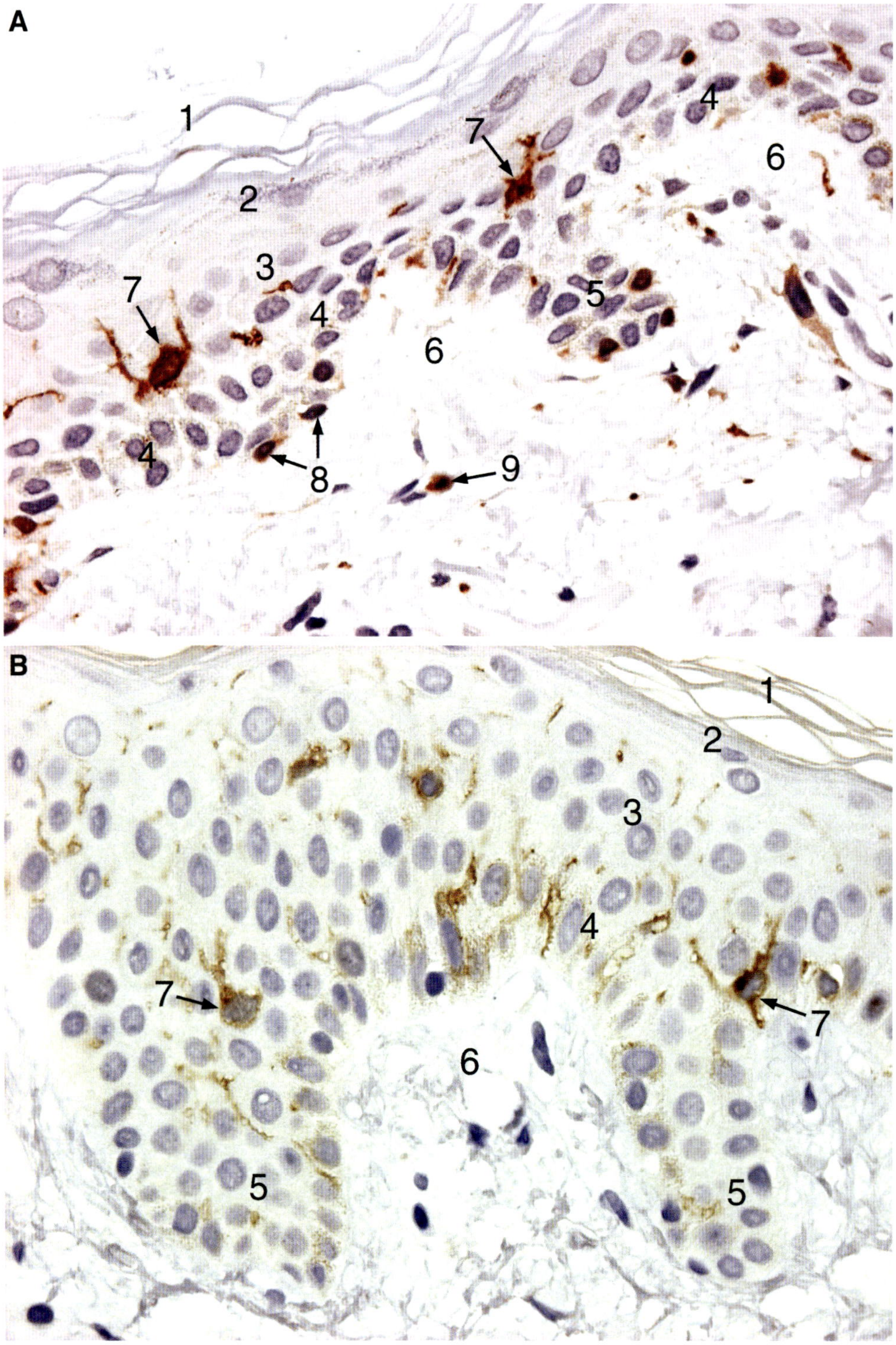

1. Stratum corneum
2. Stratum lucidum and granulosum (very thin layers in this type of skin)
3. Stratum spinosum
4. Stratum basale
5. Epithelial ridge
6. Connective tissue papilla
7. Langerhans cell
8. Merkel cell
9. S-100 immunoreactive cell of dermis (migrating precursor of Langerhans cell)

Fig. 5.31

Vascular supply of the two main types of skin. A – Hairy (hirsute) skin; B – Glabrous skin. Specimens A-B injected with red and blue india ink (from the material of K. Csányi† and J. Vajda†). Inset: Vascular glomus from the skin of human finger, HE.

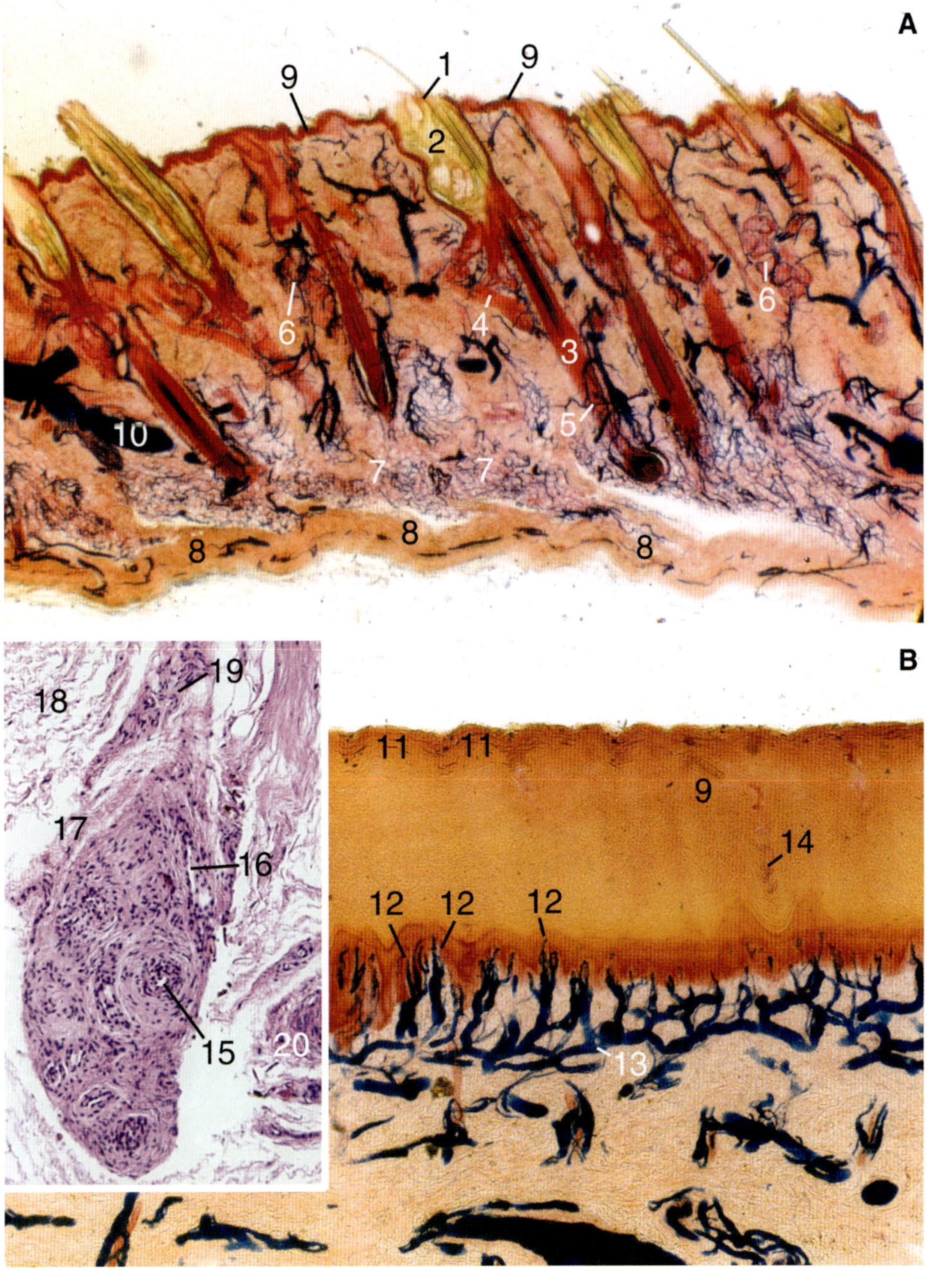

1. Hair shaft
2. Sebum released around the hair shaft
3. Hair follicle
4. Arrector pili muscle
5. Capillary network around the hair follicle
6. Capillary network of a sebaceous gland
7. Blood vessels in the hypodermis (subcutaneous plexus)
8. Galea aponeurotica
9. Epidermis, stratified squamous keratinized epithelium
10. Small artery
11. Toruli tactiles
12. Blood vessels in the connective tissue ridges of toruli tactiles
13. Blood vessels in the dermis (subpapillar plexus)
14. Sweat gland, straight segment of the excretory part
15. Arteriole in the glomus
16. Venule in the glomus
17. Common fibro-muscular capsule of glomus
18. Collagen fibers
19. Nerve fibers approaching the glomus
20. Arteriole

ATLAS—SKIN

Fig. 5.32

Hair follicle, longitudinal section, superficial part, HE. The hair shaft is not visible at this level because, in this phase of growth, it is too short to reach the surface.

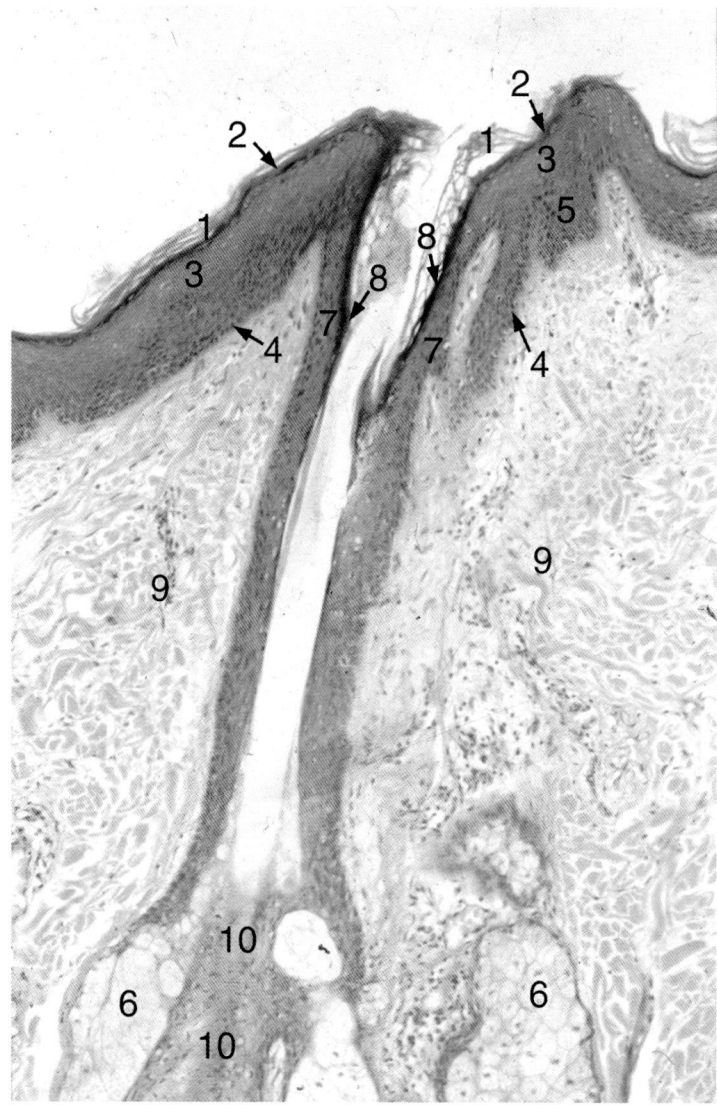

1. Stratum corneum
2. Stratum lucidum and (below) stratum granulosum
3. Stratum spinosum
4. Stratum germinativum
5. Stratum germinativum around a dermal papilla in a tangential section
6. Sebaceous gland
7. Stratum spinosum and germinativum (Malpighian layers), continuous with the wall of hair follicle
8. Stratum corneum, lucidum and granulosum, continuing inside the hair follicle
9. Collagen fibers of dermis
10. Root sheath, upper part tangentially sectioned

Fig. 5.33

Hair follicle, deep part, in longitudinal section, HE. The hair shaft is short, in the initial phase of growth.

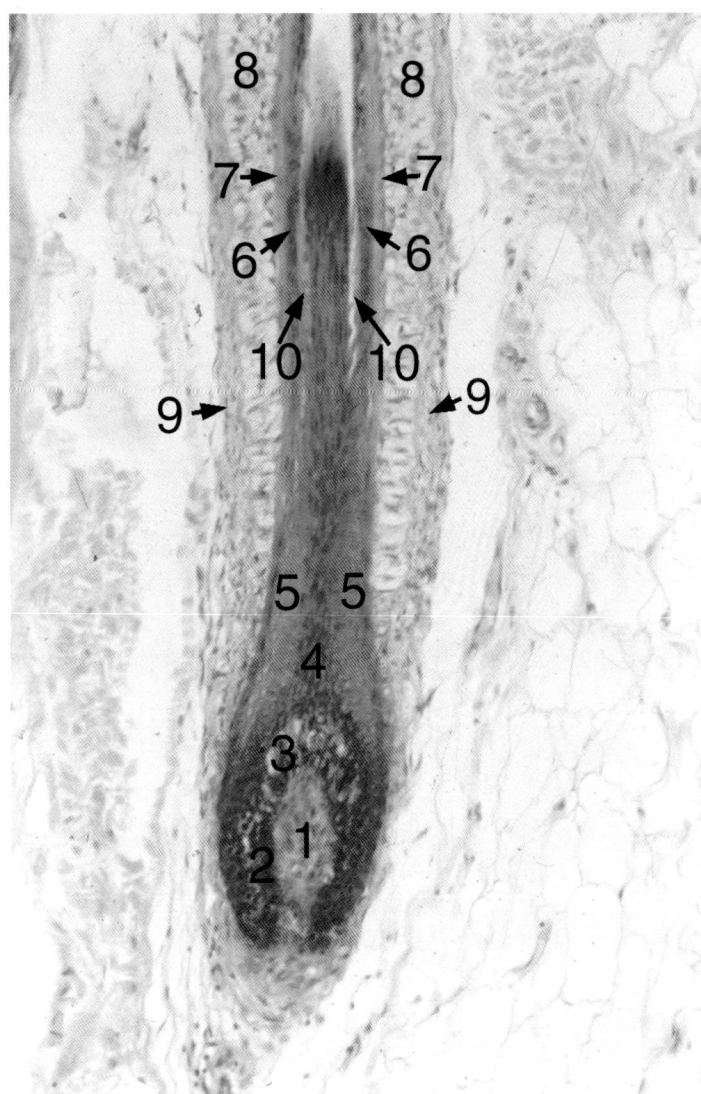

1. Hair (dermal) papilla, connective tissue
2. Matrix, proliferating cells at the junction between the epithelia of hair and the root sheath
3. Melanocytes, note the conspicuous processes
4. Medulla of hair shaft
5. Cortex of hair shaft
6. Huxley's layer of inner root sheath
7. Henle's layer of inner root sheath
8. Outer root sheath
9. Connective tissue capsule, arrow points to glassy membrane
10. Inner root sheath, overlapping cells of cuticle

Fig. 5.34

Hair follicle, Azan. Note the oblique section of hair shaft on the right side, and an approximately longitudinal section through the hair bulb on the left side.

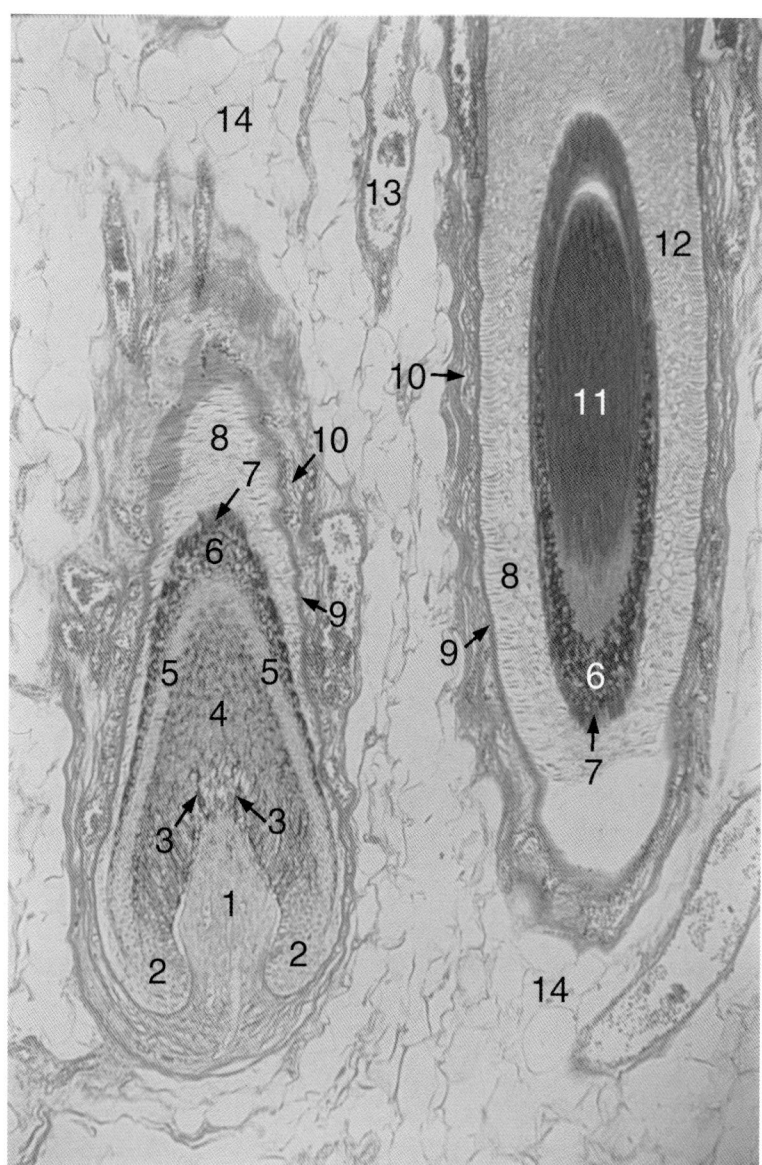

1. Dermal papilla
2. Matrix, proliferating cells at the junction between the epithelium of hair and the root sheath
3. Melanocyes
4. Medulla of hair shaft
5. Cortex of hair shaft
6. Huxley's layer of inner root sheath
7. Henle's layer of inner root sheath
8. Outer root sheath
9. Glassy membrane
10. Connective tissue capsule, arrow points at a capillary
11. The fully keratinized part of hair shaft
12. Overlapping cuticle cells of inner root sheath
13. Venule (collecting)
14. Adipocytes

Fig. 5.35

High-power micrograph demonstrating the layers of the hair follicle, Azan.

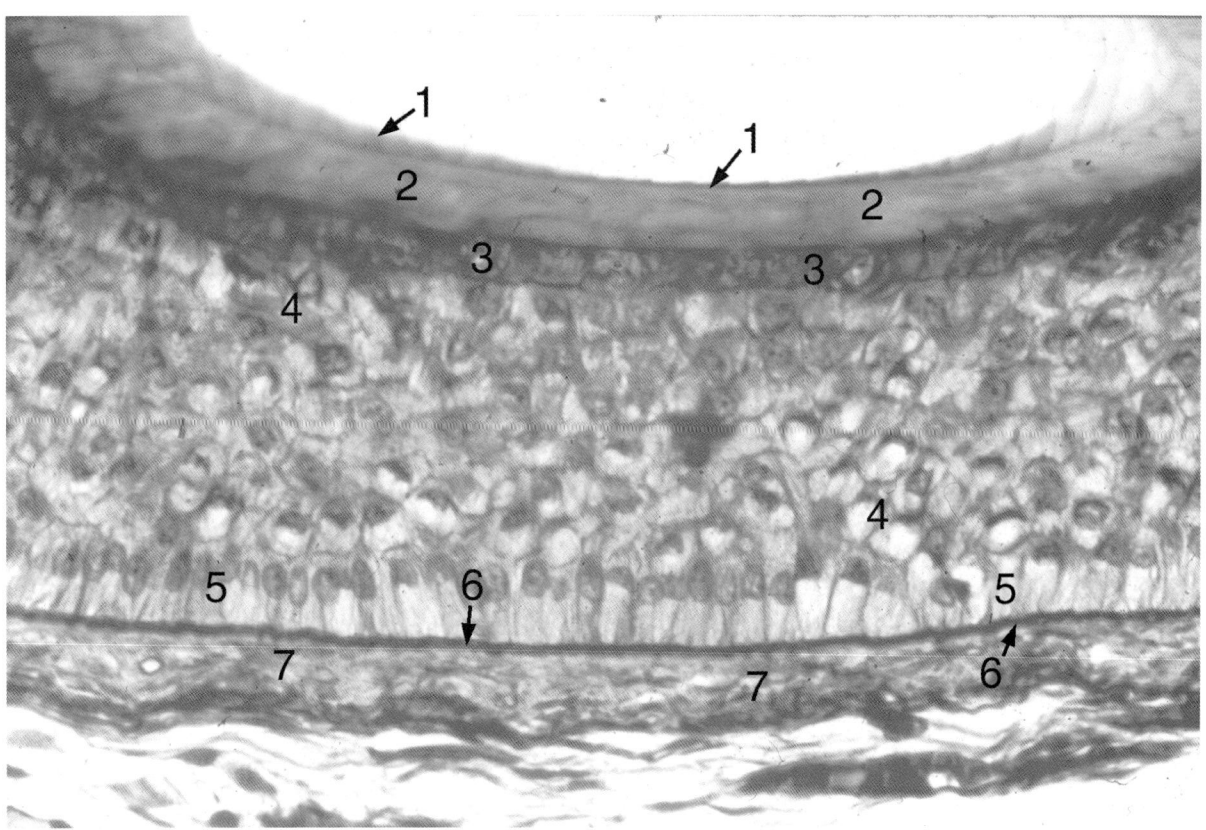

1. Cuticle of inner root sheath
2. Huxley's layer of inner root sheath
3. Henle's layer of inner root sheath
4. Outer root sheath
5. Outermost layer of outer root sheath, similar to the stratum basale of other stratified epithelia
6. Glassy membrane (corresponding to basement membrane)
7. Connective tissue capsule surrounding the hair follicle, with capillaries

Fig. 5.36

Vascular supply and innervation of hair follicle. A – Specimens injected with india ink (from the material of K. Csányi† and J. Vajda†); B – Silver impregnated specimen, photographed from the histological collection of A. Ábrahám, courtesy of Prof. K. Gulya.

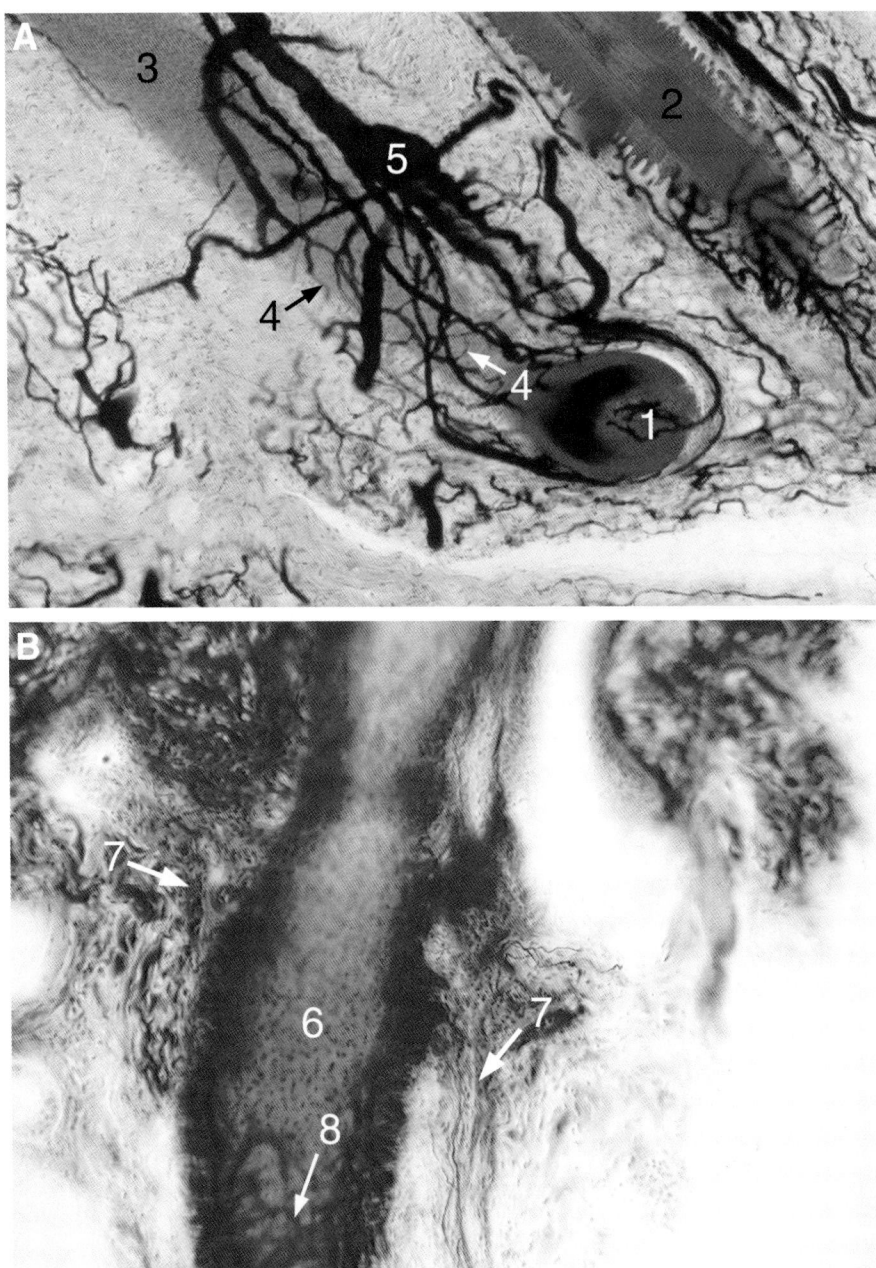

1. Dermal papilla
2. Hair follicle sectioned along the hair shaft
3. Hair follicle sectioned tangentially, along the root sheath
4. Capillary network
5. Arteriole supplying the hair follicle
6. Root sheath sectioned tangentially
7. Longitudinal nerve fibers of hair follicle
8. Circular nerve fibers of hair follicle

MYSTACIAL VIBRISSAE AND SOMATOSENSORY PATHWAYS (FIG. 5.37; 5.38; 5.39; 5.40)

This highly specialized receptor organ does not occur in humans. However, vibrissal receptors are known as the most extensively studied experimental object to exemplify tactile sensation and to elucidate somatosensory pathways in general. Several mammalian species (cats, laboratory rodents) possess a set of long hairs, called whiskers, around the nose, upper lip, eyebrow, and other exposed parts of the head, which participate in the orientation of the animal, especially in the dark. Vibrissal hairs are thick and elongated and they are embedded in deep follicles surrounded by vascular sinuses. The follicles line up in a regular array in the whisker pad, whilst they are more scattered in other regions. Tactile receptors are highly enriched in the wall of vibrissal follicles. Each follicle contains approx 300–350 lanceolate endings, sensitive to the speed of deflection, low-frequency (200 Hz) and high-frequency (1000 Hz) vibration, as well as 1500–1600 Merkel cells, detecting the direction, amplitude, and duration of vibrissal deflection. One sensory neuron innervates one to three lanceolate endings and

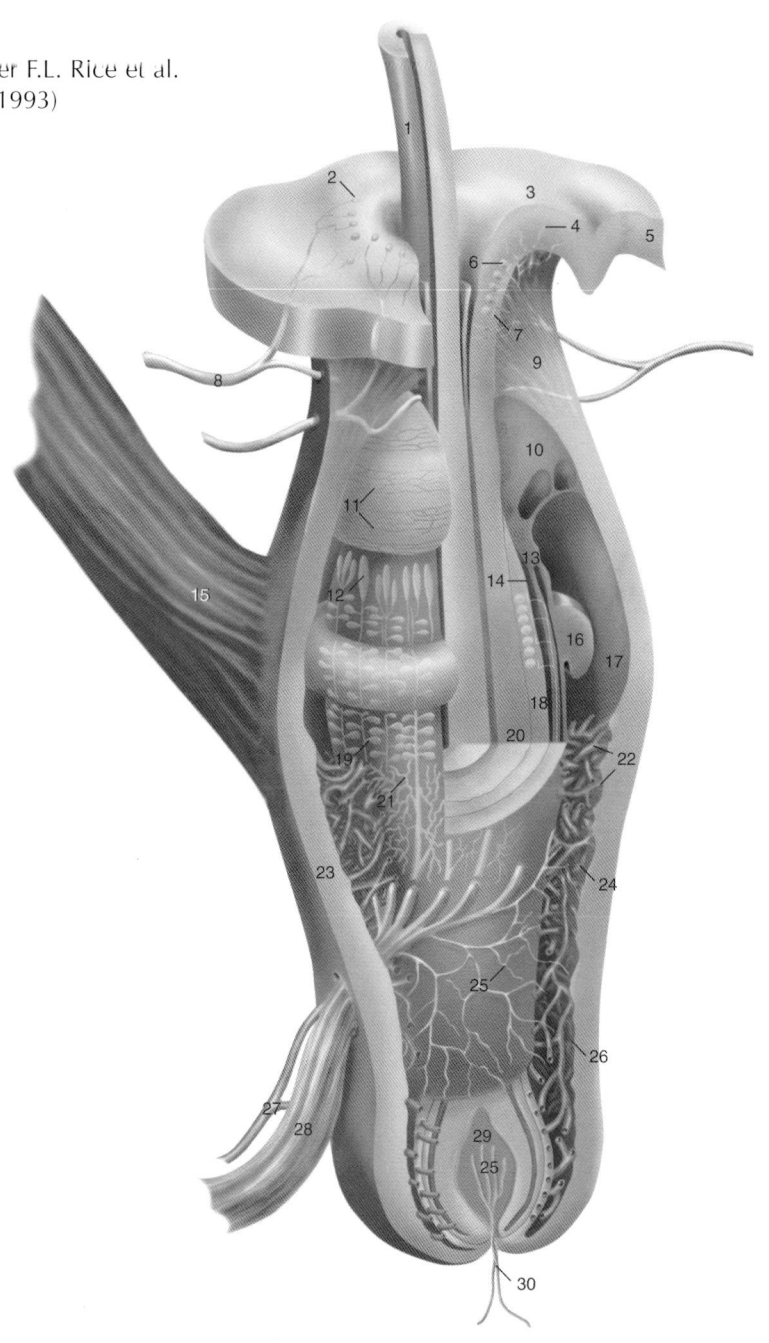

Fig. 5.37

Schematic drawing of vibrissal follicle. Redrawn after F.L. Rice et al. (Journal of Comparative Neurology, 337: 366-385, 1993)

1. Vibrissal shaft
2. Merkel ending
3. Rete ridge collar
4. Free ending
5. Epidermis
6. Merkel cell
7. Merkel ending
8. SVN (superficial vibrissal nerve)
9. Outer conical body
10. Inner conical body
11. Circular endings
12. Lanceolate endings
13. Mesenchymal sheath
14. Glassy membrane
15. Papillary muscle
16. Ringwulst
17. Ring sinus
18. External root sheath
19. Merkel endings
20. Internal root sheath
21. Reticular endings
22. Trabeculae
23. Capsule
24. Capillary network
25. Meandering endings
26. Cavernous sinus
27. Arteriole
28. DVN (deep vibrissal nerve)
29. Dermal papilla
30. Small caliber axons

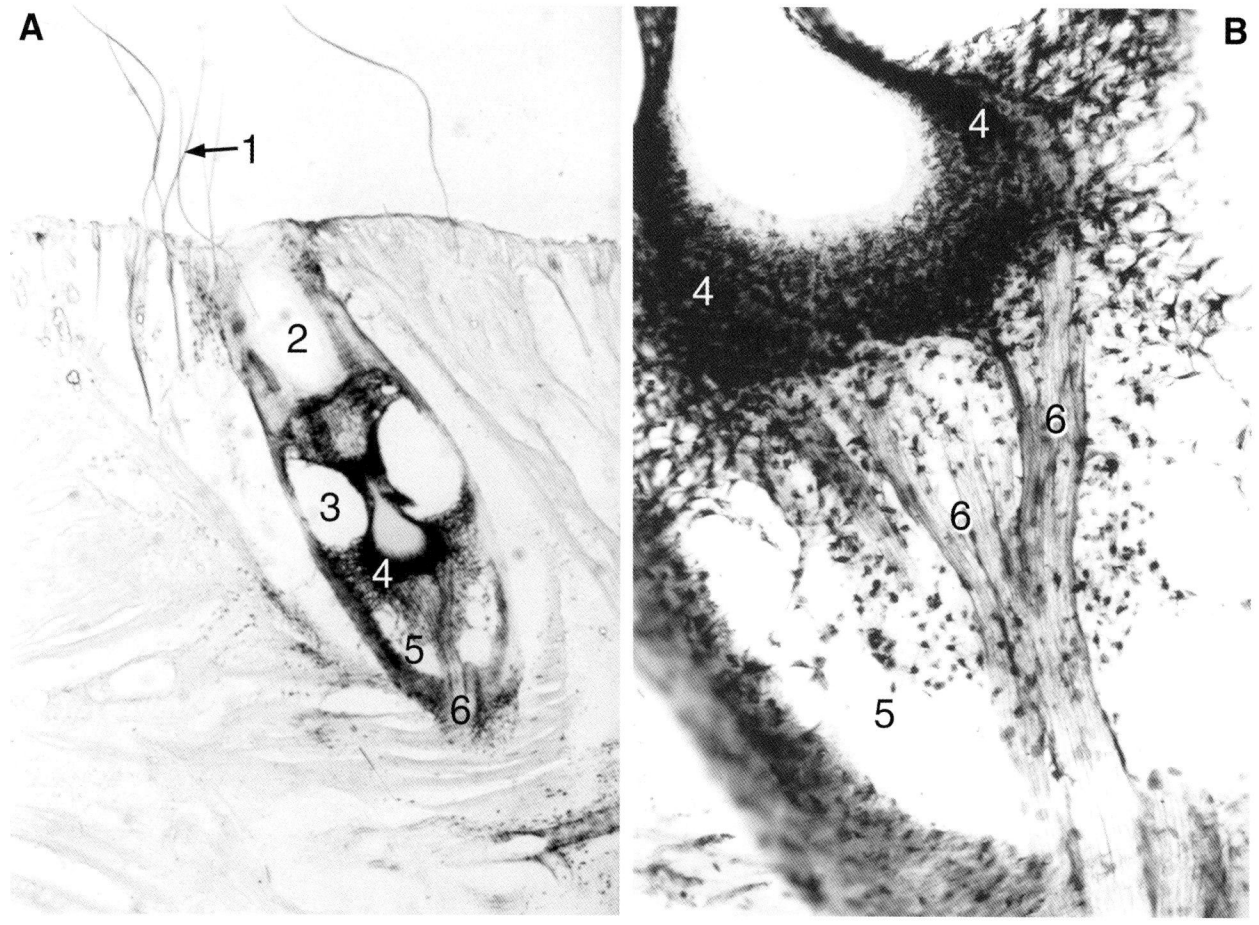

Fig. 5.38

Mystacial vibrissa of rat injected with choleratoxin B (oblique section) (A). Enlarged detail (B). The neural tracer labels the afferent nerve fibers. Courtesy of J. Páli.

1. Vellus hairs
2. Conical body
3. Ring sinus
4. Root sheaths of follicle
5. Cavernous sinus
6. Deep vibrissal nerve and its branches

three to five Merkel discs. Altogether, the Merkel and lanceolate receptors of each follicle are supplied by 250 and 150 neurons, respectively. The size of receptive fields extends to 50–100 µm on the outer cylinder, but the actual resolution is 10–15 µm when the sensitivity to the deflection of hair tips is taken into account. The Ruffini endings found in the neck region of follicles are sensitive to large-scale deflections (traction). Vibrissal hairs display an active exploratory motion of large amplitude (whisking), driven by 7 Hz septohippocampal θ activity, transmitted by the facial nerve. Ruffini and lanceolate endings (but not Merkel cells, Meissner's, or PCs) are present also in the intervibrissal skin around the fine hairs.

The area of the whisker pad of the rat is approx 1 cm^2 on either side, and it is innervated by more than 12,000 sensory neurons (Fig. 5.38). Some of these are sensitive to low-speed (3°/s), others to intermediate (100°/s) or high-speed (3000°/s) deflections of whiskers. The sensi-

tivity of neurons is further enhanced in active exploration, when the venous sinuses are engorged with the root of vibrissal hair pressed harder against a distended root sheath and its receptor endings, now more prone to deformation. Thus, even a single vibrissal hair with its abundant and topographically arranged nerve endings of sensory neurons constitutes a complex and sensitive tactile organ.

More effective still, the multiple vibrissae arranged in regular rows and the columns of the whisker pad are represented in the trigeminal ganglion, brainstem (barrelette), thalamus (barreloid), and somatosensory cortex (barrel field) according to a strict sequential somatotopy (Fig. 5.39). Such representations at the different levels correspond to functional (e.g., memory) units.

The barrel field of laboratory rodents is one of the best explored models of the somatosensory cortex, which can be visualized by conventional Nissl staining but it is also

Fig. 5.39

Neural tracing studies relevant to the innervation and central connections of vibrissae. A – Mystacial vibrissa of cat injected with the fluorescent tracer Fast Blue (longitudinal section); B -Supraorbital vibrissa of cat injected with the fluorescent tracer Diamidino Yellow (transverse section); C – Mystacial vibrissa of rat injected with choleratoxin B (transverse section); D – Horizontal section of the trigeminal (Gasserian) ganglion of rat following injection of Fast Blue into a single caudal mystacial vibrissa. Arrows indicate retrogradely labeled ganglionic cells in the infraorbital-maxillary (V2) region of the ganglion; E – Ganglionic cells of the trigeminal ganglion of rat labelled retrogradely from two neighboring mystacial vibrissae injected with Fast Blue or Diamidino Yellow, respectively. The former dye labels the cytoplasm, whereas the latter marks the nucleus. Courtesy of J. Páli.

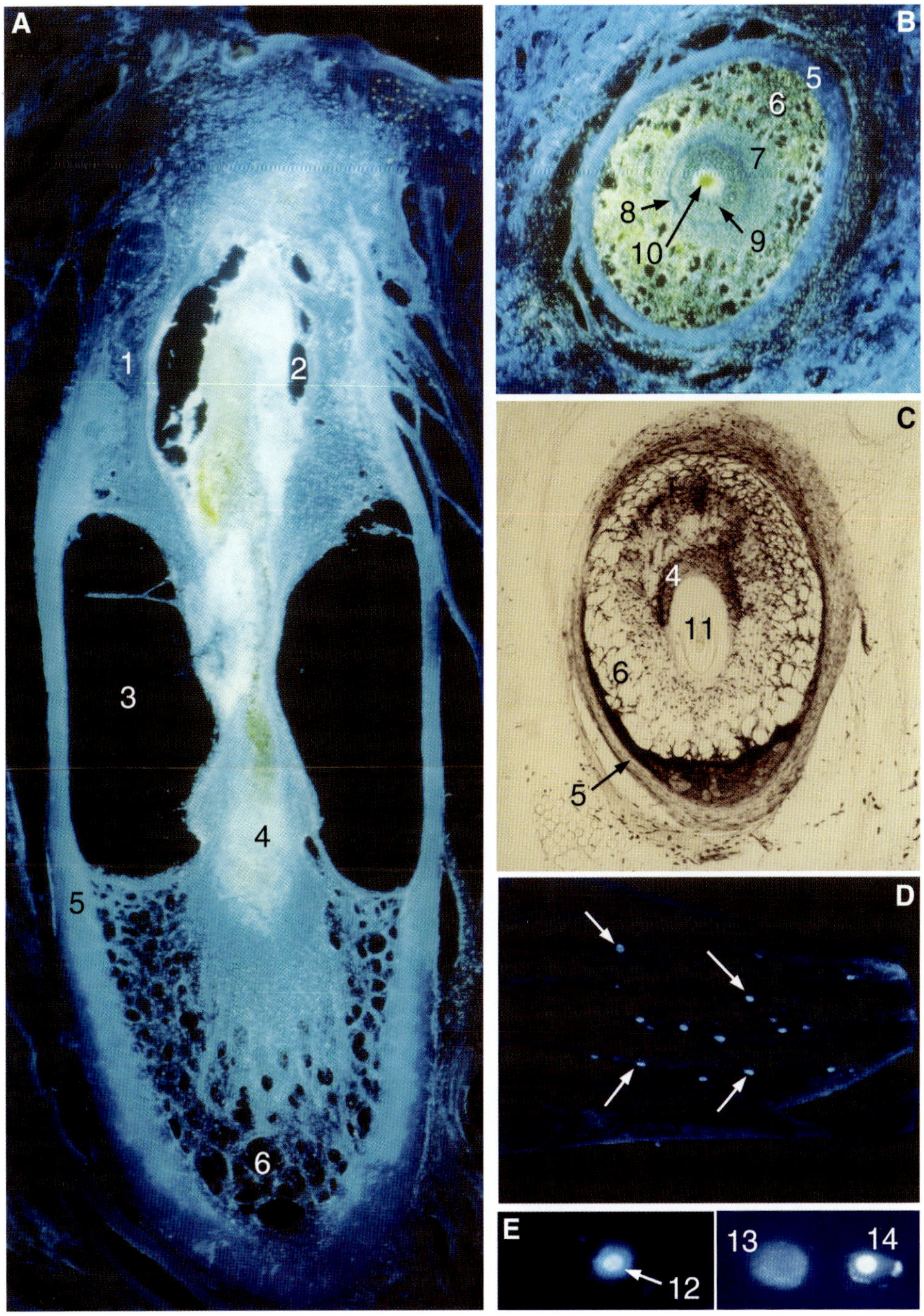

1. Conical body
2. Conical sinus
3. Ring sinus
4. Root sheaths of follicle
5. Capsule
6. Cavernous sinus
7. Mesenchymal sheath
8. External root sheath
9. Internal root sheath
10. Tracer deposit in the site of hair shaft
11. Vibrissal hair shaft
12. Double labelled neuron innervating both of the injected vibrissae
13. Fast blue labelled neuron innervating one of the injected vibrissa only
14. Diamidino yellow labelled neuron innervating the other injected vibrissa

Fig. 5.40

A-C – Brainstem sections of cat demonstrating the terminal field of the sensory nerve (barrelette), originating from a single supraorbital vibrissa, at the levels of the principal (A), interpolar (B) and caudal (C) nuclei of the trigeminal nerve. Choleratoxin B injection. D-E – Brainstem sections of rat demonstrating the terminal fields of the sensory nerves (barrelettes) originating from three mystacial vibrissae β, D2 and δ, in the principal trigeminal nucleus. Note the specificity of input in the enlarged photograph (E), where the terminal fields of non-injected vibrissae appear to be devoid of label. Choleratoxin B injection. F-G – The barrelfield of the somatosensory cortex of mouse. Nissl staining. H – Vasoactive intestinal polypeptide (VIP) immunoreactive structures in the barrelfield of mouse. The inset shows immunostained fibers (arrows) in the wall of the cortical barrel. Specimens by J. Páli (A-E) and A. Csillag (F-H).

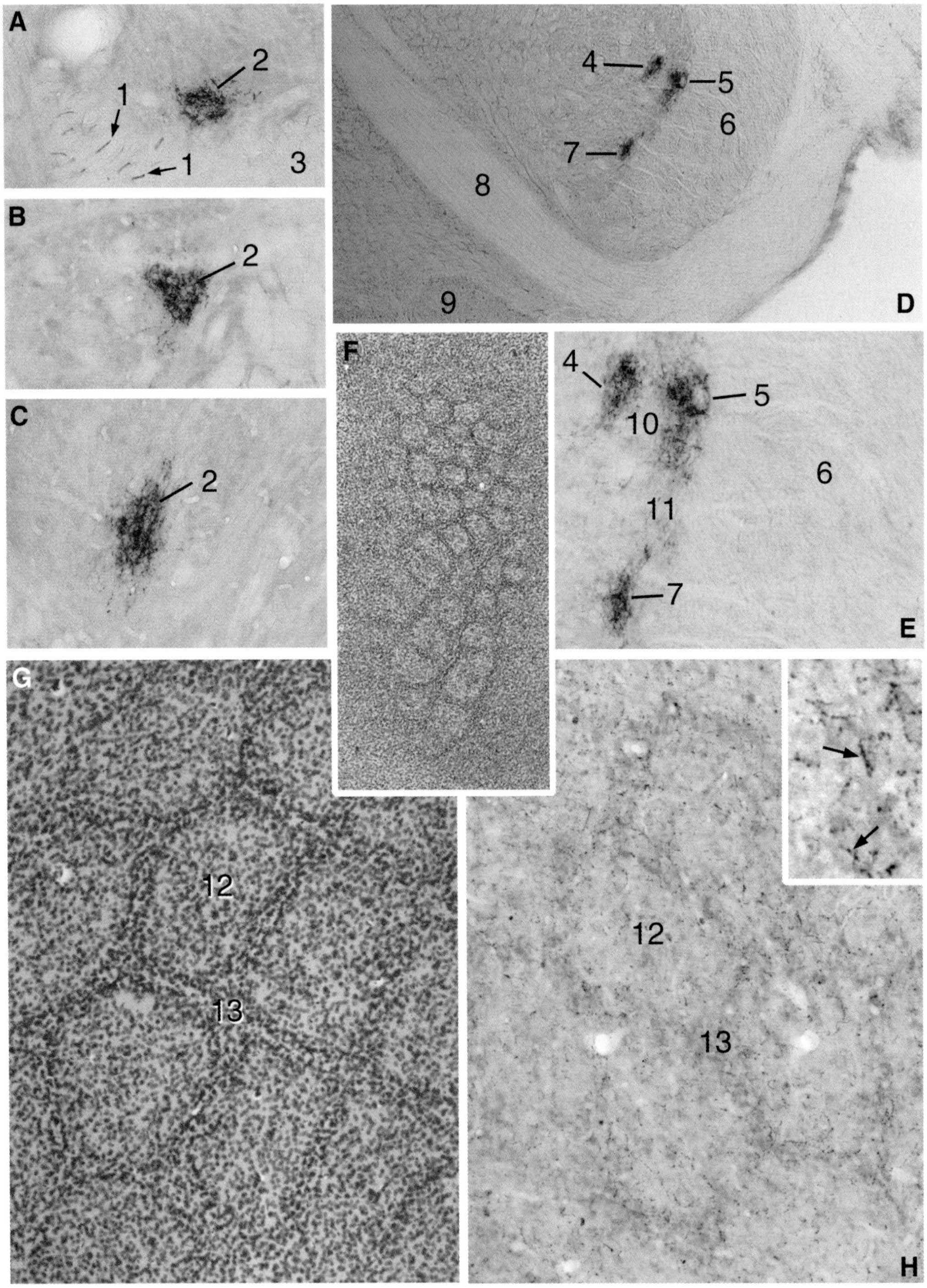

1. Myelinated fibers entering the nucleus
2. Terminal field (barrelette)
3. Spinal trigeminal tract
4. D2 barrelette
5. δ barrelette
6. Sensory root of trigeminal nerve
7. β barrelette
8. Facial nerve root
9. Superior olive
10. Unlabelled discoid area representing the non-injected D1 vibrissa
11. Unlabelled area representing the non-injected γ vibrissa
12. Hollow of cortical barrel
13. Wall (side region) of cortical barrel

known to contain a multitude of neurotransmitters in a surprisingly systematic arrangement (Fig. 5.40). For example, in the mouse the axon terminals containing vasoactive intestinal polypeptide (VIP) are distributed preferentially in the side region of barrels, whereas they are relatively scarce in the hollow of barrels (i.e., these axons seem to follow a modular distribution in the somatosensory cortex).

RECOMMENDED READINGS

Textbook and Handbooks

1. Bannister LH, Berry MM, Collins P, Dyson M, Dussek JE, Ferguson MWJ. Gray's Anatomy, 38th ed., Churchill Livingstone, Edinburgh, 1999.
2. Squire LR, Bloom FE, McConnell SK, Roberts JL, Spitzer NC, Zigmond MJ. Fundamental Neuroscience, Academic Press, Imprint of Elsevier Science, USA, 2003.3.   Hubbard JI (Ed.). The peripheral nervous system, Plenum Press, New York, London, Chapters 12-17 (Iggo A, Hunt CC, Matthews PBC, Widdicombe JG, Biscoe TJ, Sato M), 1974.

Reviews and Research Reports

1. Chambers MR, Andres KH, Duering M, Iggo A. The structure and function of the slowly adapting type II mechanoreceptor in hairy skin. Quart J Exp Physiol 1972;57:417-445.
2. Paré M, Behets C, Cornu O. Paucity of presumptive Ruffini corpuscles in the index finger pad of humans. J Comp Neurol 2003;456:260-266.
3. Potten CS. The epidermal proliferative unit: the possible role of the central basal cell. Cell Tiss Kinet 1974;7:77-88.
4. Szentágothai J. The "module-concept" in cerebral cortex architecture. Brain Res 1975;95:475-496.
5. Woolsey TA, van der Loos H. The structural organization of layer IV in the somatosensory region (SI) of mouse cerebral cortex. The description of a cortical field composed of discrete cytoarchitectonic units. Brain Res 1970;17:205-242.
6. Zilles K., Hajós F., Csillag A., Kálmán M., Sótonyi P., Schleicher A. Vasoactive intestinal polypeptide immunoreactive structures in the mouse barrel field. Brain Res 1993;618:149-154.

Index

Key: Page numbers in **bold** denote detailed, descriptive references in text; page numbers in *italics* denote references to illustrations.

A

Accommodation, 89, 91, 159
Action potential, 13, 99, 167, 168, 188, 215, 217
Adaptation, 201
    rapid, 200, 206
    slow, 200, 206
Amiloride, 190
Ampulla
    of semicircular canals 8, **43**, *44, 48, 54, 57, 62, 71*
    of semicircular ducts, 10, *11*, 24, 71
Amygdala, 181, 183, 185, 194
Angle
    of eye, *see* canthus
    iridocorneal, *see* drainage angle
Annulus
    tendineus communis, *see* Ring, tendinous
    tympanicus, 15, *23, 25, 260*
Antibodies
    to CD1A, 212, *232*
    to nCAM, 167, *175, 178*
    to S-100, 203, 212, *232*
Antihelix, 2, *2*, 3
Antitragus, 2, *2*, 3
Antrum
    mastoid, 4, *31, 46*
Apocrine sweat gland, 213
Aqueduct
    cerebral, *127*
    of cochlea, *see* cochlear canalicle
    of vestibule 8, *46, 48*
Aqueous humor, 88, **89,** 90
Arachnoid mater, 147, *148*
Arch, branchial, *21*
Area
    motor speech, of Broca, *157*
    primary visual, *see* Cortex, striate
    somatomotor, 218
    somatosensory, 198, 218, *242*, 243
    of Wernicke, *157*
Artery, arteries
    basilar, 11, *61*
    carotid, internal, 7, *31*, 100, *108*, 147, *149*, 226, 227
    central retinal, 103, *123, 124, 125,* 147, *147*
    choroidal, 88, 101, 114, *125*
    ciliary
        anterior, 101, 102, 104, 114, *142*
        long posterior, 89, 114
        short posterior 89, 100, 114
    cilioretinal, 103, *114*
    conjunctival, 105
    of the eye and orbit (summary table), **114**
    infraorbital, 113, 114, 196
    labyrinthine, 11
    lacrimal, 106, 114
    nasal, 114, *165*
    ophthalmic, **100,** *108, 128,* 147, *149*, 159
    palpebral, 114
    stapedial, *17, 18*
    transverse facial, 106, 114
    tympanic, 7, *36*
Astrocyte, 100, 147
Auricle, *1*, 2, *2, 21, 209*

B

Barrel field, *242*, 243
Barrelette, 240, *242*
Barreloid, 240
Binocular visual field, *see* Visual fields
Birbeck's granule, 212
Blind spot of Mariotte, 95
Blinking, 87, 88, **103,** 105, 106
Body
    carotid (glomus caroticum), 200, **206,** *208*, 226, 227
    ciliary, *85,* 86, 88, **89,** *90*, 93, 94, 100, 101, 102, *114, 115, 118, 119, 121, 128, 131, 133, 134*
    geniculate
        lateral, 100, *149*, 150
        medial, *149*, 151
    trapezoid, 73
    vitreous, 90, *116, 118*, 119, 120
Bone, bones
    ethmoid, 112, 167, 171, 172, *173*
    lacrimal, 106, 112
    maxilla, 106, 111, 112, 172
    nasal, 171, *173*
    of orbit, 86, 100, 106, 107, *109,* **111,** 111, 112, *112*, 113, *127, 128, 129*, 147, 159, 164, 172
    sphenoid, 107, *109*, 111, 112, 147, 172
    temporal, 2, 3, 4, 6, 7, 10, 111, *114, 127*
Border tissue of Jacoby, 147
Bowman's layer, 87, *130, 131*

Brachium
- of inferior colliculus, 76
- of superior colliculus, 76, **151**, *151, 161*

Braille reading, 217, *217*

Brodmann areas, 156, 157, *157*, 161

Bruch's membrane, **89**, 95, *96, 136, 137, 138*

Bulb
- of hair, 213
- of internal jugular vein, 5, *52*
- olfactory, 10, *149, 151*, 167, 168, *178*, **179**, 179, *179*, 180, 181, 183, 185
  - accessory, 183

Bundle
- longitudinal, medial, *see* Fasciculus longitudinalis medialis
- olivocochlear, 76
- papillo-macular, 100

C

Canal, canalicle
- carotid, **30**, *55, 56*
- cochlear, *8*, 10, **30**, *52*
- facial, 5, 6, 7, **29,** *30, 32, 41, 53, 54*
- hyaloid (Cloquet's), **93**, *118*
- infraorbital, 111, 113
- lacrimal, 105, 106, *173*
- nasolacrimal, 107
- optic, 100, 106, *108, 109*, 147
- of Petit (retrozonular space), 93
- Schlemm's, **90**, 90, *91, 131, 134*
- semicircular, *1, 4*, 5, 7, *8*, 10, **11**, 14, *29, 32, 41, 43, 44, 46, 61, 62*

Canthus, 103, *104*, 115

Capsule
- of lens, **90,** 91, *92, 93, 120, 135*
- otic, 15
- of Tenon, **106,** *111, 112,* 112, *121*

Caruncle, lacrimal, 103, *115*

Cartilage
- of Meckel, *18, 19, 25, 170,* **171,** 171, 172
- of pharyngotympanic tube, 6, *19, 56*
- of Reichert, *23, 25,* **171**
- vomeronasal, *173*, 183

Cataract, 91

Cavity (*see also* cavum, space)
- nasal, 105, *115*, 165, 166, 167, *169*, 171, *171, 172, 173, 174*
- orbital, **111**, 111, 112, 113, *114, 127, 128, 129*, 159
- trigeminal (of Meckel), *45*
- tympanic, 3, **4**, 6, 7, *25, 29, 33, 34*

Cell
- air
  - ethmoidal, *45, 114, 115, 127, 129,* 172
  - hypotympanic, *29, 31, 39*
  - mastoid, 5, *32, 35*
- basal, *145, 166,* 167, *174, 176,* 181, *182,* 188, *188, 192,* 243
- of Boettcher, 68
- of Claudius, *9,* 68
- clump, **88**, *132*
- corneocyte, **211**, *211*
- of Deiters, *9, 68, 69*
- dendritic, *see* of Langerhans
- goblet, 104, 106, *174*
- hair
  - of the membranous labyrinth, 11, 14, *70, 71*
  - of the organ of Corti, *9*, 10, 12, *12, 68, 69*
- of Hensen, *9, 68*
- hyalocyte, 94
- interdental 9, 68
- keratinocyte, 201, *202*, 210, 211, *211, 212*, 215
- of Langerhans, 104, 209, 211, 212, *232*
- mast, 88, 212
- melanocyte, 212, *212*
- of Merkel, 198, **200**, 201, *202*, 210, 211, *211*, 215, 217, *219, 232, 239,* 240
- of Müller, *96,* **98**, 99, 100, *137, 140*
- myoepithel, 105
- nerve (*see also* neuron)
  - alpha, 100
  - amacrine, *96,* 98, 99, *126*
  - bipolar
    - of retina, *96,* 98, 100, *140*
  - ganglionic
    - of retina, 95, **99**, 100, *117, 120, 137*
  - granule, 179, *179*, 180
  - horizontal, *96,* 98, 99, *140*
  - interplexiform, *96,* 98, 99
  - midget, of retina
    - bipolar, 98, 100
    - ganglionic, 100
  - mitral, 179, *179*, 180
  - periglomerular, 179, *179*, 180
  - tufted, 179, *179*, 181
- pigment
  - of retina, 89
  - of skin
- pillar (rod), 8, *9, 68*
- plasma, 106, *145*
- Purkinje, 163
- of Schwann, 167, *184*, 200, 202, 203, *221*

sustentacular, supporting, *166*, 167, *174, 176, 177, 182, 184*, 188, *188*, 207, *208, 227*
Cementum, *195*, 195, 196, *197*
Cerumen, 3
Chambers of eye (camerae bulbi), *85*, 89, 90, *90, 91*, **93–94**, *118, 121, 128, 134*
Chiasma, optic, *127*, 147, *149, 150, 151*, 157, 159, *161*
Choanae, 169, *169*, 171
Choroid of eye, 85, **89**, 101, 102, 114, 115, *118, 120, 125, 136, 147, 148*
Cilia, 95, *97, 146*, 167
  of eye (eyelashes), 103, *104, 105, 112*
  of hair cells, 10, 12
  olfactory, *166*, 167, 168, *176, 177*
Ciliary body (*see* Body, ciliary)
Ciliospinal center, 160
Circulus arteriosus, circle
  of iris, major and minor, *91, 101*, 101, 102, *102*, 105, 114, *134*
  of Haller and Zinn, 101
  of Zinn, 103, 114
Cytocriny, 212, *212*
Clivus, 94
Cochlea, *1, 4*, 7, **8**, 11, 20, 23, 30, 43, 44, 45, 46, 47, 50, 51, 53, 54, 55, 62, 66
Cog, *146*
Colliculus
  inferior, *149, 162*
  superior, *149*, 150, 151, 156, 161, *162*, 163
Color vision, 96, *141*
Concha, conchae
  of external ear, 2
  nasal, *165, 169, 170, 171*, 171, 172, *172*, 173, *174*
Cone, **95–96**, *96*, 97–100, *136–141*
Conjunctiva, 90, *91, 101*, **103–105**, 113–115, *118, 120, 121, 128, 130, 131, 144*
Convergence, 86, 87, 109, 157, 159
Corium, 200, 207, 209, 213, *222*, 228
Cornea, *85*, **86–88**, 90, *90, 91, 102*, 115, *118–122*, **121, 122**, *130, 131, 134*
Corpus callosum, *149, 157, 158*
Corpuscles, *see also* Nerve endings
  genital, 202, 203, *203*, 204, *223*
  of Golgi-Mazzoni, **204**, *204, 224*
  of Krause, **203–204**, *222*
  lamellated, **202**, *221*
  of Meissner, **201–203**, *202*, 215, 217, *219*, 240
  Pacinian (*see* lamellated)
  of Ruffini, 201, **204**, *205*, 206, 215, 217, 218, *224*, 240
    Ruffini-like spray endings, 204
    pilo-Ruffini endings, 204
  tactile, 200, 201, *202*, 215, 217, *220*, 225
  of Vater-Pacini (*see* lamellated)
Copula, *189*, 190
Corona ciliaris (pars plicata), 89
Cortex, cortical area, field
  of cerebrum
    association, 218
    auditory, *74*, **76**, *79, 80*, 81
    Broca's, *157*
    cingulate, 161
    entorhinal, *181*
    frontal, *149*, 157
    frontal eye, *157, 160*, 161, *161, 163*
    gustatory, 194
    insular, 159, 181, 185, 194
    limbic, 161
    occipital, 161, 163
    olfactory, 181
    parastriate, 156, *157*
    parietal, 159, 161, 163
    peristriate, *157*
    piriform, 181, 185
    prestriate, 156, 157
    somatomotor, 218
    somatosensory, 198, 240, *242*, 243
    striate, 154, 156, 157, *157*
    temporal, 156, 157
    visual, 151, *151*, **154**, 155, *155, 156, 158, 161*, 163
    vestibular, 81, *82*, 159
    Wernicke's, *157*
  of hair, 213, *214, 235, 236*
  of lens, 94
Corti, of
  spiral organ, **8**, *9*, 10, *66*
  tunnel, 8, 10, *68*
Cortilymph, 10
Crest, crista
  ampullary, 11, *23, 24, 71*
  lacrimal, 106, 112
  neural, 168, 171, 187, 201, 211
Cribriform plate, *see* Plate, cribriform
Crus, crura
  of antihelix, 2, *3*
  cerebri, *149*
  helicis, *3*
Crypts of iris, *132*
Crystallins, 91
Cunicula, *see* Tunnel of Corti
Cup, optic, 103, *116–125*
Cupula
  of ampullary crest, 11, *23*

Cuticle
    of hair, 213, *214, 135, 236, 237*
    of hair cell, 10
Cutis, 207
Cyclic adenosine monophosphate (cAMP), 190
Cymba conchae, 2, *3*
Cytokeratin, 203

D

Decussations
    of optic nerves, 147
    trapezoid, 73
Dermis, 204, **207**, 212, *222–224, 229, 233*
Desmosome, 195, 201, *210, 211*
Disc, optic, 86, 88, **95**, 100–102, 147
Discriminative touch, 217
Dislocation, perception of, 217
Dominance, ocular, 154
Dopamine, 99, 207
Drainage angle 90–91, *90, 91*, 93, *121, 131, 134*
Duct, ductus
    cochlear, 7, 8, *8, 20, 22*, 66–68
    endolymphatic, *18*, 19, *25*
    nasolacrimal, *105*, **105–106**, 114, 171
    reuniens, *8*, 10, *20*
    semicircular, 7, *20*
    of sweat glands, *210, 222, 231*
    utriculosaccular, 10, *25*
Dura mater, 114, *117, 118, 147–149, 178*

E

Ear
    external (*see also* Auricle, Meatus, Pinna), 2
    inner (internal), 4
    middle, 7
Eccrine sweat gland, *229*
Edinger-Westphal, accessory oculomotor nucleus of, 115, 159–160, *161*
Electroretinography (ERG), *126*
Eminence
    pyramidal, 5, *30, 35, 40, 51*
End bulb of Krause, 200, *203*, **203–204**, *222*
Endolymph, **7**, 8, 10, 11, 14, *24*
Endoneurium, 202
Endothelium, corneal, 87, *122, 130–131*
Epicanthus, 103, *104*
Epidermal Melanine Unit (EMU), 211
Epidermal Proliferative Unit (EPU) **211**, *211*
Epidermis, **207, 209–210**, *210, 214, 228, 229, 233, 239*
Epiglottis, 187, 190, 193
Epithelium
    conjunctival, **104–105**, *142, 144*
    corneal, 87, *130–131*
    of iris, *131*
    keratinized, 213, *229, 231, 233*
    of lens, 90–91, *92, 116, 118–120, 135*
    non-keratinizing, 190
    olfactory, 166, **166–167**, *173, 175–176, 178–179*
    pigment, retinal, 89, 94–95, *96, 119, 136, 138, 140*
    pigmented, *132–133*
    respiratory, *174*
    stratified squamosus, *192*, 209, 213, *222, 225, 229, 231, 233*
    vomeronasal, 181, *182, 184*
Equator
    of globe, **86**, 106, 109
    of lens, 90–91
Equatorial zone, 91, *116, 118*
Eumelanin, 212
Eustachian tube, *see* Tube, auditory
Exteroceptors, 199
Extraocular muscles (*see* Muscles, extraocular)
Eyeball (*see* Globe)
Eyebrows, 103, *104–105*, 112
Eye field
    frontal, *157, 160–161*, 161, 163, *163*
    midbrain, 161, *163*
    pontine, 161, *163*
Eyelashes, 103, *104–105*, 112, *142–143*
Eyelids, **103**, 104–105, 112, 114, *118, 120–121, 142–143*

F

Face recognition, 159
Fascia
    bulbi (Tenon's capsule), 106, *111*, 112, *112, 121*
    lacrimal, 106
    orbital, **111**
Fasciculus
    longitudinal inferior, 157
    longitudinal medial, 81, *82*, 159, 162, *162–163*
    longitudinal superior, 159, *161*
    olivocochlear (*see* Bundle, olivocochlear)
    papillomacular (*see* Bundle, papillomacular)
Fat, orbital, *112–113, 127–129, 148*
Fenestra
    cochleae (*see* Round window)
    vestibuli (*see* Oval window)
Fibers
    collagen, 87, 94, 111, *134, 142, 174, 219, 233, 234*
    of lens, **91**, *92–94, 135*
    nerve (*see* named tracts)
    of periodontal ligament, *195, 197*
    of rods, inner and outer, 8
    of Sharpey, 196

Fibroblasts, 87–88, *132*, 195, *220*
Fila olfactoria (olfactory nerve fibers), 167, *175, 178–179*
Fissure
    calcarine (of the occipital cortex), 151, 154
    choroidal, of eye, *117*
    orbital
        inferior, 111–112
        superior, 106, *108*, 111, 113–114
    palpebral, 103
    petrotympanic (of Glaser), 7, *30*
Fixation, 86, *87*, 109, 161
Fluorescence angiography (FLAG), *125*
Fold, plica
    lacrimalis, 106
    malleolar, of tympanic membrane, 3
Follicle, of hair, *173*, 200, 201, 210, 213, *214*, 228–229, 233–238
Foramen, foramina
    incisive, 171
    infraorbital, 113
    optic (*see* Canal, optic)
    singulare, 11, *23, 51*
    sphenopalatine, 171
    stylomastoid, 7
Fornix
    of brain, *149*
    of conjunctiva, 104–105, *120, 128, 144*
Fossa
    cranial
        anterior, 114
        middle, 111, 114
    hypophyseal, *127*
    incudis, 6, *24*
    interpeduncular, *127, 151*
    lacrimal, 105–106, 112
    patellar (hyaloid), 90–93
    rhomboid, *149*
    scaphoid, of auricle, 2, *3*
    subarcuate, *24*
    temporal, *127*
    triangular, of auricle, 2, *3*
    trochlear, 107
Fovea centralis of retina, 85, 86, 94, 100, *123–126*
Foveola, 94, 96, 100
Functional magnetic resonance imaging (fMRI), 159
Fundus, 123, 147
Furcula, 190

G

Gamma (γ)-aminobutyric acid (GABA), 99, 151, *152–153*
Ganglion, ganglia
    acousticofacial, *18*
    cervical, superior, 159
    ciliary, 88–89, *113*, 159, *161*
    cochlear, *see* spiral (of Corti), 10, *20*, **66**, *66*, *67*
    of facial nerve (geniculate), 7, *44*
    of glossopharyngeal nerve, 193
    parasympathetic, *208*
    pterygopalatine, *113, 165*, 193
    semilunar (Gasserian), 111, 196, 240, *241*
    sympathetic, *208*
    trigeminal, *see* semilunar (Gasserian)
    vestibular (of Scarpa), 11, *20*
Gap junction, 91, 97–99
Geniculum of facial nerve, 7, *44*
Gennari, stria of, 154, 156
Gingiva, 194, 204
Glabella, 115
Gland, glands
    apocrine, 213, *229*
    of Bowman, 87, *130–131*, 166–167, *174–175, 178*
    ceruminous, 3, *27*
    ciliary, 103
    Ebner's, 187, *192*
    eccrine, *229*
    holocrine, *229*
    of Krause, 103, *144*
    lacrimal, 103, **105–106**, 115, *118*, 144–145
    Meibomian, 103, *142–143*
    merocrine, *229*
    of Moll, 103, *143*
    mucous, 106
    nasal, *182*
    olfactory, 167–168
    palpebral, 103
    salivary, *191*
    sebaceous
        of hairy skin, 200, 209, *210*, **213**, 229, 234
        of Zeiss, 103, *143*
    serous, *192*
    sudoriferous (sweat), 209, *210, 221, 224, 233*
    tarsal, 103
    of Wolfring, 103
Glaucoma, 90
Glia, 99, 100, *117*, 147, *148*, 167, *184*
Glomerulus
    of olfactory bulb, 179, 197–180
Glomus
    carotid (carotid body), **206**, *208*, 226, *227*
    vascular (glomera), *233*
Golgi tendon organs, 109–111, 199, **206**
Golgi-Mazzoni corpuscles, **204**, *204*, 224

G proteins, 168, 183
Gustation, *see* Taste sensation
Gustducins, 190
Gyrus, gyri
    angular, 159
    cingulate, 161
    fusiform, 156–157, *158*
    lingual, 154, 158, *159*
    occipital, 154
    parahippocampal, *127*, 159

H
Hair, **213**
    blood supply of **212**
    bulb, **213**
    cortex, **213**, *214, 235, 236*
    cuticle, **213**, *214, 135, 236, 237*
    follicle, 200, *210*, **213**, *214, 228, 229, 233–238*
    innervation, 200, *201*, 212
    matrix, **213**, *214, 235–236*
    medulla, **213**, *214, 235–236*
    papilla, *214, 235*
    root, 201, 213, *214, 234–240*, 240
    root sheath, *201*, **213**, *214, 234–240*, 240, *241*
    shaft, *210*, **213**, *233, 235–236*
Haller, circle of, 101
Handle (manubrium) of malleus, 4, **6**, *21*, 35–36, *52, 58, 64*
Hasner, valve of, *105*, **106**
Hearing, mechanism of, 12
Heat-shock proteins, 91
Helicotrema, 10
Helix, 2, *2, 3*
Hemianopia, *150*, 154
Hemidesmosomes, 209
Henle's layer (of hair), 213, *214, 235–237*
Heschl, transverse temporal gyrus (convolution) of, 74, 76, 79, *80*, 81
Hippocampus, 157, 181
Horopter, 86, *87*, 157
Horner's muscle, 106
Humor
    aqueous, 88–89, **90**
    vitreous, **93**
Huxley's layer (of hair), 213, *214, 235–237*
Hypermetropia, 86
Hypothalamus, 160, 181, 194

I
Immunoglobulins, 106
Incus, *4, 5*, **6**, 13, *21*, 24–25, 29, 32–33, 36, 59
Insula, 159, 181
Integument (*see* Skin)
Interneurons, 151, *152*, 196
Intraocular pressure, 90, 106
Iodopsin, 95
Iris, *85*, 86, **88**, 90, *90, 91*, 100–103, *104, 112, 118–121, 128, 131–133, 161*
Isthmus
    of auditory tube, 6

J
Jacobson's organ, *see* Organ, vomeronasal
Jacoby, border tissue of, 147
Joints
    incudomalleolar, *33*
    incudostapedial, *33, 38, 40, 42, 59*
    receptors of, 199, 200, 203–206, **206**
Junction
    gap, *see* gap junction
    myotendinous, *see* Golgi tendon organs
    neuromuscular, 201
    tight, 89, 91, 95, 99, 167, 188

K
Keratin, 201, 211, 213
Keratohyalin, granules, 211, 213
Krause, *see* End bulb of Krause
Kuhnt, recesses of, 93

L
Labyrinth
    bony (osseous), **7**, 8, *43*, 70–71
    of ethmoid bone 172
    membranous, 7, 10, 15, *16–20, 22, 24, 70, 71*
Lacrimal apparatus, **105**
Lacrimal caruncle 103, 115
Lacus lacrimalis, 103
Lanceolate ending, 200, 215, 218, 239–240, *239*
Lamina
    anterior limiting of cornea (Bowman's), 87, *130, 131*
    basal, 87, 201, *202*, 212
    cribrosa, 103, *123*, 147, *178*
    fusca, 89, *136*
    papyracea (orbital), 112
    reticular, of spiral organ, 9, **10**, 67, 69, 71
    spiral, **8**, 9, 66–68
    terminalis, 149
    vitrea, *see* Bruch's membrane
Langerhans, cells of, 209, 211, *211*, **212**, *232*
Layer, layers, *see also* Stratum, Lamina
    basal, of skin, 210, *234*
    cornified (stratum corneum), 211, *232*
    ganglionic, of retina, 95, *96*, **99**, *136, 137*
    glomerular, of olfactory bulb, **179**, *179*

granular, external and internal, of retina, 95, *96, 136–137*
Henle's, of hair, 213, *214, 235–237*
Huxley's, of hair, 213, *214, 235–237*
Malpighian, of skin, **210**, 214, 234
mitral cell, of olfactory bulb, 179
molecular, of olfactory bulb, 179
nerve fiber, of retina, 95, *96*, **100**
neuroblast, outer and inner, *120*
neuroepithelial, of retina, **96**
nuclear, or retina, 95, *96*, **98**, *118, 136–137, 140*
papillary, of dermis, 209
pigmented, of retina, 95, *96, 118*
plexiform, of retina, 95, *96*, **98**, *137, 140*
photoreceptor (rods and cones), of retina, 95, *96*
reticular, of dermis, 209, *224*
suprachoroidal, 89–90, 101, *136*
uveal, 86
Lemniscus
    lateral, 73, *74, 77*
    medial, *82*
    trigeminal, 198
Lens, 85, 89, *90*, **90–93**, *92–94, 118, 120, 121, 128, 135*
Lens placode, *116*
Lens vesicle, *92, 116–119*
Ligament, ligamentum
    annular, of stapes, 6
    check, medial and lateral, *111*, 112, *112*, 127
    palpebral, 12, 106, *111, 127*
    pectinate, of iris, 90, *131*
    periodontal, *195*, **196**, *197*, 204
    spiral, of cochlea, 9, 10, *67, 68*
    suspensory
        of eyeball (of Lockwood), 107, **112**, *112*
        of lens, *see* Zonule, ciliary
    of Wieger (hyaloideocapsulare), 93
Light reflex, 123, **159**, 161
Limbal conjunctiva, 104
Limbal palisades of Vogt, 105
Limbus
    corneal, **87**, 90, *91, 104, 130*
    corneoscleral, 107, *107*
    palpebral, anterior, *142*
    palpebral, posterior, *143*
    scleral, 88
    of spiral lamina, 9, 10, *67, 68*
Lipofuscin, *176*
Lobe, lobes,
    frontal, *129*, 180
    occipital, 156, *158*
    temporal, *127*, 154, 159
Lockwood, *see* Ligament, suspensory
Locus Kiesselbachii, *165*

Loop of Meyer, *151*, 154
Lymphatics, lymphatic drainage, 105, 106, 167
Lymphocyte, 87, 104–105, 196, 212
Lysosome, 167
Lysozyme, 105

M

Macrophages, 87–88, 104, 212
Macula
    lutea, **94, 100**, 154
    of saccule, 8, **10**, 11, *70*
    of utricle, *8*, **10**, 11
Magnetic resonance imaging (MRI), *61, 62*, 79, 159, 181
Magnocellular cells (M-cells), 100
Magnocellular fibers (M-fibers), 100
Malleus, *4–6*, **6**, 8, 13, *21, 28, 33, 59*
Malpighian layer, *see* Layer
Mandible, *129, 170*, 171–172
Manubrium of malleus, 4, **6**, *19, 21, 25, 28–29, 35, 52, 58, 64*
Matrix, germinal of hair, 213, *214, 235–236*
Maxilla, 106, 108, 111–112, *128–129, 170*, 171–172, *173–174*
Meatus
    acoustic
        external, *1–2*, **2**, 3, 4–5, 8, *19, 26, 29–30, 45, 48, 63–64*
        internal, 1, *4*, 11, *45–46, 50–51, 57*
    nasal, 106, *170*, 181
Mechanoreceptors, 196, 200, 203–204
Meckel's cartilage, *18–19, 21, 170*, **171–172**
Medulla, of hair, **213**, *214, 235–236*
Meibomian glands, **103**, *142–143*
Meissner's corpuscles, 198, **201–202**, *202*, **215–218**, *220*
Melanin, 88, 95, *133, 138*, 211–212
Melanocyte, 88, 104, 132, **211**, *212, 214, 230*, 235
Melanosome, *212, 230*
Membrane, membrana
    basement, 88–90, *92*, 100, 167, 174, 200, 209
    basilar, **8**, *9, 66–68*
    of Bruch, 89, 95, *96, 136–138*
    of Descemet, 88, *130*
    glassy, 89, 213, *214, 235–237*, 239
    hyaloid, 90, 100
    limiting, of retina, 94–95, *96, 97*, **98–100**, *136–137, 140, 147*
    oronasal, 169, *169*, 171
    oropharyngeal (buccopharyngeal), 168–169
    tectorial, **8**, *9*, 10–11, 13–15, 22, *66–68*
    tympanic, 1, 2, **3–4**, *5*, 8, *19*, 23, *25–26*, 28, *31–32*, 34, *36, 38–39, 59, 63–64*, 209
        secondary, 10, **13**

vestibular, (of Reissner) 8, 13, *22, 66–68*
vitreous, 213
Merkel cells, 198, **200–201**, *202*, 210, **215–218**, *219, 232, 239*, 240
Merkel discs, **200–201**, 215–218, *239*, 240
Mesencephalon (midbrain), 151, 160–161, 163, 196
Mesenchyme, *16–19, 22, 26*, 88, *116–118, 120*, 168, *173*
Meyer, temporal loop of, *151*, 154
Microtubules, *97*, 167
Microvillus, microvilli, 87, 99, 105, *145–146, 166,* 167, *176–177,* 183, *184,* 187–188, 213
Midbrain, *see* Mesencephalon
Miosis, 159
Modiolus, 8, *23, 51, 66–67*
Moll, glands of, 103, *143*
Motion perception, 218
Motor end plate, 207
Movements of the eye, 106–109, 150, 161–163
Mucin, 104
Mucus, 167, *174*, 194
Müller
  cells, of retina, *96,* **98–99**, *137, 140*
  circular fibers of, 89
  orbital muscle of, 103, 112
  superior palpebral muscle of, 103
Muscle, muscles
  actions, *see* Movements of the eye
  arrector pili (piloarrector), *210,* 213, *214, 233*
  of auricle (auriculares), 2
  of Brücke, 89
  capsulopalpebral, 103
  ciliary, **89–91**, *91,* 102, 159
  dilator pupillae, 88, *132*
  extraocular, 103, **106–111**, *107–108,* 114, *142*
  Horner's, 106
  levator palpebrae superioris, 105, *108, 112–113,* 129
  oblique
    inferior, **107**, *107–108, 113, 128*
    superior, **107**, *107–108,* 112, *128*
  orbicularis oculi, *142*
  orbital, of Müller, 103, **112**
  rectus (bulbi)
    inferior, 103, **107**, *107–108, 112, 129*
    lateralis, **107**, *107, 108, 111, 113, 129,* 162
    medialis, **107**, *107–108, 111, 129, 161*
    superior, 103, **107**, *107–108, 112–113, 128–129*
  sphincter pupillae, 88, *132*
  stapedius, 6, *40, 42*
  tarsal, 103
  tensor tympani, **6**, *31, 36, 57*
Myelin, 100, 117, 147, 167, 200–201, 203–204, 206, *224, 242*

Myoepitheliocyte (myoepithelial cell), 105
Myopia, 86

N

Nail bed, 204
Nasal conchae, *see* Concha
Near reflex, 159
Neocortex, 194
Nerve endings, *see also* Corpuscles
  annulospiral, 110, 206, *207*
  bare (free), 111, 119, **200**, *224*
  corpuscular, **202–203**
  encapsulated, 111, **201–205**
  flower spray, 206, *207*
  lanceolate, 200, 215, 218, 239–240, *239*
  neurotendinous, 201, 204, **206**, *206*
  palisade, 110, **200**
  in periodontal ligaments, 198
  of vibrissae, 200, 203, 215, 218, **239–240**, *239*
Nerve (nervus), nerves (nervi), branch (ramus)
  abducent, *108,* 110, *113,* 115, *149, 163*
  alveolar, 196
  auditory, *see* vestibulocochlear
  auricular
    of facial, 2
    of vagus, 2
  auriculotemporal, 2, 3
  chorda tympani, **7**, *28–29, 33, 36, 38, 190,* 193
  ciliary
    long, 115, 160
    short, *113,* 115, 159, 161
  cochlear, 10–11, *20, 22, 61, 66*
  ethmoidal
    anterior, 115, *165,* 171
    posterior 115, *165*
  facial, 2, **6**, 25, 31, 44, 61, 106, 190, 193, *242*
  frontal, *108, 113,* 115
  glossopharyngeal, 190, *190,* 193, 207
  hypoglossal 190
  infraorbital, 115, 196
  infratrochlear, 115
  intermedius, 106
  lacrimal, 106, *108, 113,* 115
  lingual, *190*
  mandibular (V/3), 190, 193, 196
  maxillary (V/2), *108, 113,* 115, *165,* 171, 196
  nasal posterior *165*
  nasociliary, *108, 113,* 115
  nasopalatine, *165,* 193
  oculomotor, *108,* 110, *113,* 115, *149, 161–162*
  olfactory, 167–168, *175–176, 178,* **179**
  ophthalmic (V/1), *113,* 115

optic, *85*, 95, 100, 103, *107–108, 111–112, 118, 127–129,* **147**, *147–151*, 159, 161
palatine
    lesser, 193
palpebral, 115
petrosal
    greater, 7, *44*, 106, 193
    lesser, 7
of pterygoid canal (Vidian), 106
to stapedius muscle, 6
to tensor tympani muscle, 6
trigeminal, 111, *149*, 193, 196, *242*
trochlear, *108*, 110, *113*, *149*
tympanic, 4, 7, *39, 60*
vagus, *190*, 193, 207
vestibular, 11, *20*, *44*, 162
vestibulocochlear, 11, *20*
vomeronasal, 182–183, *184*
zygomatic, 106
zygomaticofacial, 115
vibrissal
    deep, *239*
    superficial, *239*
Neuroglia, 117
Neuron, *see* Cell, nerve
Neurotendinous ending, *see* Nerve endings
Neurotransmitters, 98–99, 152, 207, 243
Nexus, *see* Gap junction
Nociceptors, 200
Nucleus, nuclei
    of abducent nerve, 162, *162–163*
    accessory optic, 163
    amygdaloid, 194
    arcuate, 150
    bed, of stria terminalis, 194
    of Cajal, interstitial, 162, *162–163*
    caudate, 181
    of cochlear nerve, 73, *74–75*
    of corpus trapezoideum (trapezoid body), 73
    of Darkschewitsch, *162–163*
    of Deiters, *82*, *163*
    of Edinger-Westphal, 115, 159, 160, *161*
    hypothalamic, 183, 194
    lateral geniculate, 147, **150**, *151–1582*, 154, *161*
    of lateral lemniscus, 76
    mesencephalic, of trigeminal, 196
    of oculomotor nerve, 115, *162–163*, *161–163*
    olfactory, anterior, **180**
    parabrachial, 194
    perihypoglossal, 161–163, *162*
    pregeniculate, *152*
    pretectal, 150, 159, *161*
    salivatory
        superior, 106
    septal, 165
    suprachiasmatic, 150
    thalamic, 151, 153, 181, 198
    of tractus solitarius (solitary tract), 193–194
    of trigeminal nerve
        mesencephalic, 111
        pontine (principal), 198, 215, *242*
    trochlear, 115, 162, *162–163*
    of vestibular nerve, *82*
        inferior (of Roller), *163*
        medial (of Schwalbe), *163*
        superior (of Bechterew), *163*
Nystagmus
    optokinetic, 162–163

O

Object vision, 86, 89
Ocular dominance columns, 154
Olfaction, mechanism of, 168, 181
Olfactory binding proteins (Odor-binding proteins), 168
Oligodendrocyte, oligodendroglia, 147
Opening (Aditus, Aperture, Hiatus, Orifice)
    of auditory tube, 6, *28, 35*
    of cochlear canaliculus, 19
    of Eustachian tube, 6, *171*
    of greater petrosal nerve, *50*
    orbital, 111
    pupillary, 89, *90*, 93
    of vestibular aqueduct, 8
Operculum
    temporal, 76, 79
    parietal, *79*
Opsins, 95–96
Optic chiasma, *see* Chiasma
Optic cup, 103, *116–125*
Optic disc, 86, 88, 93, 95, 101, 147
Ora serrata, 89, 94, 98, 118, 120
Orbiculus ciliaris (pars plana), 89
Orbit
    arteries, 100, **114**
    bony structure, 111, *127–129*
    nerves, 113, **115**
Organ
    neurotendinous, of Golgi 204, **206**, *206*
    spiral, of Corti, **8**, *9, 66–68*
    vomeronasal, of Jacobson, 169, 173, **181–185**, *182, 184*
Ossicles, auditory, 5, *13*
Otoacoustic emission, 14
Otocyst, 15
Otolith, 11, *70*

Owls, 76
Oxytalan, 195–196

P

Palate
    development, 196–171
    hard, *165*, 187
    primary, secondary, *169*, 169–171
    soft, 187
    nerves, 193–194
Palm skin, 201, **209**, 215, *219–221*, 224
Palpebrae, *see* Eyelids
Panniculus adiposus, 209
Papilla
    dermal, 201, 209, *230, 234, 236, 238*
    filiform, 187, *191*
    foliate, 187, *191*, 193
    fungiform, 187, *191*, 193, *225*
    of hair, 213, *214, 235, 240*
    lacrimal, 106
    optic, 95
    vallate, 187, 190, *192*
Paradoxical cold, 217
Parafovea, **94**, 100
Paraneurons, 201, 207
Parinaud's syndrome, 159
Part, pars
    caeca retinae, 94
    ciliaris of retina, 94
    iridica of retina, 88
    optica of retina, 94, **95**
    plana of ciliary body, 89, *131*
    plicata of ciliary body, 89, 102, *131, 133*
Parvocellular cells (P-cells), 100, 151
Parvocellular fibers (P-fibers), 100
Pathways
    auditory, **73**, *74–75, 77–80*
    gustatory, **193–194**
    olfactory, **179–181**, *180*
    retinotectal, 161
    visual, **147**, *151*
Pedicle of cone, 96–97, *98–99, 140*
Perifovea, 94, 98, 100, *141*
Perilymph, 7, *9*, 10
Perineurium, *175*, 204, *224*
Periodontium, **194–198**, *195, 197*, 204
Periorbita, 106, *111*, **112**
Petrous temporal, *see* Bone, temporal
Phaeomelanin, 212
Pharynx, *169*, 171, 187, 194
Pheromones, 183, 185
Phosphoinositol, 168

Photoreceptors, 96, 98–100, *139–141*
Pia mater, 114, *117–118*, 147, *148, 178*
Pigment
    in epidermis, 211–212, *230*
    in hair, 212
    visual, 95
Pigment epithelium, 88–89, 94, **95**, *96, 116, 118–120, 163, 138, 140*
Pilosebaceus unit, 209, **213**
Pinna, **2**, 3, *21, 27*
Pits
    nasal, 171
    olfactory, *116*, 168, *169*
    optic, *116*
    otic, 15
    vomeronasal, 183
Placodes
    auditory, 15
    epibranchial, 168, 187
    lens, *116*, **168**
    nasal, 168
    olfactory, 168
    otic, 19
    trigeminal, 168
Plate
    cribriform, *85*, 88, 100, 147, 166–167, *175, 178–179*
    orbital, of ethmoid bone, 112
    orbital, of frontal bone, 111, *128–129*
    prechordal, 168
Plexus
    artery
        choroid of eye, 100–103
    nerve
        carotid, internal, 159
        dental, inferior and superior, 196
        dermal, *210*, 212
    vein
        episcleral, 90, 105
        pterygoid, 113
        scleral, 100–103
        tympanic, 4, 7
Plica, *see* Fold
Poles
    of eyeball, 86
    of lens, 90
Pons, 106, *149, 162–163*, 193–194
Ponticulus, *35, 38, 40*
Pouch of Rathke, 168
Positron emission tomography (PET), 157, 181
Premaxilla, 196
Presbyopia, 89

Process, processus
　　ciliary, **89**, 90–91, *91*, 93, *131–133*
　　clinoid, *127*, 172
　　cochleariform, **5**, *30, 51*
　　pterygoid, 171
　　uncinate, *172*
　　zygomatic, 111
Prominence, swelling
　　of facial canal, 168
　　frontonasal, 171
　　mandibular, 168
　　maxillary, 168
　　nasal, lateral, 116, 168
　　nasal, medial, 169
　　spiral, *67–68*
Promontory, **5**, *35, 39, 41, 54–56, 58*
Prosopagnosia, **157**, 159
Pulvinar, *149*, 150–151
Punctum lacrimale, *105*
Pupil of eye, **88**, 89, 93, *104, 118*, 159
Pupillary ruff, 88
Purple, visual, 95

R

Radiation
　　acoustic, 73, 154
　　optic, 150, **151**, *151*, 154, *161*
Rathke's pouch, 168
Receptive field
Receptor, receptors
　　auditory, 7–8
　　chemoreceptors, 200
　　classification, 199–200
　　gustatory, 187
　　interoceptors, 199
　　joint, 206
　　proprioceptors, 199
　　mechanoreceptors, 188, 196, 200, 203–204
　　nociceptors, 200
　　olfactory, 166–168, *174, 176, 179*
　　thermoreceptors, 200, 217
　　vestibular, 7
　　visual, 95–96, 98–100, *139–141*
Receptor molecules, 95, 168, 183, 190
Recess, recessus, recesses
　　cochlear, 8
　　elliptical, 8
　　epitympanic, *5*, 6
　　infundibular, 147, *149*
　　of Kuhnt, 93
　　optic, 149
　　sphenoethmoidal, 166
　　spherical, 8
Reflexes
　　accomodation, 89, 91, 159
　　consensual pupillary, **159**
　　corneal, 88
　　light, 123, **159**, *161*
　　near, 159
　　optokinetic, 162–163
　　proprioceptive, 206
　　vestibulo-ocular, 162, *163*
Refractive media of the eye, 86
Regional cerebral blood flow (rCBF), 157
Retina
　　circulation, 100–114, *123–125*
　　development, *116–120*
　　parts, 94–100, *136–137*
　　structure (architecture), 94–100, *131, 137–139*
Retinal, 95
Retinal detachment, 94
Retinol, 95
Rhodopsin, 95
Ring, annulus
　　fibrocartilagineous, **3**, *23, 25–26, 28, 34*
　　tendinous, common, 106–107, *108*, 109, 112–113
　　tympanic, 15, *23, 25, 260*
Rods
　　of Corti (pillar cells), 8, *68*
　　olfactory, 167
　　of retina, 95–96, *96–97, 136–139*, 140
Root of hair, 201, 213, *214, 234–240*, 240
Roughness, perception of, 217

S

Sac
　　conjunctival, 105, 107,*118, 121*
　　endolymphatic, 10
　　lacrimal, 103–106, *105, 111–112*, 114, *146*
Saccades, 163
Saccule, 7, **10**, 19–20, 25, 70
Second messengers, 168
Scala
　　tympani, **8**, *8*, 15, *61, 66–68*
　　vestibuli, **8**, *8*, 15, *61, 66–68*
Schlemm's canal, 90, *90–91, 131, 134*
Schwann cell, 167, *184*, 200, 202, *205, 221*
Sclera, 85, **88**, *90–91*, 100–104, *118–121, 136, 142, 147–148*
Scleral spur, 89, *91*
Sella turcica, *127, 149*
Sensory modalities, 159, 167, 181, 194, 200
Septum, septa
　　nasal, *127, 129, 165*, 166, *170*, 171, *172–173*, 181, 183

orbital, 103–105, *111–112*, 112–113, *128*
pellucidum, *149*
Sinus, sinuses
 cavernous, 113
 paranasal
  ethmoidal, *see* Cell, air
  frontal, 113–114, *149, 165, 172*
  lacrimal, of Meier, 106
  maxillary, 105, *108, 128–129*, 172, *172, 174*
  sphenoidal, 114, *149, 165, 172*
 venous of sclera, 90
 of vibrissae, 239–240
  cavernous, *239–241*
  ring, *239–241*
Skin
 of auricle, 2, *27*, 209
 blood supply, 212–213
 development, 209–210
 end organs, 200–207, *220–224*
 of external acoustic meatus, 2, *26*, 209
 of eyelids, 103
 functions, 209
 glands, 209, *210*
 innervation, 212
 pigmentation, 211–212
 structure, 210–212
 types, regions, 209
Skin-associated lymphatic tissue (SALT), 212
Smooth pursuit, 163
Space, spaces
 episcleral, *142*
 of Fontana, 90, *134*
 intraretinal, *118–119*
 of Nuel, **10**, *68*
 perilymphatic, **10**, *25, 66, 70–71*
 of Petit, 93
 periodontal, 195–196
 retrolental, 93, *121*
 retrozonular, 93
 subarachnoid, *117*, 147, *148, 178*
Speech areas, *157*
Spherule of rod, 96–97, **98**
Sphincter pupillae, *see* Muscle, sphincter
Spindle, muscle (neuromuscular), 109–110, **206**, *207, 224*
Stapes, 4–5, **6**, *21, 25, 29, 32, 40–42, 48, 59*
Statoconia, 11, *70*
Stereocilia, 10–14, *69, 71*
Stereognosis, 218
Stratum, strata (layer, zone)
 basale, of skin, 210, *220, 230–232*
 corneum, of skin, **211**, *220, 230–232, 234*
 fibrous, of tympanic membrane, 4, *26*
 germinativum, of skin, 210, *230–231, 234*
 granulosum, of skin, **211**, *220, 230–231, 234*
 of hair follicle, 213
 lucidum, of skin, **211**, *220, 230–232, 234*
 spinosum, of skin, **211**, *220, 230–232, 234*
Stria, striae (stripe)
 of Gennari, 154
 olfactory, 180–181
 vascularis, of cochlea, **8**, *9, 66–68*
Stroma
 of ciliary body, 89, *133*
 of iris, 88, *132*
Subcutis (hypodermis), 204, 209, *210, 221–222, 228–229, 233*
Substance, perforated anterior, 147
Substantia propria, 87–88, *130–131*
Sulcus, sulci (groove, fissure)
 calcarine, 152, 154
 of the optic chiasma (groove of sphenoid), 147
 intraparietal, 156, 159
 olfactory, 180
 spiralis, of cochlea, 8, *9, 67–68*
 tympanic, 3
Supercilii, *see* Eyebrows
Synapse, synaptic
 dendrodendritic, 179–180, 207
 dyad, 99
 serial, 99
 triad, 98–99, *140*
Synaptic ribbons, *97*, 98–99, *140*

T

Tactile acuity (resolution), 217
Tactile sensation, mechanism, 200–207
Tarsus (tarsal plate), 103
Taste buds, 187–188, *191–192*
Taste receptor cells, 188
Taste sensation, 193–194
Tears, 103, 105
Tectum of midbrain, 161, 163
Tegmen tympani, **5**, *31–32, 37*
Tegmentum
 of mesencephalon, 160
Temporal bone, *see* Bone, temporal
Tenon's capsule, *see* Fascia bulbi
Thalamus, 150, 161, 194, 240
Thermoreceptor, 200, 217
Tip links of cilia, 12–13
Tongue
 development, *189*, 190

INDEX

nerves, 190, *190*
papillae, 187–188, *191–192*
Tonofibrils, tonofilaments, 211
Touch dome, **201**, 210, 215
Trabecular meshwork, 90, *91*, 93
    corneoscleral, 90, *134*
    uveal meshwork, 90, *131, 134*
Tract, tracts, tractus (*see also* Bundle and Fasciculus)
    olfactory, *149, 151*, 179, **180**, *180*
    optic, 100, **147, 149**, 150, *151, 161*
    spinal of trigeminal, *242*
    spiralis foraminosus, 11
    trigeminothalamic, 194, 198
    uveal, 86, 89, 100
    vestibulocerebellar, *82*
    vestibulospinal, *82*
Tragus, 2, *2–3*
Transducin, 95
Trichohyalin, 213
Trochlea of superior oblique, 107, *107–108*, 112
Trunk, sympathetic, 159
Tube, auditory (pharyngotympanic), 5, **6**, *19, 31*
Tuber cinereum, 149, *151*
Tunica, tunicae, tunics
    of eyeball, 118
        fibrosa, 147
        nervosa, 86
        vasculosa (uvea), 147
Tunnel of Corti
    inner (cuniculum internum), 9, **10**, *67–68*
    intermediate (cuniculum medium, spaces of Nuel), 9, **10**, 68
    outer (cuniculum externum), 9, **10**, 68
Two-point limen, 217

U

Ultrasonography, 121
Umami, 188
Umbo of tympanic membrane, **6**, *28, 34, 39*
Uncus, *79*
Uniocular visual field, *see* Visual fields
Utricle, 7, **10**, *19–20, 25*
Uveal tract, *see also* Choroid 86, 89, 100, 103

V

Valve of nasolacrimal duct (Hasner's), 106
Vas spirale, *9*, 10
Vater-Pacinian corpuscle, *see* Corpuscle, Pacinian

Vein, veins
    angular, 106, 113
    aqueous, 90
    central, of retina, 103, 113, *117–118, 123–124, 147*
    choroidal, of eye, 101–102
    ciliary, 90, 102
    conjunctival, 105
    episcleral, 90, 105
    facial, 113
    hyaloid, *117*
    infraorbital, **113**
    jugular, internal, 11
    labyrinthine, 11
    ophthalmic, 103, *108*, **113**
    palpebral, 105
    pterygoid (plexus), 113
    supraorbital, **113**
    vortex (venae vorticosae), 88, 102–103, 113, *125, 148*
Vellus hairs, 213, *240*
Ventricles
    of brain
        fourth, *61*
        lateral, 154
        third, *79*, *116*, 147, *149*
    optic, *116–120*
Vergence movements, 109, **163**
Vesicle,
    auditory (otic), 15, *16–17*
    optic, *116–120*
Vestibule, 1, *4*, 7, **8**, *10, 18, 25, 46–48, 50–54, 57–59, 61–62*
Vibration detection, **218**
Vibrissae, mystacial, 215, 218, **239–243**, *239–242*
Vieth-Müller circle, *see* Horopter
Visual acuity, 96
Visual axis, *85*, 86
Visual fields, 86, 151, **154**, **157**, 161
Visual purple, *see* Rhodopsin
Vitreous body, *see* Body, vitreous
Vomer, 171, *182*
Vomeronasal cartilage, *173*, 183
Vomeronasal organ,
    see Organ, vomeronasal

W

Whiskers, *see* Vibrissae
Willis, arterial circle of, *see* Circulus arteriosus
Wolfring, glands of, *see* Glands

X
X-cells, 100

Y
Y-cells, 100

Z
Zone, zona
    arcuata of cochlea, 9, **10**
    pectinata of cochlea, 9, **10**
Zonula
    adherens, 98–99, *140*, 167
    occludens, *177, 184*, 188
Zonule, ciliary (of Zinn), 89–91